Annals of Mathematics Studies
Number 200

The Norm Residue Theorem
in Motivic Cohomology

Christian Haesemeyer

Charles A. Weibel

PRINCETON UNIVERSITY PRESS

PRINCETON AND OXFORD

2019

Published by Princeton University Press
41 William Street, Princeton, New Jersey 08540
6 Oxford Street, Woodstock, Oxfordshire OX20 1TR

press.princeton.edu

LCCN 2018967600
ISBN: 978-0-691-18182-0
ISBN (pbk.): 978-0-691-19104-1

British Library Cataloging-in-Publication Data is available

Editorial: Vickie Kearn, Susannah Shoemaker, and Lauren Bucca
Production Editorial: Nathan Carr
Production: Erin Suydam
Publicity: Matthew Taylor and Kathryn Stevens

This book has been composed in LaTeX

Printed on acid-free paper. ∞

Printed in the United States of America

10 9 8 7 6 5 4 3 2 1

This book is dedicated to

Vladimir Voevodsky (1966–2017)

and

Andrei Suslin (1950–2018)

Contents

Preface

LET k BE a field and ℓ a prime number different from $\mathrm{char}(k)$. The étale cohomology ring $\oplus H^n_{\text{ét}}(k, \mu_\ell^{\otimes n})$ is the Galois cohomology of the group μ_ℓ of ℓth roots of unity and its twists $\mu_\ell^{\otimes n}$. In degree 0 it is $H^0_{\text{ét}}(k, \mathbb{Z}/\ell) \cong \mathbb{Z}/\ell$, and in degree 1 it is given by the Kummer isomorphism $k^\times/\ell \xrightarrow{\sim} H^1_{\text{ét}}(k, \mu_\ell)$. If k contains μ_ℓ then $H^2_{\text{ét}}(k, \mu_\ell^{\otimes 2}) \cong \mu_\ell \otimes \mathrm{Br}_\ell(k)$; Tate realized that in the general case we should have $H^2_{\text{ét}}(k, \mu_\ell^{\otimes 2}) \cong K_2(k)/\ell$, and this was proven by Suslin and Merkurjev [MeS82]; this identification evolved from the classical ℓ^{th} power norm residue symbol of Hilbert's 9th Problem. The higher cohomology groups $H^n_{\text{ét}}(k, \mu_\ell^{\otimes n})$ have seemed difficult to describe in general.

The Kummer isomorphism induces a ring homomorphism from $\wedge^*(k^\times)$, the exterior algebra of the abelian group $k^\times$, to the étale cohomology ring, and Tate observed that the cup product $[a] \cup [1-a]$ vanishes in $H^2_{\text{ét}}(k, \mu_\ell^{\otimes 2})$. Inspired by this, Milnor defined a graded ring $K^M_*(k)$, called the *Milnor K-theory* of k, as the quotient of $\wedge^*(k^\times)$ by the ideal generated by elements of the form $\{a, 1-a\}$, $a \in k - \{0, 1\}$. By construction, there is a canonical ring homomorphism

$$K^M_*(k)/\ell \longrightarrow \oplus_n H^n_{\text{ét}}(k, \mu_\ell^{\otimes n}),$$

called the *norm residue homomorphism* to reflect its origins in Hilbert's Problem. The main theorem of this book is that the norm residue homomorphism is an isomorphism:

Theorem A. *For all fields k containing $1/\ell$, and all n, the norm residue is an isomorphism.*

As a consequence, we obtain a presentation of the étale cohomology ring in terms of generators (the $[a]$ in $H^1_{\text{ét}}$) and relations (the $\{a, 1-a\}$ in $H^2_{\text{ét}}$).

The proof of Theorem A was completed by Voevodsky in the 2010–11 papers [Voe10c] and [Voe11], but depends on the work of many other people. See the historical notes at the end of chapter 1 for details.

For any smooth variety X over k, we can form the bigraded motivic cohomology ring $H^{*,*}(X, \mathbb{Z}/\ell)$ and there is a ring homomorphism which in bidegree (p, q) is $H^{p,q}(X, \mathbb{Z}/\ell) \to H^p_{\text{ét}}(X, \mu_\ell^{\otimes q})$. When $X = \mathrm{Spec}(k)$, the diagonal entry $H^{p,p}(k, \mathbb{Z})$ is isomorphic to $K^M_p(k)$, and $H^{p,p}(k, \mathbb{Z}/\ell) \cong K^M_p(k)/\ell$. Theorem A is a special case of the following more sweeping result, which we also prove.

Theorem B. *Let X be a smooth variety over a field containing $1/\ell$. Then the map $H^{p,q}(X, \mathbb{Z}/\ell) \to H^p_{\text{ét}}(X, \mu_\ell^{\otimes q})$ is an isomorphism for all $p \le q$.*

Theorem B has a homological mirror in the derived category of Nisnevich sheaves. To formulate it, let π_* denote the direct image functor from étale sheaves to Nisnevich sheaves on the category $\mathbf{Sm}/k$ of smooth schemes over k, and recall that $H^n_{\text{ét}}(X, \mu_\ell^{\otimes q})$ is the Nisnevich hypercohomology $H^n_{\text{nis}}(X, \mathbf{R}\pi_*\mu_\ell^{\otimes q})$. The map $H^{p,q}(X, \mathbb{Z}/\ell) \to H^p_{\text{ét}}(X, \mu_\ell^{\otimes q})$ is just the cohomology on X of a natural map $\mathbb{Z}/\ell(q) \to \mathbf{R}\pi_*\mu_\ell^{\otimes q}$ in the derived category. It factors through the good truncation at q of $\mathbf{R}\pi_*\mu_\ell^{\otimes q}$, and we have

Theorem C. *For all q, the map $\mathbb{Z}/\ell(q) \to \tau^{\le q}\mathbf{R}\pi_*\mu_\ell^{\otimes q}$ is an isomorphism in the derived category of Nisnevich sheaves on the category of smooth simplicial schemes over k.*

As noted above, it is easy to see that Theorem C implies Theorem B, which in turn implies Theorem A. We will see in chapter 2 that all three theorems are equivalent. This equivalence is due to Suslin and Voevodsky [SV00a].

The history of these theorems is quite interesting. In the late 1960s, Milnor and Tate verified that the norm residue homomorphism is always an isomorphism for local and global fields, fields for which the only issue for $n > 2$ is torsion for $\ell = 2$. Inspired by these calculations, Milnor stated in [Mil70] that

> Bass and Tate also consider the more general [norm residue] homomorphism [for ℓ odd] ... but we will only be interested in the case $\ell = 2$. I do not know of any examples for which the [norm residue] homomorphism fails to be bijective.

In the 1982 papers [Me81, MeS82], Merkurjev and Suslin showed that the norm residue homomorphism is an isomorphism for $n = 2$ and all ℓ. By the mid-1990s, Milnor's statement for $\ell = 2$ had become known as the *Milnor Conjecture*.

The parallel conjecture when ℓ is odd, dubbed the *Bloch–Kato conjecture* in [SV00a], was first clearly formulated by Kazuya Kato in [Kat80, p. 608]:

> Concerning this homomorphism, the experts perhaps have the following Conjecture in mind. (cf. [Mil70, §6])

> *Conjecture.* The [norm residue] homomorphism is bijective for any field k and any integer ℓ which is invertible in k.

The version stated by Spencer Bloch was: "I wonder whether the whole cohomology algebra $\oplus H^r(F, \mu_{\ell^\nu}^{\otimes r})$ might not be generated by H^1?" [Blo80, p. 5.12].

We now turn to the rise of motivic cohomology. In the early 1980s, S. Lichtenbaum [Lic84, §3] and A. Beilinson [Beĭ87, 5.10.D] formulated a set of conjectures describing the (then-hypothetical) complexes of sheaves $\mathbb{Z}(n)$ and properties they should enjoy. These complexes were later constructed by Voevodsky and others, and their cohomology is the motivic cohomology developed in [Voe00b]. Among these properties is the assertion BL(n) that Theorems B and C hold for $n = q$; this has often been referred to as the *Beilinson–Lichtenbaum conjecture*. Another is the assertion H90(n) that $H_{\text{ét}}^{n+1}(k, \mathbb{Z}(n))$ should vanish.

In 1994, Suslin and Voevodsky showed that the Bloch–Kato conjecture was equivalent to the Beilinson–Lichtenbaum conjecture, and to H90(n). Their proof required resolution of singularities over k, a restriction that was later removed in [GL01] and [Sus03]. Our chapter 2 provides a shorter proof of these equivalences. This shows that Theorems A, B, and C are equivalent.

For $\ell = 2$, the proof of Theorems A, B, and C was announced by Voevodsky in 1996, and published in the 2003 paper [Voe03a].

A proof of the Bloch–Kato conjecture (Theorem A) was announced by Voevodsky in 1998, assuming the existence of what we now call a *Rost variety* (see Definition 1.24). Rost produced such a variety that same year, in [Ros98a], but the complete proof that Rost's variety had the properties required by Voevodsky did not appear until much later ([SJ06], [Ros06] and [HW09]). The proof of Theorem A appeared in the 2003 preprint [Voe03b]—modulo the assumption that Rost varieties exist and two other assertions. One of these assertions, concerning the motivic cohomology operations on $H^{*,*}(X, \mathbb{Z}/\ell)$, was incorrect; happily, it was found to be avoidable [Wei09]. The full proof of Theorem A was published by Voevodsky in the 2010–11 papers [Voe10c] and [Voe11].

In this book we shall prove Theorems A, B, and C for all ℓ, following the lines of [Voe11]. We will also establish the appropriate replacement assertions concerning the motivic cohomology groups $H^{*,*}(X, \mathbb{Z})$.

Prerequisite material

Our proof will use the machinery of motivic cohomology. In order to keep the book's length reasonable (and preserve our sanity), we need to assume a certain amount of material. Primarily, this means:

1. the material on the pointed motivic homotopy category of spaces, due to Morel–Voevodsky and found in [MV99] (see sections 12.7–12.9);
2. the construction of reduced power operations P^i in [Voe03c] (see section 13.3);
3. the theory of presheaves with transfers, as presented in [MVW] (the original source for this material is Voevodsky's paper [Voe00b]);
4. the main facts about algebraic cobordism, due to Levine and Morel and found in the book [LM07]; we have summarized the facts we need about algebraic cobordism, especially the degree formulas, in chapter 8.

As should be clear from this list, the material we assume does not arise in a vacuum. These topics imply, for example, that the reader is at least comfortable with the basic notions of Algebraic Geometry, including étale cohomology, and basic notions in homotopy theory, including model categories (for chapter 12).

Acknowledgments

THE AUTHORS ARE deeply indebted to Vladimir Voevodsky, who not only had the vision to carry out the mathematics in this book, but who encouraged the authors and provided explanations when needed. They are also indebted to Andrei Suslin and Markus Rost, who not only provided much of the mathematics in this book, especially part II, but also gave many lectures on this material.

The chapter on Model Categories could not have been written without the assistance of Philip Hirschhorn and Rick Jardine, who provided significant guidance in the writing of this chapter. Sasha Merkurjev made several very useful and insightful comments, Joël Riou very helpfully commented on several chapters, especially chapters 9 and 12, and the referees did an amazing job. We thank all of them.

Finally, the authors were supported by numerous grants over the period (2007–2018) that this book was written. This includes support from the IAS, NSF, NSA, and their host institutions.

Christian Haesemeyer
Charles A. Weibel

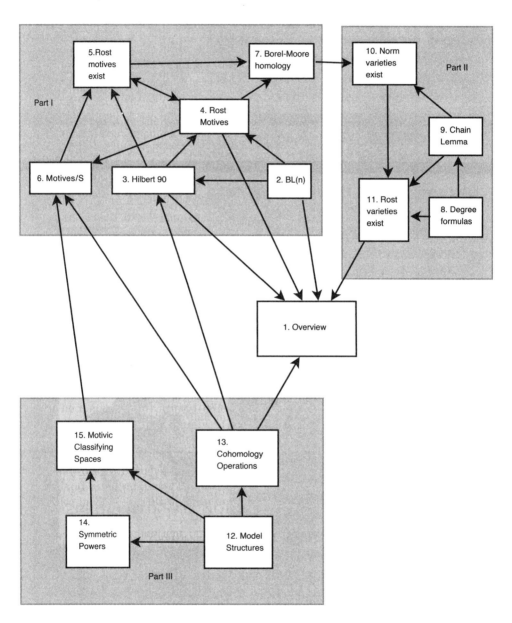

Dependency graph of the chapters

Part I

Chapter One

An Overview of the Proof

THE PURPOSE OF this chapter is to give the main steps in the proof of Theorems A and B (stated in the introduction) that for each n the norm residue homomorphism

$$K_n^M(k)/\ell \longrightarrow H_{\text{ét}}^n(k, \mu_\ell^{\otimes n}) \tag{1.1}$$

is an isomorphism, and $H^{p,n}(X, \mu_\ell^{\otimes n}) \cong H_{\text{ét}}^p(X, \mu_\ell^{\otimes n})$ for $p \leq n$. We proceed by induction on n. It turns out that in order to prove Theorems A, B, and C, we must simultaneously prove several equivalent (but more technical) assertions, H90(n) and BL(n), which are defined in 1.5 and 1.28.

1.1 FIRST REDUCTIONS

We fix a prime ℓ and a positive integer n. In this section we reduce Theorems A and B to H90(n), an assertion (defined in 1.5) about the étale cohomology of the ℓ-local motivic complex $\mathbb{Z}_{(\ell)}(n)$. We begin with a series of reductions, the first of which is a special case of the *transfer argument*.

The transfer argument 1.2. Let F be a covariant functor on the category of fields which are algebraic over some base field, taking values in $\mathbb{Z}/\ell$-modules and commuting with direct limits. We suppose that F is also contravariant for finite field extensions k'/k, and that the evident composite from $F(k)$ to itself is multiplication by $[k':k]$. The contravariant maps are commonly called *transfer maps*. If $[k':k]$ is prime to ℓ, the transfer hypothesis implies that $F(k)$ injects as a summand of $F(k')$. More generally, $F(k)$ injects into $F(k')$ for any algebraic extension k' consisting of elements whose degree is prime to ℓ. Thus to prove that $F(k) = 0$ it suffices to show that $F(k') = 0$ for the field k'.

Both $k \mapsto K_n^M(k)/\ell$ and $k \mapsto H_{\text{ét}}^n(k, \mu_\ell^{\otimes n})$ satisfy these hypotheses, and so do the kernel and cokernel of the norm residue map (1.1), because the norm residue commutes with these transfers. Thus if the norm residue is an isomorphism for k' it is an isomorphism for k, by the transfer argument applied to the kernel and cokernel of (1.1). For this reason, we may assume that k contains all ℓ^{th} roots of unity, that k is a perfect field, and even that k has no field extensions of degree prime to ℓ.

The second reduction allows us to assume that we are working in characteristic zero, where, for example, the resolution of singularities is available.

Lemma 1.3. *If* (1.1) *is an isomorphism for all fields of characteristic* 0, *then it is an isomorphism for all fields of characteristic* $\neq \ell$.

Proof.[1] Let R be the ring of Witt vectors over k and K its field of fractions. By the standard transfer argument 1.2, we may assume that k is a perfect field, so that R is a discrete valuation ring. In this case, the specialization maps "sp" are defined and compatible with the norm residue maps in the sense that

$$
\begin{array}{ccc}
K_n^M(K)/\ell & \longrightarrow & H_{\text{ét}}^n(K, \mu_\ell^{\otimes n}) \\
\text{sp} \downarrow & & \text{sp} \downarrow \\
K_n^M(k)/\ell & \longrightarrow & H_{\text{ét}}^n(k, \mu_\ell^{\otimes n})
\end{array}
$$

commutes (see [Wei13, III.7.3]). Both specialization maps are known to be split surjections. Since $\text{char}(K) = 0$, the result follows. $\square$

Our third reduction translates the problem into the language of motivic cohomology, as the condition H90(n) of Definition 1.5.

The (integral) motivic cohomology of a smooth variety X is written as $H^{n,i}(X, \mathbb{Z})$ or $H^n(X, \mathbb{Z}(i))$; it is defined to be the Zariski hypercohomology on X of $\mathbb{Z}(i)$; see [MVW, 3.4]. Here $\mathbb{Z}(i)$ is a cochain complex of étale sheaves which is constructed, for example, in [MVW, 3.1]. By definition, $\mathbb{Z}(i) = 0$ for $i < 0$ and $\mathbb{Z}(0) = \mathbb{Z}$, so $H^n(X, \mathbb{Z}(i)) = 0$ for $i < 0$ and even $i = 0$ when $n \neq 0$. There are pairings $\mathbb{Z}(i) \otimes \mathbb{Z}(j) \to \mathbb{Z}(i+j)$ making $H^*(X, \mathbb{Z}(*))$ into a bigraded ring. When k is a field, we often write $H^*(k, \mathbb{Z}(*))$ for $H^*(\text{Spec } k, \mathbb{Z}(*))$.

There is a quasi-isomorphism $\mathbb{Z}(1) \xrightarrow{\simeq} \mathcal{O}^\times[-1]$; see [MVW, 4.1]. This yields an isomorphism $H^1(X, \mathbb{Z}(1)) \cong \mathcal{O}_X^\times$. When $X = \text{Spec}(k)$ for a field k, the Steinberg relation holds in $H^2(X, \mathbb{Z}(2))$: if $a \neq 0, 1$ then $a \cup (1-a) = 0$. The presentation of $K_*^M(k)$ implies that we have a morphism of graded rings $K_*^M(k) \to H^*(k, \mathbb{Z}(*))$ sending $\{a_1, ..., a_n\}$ to $a_1 \cup \cdots \cup a_n$. It is a theorem of Totaro and Nesterenko–Suslin that $K_n^M(k) \cong H^n(\text{Spec } k, \mathbb{Z}(n))$ for each n; proofs are given in [NS89], [Tot92], and [MVW, Thm. 5.1].

We can of course vary the coefficients in this construction. Given any abelian group A, we may consider $H^n(X, A(i))$, where $A(i)$ denotes $A \otimes \mathbb{Z}(i)$; $H^*(X, A(*))$ is a ring if A is. Because Zariski cohomology commutes with direct limits, we have $H^n(X, \mathbb{Z}(i)) \otimes \mathbb{Q} \xrightarrow{\simeq} H^n(X, \mathbb{Q}(i))$ and $H^n(X, \mathbb{Z}(i)) \otimes \mathbb{Z}_{(\ell)} \xrightarrow{\simeq} H^n(X, \mathbb{Z}_{(\ell)}(i))$. Because $H_{\text{zar}}^{n+1}(\text{Spec } k, \mathbb{Z}(n)) = 0$ [MVW, 3.6], this implies that we have

$$K_n^M(k)/\ell \cong H_{\text{zar}}^n(\text{Spec } k, \mathbb{Z}/\ell(n)). \tag{1.4}$$

Since each $A(i)$ is a complex of étale sheaves, we can also speak about the étale motivic cohomology $H_{\text{ét}}^*(X, A(i))$. There is a motivic-to-étale map

---

1. Taken from [Voe96, 5.2].

$H^*(X, A(i)) \to H^*_{\text{ét}}(X, A(i))$; it is just the change-of-topology map $H^*_{\text{zar}} \to H^*_{\text{ét}}$. For $A = \mathbb{Z}/\ell$ we have isomorphisms $H^n_{\text{ét}}(X, \mathbb{Z}/\ell(i)) \cong H^n_{\text{ét}}(X, \mu_\ell^{\otimes i})$ for all $n, i \geq 0$; see [MVW, 10.2]. We also have $H^n_{\text{ét}}(k, \mathbb{Z}(i))_{(\ell)} = H^n_{\text{ét}}(k, \mathbb{Z}_{(\ell)}(i))$ and $H^n_{\text{ét}}(k, \mathbb{Z}(i)) \otimes \mathbb{Q} = H^n_{\text{ét}}(k, \mathbb{Q}(i))$.

The condition H90(n)

Definition 1.5. Fix n and ℓ. We say that *H90(n) holds* if $H^{n+1}_{\text{ét}}(k, \mathbb{Z}_{(\ell)}(n)) = 0$ for any field k with $1/\ell \in k$. Note that H90(0) holds as $H^1_{\text{ét}}(k, \mathbb{Z}) = 0$, and that H90(n) implicitly depends on the prime ℓ.

The name "H90(n)" comes from the observation that H90(1) is equivalent to the localization at ℓ of the classical Hilbert's Theorem 90:

$$H^2_{\text{ét}}(k, \mathbb{Z}(1)) \cong H^2_{\text{ét}}(k, \mathbb{G}_m[-1]) = H^1_{\text{ét}}(k, \mathbb{G}_m) = 0.$$

We now connect H90(n) to $K^M_n(k)$.

Lemma 1.6. *For all $n > i$, $H^n_{\text{ét}}(k, \mathbb{Z}(i))$ is a torsion group, and its ℓ-torsion subgroup is $H^n_{\text{ét}}(k, \mathbb{Z}_{(\ell)}(i))$. When $1/\ell \in k$ and $n \geq i+1$ we have $H^{n+1}_{\text{ét}}(k, \mathbb{Z}_{(\ell)}(i)) \cong H^n_{\text{ét}}(k, \mathbb{Q}/\mathbb{Z}_{(\ell)}(i))$, while there is an exact sequence*

$$K^M_n(k) \otimes \mathbb{Q}/\mathbb{Z}_{(\ell)} \to H^n_{\text{ét}}(k, \mathbb{Q}/\mathbb{Z}_{(\ell)}(n)) \to H^{n+1}_{\text{ét}}(k, \mathbb{Z}_{(\ell)}(n)) \to 0.$$

Proof. We have $H^n_{\text{ét}}(k, \mathbb{Q}(i)) \cong H^n(k, \mathbb{Q}(i))$ for all n by [MVW, 14.23]. If $n > i$, $H^n(k, \mathbb{Q}(i))$ vanishes (by [MVW, 3.6]) and hence $H^n_{\text{ét}}(k, \mathbb{Z}(i))$ is a torsion group. Its ℓ-torsion subgroup is $H^n_{\text{ét}}(k, \mathbb{Z}(i))_{(\ell)} = H^n_{\text{ét}}(k, \mathbb{Z}_{(\ell)}(i))$. Set $D(i) = \mathbb{Q}/\mathbb{Z}_{(\ell)}(i)$. The étale cohomology sequence for the exact sequence $0 \to \mathbb{Z}_{(\ell)}(i) \to \mathbb{Q}(i) \to D(i) \to 0$ yields the second assertion (for $n \geq i+1$), and (taking $n = i$) yields the commutative diagram:

$$
\begin{array}{ccccccc}
H^n(k, \mathbb{Z}_{(\ell)}(n)) & \longrightarrow & H^n(k, \mathbb{Q}(n)) & \longrightarrow & H^n(k, D(n)) & \longrightarrow & 0 \\
\downarrow & & \cong \downarrow & & \downarrow & & \\
H^n_{\text{ét}}(k, \mathbb{Z}_{(\ell)}(n)) & \longrightarrow & H^n_{\text{ét}}(k, \mathbb{Q}(n)) & \longrightarrow & H^n_{\text{ét}}(k, D(n)) & \xrightarrow{\text{onto}} & H^{n+1}_{\text{ét}}(k, \mathbb{Z}_{(\ell)}(n)).
\end{array}
$$

The bottom right map is onto because $H^{n+1}_{\text{ét}}(k, \mathbb{Q}(i)) = 0$. Since $H^n(k, D(n)) \cong K^M_n(k) \otimes \mathbb{Q}/\mathbb{Z}_{(\ell)}$, a diagram chase yields the exact sequence. $\square$

The example $\text{Br}(k)_{(\ell)} = H^2_{\text{ét}}(k, \mathbb{Q}/\mathbb{Z}_{(\ell)}(1)) \cong H^3_{\text{ét}}(k, \mathbb{Z}_{(\ell)}(1))$ shows that the higher étale cohomology of $\mathbb{Z}(n)$ and $\mathbb{Z}_{(\ell)}(n)$ need not vanish.

Theorem 1.7. *Fix n and ℓ. If $K^M_n(k)/\ell \xrightarrow{\simeq} H^n_{\text{ét}}(k, \mu_\ell^{\otimes n})$ holds for every field k containing $1/\ell$, then H90(n) holds.*

Of course, the weaker characteristic 0 hypothesis suffices by Lemma 1.3.

Proof. Recall that $K_n^M(k) \cong H_{\text{zar}}^n(\operatorname{Spec} k, \mathbb{Z}(n))$. The change of topologies map $H_{\text{zar}}^n \to H_{\text{ét}}^n$ yields a commutative diagram:

$$
\begin{array}{ccccccc}
K_n^M(k) & \xrightarrow{\ell} & K_n^M(k) & \longrightarrow & K_n^M(k)/\ell & \longrightarrow & 0 \\
\downarrow & & \downarrow & & \text{Norm} \downarrow \text{residue} & & \\
H_{\text{ét}}^n(k, \mathbb{Z}(n)) & \xrightarrow{\ell} & H_{\text{ét}}^n(k, \mathbb{Z}(n)) & \longrightarrow & H_{\text{ét}}^n(k, \mu_\ell^{\otimes n}) & \longrightarrow & H_{\text{ét}}^{n+1}(k, \mathbb{Z}(n)) \xrightarrow{\ell}
\end{array}
$$

The right vertical map is the Norm residue homomorphism, because the left vertical maps are multiplicative, and $H_{\text{ét}}^1(k, \mathbb{Z}(1)) = k^\times$. If the norm residue is a surjection, then $H_{\text{ét}}^{n+1}(k, \mathbb{Z}(n))$ has no ℓ-torsion. But it is a torsion group, and its ℓ-primary subgroup is $H_{\text{ét}}^{n+1}(k, \mathbb{Z}_{(\ell)}(n))$ by Lemma 1.6. As this must be zero for all k, H90(n) holds. $\qquad\square$

The converse of Theorem 1.7 is true, and will be proven in chapter 2 as Theorem 2.38 and Corollary 2.42. For reference, we state it here. Note that parts a) and b) are the conclusions of Theorems A and B (stated in the introduction).

Theorem 1.8. *Fix n and ℓ. Suppose that H90(n) holds. If k is any field containing $1/\ell$, then:*

a) *the norm residue $K_n^M(k)/\ell \to H_{\text{ét}}^n(k, \mu_\ell^{\otimes n})$ is an isomorphism;*
b) *for every smooth X over k and all $p \leq n$, the motivic-to-étale map $H^p(X, \mathbb{Z}/\ell(n)) \to H_{\text{ét}}^p(X, \mu_\ell^{\otimes n})$ is an isomorphism.*

1.2 THE QUICK PROOF

With these reductions behind us, we can now present the proof that the norm residue is an isomorphism. In order to keep the exposition short, we defer definitions and proofs to later sections.

We will proceed by induction on n, assuming H90(n $-$ 1) holds. By Theorems 1.7 and 1.8, this is equivalent to assuming that $K_{n-1}^M(k)/\ell \cong H_{\text{ét}}^{n-1}(k, \mu_\ell^{\otimes n-1})$ for all fields k containing $1/\ell$.

Definition 1.9. We say that a field k containing $1/\ell$ is *ℓ-special* if k has no finite field extensions of degree prime to ℓ. This is equivalent to the assertion that every finite extension is a composite of cyclic extensions of degree ℓ, and hence that the absolute Galois group of k is a pro-ℓ-group.

If k is a field containing $1/\ell$, any maximal prime-to-ℓ algebraic extension is ℓ-special. These extensions correspond to the Sylow ℓ-subgroups of the absolute Galois group of k.

The following theorem first appeared as [Voe03a, 5.9]; it will be proven in section 3.1 as Theorem 3.11.

Theorem 1.10. *Suppose that H90(n − 1) holds. If k is an ℓ-special field and* $K_n^M(k)/\ell = 0$, *then* $H_{\text{ét}}^n(k, \mu_\ell^{\otimes n}) = 0$ *and hence* $H_{\text{ét}}^{n+1}(k, \mathbb{Z}_{(\ell)}(n)) = 0$.

The main part of this book is devoted to proving the following deep theorem.

Theorem 1.11. *Suppose that H90(n − 1) holds. Then for every field k of characteristic 0 and every nonzero symbol* $\underline{a} = \{a_1, \ldots, a_n\}$ *in* $K_n^M(k)/\ell$ *there is a smooth projective variety* $X_{\underline{a}}$ *whose function field* $K_{\underline{a}} = k(X_{\underline{a}})$ *satisfies:*

(a) $\underline{a}$ *vanishes in* $K_n^M(K_{\underline{a}})/\ell$; *and*
(b) *the map* $H_{\text{ét}}^{n+1}(k, \mathbb{Z}_{(\ell)}(n)) \to H_{\text{ét}}^{n+1}(K_{\underline{a}}, \mathbb{Z}_{(\ell)}(n))$ *is an injection.*

Outline of proof. (See Figure 1.1.) The varieties $X_{\underline{a}}$ we use to prove Theorem 1.11 are called *Rost varieties for* $\underline{a}$; they are defined in section 1.3 (see 1.24). Part of the definition is that any Rost variety satisfies condition (a). The proof that a Rost variety exists for every $\underline{a}$, which is due to Markus Rost, is postponed until part II of this book, and is given in chapter 11 (Theorem 11.2).

The proof that Rost varieties satisfy condition (b) of Theorem 1.11 will be given in chapter 4 (in Theorem 4.20). The proof requires the motive of the Rost variety to have a special summand called a *Rost motive*; the definition of Rost motives is given in section 4.3 (see 4.11).

The remaining difficult step in the proof of Theorem 1.11, due to Voevodsky, is to show that there is always a Rost variety for $\underline{a}$ which has a Rost motive. We give the proof of this in chapter 5, using the simplicial scheme $\mathfrak{X}$ which is defined in 1.32. The input to the proof is a cohomology class $\mu \in H^{2b+1,b}(\mathfrak{X}, \mathbb{Z})$; μ will be constructed in chapter 3, starting from $\underline{a}$; see Corollary 3.16. The class μ is used to construct a motivic cohomology operation ϕ and chapter 6 is devoted to showing that ϕ coincides with the operation βP^b ($b = (\ell^{n-1} - 1)/(\ell - 1)$); see Theorem 6.34. The proof requires facts about motivic cohomology operations which are developed in part III. □

The quick proof

Assuming Theorems 1.8, 1.10 and 1.11, we can now prove Theorems A and B of the introduction. This argument originally appeared on p. 97 of [Voe03a].

Theorem 1.12. *If H90(n − 1) holds, then H90(n) holds. By Theorem 1.8, this implies that for every field k containing* $1/\ell$:

a) *the norm residue* $K_n^M(k)/\ell \to H_{\text{ét}}^n(k, \mu_\ell^{\otimes n})$ *is an isomorphism;*
b) *for every smooth X over k and all* $p \le n$, *the motivic-to-étale map* $H^p(X, \mathbb{Z}/\ell(n)) \to H_{\text{ét}}^p(X, \mu_\ell^{\otimes n})$ *is an isomorphism.*

Since H90(1) holds, it follows by induction on n that H90(n) holds for every n. Note that Theorem A is 1.12(a) and Theorem B is 1.12(b).

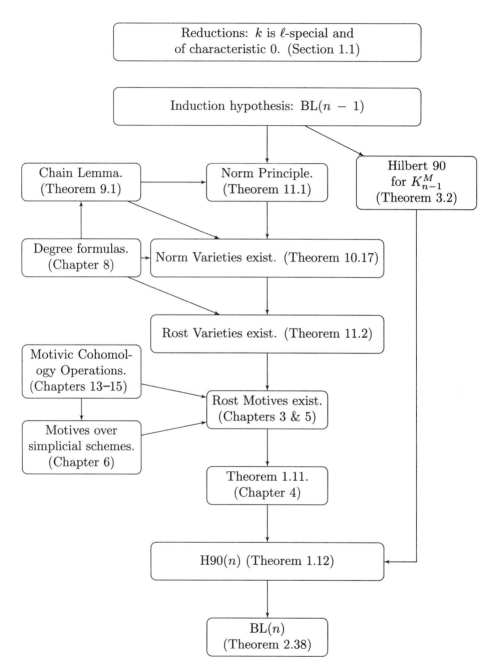

Figure 1.1: Dependency chart of Main Theorem 1.11

Proof of Theorem 1.12. Fix k, and an algebraically closed overfield Ω of infinite transcendence degree $> |k|$ over k. We first use transfinite recursion to produce an ℓ-special field k' ($k \subset k' \subset \Omega$) such that $K_n^M(k)/\ell \to K_n^M(k')/\ell$ is zero and $H_{\text{ét}}^{n+1}(k, \mathbb{Z}_{(\ell)}(n))$ embeds into $H_{\text{ét}}^{n+1}(k', \mathbb{Z}_{(\ell)}(n))$.

Well-order the symbols in $K_n^M(k)$: $\{\underline{a}_\lambda\}_{\lambda < \kappa}$. Fix $\lambda < \kappa$; inductively, there is an intermediate field k_λ such that $\underline{a}_\mu$ vanishes in $K_n^M(k_\lambda)/\ell$ for all $\mu < \lambda$ and $H_{\text{ét}}^{n+1}(k, \mathbb{Z}_{(\ell)}(n))$ embeds into $H_{\text{ét}}^{n+1}(k_\lambda, \mathbb{Z}_{(\ell)}(n))$. If $\underline{a}_\lambda$ vanishes in $K_n^M(k_\lambda)/\ell$, set $k_{\lambda+1} = k_\lambda$. Otherwise, Theorem 1.11 states that there is a variety X_λ over k_λ whose function field $K = k_\lambda(X_\lambda)$ splits a_λ, and such that $H_{\text{ét}}^{n+1}(k_\lambda, \mathbb{Z}_{(\ell)}(n))$ embeds into $H_{\text{ét}}^{n+1}(K, \mathbb{Z}_{(\ell)}(n))$; set $k_{\lambda+1} = K$. If λ is a limit ordinal, set $k_\lambda = \cup_{\mu < \lambda} k_\mu$. Finally, let k' be a maximal prime-to-ℓ algebraic extension of k_κ. Then $H_{\text{ét}}^{n+1}(k, \mathbb{Z}_{(\ell)}(n))$ embeds into $H_{\text{ét}}^{n+1}(k_\kappa, \mathbb{Z}_{(\ell)}(n))$, which embeds in $H_{\text{ét}}^{n+1}(k', \mathbb{Z}_{(\ell)}(n))$ by the usual transfer argument 1.2. By construction, k' splits every symbol in $K_n^M(k)$.

Iterating this construction, we obtain an ascending sequence of field extensions $k^{(m)}$; let L denote the union of the $k^{(m)}$. Then L is ℓ-special and $K_n^M(L)/\ell = 0$ by construction, so $H_{\text{ét}}^{n+1}(L, \mathbb{Z}_{(\ell)}(n)) = 0$ by Theorem 1.10. Since $H_{\text{ét}}^{n+1}(k, \mathbb{Z}_{(\ell)}(n))$ embeds into $H_{\text{ét}}^{n+1}(L, \mathbb{Z}_{(\ell)}(n))$, we have $H_{\text{ét}}^{n+1}(k, \mathbb{Z}_{(\ell)}(n)) = 0$. Since this holds for any k, H90(n) holds. $\square$

In the remainder of this chapter, we introduce the ideas and basic tools we will use in the rest of the book.

1.3 NORM VARIETIES AND ROST VARIETIES

In this section we give the definition of norm varieties and Rost varieties; see Definitions 1.13 and 1.24. These varieties are the focus of the main theorem 1.11, and will be shown to exist in chapters 10 and 11 in part II.

We begin with the notions of a splitting variety and a norm variety for a symbol $\underline{a} \in K_n^M(k)/\ell$. Norm varieties will be the focus of chapter 10.

Definition 1.13. Let $\underline{a}$ be a symbol in $K_n^M(k)/\ell$. A field F over k is said to *split* $\underline{a}$, and be a *splitting field* for $\underline{a}$, if $\underline{a} = 0$ in $K_n^M(F)/\ell$. A variety X over k is called a *splitting variety* for $\underline{a}$ if its function field splits $\underline{a}$ (i.e., if $\underline{a}$ vanishes in $K_n^M(k(X))/\ell$).

A splitting variety X is called an *ℓ-generic* splitting variety if any splitting field F has a finite extension E of degree prime to ℓ with $X(E) \neq \emptyset$.

A *norm variety* for a nonzero symbol $\underline{a}$ in $K_n^M(k)/\ell$ is a smooth projective ℓ-generic splitting variety of dimension $\ell^{n-1} - 1$.

We will show in Theorem 10.17 that norm varieties always exist for all n when $\text{char}(k) = 0$. When $n = 1$, the 0-dimensional variety $X = \text{Spec } k(\sqrt[\ell]{a})$ is a norm variety for a because $K_n^M(k)/\ell = k^\times / k^{\times \ell}$. When $n = 2$, Severi–Brauer varieties are norm varieties by Proposition 1.25.

Remark 1.13.1. (Specialization) Let Y be a reduced subscheme of X, not contained in the singular locus of X. If X is a splitting variety for $\underline{a}$ then so is Y. When X is a smooth splitting variety, such as a norm variety for $\underline{a}$, this implies that $\underline{a}$ is split by every field E with $X(E) \neq \emptyset$.

To see this, pick a closed nonsingular point x lying on Y. By specialization [Wei13, III.7.3], there is a map $K_n^M(k(X)) \to K_n^M(k(Y))$ sending the class of $\underline{a}$ on $k(X)$ to the class of $\underline{a}$ on $k(Y)$.

Severi–Brauer varieties

Recall that the set of minimal left ideals of the matrix algebra $M_\ell(k)$ correspond to the k-points of the projective space $\mathbb{P}_k^{\ell-1}$; if I is a minimal left ideal corresponding to a line L of k^ℓ then the rows of matrices in I all lie on L.

Now fix a symbol $\underline{a} = \{a_1, a_2\}$ and a primitive ℓ^{th} root of unity in k, ζ. Let $A = A(\underline{a})$ denote the central simple algebra $k\{x, y\}/(x^\ell = a_1, y^\ell = a_2, xy = \zeta yx)$. It is well known that there is a smooth projective variety X of dimension $\ell-1$, defined over k, such that for every field F over k, $X(F)$ is the set of (nonzero) minimal ideals of $A \otimes_k F$: $X(F) \neq \emptyset$ if and only if $A \otimes_k F \cong M_\ell(F)$. The variety X is called the *Severi–Brauer* variety of A.

Here is one way to construct the Severi–Brauer variety X. If $E = k(\sqrt[\ell]{a_1})$ then $A \otimes_k E \cong M_\ell(E)$; the Galois group of E/k acts on the set of minimal ideals of $A \otimes_k E$ and hence on $\mathbb{P}_E^{\ell-1}$ and $X \times_k E$ is $\mathbb{P}_E^{\ell-1}$ with this Galois action. Now apply Galois descent. This method originated in [Ser63]; see [KMRT98].

Definition 1.14. If k contains a primitive ℓ^{th} root of unity, ζ, the *Severi–Brauer variety* X associated to a symbol $\underline{a} = \{a_1, a_2\}$ is defined to be the Severi–Brauer variety of $A = A(\underline{a})$. (The variety is independent of the choice of ζ.) If k does not contain a primitive ℓ^{th} root of unity, we will mean the Severi–Brauer variety for $\{a_1, a_2\}$ defined over $k(\zeta)$.

If $\zeta \in F$, there is a canonical map $K_2^M(F)/\ell \longrightarrow {}_\ell\mathrm{Br}(k)$, sending $\{a_1, a_2\}$ to its associated central simple algebra A. The Merkurjev–Suslin Theorem [MeS82] states that this is an isomorphism. Since $A \otimes_k k(X)$ is a matrix algebra by construction, the Merkurjev–Suslin Theorem implies that $k(X)$ splits $\underline{a}$. Here is a more elementary proof.

Lemma 1.15. *Every symbol* $\underline{a} = \{a_1, a_2\}$ *is split by its Severi–Brauer variety.*

Proof. (Merkurjev) Fix $\alpha = \sqrt[\ell]{a_1}$ and set $E = k(\alpha)$. Recall from [Wei82] (or 11.12) that the *Weil restrictions* of $\mathbb{A}^1$ along E and k are isomorphic to the affine spaces $\mathbb{A}^\ell$ and $\mathbb{A}^1$ over k, and the Weil restriction of the norm map $N_{E/k}$ is a map $N : \mathbb{A}^\ell \to \mathbb{A}^1$. Then the Severi–Brauer variety X is birationally equivalent to the subvariety of $\mathbb{A}^\ell$ defined by $N(X_0, \ldots, X_{\ell-1}) = a_2$.

In the function field $k(X)$, we set $x_i = X_i/X_0$ and $c = N(1, x_1, \ldots, x_{\ell-1})$, so that $c X_0^\ell = a_2$. Then $k(X) = k(x_1, \ldots, x_{\ell-1})(\beta)$, $\beta^\ell = a_2/c$. By construction, the element $y = 1 + \sum x_i \alpha^i$ of $E(X) = E(x_1, \ldots)$ has $Ny = c = a_2/\beta^\ell$ so in $K_2^M(k(X))/\ell$ we have

$$\{a_1, a_2\} = \{a_1, a_2/\beta^\ell\} = \{a_1, Ny\} = N\{\alpha^\ell, y\} = N(0) = 0.$$

Thus the field $k(X)$ splits the symbol $\underline{a}$.   □

Corollary 1.16. *The Severi–Brauer variety X of a symbol $\underline{a} = \{a_1, a_2\}$ is a norm variety for $\underline{a}$.*

Proof. Since any norm variety for $k(\zeta)$ is also a norm variety for k, and a field F splits $\underline{a}$ iff $F(\zeta)$ splits $\underline{a}$, we may assume that k contains a primitive ℓ^{th} root of unity. Thus X exists and is a smooth projective variety of dimension $\ell-1$. By Lemma 1.15, $k(X)$ splits the symbol. Finally, suppose that a field F/k splits $\underline{a}$. Then the associated central simple algebra is trivial $(A \otimes_k F \cong M_\ell(F))$ and hence $X(F) \neq \emptyset$.   □

The characteristic number s_d(x)

The definition of a Rost variety also involves the notion of a ν_i-variety, which is defined using the classical characteristic number $s_d(X)$.

Let X be a smooth projective variety of dimension $d > 0$. Recall from [MS74, §16] that there is a characteristic class $s_d : K_0(X) \to CH^d(X)$ corresponding to the symmetric polynomial $\sum t_j^d$ in the Chern roots t_j of a bundle; the characteristic number is the degree of the characteristic class. We shall write $s_d(X)$ for the characteristic number of the tangent bundle T_X, i.e., $s_d(X) = \deg(s_d(T_X))$. When $d = \ell^\nu - 1$, we know that $s_d(X) \equiv 0 \pmod{\ell}$; see [MS74, 16.6 and 16-E] and [Sto68, pp. 128–29] or [Ada74, II.7].

Definition 1.17. A ν_i-*variety* over a field k is a smooth projective variety X of dimension $d = \ell^i - 1$, with $s_d(X) \not\equiv 0 \pmod{\ell^2}$.

Remark. In topology, a smooth complex variety X of dimension $d = \ell^i - 1$ for which $s_d(X) \equiv \pm \ell \pmod{\ell^2}$ is called a *Milnor manifold*. In complex cobordism theory, the bordism classes of Milnor manifolds in MU_d are among the generators of the complex cobordism ring MU_* of stably complex manifolds.

Examples 1.18. (1) It is well known that $s_d(\mathbb{P}^d) = d + 1$; see [MS74, 16.6]. Setting $d = \ell - 1$, we see that $\mathbb{P}^{\ell-1}$ (and any form of it) is a ν_1-variety. In particular, the Severi–Brauer variety of a symbol $\{a_1, a_2\}$ is a ν_1-variety, since it is a form of $\mathbb{P}^{\ell-1}$.

(2) A smooth hypersurface X of degree ℓ in $\mathbb{P}^{d+1}$ has $s_d(X) = \ell(d+2-\ell^d)$ by [MS74, 16-D], so if $d = \ell^i - 1$ we see that X is a ν_i-variety and $X(\mathbb{C})$ is a Milnor manifold.

(3) We will see in Proposition 10.14 that if $\mathrm{char}(k) = 0$, any norm variety for a symbol $\{a_1, \ldots, a_n\}$ $(n \geq 2)$ is a ν_{n-1}-variety.

Borel–Moore homology

The Borel–Moore homology group $H^{BM}_{-1,-1}(X)$ of a scheme X is defined as $\mathrm{Hom}_{\mathbf{DM}}(\mathbb{Z}, M^c(X)(1)[1])$ if $\mathrm{char}(k) = 0$ (resp., $\mathrm{Hom}_{\mathbf{DM}}(\mathbb{Z}[1/p], \mathbb{Z}[1/p] \otimes M^c(X)$ $(1)[1])$ if $\mathrm{char}(k) = p > 0$ and k is perfect); see [MVW, 16.20]. Here $M^c(X)$ is the motive of X with compact supports. $H^{BM}_{-1,-1}(X)$ is a covariant functor in X for proper maps, and contravariant for finite flat maps, because $M^c(X)$ has these properties; see [MVW, 16.13]. When X is projective, the natural map from $M(X) = \mathbb{Z}_{\mathrm{tr}}(X)$ to $M^c(X)$ is an isomorphism in $\mathbf{DM}$, so the Borel–Moore homology group agrees with the usual motivic homology group $H_{-1,-1}(X, R)$, which is defined as $\mathrm{Hom}_{\mathbf{DM}}(R, R_{\mathrm{tr}}(X)(1)[1])$, where R is $\mathbb{Z}$ (resp., $\mathbb{Z}[1/p]$); see [MVW, 14.17].

Proposition 1.19. *Let X be a smooth variety over a perfect field k. Then $H^{BM}_{-1,-1}(X)$ is the group generated by symbols $[x, \alpha]$, where x is a closed point of X and $\alpha \in k(x)^\times$, modulo the relations*

(i) $[x, \alpha][x, \alpha'] = [x, \alpha\alpha']$ and
(ii) for every point y of X such that $\dim(\overline{\{y\}}) = 1$, the image of the tame symbol $K_2(k(y)) \to \oplus k(x)^\times$ is zero.

   That is, we have an exact sequence

$$\bigoplus_y K^M_2(k(y)) \xrightarrow{\ tame\ } \bigoplus_x k(x)^\times \xrightarrow{\ \oplus[x,-]\ } H^{BM}_{-1,-1}(X) \to 0.$$

Proof. Let A denote the abelian group presented in the Proposition, and set $d = \dim(X)$. Note that A is uniquely p-divisible when k is a perfect field of characteristic $p > 0$, because each $k(x)^\times$ is uniquely p-divisible, and the group $K^M_2(k(y))$ is also uniquely p-divisible by Lemma 1.20.

   We first show that A is isomorphic to $H^{2d+1,d+1}(X, \mathbb{Z})$. To this end, consider the hypercohomology spectral sequence $E^{p,q}_2 = H^p(X, \mathcal{H}^q) \Rightarrow H^{p+q,d+1}(X, \mathbb{Z})$, where $\mathcal{H}^q$ denotes the Zariski sheaf associated to the presheaf $H^{q,d+1}(-, \mathbb{Z})$. Since $H^{q,d+1} = 0$ for $q > d+1$, the terms $E^{p,q}_2$ are zero unless $p \leq d$ and $q \leq d+1$. From this we deduce that $H^{2d+1,d+1}(X, \mathbb{Z}) \cong H^d(X, \mathcal{H}^{d+1})$.

   For each n, $\mathcal{H}^n$ is a homotopy invariant Zariski sheaf, by [MVW, 24.1]. Moreover, it has a canonical flasque "Gersten" resolution on each smooth X, given in [MVW, 24.11], whose c^{th} term is the coproduct of the skyscraper sheaves $H^{n-c,d+1-c}(k(z))$ for which z has codimension c in X. Taking $n = d+1$, and recalling that $K^M_n \cong H^{n,n}$ on fields, we see that the skyscraper sheaves in the $(d-1)^{st}$ and d^{th} terms take values in $K^M_2(k(y))$ and $K^M_1(k(x))$. Moreover, by [Wei13, V.9.2 and V(6.6.1)], the map $K^M_2(k(y)) \to K^M_1(k(x))$ is the tame symbol if $x \in \overline{\{y\}}$, and zero otherwise. As $H^d(X, \mathcal{H}^{d+1})$ is obtained by taking global sections of the Gersten resolution and then cohomology, we see that it is isomorphic to A.

   Now suppose that $\mathrm{char}(k) = 0$. Using motivic duality with $d = \dim(X)$ (see [MVW, 16.24] or [FV00, 7.1]), the proof is finished by the duality calculation:

$$H^{BM}_{-1,-1}(X,\mathbb{Z}) = \mathrm{Hom}(\mathbb{Z}, M^c(X)(1)[1])$$
$$= \mathrm{Hom}(\mathbb{Z}(d)[2d], M^c(X)(d+1)[2d+1]) \qquad (1.19\mathrm{a})$$
$$= \mathrm{Hom}_{\mathbf{DM}}(M(X), \mathbb{Z}(d+1)[2d+1]) = H^{2d+1,d+1}(X).$$

Now suppose that $\mathrm{char}(k) > 0$. Since $H^{2d+1,d+1}(X,\mathbb{Z}) \cong A$ is uniquely divisible, the duality calculation (1.19a) goes through with $\mathbb{Z}$ replaced by $\mathbb{Z}[1/p]$, using the characteristic p version of motivic duality (see [Kel13, 5.5.14]). □

Lemma 1.20. *(Bloch–Kato–Gabber)* If F is a field of transcendence degree 1 over a perfect field k of characteristic p, $K^M_2(F)$ is uniquely p-divisible.

Proof. For any field F of characteristic p, the group $K_2(F)$ has no p-torsion (see [Wei13, III.6.7]), and the $d\log$ map $K_2(F)/p \to \Omega^2_F$ is an injection with image $\nu(2)$; see [Wei13, III.7.7.2]. Since k is perfect, $\Omega^1_k = 0$ and Ω^1_F is 1–dimensional, so $\Omega^2_F = 0$ and hence $K_2(F)/p = 0$. □

The motivic homology functor $H^{BM}_{-1,-1}(X)$ has other names in the literature. It is isomorphic to the K-cohomology groups $H^d(X, \mathcal{K}_{d+1})$ [Qui73] and $H^d(X, \mathcal{K}^M_{d+1})$, where $d = \dim(X)$, and to Rost's Chow group with coefficients $A_0(X, \mathcal{K}_1)$ [Ros96]. Since we will only be concerned with smooth projective varieties X and integral coefficients, we will omit the superscript "BM" and the coefficients and just write $H_{-1,-1}(X)$.

Examples 1.21. (i) $H_{-1,-1}(\mathrm{Spec}\, E) = E^\times$ for every field E over k. This is immediate from the presentation in 1.19.
(ii) If E is a finite extension of k, the proper pushforward from $E^\times = H_{-1,-1}$ ($\mathrm{Spec}\, E$) to $k^\times = H_{-1,-1}(\mathrm{Spec}\, k)$ is just the norm map $N_{E/k}$.
(iii) For any proper variety X over k, the pushforward map

$$N_{X/k} : H_{-1,-1}(X) \to H_{-1,-1}(\mathrm{Spec}\, k) = k^\times$$

is induced by the composites $\mathrm{Spec}\, k(x) \to X \to \mathrm{Spec}\, k$, $x \in \overset{\cdot}{X}$. By (ii), we see that $N_{X/k}$ sends $[x, \alpha]$ to the norm $N_{k(x)/k}(\alpha)$.

Definition 1.22. When X is proper, the projections $X \times X \to X$ are proper and we may define the reduced group $\overline{H}_{-1,-1}(X)$ to be the coequalizer of $H_{-1,-1}(X \times X) \rightrightarrows H_{-1,-1}(X)$, i.e., the quotient of $H_{-1,-1}(X)$ by the difference of the two projections.

Example 1.23. When $E = k(\sqrt[\ell]{a})$ is a cyclic field extension of k, with Galois group generated by σ, then $\overline{H}_{-1,-1}(\mathrm{Spec}\, E)$ is the cokernel of $E^\times \xrightarrow{1-\sigma} E^\times$, and Hilbert's Theorem 90 induces an exact sequence

$$0 \to \overline{H}_{-1,-1}(\mathrm{Spec}\, E) \xrightarrow{N_{E/k}} k^\times \xrightarrow{a\cup} \mathrm{Br}(E/k) \to 0.$$

Note that $\mathrm{Br}(E/k)$ is a subgroup of $K_2^M(k)/\ell$ when $\mu_\ell \subset k^\times$. We will generalize this in Proposition 7.7, using $K_{n+1}^M(k)/\ell$.

Rost varieties

Definition 1.24. A *Rost variety* for a sequence $\underline{a} = (a_1, \ldots, a_n)$ of units of k is a ν_{n-1}-variety X satisfying:

(a) X is a splitting variety for $\underline{a}$, i.e., $\underline{a}$ vanishes in $K_n^M(k(X))/\ell$;
(b) for each integer i, $1 \le i < n$, there is a ν_i-variety mapping to X;
(c) the map $N : \overline{H}_{-1,-1}(X) \to k^\times$ is an injection.

When $n = 1$, $\mathrm{Spec}(k(\sqrt[\ell]{a}))$ is a Rost variety for a. When $n = 2$, Proposition 1.25 shows that Severi–Brauer varieties of dimension $\ell-1$ are Rost varieties. In chapter 11 we will show that Rost varieties exist over ℓ-special fields for all n, ℓ and $\underline{a}$, at least when $\mathrm{char}(k) = 0$. More specifically, Theorem 11.2 shows that norm varieties for $\underline{a}$ are Rost varieties for $\underline{a}$.

Proposition 1.25. *The Severi–Brauer variety X of a symbol $\underline{a} = \{a_1, a_2\}$ is a Rost variety for $\underline{a}$.*

Proof. By Lemma 1.15, X splits $\underline{a}$; by Example 1.18(1), X is a ν_1-variety. Finally, Quillen proved that $H_{-1,-1}(X) = H^1(X, \mathcal{K}_2)$ is isomorphic to $K_1(A)$, and it is classical that $K_1(A)$ is the image of $A^\times \to k^\times$; see [Wan50, p. 327]. $\square$

1.4   THE BEILINSON–LICHTENBAUM CONDITIONS

Our approach to Theorems A and B (for n) will use their equivalence with a more general condition, which we call the *Beilinson–Lichtenbaum* condition BL(n). In this section, we define BL(n) (in 1.28); in section 2.1 we show that it implies the corresponding condition BL(p) for all $p < n$.

Consider the morphism of sites $\pi : (\mathbf{Sm}/k)_{\text{ét}} \to (\mathbf{Sm}/k)_{\text{zar}}$, where π_* is restriction and π^* sends a Zariski sheaf $\mathcal{F}$ to its associated étale sheaf $\mathcal{F}_{\text{ét}}$. The total direct image $\mathbf{R}\pi_*$ sends an étale sheaf (or complex of sheaves) $\mathcal{F}$ to a Zariski complex such that $H_{\text{zar}}^*(X, \mathbf{R}\pi_*\mathcal{F}) = H_{\text{ét}}^*(X, \mathcal{F})$. In particular, the Zariski cohomology of $\mathbf{R}\pi_* \mu_\ell^{\otimes n}$ agrees with the étale cohomology of $\mu_\ell^{\otimes n}$.

Recall [Wei94, 1.2.7] that the good truncation $\tau^{\le n}\mathcal{C}$ of a cochain complex $\mathcal{C}$ is the universal subcomplex which has the same cohomology as $\mathcal{C}$ in degrees $\le n$ but is acyclic in higher degrees. Applying this to $\mathbf{R}\pi_*\mathcal{F}$ leads to the following useful complexes.

Definition 1.26. The cochain complexes of Zariski sheaves $L(n)$ and $L/\ell^\nu(n)$ are defined to be

$$L(n) = \tau^{\le n}\mathbf{R}\pi_*[\mathbb{Z}_{(\ell)}(n)] \quad \text{and} \quad L/\ell^\nu(n) = \tau^{\le n}\mathbf{R}\pi_*[\mathbb{Z}/\ell^\nu(n)].$$

We know by [MVW, 10.3] that for each n (and all ν) there is a quasi-isomorphism of complexes of étale sheaves $\mu_{\ell^\nu}^{\otimes n} \xrightarrow{\sim} \mathbb{Z}/\ell^\nu(n)$. When X is a Zariski local scheme this implies that $H^n(X, L(n))$ is: $H_{\text{ét}}^n(X, \mathbb{Z}_{(\ell)}(n))$ for $p \leq n$ and zero for $p > n$; while $H^n(X, L/\ell^\nu(n))$ is: $H_{\text{ét}}^p(X, \mu_{\ell^\nu}^{\otimes n})$ for $p \leq n$ and zero for $p > n$.

Now $\mathbb{Z}_{(\ell)}(n)$ and the $\mathbb{Z}/\ell^\nu(n)$ are étale sheaves with transfers, so their canonical flasque resolutions $E^\bullet$ are complexes of étale sheaves with transfers by [MVW, 6.20]. The restriction $\pi_* E^\bullet$ to the Zariski site inherits the transfer structure, so the truncations $L(n)$ and $L/\ell^\nu(n)$ are complexes of Zariski sheaves with transfers.

The adjunction $1 \to \mathbf{R}\pi_*\pi^*$ gives a natural map of Zariski complexes $\mathbb{Z}/\ell^\nu(n) \to \mathbf{R}\pi_*[\mathbb{Z}/\ell^\nu(n)]$. Since the complexes $\mathbb{Z}(n)$ and $\mathbb{Z}/\ell^\nu(n)$ are zero above degree n by construction ([MVW, 3.1]), we may apply $\tau^{\leq n}$ to obtain morphisms of sheaves on $\mathbf{Sm}/k$:

$$\mathbb{Z}_{(\ell)}(n) \xrightarrow{\tilde{\alpha}_n} L(n), \quad \mathbb{Z}/\ell^\nu(n) \xrightarrow{\alpha_n} L/\ell^\nu(n). \tag{1.27}$$

Definition 1.28. We will say that $BL(n)$ *holds* if the map $\mathbb{Z}/\ell(n) \xrightarrow{\alpha_n} L/\ell(n)$ is a quasi-isomorphism for any field k containing $1/\ell$. This is equivalent to the seemingly stronger but analogous assertion with coefficients $\mathbb{Z}/\ell^\nu$; see 1.29(a).

Beilinson and Lichtenbaum had conjectured that $BL(n)$ holds for all n, whence the name; see [Lic84, §3] and [Beĭ87, 5.10.D].

Lemma 1.29. *If $BL(n)$ holds then:*

(a) $\alpha_n : \mathbb{Z}/\ell^\nu(n) \xrightarrow{\sim} \tau^{\leq n}\mathbf{R}\pi_*\mu_{\ell^\nu}^{\otimes n}$ is a quasi-isomorphism for all $\nu \geq 1$;
(b) $\mathbb{Q}/\mathbb{Z}_{(\ell)}(n) \xrightarrow{\sim} \tau^{\leq n}\mathbf{R}\pi_*[\mathbb{Q}/\mathbb{Z}_{(\ell)}(n)]$ is a quasi-isomorphism;
(c) $\tilde{\alpha}_n : \mathbb{Z}_{(\ell)}(n) \xrightarrow{\sim} L(n) = \tau^{\leq n}\mathbf{R}\pi_*[\mathbb{Z}_{(\ell)}(n)]$ is also a quasi-isomorphism;
(d) $K_n^M(k)_{(\ell)} \to H_{\text{ét}}^n(k, \mathbb{Z}_{(\ell)}(n))$ is an isomorphism for all k containing $1/\ell$.

Proof. The statement for $\mathbb{Z}/\ell^\nu$ coefficients follows by induction on ν using the morphism of distinguished triangles:

$$\begin{array}{ccccccccc} \mathbb{Z}/\ell(n)[-1] & \longrightarrow & \mathbb{Z}/\ell^{\nu-1}(n) & \longrightarrow & \mathbb{Z}/\ell^\nu(n) & \longrightarrow & \mathbb{Z}/\ell(n) & \longrightarrow & \mathbb{Z}/\ell^{\nu-1}(n)[1] \\ \downarrow\cong & & \downarrow\cong & & \alpha_n\downarrow & & \downarrow\cong & & \downarrow\cong \\ L/\ell(n)[-1] & \longrightarrow & L/\ell^{\nu-1}(n) & \longrightarrow & L/\ell^\nu(n) & \longrightarrow & L/\ell(n) & \longrightarrow & L/\ell^{\nu-1}(n)[1]. \end{array}$$

Taking the direct limit over ν in part (a) yields part (b).

Since $\tilde{\alpha}_n$ is also an isomorphism for $\mathbb{Q}$ coefficients by [MVW, 14.23], the coefficient sequence for $0 \to \mathbb{Z}_{(\ell)}(n) \to \mathbb{Q}(n) \to \mathbb{Q}/\mathbb{Z}_{(\ell)}(n) \to 0$ shows that $\mathbb{Z}_{(\ell)}(n) \xrightarrow{\sim} L(n)$ is also a quasi-isomorphism. Part (d) is immediate from (c) and $K_n^M(k)_{(\ell)} \cong H_{\text{zar}}^n(k, \mathbb{Z}(n))_{(\ell)} = H_{\text{zar}}^n(k, \mathbb{Z}_{(\ell)}(n))$. $\square$

The main result in chapter 2 is that BL(n) is equivalent to H90(n) and hence Theorem A, that $K_n^M(k)/\ell \cong H_{\text{ét}}^n(k, \mu_\ell^{\otimes n})$. The fact that H90(n) implies BL(n) is proven in Theorem 2.38. Here is the easier converse, that BL(n) implies H90(n).

Lemma 1.30. *If BL(n) holds then H90(n) holds.*
In addition, if BL(n) holds then for any field k containing $1/\ell$:

(a) $K_n^M(k)/\ell \cong H^n(k, \mathbb{Z}/\ell(n)) \cong H_{\text{ét}}^n(k, \mu_\ell^{\otimes n})$.
(b) *For all $p \leq n$, $H^p(k, \mathbb{Z}/\ell(n)) \cong H_{\text{ét}}^p(k, \mu_\ell^{\otimes n})$.*

Proof. Applying $H^p(k, -)$ to α_n yields (b). Setting $p = n$ in (b) proves (a), because $K_n^M(k)/\ell \cong H^n(k, \mathbb{Z}/\ell(n))$. By Theorem 1.7, (a) implies H90(n). □

Corollary 1.31.[2] *If BL(n) holds then for every smooth simplicial scheme $X_\bullet$ the map $H^{p,n}(X_\bullet, \mathbb{Z}/\ell) \to H_{\text{ét}}^p(X_\bullet, \mu_\ell^{\otimes n})$ is an isomorphism for all $p \leq n$. It is an injection when $p = n + 1$.*

Proof. First, suppose that X is a smooth scheme. A comparison of the hypercohomology spectral sequences $H^p(X, \mathcal{H}^q) \Rightarrow H^{p+q}(X)$ for coefficient complexes $L/\ell(n)$ and $R\pi_*[\mathbb{Z}/\ell(n)]$ shows that $\alpha_n : H^{p,n}(X, \mathbb{Z}/\ell) \to H_{\text{ét}}^p(X, \mu_\ell^{\otimes n})$ is an isomorphism for $p \leq n$ and an injection for $p = n + 1$.

For $X_\bullet$, the assertion follows from a comparison of the spectral sequences $E_2^{p,q} = H^q(X_p) \Rightarrow H^{p+q}(X_\bullet)$ for the Zariski and étale topologies, and the result for each smooth scheme X_p. □

1.5   SIMPLICIAL SCHEMES

In this section, we construct a certain simplicial scheme $\mathfrak{X}$ which will play a crucial role in our constructions, and introduce some features of its cohomology.

It is well known that the hypercohomology of a simplicial scheme $X_\bullet$ agrees with the group of morphisms in the derived category of sheaves of abelian groups, from the representable simplicial sheaf $\mathbb{Z}[X_\bullet]$ (regarded as a complex of sheaves via the Dold–Kan correspondence) to the coefficient sheaf complex. Applying this to the coefficient complex $A(q)$, we obtain the original definition of the motivic cohomology of $X_\bullet$: $H^{p,q}(X_\bullet, A) = H_{\text{zar}}^p(X_\bullet, A(q))$; see [MVW, 3.4].

For our purposes, it is more useful to work in the triangulated category **DM**, which is a quotient of the derived category of Nisnevich sheaves with transfers, or its triangulated subcategory $\mathbf{DM}_{\text{nis}}^{\text{eff}}$, where we have

$$H^{p,q}(X_\bullet, A) \cong \text{Hom}_{\mathbf{DM}_{\text{nis}}^{\text{eff}}}(\mathbb{Z}_{\text{tr}}(X_\bullet), A(q)[p]) = \text{Hom}_{\mathbf{DM}}(\mathbb{Z}_{\text{tr}}(X_\bullet), A(q)[p]).$$

---

2. Taken from [Voe03a, 6.9]. It is needed for Lemma 3.13.

See [MVW, 14.17]. Similarly, the étale motivic cohomology $H_{\text{ét}}^*(X_\bullet, A(q))$ is the étale hypercohomology of the étale sheaf $A(q)_{\text{ét}}$ underlying $A(q)$, and agrees with $\text{Hom}_{\mathbf{DM}_{\text{ét}}^-}(\mathbb{Z}_{\text{tr}}(X_\bullet), A(q)[p])$; see [MVW, 10.1, 10.7].

We begin with a simplicial set construction. Associated to any nonempty set S there is a contractible simplicial set $\check{C}(S): n \mapsto S^{n+1}$; the face maps are projections (omit a term) and the degeneracy maps are diagonal maps (duplicate a term). In fact, $\check{C}(S)$ is the 0-coskeleton of S; see Lemma 12.6. More generally, for any set T, the projection $T \times \check{C}(S) \to T$ is a homotopy equivalence; it is known as the canonical cotriple resolution of T associated to the cotriple $\bot(T) = T \times S$; see [Wei94, 8.6.8].

Definition 1.32. Let X be a (nonempty) smooth scheme over k. We write $\mathfrak{X} = \check{C}(X)$ for the simplicial scheme $\mathfrak{X}_n = X^{n+1}$, whose face maps are given by projection:

$$X \Leftarrow X \times X \Lleftarrow X^3 \Lleftarrow X^4 \cdots .$$

That is, $\mathfrak{X}$ is the 0-coskeleton of X.

We may regard $\mathfrak{X}$ and $\mathfrak{X} \times Y$ as simplicial representable presheaves on $\mathbf{Sm}/k$; for any smooth U, $\mathfrak{X}(U) = \check{C}(X(U))$. Thus if $X(Y) = \text{Hom}(Y, X) \neq \emptyset$ then the projection $(\mathfrak{X} \times Y)(U) \to Y(U)$ is a homotopy equivalence for all U by the cotriple remarks above. In particular, $\mathfrak{X}(k)$ is either contractible or $\emptyset$, according to whether or not X has a k-rational point.

Remark 1.32.1. A map of simplicial presheaves is called a *global weak equivalence* if its evaluation on each U is a weak equivalence of simplicial sets. It follows that $\mathfrak{X} \to \text{Spec}(k)$ is a global weak equivalence if and only if X has a k-rational point, and more generally that the projection $\mathfrak{X} \times Y \to Y$ is a global weak equivalence if and only if $\text{Hom}(Y, X) \neq \emptyset$.

We will frequently use the following standard fact. We let R denote $\mathbb{Z}$ if $\text{char}(k) = 0$, and $\mathbb{Z}[1/\text{char}(k)]$ if k is a perfect field of positive characteristic.

Lemma 1.33. *For all smooth Y and $p > q$, $\text{Hom}_{\mathbf{DM}}(R, R_{\text{tr}}(Y)(q)[p]) = 0$.*

Proof. By definition [MVW, 3.1], $R_{\text{tr}}(Y)(q)[q]$ is a chain complex $C_*(Y \times \mathbb{G}_m^{\wedge q})$ of sheaves which is zero in positive cohomological degrees. By [MVW, 14.16],

$$\text{Hom}(R, R_{\text{tr}}(Y)(q)[p]) \cong H_{\text{zar}}^{p-q}(k, R_{\text{tr}}(Y)(q)[q]) = H^{p-q}R_{\text{tr}}(Y)(q)[q](k). \quad \square$$

Lemma 1.34. *For every smooth X, $H_{-1,-1}(\mathfrak{X}) \cong \overline{H}_{-1,-1}(X)$.*

Proof. For all p and $n > 1$, Lemma 1.33 yields $\text{Hom}_{\mathbf{DM}}(R, R_{\text{tr}}X^p(1)[n]) = 0$. Therefore every row below $q = -1$ in the spectral sequence

$$E_{pq}^1 = \text{Hom}(R[q], R_{\text{tr}}X^{p+1}(1)) \Rightarrow \text{Hom}(R, R_{\text{tr}}\mathfrak{X}(1)[p - q]) = H_{q-p,-1}(\mathfrak{X})$$

is zero. The homology at $(p, q) = (0, -1)$ yields the exact sequence

$$0 \longleftarrow H_{-1,-1}(\mathfrak{X}) \longleftarrow H_{-1,-1}(X) \longleftarrow H_{-1,-1}(X \times X).$$

Since $\overline{H}_{-1,-1}(X)$ is the cokernel of the right map, the result follows.   □

Lemma 1.35.[3] *For every smooth* X, $H^{0,0}(\mathfrak{X}, R) = R$ *and* $H^{p,0}(\mathfrak{X}, R) = 0$ *for* $p > 0$; $H^{0,1}(\mathfrak{X}, \mathbb{Z}) = H^{2,1}(\mathfrak{X}, \mathbb{Z}) = 0$ *and* $H^{1,1}(\mathfrak{X}; \mathbb{Z}) \cong H^{1,1}(\operatorname{Spec} k; \mathbb{Z}) \cong k^{\times}$.

Proof. The spectral sequence $E_1^{p,q} = H^q(X^{p+1}; R) \Rightarrow H^{p+q,0}(\mathfrak{X}; R)$ degenerates at E_2 for X smooth, being zero for $q > 0$, and the R-module cochain complex of the contractible simplicial set $\check{C}(\pi_0(X))$ for $q = 0$.

The spectral sequence $E_1^{p,q} = H^q(X^{p+1}; \mathbb{Z}(1)) \Rightarrow H^{p+q,1}(\mathfrak{X}; \mathbb{Z})$ degenerates at E_2, all rows vanishing except for $q = 1$ and $q = 2$, because $\mathbb{Z}(1) \cong \mathcal{O}^{\times}[-1]$; see [MVW, 4.2]. We compare this with the spectral sequence converging to $H_{\text{ét}}^{p+q}(\mathfrak{X}; \mathbb{G}_m)$; $H_{\text{zar}}^q(Y, \mathcal{O}^{\times}) \to H_{\text{ét}}^q(Y, \mathbb{G}_m)$ is an isomorphism for $q = 0, 1$ (and an injection for $q = 2$). Hence we have $H^{q,1}(\mathfrak{X}) = H_{\text{ét}}^{q,1}(\mathfrak{X})$ for $q \leq 2$, and $H_{\text{ét}}^{q,1}(\mathfrak{X}) \cong H_{\text{ét}}^{q,1}(k) = H_{\text{ét}}^{q-1}(k, \mathbb{G}_m)$ by Lemma 1.37.   □

Recall that if $f : X_{\bullet} \to Y_{\bullet}$ is a morphism of simplicial objects in any category with coproducts and a final object, the cone of f is also a simplicial object. It is defined in [Del74, 6.3.1].

Definition 1.36. The suspension $\Sigma X_{\bullet}$ of a simplicial scheme $X_{\bullet}$ is the cone of $(X_{\bullet})_+ \to \operatorname{Spec}(k)_+$. The reduced suspension $\widetilde{\Sigma} X_{\bullet}$ of any simplicial scheme $X_{\bullet}$ is the pointed pair $(\Sigma X_{\bullet}, \text{point})$, where "point" is the image of $\operatorname{Spec}(k)$ in $\Sigma X_{\bullet}$.

If $X_{\bullet}$ is pointed then $H^{p,q}(\widetilde{\Sigma} X_{\bullet}) = \widetilde{H}^{p,q}(\Sigma X_{\bullet})$, but this makes little sense when $X_{\bullet}$ has no k-points. The pointed pair is chosen to avoid this problem. By construction there is a long exact sequence on cohomology:

$$\cdots \to H^{p-1,q}(X_{\bullet}) \to H^{p,q}(\widetilde{\Sigma} X_{\bullet}) \to H^{p,q}(\operatorname{Spec} k) \to H^{p,q}(X_{\bullet}) \to \cdots.$$

In particular, if $X_{\bullet}$ is pointed then we have the suspension isomorphism $\sigma_s :$ $\widetilde{H}^{p-1,q}(X_{\bullet}) \to H^{p,q}(\widetilde{\Sigma} X_{\bullet})$. If $p > q$ then $H^{p,q}(X_{\bullet}) \xrightarrow{\cong} H^{p+1,q}(\widetilde{\Sigma} X_{\bullet})$, because in this range $H^{p,q}(\operatorname{Spec} k) = 0$.

Lemma 1.37.[4] *If* X *has a point* x *with* $[k(x):k] = e$ *then for each* (p, q) *the group* $H^p(\widetilde{\Sigma} \mathfrak{X}, \mathbb{Z}(q))$ *has exponent* e. *Hence the kernel and cokernel of each* $H^p(k, \mathbb{Z}(q)) \to H^p(\mathfrak{X}, \mathbb{Z}(q))$ *has exponent* e.

The maps $H_{\text{ét}}^{p,q}(k, \mathbb{Z}) \xrightarrow{\cong} H_{\text{ét}}^{p,q}(\mathfrak{X}, \mathbb{Z})$ *are isomorphisms for all* (p, q). *Therefore* $H_{\text{ét}}^{*,*}(\widetilde{\Sigma} \mathfrak{X}, \mathbb{Z}) = 0$ *and* $H_{\text{ét}}^{*,*}(\widetilde{\Sigma} \mathfrak{X}, \mathbb{Z}/\ell) = 0$.

---

3. $H^{0,0}(\mathfrak{X})$ and $H^{0,1}(\mathfrak{X})$ are used in 4.5 and 4.15.
4. Based on Lemmas 9.3 and 7.3 of [Voe03a], respectively.

Proof. Set $F(Y) = H^p(\widetilde{\Sigma}\mathfrak{X} \times Y, \mathbb{Z}(q))$; this is a presheaf with transfers which vanishes on $\mathrm{Spec}(k(x))$. As with any presheaf with transfers, the composition $F(k) \to F(k(x)) \to F(k)$ is multiplication by e. It follows that $e \cdot F(k) = 0$.

Now any nonempty X has a point x with $k(x)/k$ étale, and $\mathfrak{X}_x = \mathfrak{X} \times \mathrm{Spec}\, k(x)$ is an étale cover of $\mathfrak{X}$. Since the map from $\mathfrak{X}_x$ to the étale cover x of $\mathrm{Spec}(k)$ is a global weak equivalence, the second assertion follows from a comparison of the descent spectral sequences for the covers of $\mathfrak{X}$ and $\mathrm{Spec}\, k$. □

As in topology, the integral Bockstein $\tilde{\beta}: H^{p,q}(Y, \mathbb{Z}/\ell) \to H^{p+1,q}(Y, \mathbb{Z})$ is the boundary map in the cohomology sequence for the coefficient sequence $0 \to \mathbb{Z}(q) \xrightarrow{\ell} \mathbb{Z}(q) \to \mathbb{Z}/\ell(q) \to 0$; the usual Bockstein $\beta: H^{p,q}(Y, \mathbb{Z}/\ell) \to H^{p+1,q}(Y, \mathbb{Z}/\ell)$ is the boundary map for $0 \to \mathbb{Z}(q) \xrightarrow{\ell} \mathbb{Z}/\ell^2(q) \to \mathbb{Z}/\ell(q) \to 0$. Both are natural in Y; see 1.42(3) and section 13.1 for more information.

Corollary 1.38. *Suppose that X has a point of degree ℓ. Then the motivic cohomology groups $H^{*,*}(\widetilde{\Sigma}\mathfrak{X}, \mathbb{Z})$ have exponent ℓ, and we have exact sequences:*

$$0 \to H^{p,q}(\widetilde{\Sigma}\mathfrak{X}, \mathbb{Z}) \to H^{p,q}(\widetilde{\Sigma}\mathfrak{X}, \mathbb{Z}/\ell) \xrightarrow{\tilde{\beta}} H^{p+1,q}(\widetilde{\Sigma}\mathfrak{X}, \mathbb{Z}) \to 0,$$

$$H^{p-1,q}(\widetilde{\Sigma}\mathfrak{X}, \mathbb{Z}/\ell) \xrightarrow{\beta} H^{p,q}(\widetilde{\Sigma}\mathfrak{X}, \mathbb{Z}/\ell) \xrightarrow{\beta} H^{p+1,q}(\widetilde{\Sigma}\mathfrak{X}, \mathbb{Z}/\ell).$$

Corollary 1.39. *If $BL(n-1)$ holds and X is smooth then $H^{p,n-1}(\widetilde{\Sigma}\mathfrak{X}, \mathbb{Z}/\ell) = 0$ for all $p \le n$, and $H^{p,q}(\mathrm{Spec}\, k, \mathbb{Z}/\ell) \xrightarrow{\cong} H^{p,q}(\mathfrak{X}, \mathbb{Z}/\ell)$ for all $p \le q < n$.*

Proof. As $H^{p,n-1}_{\text{ét}}(\widetilde{\Sigma}\mathfrak{X}, \mathbb{Z}/\ell) = 0$ by Lemma 1.37, the first assertion follows from 1.31. The second assertion follows from the cohomology sequence in Definition 1.36, and Lemma 1.30. □

Example 1.40. Assume that $BL(n-1)$ holds, and that X has a point of degree ℓ. Then $H^{n,n-1}(\widetilde{\Sigma}\mathfrak{X}, \mathbb{Z}/\ell) = 0$ by 1.39. From the first sequence in 1.38, and naturality of $\tilde{\beta}$, we see that $H^{n+1,n-1}(\widetilde{\Sigma}\mathfrak{X}, \mathbb{Z}) = 0$ and hence the integral Bockstein

$$H^{n+1,n-1}(\widetilde{\Sigma}\mathfrak{X}, \mathbb{Z}/\ell) \xrightarrow{\tilde{\beta}} H^{n+2,n-1}(\widetilde{\Sigma}\mathfrak{X}, \mathbb{Z})$$

is injective. It follows that the integral Bockstein $\tilde{\beta}: H^{n,n-1}(\mathfrak{X}, \mathbb{Z}/\ell) \to H^{n+1,n-1}(\mathfrak{X}, \mathbb{Z})$ is an injection because, as noted in 1.36, $H^{n,n-1}(\mathfrak{X}, \mathbb{Z}/\ell) \cong H^{n+1,n-1}(\widetilde{\Sigma}\mathfrak{X}, \mathbb{Z}/\ell)$ and $H^{n+1,n-1}(\mathfrak{X}, \mathbb{Z}) \cong H^{n+2,n-1}(\widetilde{\Sigma}\mathfrak{X}, \mathbb{Z})$.

1.6 MOTIVIC COHOMOLOGY OPERATIONS

Cohomology operations are another fundamental tool we shall need, both in section 3.4 (to construct the element μ of Corollary 3.16), and in chapter 5 (to show that Rost motives exist). We refer the reader to chapter 13 for more discussion.

Recall that for each coefficient group A, and all $p, q \geq 0$, the motivic cohomology groups $H^{p,q}(-, A) = H^p(-, A(q))$ are contravariant functors from the category $\Delta^{\mathrm{op}}\mathbf{Sm}/k$ of smooth simplicial schemes over k to abelian groups. For each set of integers n, i, p, q and every two groups A and B, a *cohomology operation* ϕ from $H^{n,i}(-, A)$ to $H^{p,q}(-, B)$ is just a natural transformation. The *bidegree* of ϕ is $(p - n, q - i)$.

There is a *twist isomorphism* $\sigma_t : H^{n,i}(X, A) \xrightarrow{\sim} H^{n+1,i+1}(X_+ \wedge \mathbb{G}_m, A)$ of bidegree $(1, 1)$ in motivic cohomology; see [Voe03c, 2.4] or [MVW, 16.25].

Definition 1.41. A family of operations $\phi_{(n,i)} : H^{n,i}(-, A) \to H^{n+p,i+q}(-, B)$ with a fixed bidegree (p, q) is said to be *bi-stable* if it commutes with the suspension and twist isomorphisms, σ_s and σ_t.

Examples 1.42. There are several kinds of bi-stable operations.

1. Any homomorphism $A \to B$ induces a bi-stable operation of bidegree $(0, 0)$, the change of coefficients map $H^{*,*}(-, A) \to H^{*,*}(-, B)$.
2. If R is a ring and A is an R-module then multiplication by $\lambda \in H^{p,q}(k, R)$ is a bi-stable operation of bidegree (p, q) from $H^{*,*}(-, A)$ to itself.
3. The *integral Bockstein* $\tilde{\beta} : H^{n,i}(X, \mathbb{Z}/\ell) \to H^{n+1,i}(X, \mathbb{Z})$ and its reduction modulo ℓ, the usual *Bockstein* $\beta : H^{n,i}(X, \mathbb{Z}/\ell) \to H^{n+1,i}(X, \mathbb{Z}/\ell)$ are both bi-stable operations. They are the boundary maps in the long exact cohomology sequence associated to the coefficient sequences

$$0 \to \mathbb{Z}(q) \xrightarrow{\ell} \mathbb{Z}(q) \to \mathbb{Z}/\ell(q) \to 0, \quad \text{and}$$

$$0 \to \mathbb{Z}/\ell(q) \xrightarrow{\ell} \mathbb{Z}/\ell^2(q) \to \mathbb{Z}/\ell(q) \to 0.$$

4. In [Voe03c, p. 33], Voevodsky constructed the *reduced power* operations

$$P^i : H^{p,q}(X, \mathbb{Z}/\ell) \to H^{p+2i(\ell-1), q+i(\ell-1)}(X, \mathbb{Z}/\ell)$$

and proved that they are bi-stable. If $\ell = 2$ it is traditional to write Sq^{2i} for P^i and Sq^{2i+1} for βP^i.

We may compose bi-stable operations if the coefficient groups match: $\phi' \circ \phi$ is a bi-stable operation whose bidegree is bidegree(ϕ') + bidegree(ϕ). It follows that the stable cohomology operations with $A = B = R$ form a bigraded ring, and that $H^{*,*}(k, R)$ is a subring.

Definition 1.43. (Milnor operations). There is a family of motivic operations Q_i on $H^{*,*}(X, \mathbb{Z}/\ell)$ constructed in [Voe03c, §13], called the *Milnor operations*. The bidegree of Q_i is $(2\ell^i - 1, \ell^i - 1)$, Q_0 is the Bockstein β, Q_1 is $P^1\beta - \beta P^1$, and the other Q_i are defined inductively.

If $\ell > 2$ the inductive formula is $Q_{i+1} = [P^{\ell^i}, Q_i]$. If $\ell = 2$ the inductive formula is $Q_i = [\beta, P^{\mathbf{r}_i}]$; this differs from $[P^{2^{i-1}}, Q_i]$ by correction terms involving $[-1] \in k^\times / k^{\times 2} = H^{1,1}(k, \mathbb{Z}/2)$. See section 13.4 in part III.

We list a few properties of these operations here, referring the reader to section 13.4 for a fuller discussion. The Q_i satisfy $Q_i^2 = 0$ and $Q_i Q_j = -Q_j Q_i$, are $K_*^M(k)$-linear and generate an exterior algebra under composition.

The following theorem concerns the vanishing of a motivic analogue of the classical Margolis homology; see section 13.6 in part III. It was established for $i = 0$ in 1.38, and will be proven for all i in Theorem 13.24. This exact sequence will be used in Propositions 3.15 and 3.17 to show that the Q_i are injections in an appropriate range.

Theorem 1.44. *If X is a Rost variety for $(a_1, ..., a_n)$, the following sequence is exact for all $i < n$ and all (p, q).*

$$H^{p-2\ell^i+1, q-\ell^i+1}\left(\widetilde{\Sigma}\mathfrak{X}, \mathbb{Z}/\ell\right) \xrightarrow{Q_i} H^{p,q}\left(\widetilde{\Sigma}\mathfrak{X}, \mathbb{Z}/\ell\right) \xrightarrow{Q_i} H^{p+2\ell^i-1, q+\ell^i-1}\left(\widetilde{\Sigma}\mathfrak{X}, \mathbb{Z}/\ell\right)$$

Remark. In Theorem 1.44, it suffices that for each $i < n$ there is a ν_i-variety X_i and a map $X_i \to X$. This is the formulation given in Theorem 13.24.

1.7   HISTORICAL NOTES

As mentioned in the introduction, the question of whether the norm residue is always an isomorphism was first raised by Milnor in his 1970 paper [Mil70] defining what we now call "Milnor K-theory." For local and global fields, Tate had already checked that it was true for $n = 2$ (i.e., for K_2) and all primes ℓ (published in [Tat76]), and Milnor checked in his paper that it was true for all $n > 2$ (where the groups have exponent 2). Kato verified that the norm residue was an isomorphism for fields arising in higher class field theory, and stated the question as a conjecture in [Kat80]. Bloch also asked about it in [Blo80, p. 5.12].

Originally, *norm residue homomorphism* referred to the symbol $(a, b)_k$ of a central simple algebra in the group $\mu_\ell(k)$ of a local field, arising in Hilbert's 9th Problem. Later it was realized that the symbol should take values in the Brauer group, or more precisely $\mu_\ell \otimes \mathrm{Br}_\ell(k)$, and that this map factored through $K_2(k)/\ell$; see [Mil71, 15.5]. The use of this term for the map from $K_*^M(F)/m$ to $H_{\text{ét}}^2(F, \mu_\ell^{\otimes 2})$ seems to have originated in Suslin's 1986 ICM talk [Sus87, 4.2].

The question was completely settled for $n = 2$ by Merkurjev and Suslin in the 1982 paper [MeS82]. Their key geometric idea was the use of Severi–Brauer varieties, which we now recognize as the Rost varieties for $n = 2$. The case $n = 3$ for $\ell = 2$ was settled independently by Rost and Merkurjev–Suslin in the late 1980s. In 1990, Rost studied Pfister quadrics (Rost varieties for $\ell = 2$) and constructed what we now call its Rost motive; see [Ros90] and [Voe03a, 4.3].

In 1994, Suslin and Voevodsky noticed that this conjecture about the norm residue being an isomorphism would imply a circle of conjectures due to Beilinson [Beĭ87] and Lichtenbaum [Lic84] regarding the (then hypothetical) complexes of sheaves $\mathbb{Z}(n)$; the preprint was posted in 1995 and an expanded version was eventually published in [SV00a]. This is the basis of our chapter 2.

In 1996, Voevodsky announced the proof of Milnor's conjecture for $\ell = 2$, using work of Rost on the motive of a Pfister quadric. The 1996 preprint [Voe96] was expanded into [Voe03a] and [Voe03c], which appeared in 2003.

In 1998, Voevodsky announced the proof of the Bloch–Kato conjecture, i.e., Milnor's conjecture for $\ell > 2$, assuming the existence of what we call Rost varieties (1.24). Details of this program appeared in the 2003 preprint [Voe03b], and the complete proof was published in 2011 [Voe11].

Later in 1998, Rost announced the construction of norm varieties; the construction was released in the preprints [Ros98a] and [Ros98b], but did not contain the full proof that his norm varieties were "Rost varieties," i.e., satisfied the properties (1.24) required by Voevodsky's program. Most of those details appeared in [SJ06]; Rost's informal notes [Ros06] provided other details, and the final details were published in [HW09].

The combination of Rost's construction and Voevodsky's work combines to verify not only the Bloch–Kato conjecture (proving Theorem A) but also proving Theorems B and C, which are stated in the overview of this book.

The material in section 1.5 is taken from the appendix of [Voe03a].

Chapter Two

Relation to Beilinson–Lichtenbaum

RECALL FROM 1.26 that $L/\ell(n)$ is the good truncation complex $\tau^{\leq n}\mathbf{R}\pi_*(\mu_\ell^{\otimes n})$, where π is the change-of-topology morphism $(\mathbf{Sm}/k)_{\text{ét}} \to (\mathbf{Sm}/k)_{\text{zar}}$. We say that "BL(n) holds" if the map $\alpha_n : \mathbb{Z}/\ell(n) \to L/\ell(n)$ is a quasi-isomorphism for any field k containing $1/\ell$.

In this chapter, we show how the Beilinson–Lichtenbaum condition (that BL(n) holds) is equivalent to the assertion that the norm residue is an isomorphism, and also to the property that H90(n) holds, a fact that is important to make the inductive step in the proofs of Theorems A, B, and C work. Most of this material is taken from [SV00a], which assumes resolution of singularities, and from [Sus03], which is based upon [GL01] and removes this assumption. The content of this chapter is captured in the following diagram of implications:

$$
\begin{array}{ccccc}
BL(n) & \xrightarrow[\;2.37\;]{\;1.30\;} & BK(n) & \xrightarrow[\;2.38\;]{\;1.7\;} & H90(n) \\
\Big\downarrow{\scriptstyle 2.9} & & \Big\downarrow{\scriptstyle 2.12} & & \Big\downarrow{\scriptstyle 2.11} \\
BL(n-1) & & BK(n-1) & & H90(n-1)
\end{array}
$$

Figure 2.1: Scheme of the proof

The preparatory vertical implications are demonstrated in sections 2.1 and 2.2. The crucial ingredient in showing that the left vertical implication holds is an analysis of the contractions of the motivic complex $\mathbb{Z}/\ell(n)$ (see Example 2.4(2)) and the complex $L/\ell(n)$ (see Corollary 2.7). The middle and right vertical implications are consequences of 2.10, which shows that the Gysin triangle associated to an open subset of $\mathbb{A}^1-\{0\}$ splits. The implications from left to right are proved in chapter 1. The remainder of this chapter is taken up with the proofs of the implications from right to left; the idea is to use a dimension shift argument relying on an analysis of an appropriately defined cohomology of the boundary of an algebraic simplex, combined with the vertical implications proved in the first two sections to set up an inductive proof of BL(n) from its special cases BK(n) and H90(n).

2.1 BL(n) IMPLIES BL(n − 1)

In this section, we prove the useful reduction that if the Bloch–Lichtenbaum condition BL(n) holds then so does BL(n − 1); see Theorem 2.9. For the proof, we will need the following construction.

If F is a homotopy invariant presheaf, we define the presheaf F_{-1} to be $X \mapsto F(X \times (\mathbb{A}^1 - \{0\}))/F(X)$, as in [MVW]. If F is the presheaf underlying a homotopy invariant sheaf with transfers $\mathcal{F}$, then $F_{-1}(U) = H^0(U, \mathcal{F}_{-1})$. More generally, $H^p(U, \mathcal{F})_{-1} \cong H^p(U, \mathcal{F}_{-1})$; see [MVW, 24.8].

This generalizes to complexes using $\underline{RHom}$, the internal Hom complex in $\mathbf{DM}^{\mathrm{eff}}_{\mathrm{nis}}$ which is constructed in [MVW, 14.12].

Definition 2.1. Let $\mathcal{F}$ be a bounded above complex of sheaves with transfers, with homotopy invariant cohomology. Then $\mathcal{F}_{-1}$ denotes $\underline{RHom}(\mathbb{L}, \mathcal{F})[1]$, where $\mathbb{L}$ is $\mathbb{Z}(1)[2] \cong (\mathbb{Z}_{\mathrm{tr}}(\mathbb{A}^1 - \{0\})/\mathbb{Z})[1] \cong \mathbb{Z}_{\mathrm{tr}}(\mathbb{P}^1)/\mathbb{Z}$; see [MVW, 15.2]. That is, $\mathcal{F}_{-1}(X)$ is the complex $\mathbf{R}\operatorname{Hom}(\mathbb{Z}_{\mathrm{tr}}(X \times (\mathbb{A}^1 - \{0\})), \mathcal{F})/\mathbf{R}\operatorname{Hom}(\mathbb{Z}_{\mathrm{tr}}(X), \mathcal{F})$ or, equivalently, $\mathbf{R}\operatorname{Hom}(\mathbb{Z}_{\mathrm{tr}}(X \times \mathbb{P}^1), \mathcal{F}[1])/\mathbf{R}\operatorname{Hom}(\mathbb{Z}_{\mathrm{tr}}(X), \mathcal{F}[1])$.

Lemma 2.2. *Let $\mathcal{F}$ be a bounded above complex of homotopy invariant sheaves with transfers. Then for smooth X, $H^p(X, \mathcal{F})_{-1} \cong H^p(X, \mathcal{F}_{-1})$.*

Proof. Because $H^p(X \times (\mathbb{A}^1 - \{0\}), \mathcal{F}) \cong H^p(X, \mathcal{F}) \oplus \operatorname{Hom}(\mathbb{Z}_{\mathrm{tr}}(X)(1)[1], \mathcal{F}[p])$,

$$H^p(X, \mathcal{F})_{-1} \cong \operatorname{Hom}(\mathbb{Z}_{\mathrm{tr}}(X)(1)[2], \mathcal{F}[p+1]) \cong H^{p+1}(X \otimes \mathbb{L}, \mathcal{F})$$
$$\cong \operatorname{Hom}(\mathbb{Z}_{\mathrm{tr}}(X), \underline{RHom}(\mathbb{L}, \mathcal{F})[p+1]) \cong H^p(X, \mathcal{F}_{-1}). \quad (2.2.1)$$

Here we have used the adjunction $\operatorname{Hom}(K \otimes M, \mathcal{F}) \cong \operatorname{Hom}(K, \underline{RHom}(M, \mathcal{F}))$ of [MVW, 14.12], which is valid for any geometric motive M. $\qquad\square$

Lemma 2.3. *There is a natural map $\delta : \mathcal{F}[-1] \to \mathcal{F}(1)_{-1}$ in $\mathbf{DM}^{\mathrm{eff}}_{\mathrm{nis}}$. Applying $\operatorname{Hom}(\mathbb{Z}_{\mathrm{tr}}(X)[-p], -)$ yields natural maps $H^{p-1}(X, \mathcal{F}) \to H^p(X, \mathcal{F}(1)_{-1})$.*

Proof. Since $\mathcal{F}(1)_{-1}[1] = \underline{RHom}(\mathbb{L}, \mathcal{F}(1)[2]) = \underline{RHom}(\mathbb{L}, \mathcal{F} \otimes \mathbb{L})$, and $\operatorname{Hom}(\mathcal{F} \otimes \mathbb{L}, \mathcal{G}) \simeq \operatorname{Hom}(\mathcal{F}, \underline{RHom}(\mathbb{L}, \mathcal{G}))$ [MVW, 14.12] we have a canonical isomorphism:

$$\operatorname{Hom}(\mathcal{F} \otimes \mathbb{L}, \mathcal{F} \otimes \mathbb{L}) \xrightarrow{\sim} \operatorname{Hom}(\mathcal{F}, \underline{RHom}(\mathbb{L}, \mathcal{F} \otimes \mathbb{L})) \cong \operatorname{Hom}(\mathcal{F}, \mathcal{F}(1)_{-1}[1]).$$

The natural map δ is the image of the identity map on $\mathcal{F} \otimes \mathbb{L}$. $\qquad\square$

Examples 2.4. (1) If A is a constant sheaf, $A_{-1} = 0$: $H^*(X \times \mathbb{P}^1, A) \cong H^*(X, A)$.
(2) When $n > 0$, the map δ of 2.3 induces an isomorphism $\mathbb{Z}(n-1)[-1] \xrightarrow{\sim} \mathbb{Z}(n)_{-1}$, and isomorphisms $H^p(X, \mathbb{Z}(n))_{-1} \cong H^{p-1}(X, \mathbb{Z}(n-1))$; this is a consequence of the Localization Theorem, and is established in [MVW, 23.1]. By the 5-lemma, we also have $\delta : \mathbb{Z}/\ell(n-1)[-1] \xrightarrow{\sim} \mathbb{Z}/\ell(n)_{-1}$.

(3) The map δ induces an isomorphism $\mathbf{R}\pi_*(\mu_\ell^{\otimes n-1})[-1] \xrightarrow{\simeq} \mathbf{R}\pi_*(\mu_\ell^{\otimes n})_{-1}$, and $H^p_{\text{ét}}(X \times \mathbb{P}^1, \mu_\ell^{\otimes n}) \cong H^p_{\text{ét}}(X, \mu_\ell^{\otimes n}) \oplus H^{p-2}_{\text{ét}}(X, \mu_\ell^{\otimes n-1})$; see [MVW, 23.3].

(4) By iteration, $\mathcal{F}_{-c} = \underline{RHom}(\mathbb{L}^c, \mathcal{F})[c]$. Thus from (1) and (2) we get

$$\mathbb{Z}(n)_{-c} \cong \begin{cases} \mathbb{Z}(n-c)[-c], & c \leq n; \\ 0, & c > n. \end{cases}$$

Replacing $\mathbb{Z}$ by $\mathbb{Z}/\ell^\nu$ yields a parallel formula for $\mathbb{Z}/\ell^\nu(n)_{-c}$.

Let $\mathcal{H}^p(\mathcal{F})$ denote the pth cohomology sheaf of $\mathcal{F}$.

Lemma 2.5. *The pth cohomology sheaf of the complex $\mathcal{F}_{-1}$ is*

$$\mathcal{H}^p(\mathcal{F}_{-1})(X) = \mathcal{H}^p(\mathcal{F})(X \times (\mathbb{A}^1 - \{0\}))/\mathcal{H}^p(\mathcal{F})(X).$$

Proof. Set $H = H^p(-, \mathcal{F})$; it is a homotopy invariant presheaf with transfers by [MVW, 13.3, 13.8], with associated sheaf $\mathcal{H} = \mathcal{H}^p(\mathcal{F})$. By Lemma 2.2, $H^p(-, \mathcal{F}_{-1}) \cong H_{-1}$, so $\mathcal{H}^p(\mathcal{F}_{-1})$ is the sheafification $(H_{-1})_{\text{nis}}$ of H_{-1}. But by [MVW, 23.5], $(H_{-1})_{\text{nis}}$ is the sheaf $\mathcal{H}_{-1}(X) = \mathcal{H}(X \times (\mathbb{A}^1 - \{0\}))/\mathcal{H}(X)$. □

Recall that $\tau^{\leq n}\mathcal{F}$ denotes the good truncation of a cochain complex $\mathcal{F}$ at level n. By construction, $\mathcal{H}^p(\tau^{\leq n}\mathcal{F})$ is $\mathcal{H}^p(\mathcal{F})$ if $p \leq n$, and 0 otherwise.

Proposition 2.6. $(\tau^{\leq n}\mathcal{F})_{-1} \xrightarrow{\simeq} \tau^{\leq n}(\mathcal{F}_{-1})$ *is a quasi-isomorphism.*

Proof. By Lemma 2.5, the pth cohomology sheaf of $(\tau^{\leq n}\mathcal{F})_{-1}$ is

$$X \mapsto \begin{cases} \mathcal{H}^p(\mathcal{F})(X \times (\mathbb{A}^1 - \{0\}))/\mathcal{H}^p(\mathcal{F})(X), & p \leq n \\ 0, & p > n \end{cases} = \begin{cases} \mathcal{H}^p(\mathcal{F}_{-1})(X), & p \leq n \\ 0, & p > n. \end{cases}$$

Hence the natural map $(\tau^{\leq n}\mathcal{F})_{-1} \to \mathcal{F}_{-1}$ factors through $\tau^{\leq n}(\mathcal{F}_{-1})$, and $\mathcal{H}^p((\tau^{\leq n}\mathcal{F})_{-1}) \to \mathcal{H}^p(\tau^{\leq n}(\mathcal{F}_{-1}))$ is an isomorphism for all p. □

Corollary 2.7. *The map $\delta : L/\ell(n-1)[-1] \xrightarrow{\simeq} L/\ell(n)_{-1}$ is an isomorphism. In particular, for smooth X we have*

$$H^{p-1}(X, L/\ell(n-1)) \cong H^p(X, L/\ell(n)_{-1}) \cong H^p(X, L/\ell(n))_{-1}.$$

Proof. Since $L/\ell(n) = \tau^{\leq n}\mathbf{R}\pi_*(\mu_\ell^{\otimes n})$, the first assertion is immediate from 2.6 and Example 2.4(3). The second assertion follows from this and Lemma 2.2. □

Proposition 2.8. *For smooth X, we have a commutative diagram*

$$
\begin{array}{ccc}
H^{p-1}(X, \mathbb{Z}/\ell(n-1)) & \xrightarrow[\delta]{\simeq} & H^p(X, \mathbb{Z}/\ell(n)_{-1}) \\
\downarrow{\scriptstyle \alpha_{n-1}} & & \downarrow{\scriptstyle \alpha_n} \\
H^{p-1}(X, L/\ell(n-1)) & \xrightarrow[\delta]{\simeq} & H^p(X, L/\ell(n)_{-1}) \\
\downarrow & & \downarrow \\
H^{p-1}_{\text{ét}}(X, \mu_\ell^{\otimes n-1}) & \xrightarrow[\delta]{\simeq} & H^p_{\text{ét}}(X, (\mu_\ell^{\otimes n})_{-1}).
\end{array}
$$

Proof. By Lemma 2.3, $H^p(X, \delta) = \operatorname{Hom}(\mathbb{Z}_{\text{tr}}(X)[-p], \delta)$ is a natural transformation. Applying it to $\mathbb{Z}/\ell(n) \xrightarrow{\alpha} L/\ell \to \mathbf{R}\pi_* \mu_\ell^{\otimes n}$ yields the commutative diagram. The rows are isomorphisms by 2.4(2), 2.7 and 2.4(3). □

Theorem 2.9.[1] *If BL(n) holds then BL(n − 1) holds.*

Proof. Consider the diagram of Proposition 2.8. If BL(n) holds then for every local X the top right vertical is an isomorphism in the diagram, and hence the top left vertical is an isomorphism, i.e., α_{n-1} is a quasi-isomorphism on the stalks of X. It follows that α_{n-1} is a quasi-isomorphism, i.e., BL(n − 1) holds. □

2.2   H90(n) IMPLIES H90(n − 1)

The fact that H90(n) implies H90(n − 1) will be used in Theorem 2.37 to show that $H^{p,n}(k, \mathbb{Z}/\ell) \to H^p_{\text{ét}}(k, \mu_\ell^{\otimes n})$ is an isomorphism for all $p \leq n$. Since the proof is elementary, we give it here.

By a *motivic complex* we shall mean a cochain complex of Zariski sheaves with transfers, having homotopy invariant cohomology. One source of motivic complexes comes from the total direct image $\mathbf{R}\pi_*$ associated to the morphism of sites $\pi : (\mathbf{Sm}/k)_{\text{ét}} \to (\mathbf{Sm}/k)_{\text{zar}}$. If $\mathcal{F}$ is an étale sheaf with transfers, having homotopy invariant cohomology, then $\mathbf{R}\pi_*(\mathcal{F})$ is a motivic complex by [MVW, 6.20], and $H^*_{\text{zar}}(X, \mathbf{R}\pi_* \mathcal{F}) = H^*_{\text{ét}}(X, \mathcal{F})$.

Lemma 2.10. *For all dense open $U \subseteq (\mathbb{A}^1 - \{0\})$ the localization triangle*

$$
\oplus_{x \notin U} M(x)(1)[1] \to M(U) \to M(\operatorname{Spec} k) \to
$$

is split exact in **DM**, *where the sum is over all closed points of $\mathbb{A}^1$ not in U (see [MVW, 15.15]). Hence for every motivic complex $\mathcal{C}$ we have a split exact*

---

1. Theorem 2.9 is used in 2.24 and 3.15.

localization sequence, with the splitting natural in U and $\mathcal{C}$:

$$0 \to H^n(k, \mathcal{C}) \to H^n(U, \mathcal{C}) \xrightarrow{\partial} \oplus_{x \notin U} H^{n-1}(x, \mathcal{C}(-1)) \to 0.$$

Taking the direct limit over all U yields the split exact sequence

$$0 \to H^n(k, \mathcal{C}) \to H^n(k(t), \mathcal{C}) \xrightarrow{\partial} \oplus_{x \in \mathbb{A}^1} H^{n-1}(x, \mathcal{C}(-1)) \to 0.$$

Proof. The group of finite correspondences from $\operatorname{Spec} k$ to X is the same as the group of zero-cycles on X; see [MVW, 1.10]. As U is dense in $\mathbb{A}^1$, it contains a zero-cycle of degree 1, yielding a finite correspondence $\operatorname{Spec} k \to U$ of degree 1. Passing to motives via $M(X) = \mathbb{Z}_{\mathrm{tr}}(X)$, we get a morphism $M(\operatorname{Spec} k) \to M(U)$ whose composition with $M(U) \to M(\mathbb{A}^1)$ is the canonical isomorphism $M(\operatorname{Spec} k) \xrightarrow{\cong} M(\mathbb{A}^1)$. This splits the map $M(U) \to M(\operatorname{Spec} k)$, as required. $\qquad\square$

Remark 2.10.1. There is a notational ambiguity with $H^{*,*}(k(t), A)$, since it could be computed relative to $\mathbf{Sm}/k(t)$ or as the direct limit of the $H^{*,*}(U, A)$ relative to $\mathbf{Sm}/k$, as in Lemma 2.10. Happily, the two coincide by [MVW, 3.9].

Example 2.11. Applying Lemma 2.10 to the motivic complex $\mathcal{C} = \mathbb{Z}(n)$ yields the split exact sequence

$$0 \to K_n^M(k) \to K_n^M(k(t)) \xrightarrow{\partial} \oplus_{x \in \mathbb{A}^1} K_{n-1}^M(k(x)) \to 0$$

of [Mil70, 2.3]. Applying 2.10 to the motivic complexes $\mathbb{Z}_{(\ell)}(n)$ and $\mathbf{R}\pi_* \mu_\ell^{\otimes n}$ yields the corresponding sequences for $H_{\text{ét}}^n(k, \mathbb{Z}_{(\ell)})$ and $H_{\text{ét}}^n(k, \mu_\ell^{\otimes n})$. Finally, applying 2.10 to $\mathbf{R}\pi_* \mathbb{Z}(n)_{\text{ét}}$ yields the split exact sequence

$$0 \to H_{\text{ét}}^{n+1}(k, \mathbb{Z}(n)) \to H_{\text{ét}}^{n+1}(k(t), \mathbb{Z}(n)) \xrightarrow{\partial} \oplus_{x \in \mathbb{A}^1} H_{\text{ét}}^n(k(x), \mathbb{Z}(n-1)) \to 0.$$

The next result shows that if $K_n^M(F)/\ell \xrightarrow{\sim} H_{\text{ét}}^n(F, \mu_\ell^{\otimes n})$ for all fields F over k then $K_{n-1}^M(k)/\ell \xrightarrow{\sim} H_{\text{ét}}^{n-1}(k, \mu_\ell^{\otimes n-1})$. It will be used in Theorem 2.37.

Corollary 2.12. *If $H90(n)$ holds then $H90(n-1)$ also holds.*
If $K_n^M(k(t))/\ell \to H_{\text{ét}}^n(k(t), \mu_\ell^{\otimes n})$ is an isomorphism (resp., onto) then so is $K_{n-1}^M(k)/\ell \to H_{\text{ét}}^{n-1}(k, \mu_\ell^{\otimes n-1})$.

Proof. By Example 2.11, $H_{\text{ét}}^n(k, \mathbb{Z}_{(\ell)}(n-1))$ is a summand of $H_{\text{ét}}^{n+1}(k(t), \mathbb{Z}_{(\ell)}(n))$, whence the first assertion. For the second, we know by Example 2.11 that there is a commutative diagram

$$\begin{array}{ccc}
K_n^M(k(t))/\ell & \xrightarrow{\ \partial\ } & K_{n-1}^M(k)/\ell \\
\Big\downarrow{\scriptstyle\cong} & & \Big\downarrow \\
H_{\text{ét}}^n(k(t), \mu_\ell^{\otimes n}) & \xrightarrow{\ \partial\ } & H_{\text{ét}}^{n-1}(k, \mu_\ell^{\otimes n-1}).
\end{array}$$

By Example 2.11, both horizontal arrows are split surjections, and the splittings are compatible. If we assume the left vertical is an isomorphism (resp., a surjection), then so is the right vertical morphism.   □

2.3   COHOMOLOGY OF SINGULAR VARIETIES

Recall that the schemes $\Delta^m = \text{Spec}(k[t_0, \ldots, t_m]/(1 - \sum t_i)$ fit together to form a cosimplicial scheme $\Delta^\bullet$, where the ith coface map $\partial_i : \Delta^{m-1} \to \Delta^m$ is given by setting $t_i = 0$, and codegeneracies are given by sending t_i to $t_i + t_{i+1}$. The $m+1$ images of the point Δ^0 in Δ^m are called its *vertices*.

In order to shift cohomological indices, we shall modify the topologist's sphere trick, replacing S^{m-1} with $\partial\Delta^m$, the closed subvariety of Δ^m defined as the union of all of its maximal proper faces $\partial_i\Delta^{m-1}$. Our first task is to define the cohomology of a singular algebraic variety, such as $\partial\Delta^m$, with coefficients in a sheaf defined only on $\mathbf{Sm}/k$.

Definition 2.13. For each scheme X over k, let $\mathbb{Z}[X]$ denote the Nisnevich sheaf associated to the presheaf $U \mapsto \mathbb{Z}[\text{Hom}_{\mathbf{Sch}/k}(U, X)]$ on $\mathbf{Sch}/k$, and let $\mathbb{Z}_{\mathbf{Sm}}[X]$ denote the restriction of $\mathbb{Z}[X]$ to $\mathbf{Sm}/k$.

If X is a scheme of finite type over a field k, and $\mathcal{F}$ is any complex of Nisnevich sheaves on $\mathbf{Sm}/k$, we define the cohomology groups $H^p(X_{\mathbf{Sm}}, \mathcal{F})$ to be $\text{Hom}_{\mathbf{D}(\mathbf{Sm}/k)}(\mathbb{Z}_{\mathbf{Sm}}[X], \mathcal{F}[p])$.

If X is smooth, $\mathbb{Z}_{\mathbf{Sm}}[X]$ is the free abelian group sheaf generated by a representable sheaf, and our definition agrees with the usual definition of (Nisnevich) sheaf hypercohomology. However, if X is singular and $\mathcal{F}$ is the restriction of a complex $\mathcal{F}'$ defined on all schemes then $H^*(X_{\mathbf{Sm}}, \mathcal{F})$ will not in general equal $H^*(X, \mathcal{F}')$. For example, if $\mathcal{O}_{\mathbf{Sm}}$ is the restriction of $\mathcal{O}$ to $\mathbf{Sm}/k$ and X is the affine cusp then $H^0(X_{\mathbf{Sm}}, \mathcal{O}_{\mathbf{Sm}}) = k[x]$ while $H^0(X, \mathcal{O}) = k[x^2, x^3]$. If k admits resolution of singularities and $\mathcal{F}$ is a homotopy invariant complex of sheaves with transfers, the cohomology $H^p(X_{\mathbf{Sm}}, \mathcal{F})$ agrees with the *cdh* cohomology of X with coefficients in $\mathcal{F}'$ by [MVW, 13.27].

Note that $\mathbb{Z}_{\mathbf{Sm}}[X_{\text{red}}] = \mathbb{Z}_{\mathbf{Sm}}[X]$, so that $H^*(X_{\mathbf{Sm}}, \mathcal{F}) = H^*((X_{\text{red}})_{\mathbf{Sm}}, \mathcal{F})$. If $X = X_1 \cup X_2$ is the union of closed subschemes we have an exact sequence of sheaves

$$0 \to \mathbb{Z}_{\mathbf{Sm}}[X_1 \cap X_2] \to \mathbb{Z}_{\mathbf{Sm}}[X_1] \times \mathbb{Z}_{\mathbf{Sm}}[X_2] \to \mathbb{Z}_{\mathbf{Sm}}[X] \to 0;$$

applying $\mathbf{R}\operatorname{Hom}(-, \mathcal{F})$, it follows that the cohomology groups $H^*(-_{\mathbf{Sm}}, \mathcal{F})$ satisfy Mayer–Vietoris for closed covers. The generalization of Mayer–Vietoris to larger closed covers is the Čech spectral sequence.

Example 2.14. Suppose we are given a scheme which is the union $X = \cup X_i$ of finitely many closed subschemes X_i. Then there is a quasi-isomorphism from the Čech complex

$$\mathbb{Z}_{\mathbf{Sm}}[\check{C}\{X_i\}] : \quad 0 \to \mathbb{Z}_{\mathbf{Sm}}[\cap X_i] \to \cdots \to \oplus_{i<j}\mathbb{Z}_{\mathbf{Sm}}[X_i \cap X_j] \to \oplus_i \mathbb{Z}_{\mathbf{Sm}}[X_i] \to 0$$

to $\mathbb{Z}_{\mathbf{Sm}}[\cup X_i]$. Thus $H^*(X_{\mathbf{Sm}}, \mathcal{F})$ is the cohomology of $\mathbf{R}\operatorname{Hom}(\mathbb{Z}_{\mathbf{Sm}}[\check{C}\{X_i\}], \mathcal{F})$ for each complex of sheaves $\mathcal{F}$ on $\mathbf{Sm}/k$, and we can compute it using the associated Čech spectral sequence

$$E_1^{p,q} = \check{C}^p(\{X_i\}, H^q(-_{\mathbf{Sm}}, \mathcal{F})) = \prod_{i_0 < \cdots < i_p} H^q((\cap X_{i_r})_{\mathbf{Sm}}, \mathcal{F}) \Rightarrow H^{p+q}(X_{\mathbf{Sm}}, \mathcal{F}).$$

The rows are the Čech complexes $\check{C}(\{X_i\}; H) = H(\check{C}\{X_i\})$ for the presheaves $H = H^q(-_{\mathbf{Sm}}, \mathcal{F})$. Because the spectral sequence is bounded in p, it is convergent even if $\mathcal{F}$ is an unbounded complex. If the X_i and all intersections $\cap X_{i_r}$ are smooth, then the $E_1^{p,q}$ terms are the sheaf hypercohomology $H_{\mathrm{nis}}^p(X, \mathcal{F})$.

As an application, we consider $\partial \Delta^m$, the union of the maximal face subschemes $X_i = \partial_i \Delta^m$ of Δ^m. If $\mathcal{F}$ is a complex on $\mathbf{Sm}/k$ with homotopy invariant cohomology then the Čech spectral sequence for $\partial \Delta^m$ collapses at E_2, because each $H^q(\cap X_{i_r \, \mathbf{Sm}}, \mathcal{F})$ is $H^q(k, \mathcal{F})$ when $\cap X_{i_r} \neq \emptyset$, $E_1^{p,q} = 0$ for $p \geq m$, and the rows of E_1 compute the homology of the $(m-1)$-sphere with constant coefficients $H^q(k, \mathcal{F})$. This yields the natural isomorphism[2]

$$H^p(\partial \Delta_{\mathbf{Sm}}^m, \mathcal{F}) \xrightarrow{\simeq} H^p(k, \mathcal{F}) \oplus H^{p+1-m}(k, \mathcal{F}). \tag{2.15}$$

We may repeat the above discussion for the étale topology, using the observation that $\mathbb{Z}[X]$ is already an étale sheaf, and writing $\mathbb{Z}_{\mathbf{Sm}}[X]$ for the restriction of this étale sheaf to $(\mathbf{Sm}/k)_{\mathrm{\acute{e}t}}$ by abuse of notation. As in Example 2.14, we have a quasi-isomorphism $\mathbb{Z}_{\mathbf{Sm}}[\check{C}\{X_i\}] \to \mathbb{Z}_{\mathbf{Sm}}[\cup X_i]$ of étale sheaves and isomorphisms $H_{\mathrm{\acute{e}t}}^p((\cup X_i)_{\mathbf{Sm}}, \mathcal{F}_{\mathbf{Sm}}) \cong \operatorname{Hom}_{\mathrm{\acute{e}t}}(\mathbb{Z}_{\mathbf{Sm}}[\check{C}\{X_i\}], \mathcal{F}_{\mathbf{Sm}}[p])$ for every complex $\mathcal{F}_{\mathbf{Sm}}$ of étale sheaves on $\mathbf{Sm}/k$. For locally constant torsion complexes $\mathcal{F}$ defined on $\mathbf{Sch}/k$, such as $\mu_\ell^{\otimes n}$, the usual étale cohomology of $\mathcal{F}$ is compatible with the cohomology of the restriction $\mathcal{F}_{\mathbf{Sm}}$ of $\mathcal{F}$ to $\mathbf{Sm}$, at least in the following sense.

Lemma 2.16. *Suppose $X = \cup X_i$ is a finite union of closed subschemes X_i. Let $\mathcal{F}$ be a complex with locally constant torsion étale cohomology sheaves on $\mathbf{Sch}/k$,*

---

2. The shifting trick (2.15) is based on Lemma 9.2 of [SV00a].

whose torsion is prime to $\mathrm{char}(k)$. *Then there is a natural quasi-isomorphism:*

$$\mathbf{R}\,\mathrm{Hom}_{\mathbf{Sch}_{\acute{e}t}}(\mathbb{Z}[X], \mathcal{F}) \xrightarrow{\simeq} \mathbf{R}\,\mathrm{Hom}_{\mathbf{Sch}_{\acute{e}t}}(\mathbb{Z}[\check{C}\{X_i\}], \mathcal{F}).$$

Proof. It suffices to consider the case $X = X_1 \cup X_2$, as the general case follows by induction. Let $p: X \to \mathrm{Spec}(k)$ denote the structure map so that $\mathbf{R}p_*\mathcal{F} = \mathbf{R}\,\mathrm{Hom}_{\mathbf{Sch}}(\mathbb{Z}[X], \mathcal{F})$. If $\iota^i: X_i \to X$ is the inclusion, $\mathbf{R}\iota_*^i\mathcal{F} = \iota_*^i\mathcal{F}$ by proper base-change [Mil80, VI.2.5], and $\mathbf{R}p_*\mathbf{R}\iota_*^i\mathcal{F}$ is quasi-isomorphic to $\mathbf{R}\,\mathrm{Hom}(\mathbb{Z}[X_i], \mathcal{F})$. Applying $\mathbf{R}p_*$ to the triangle $\mathcal{F} \to \oplus\iota_*^i(\mathcal{F}) \to \iota_*^{12}(\mathcal{F})$ yields the triangle

$$\mathbf{R}\,\mathrm{Hom}_{\acute{e}t}(\mathbb{Z}[X], \mathcal{F}) \to \oplus_{i=1}^2 \mathbf{R}\,\mathrm{Hom}_{\acute{e}t}(\mathbb{Z}[X_i], \mathcal{F}) \to \mathbf{R}\,\mathrm{Hom}_{\acute{e}t}(\mathbb{Z}[X_{12}], \mathcal{F}).$$

Since $\mathbf{R}\,\mathrm{Hom}_{\acute{e}t}(\mathbb{Z}[\check{C}\{X_i\}], \mathcal{F})$ also fits into this triangle, the result follows. $\square$

We now consider the change-of-topology morphism $\pi: \mathbf{Sm}_{\acute{e}t} \to \mathbf{Sm}_{\mathrm{nis}}$. If $\mathcal{F}$ is a complex of étale sheaves on $\mathbf{Sch}/k$, and $\mathcal{F}_{\mathbf{Sm}}$ is its restriction to $(\mathbf{Sm}/k)_{\acute{e}t}$, the total direct image $\mathbf{R}\pi_*\mathcal{F}_{\mathbf{Sm}}$ is a complex of Nisnevich sheaves on $(\mathbf{Sm}/k)_{\mathrm{nis}}$. We can take the cohomology of $\mathbf{R}\pi_*\mathcal{F}_{\mathbf{Sm}}$ in the sense of Definition 2.13.

Proposition 2.17. *Suppose* $X = \cup X_i$ *is a finite union of smooth closed sub-schemes* X_i, *and that all finite intersections of the* X_i *are smooth. Let* $\mathcal{F}$ *be a complex with locally constant torsion étale cohomology sheaves on* $\mathbf{Sch}/k$, *whose torsion is prime to* $\mathrm{char}(k)$. *Then*

$$H_{\acute{e}t}^*(X, \mathcal{F}) \xrightarrow{\simeq} H_{\acute{e}t}^*(X_{\mathbf{Sm}}, \mathcal{F}_{\mathbf{Sm}}) \xrightarrow{\simeq} H_{\mathrm{nis}}^*(X_{\mathbf{Sm}}, \mathbf{R}\pi_*\mathcal{F}_{\mathbf{Sm}}).$$

Proof. By Lemma 2.16, $H_{\acute{e}t}^p(X, \mathcal{F})$ is isomorphic to $\mathrm{Hom}_{\mathbf{Sch}_{\acute{e}t}}(\mathbb{Z}[\check{C}\{X_i\}], \mathcal{F}[p])$. Since the X_i and their intersections are smooth, this is the same as

$$\mathrm{Hom}_{\mathbf{Sm}_{\acute{e}t}}(\mathbb{Z}[\check{C}\{X_i\}], \mathcal{F}_{\mathbf{Sm}}[p]) = \mathrm{Hom}_{\mathbf{Sm}_{\mathrm{nis}}}(\mathbb{Z}[\check{C}\{X_i\}], \mathbf{R}\pi_*(\mathcal{F}_{\mathbf{Sm}})[p]).$$

Now $\mathbb{Z}[\check{C}\{X_i\}] \simeq \mathbb{Z}_{\mathbf{Sm}}[X]$, by Example 2.14. Therefore the left side is $H_{\acute{e}t}^p(X_{\mathbf{Sm}}, \mathcal{F}_{\mathbf{Sm}})$, and the right side is the same as $H_{\mathrm{nis}}^p(X_{\mathbf{Sm}}, \mathbf{R}\pi_*\mathcal{F}_{\mathbf{Sm}}) = \mathrm{Hom}_{\mathbf{Sm}_{\mathrm{nis}}}(\mathbb{Z}_{\mathbf{Sm}}[X], \mathbf{R}\pi_*\mathcal{F}_{\mathbf{Sm}}[p])$. $\square$

Corollary 2.18. *For* $X = \cup X_i$ *as in Proposition 2.17 and any* $p \le n$ *we have natural isomorphisms*

$$H^p(X_{\mathbf{Sm}}, L/\ell(n)) \cong H_{\acute{e}t}^p(X, \mu_\ell^{\otimes n}).$$

Proof. It is enough to show that $H^p(X_{\mathbf{Sm}}, \tau^{\le n}\mathcal{F}) \cong H^p(X_{\mathbf{Sm}}, \mathcal{F})$ for every complex $\mathcal{F}$ of Nisnevich sheaves, as the corollary is just the special case $\mathcal{F} = \mathbf{R}\pi_*(\mu_\ell^{\otimes n})$. Without loss of generality, we can replace $\tau^{>n}\mathcal{F}$ by a complex $\mathcal{E}$ of injective sheaves, zero in degrees $\le n$, to assume that $H^p(X_{\mathbf{Sm}}, \tau^{>n}\mathcal{F}) =$

$\text{Hom}(\mathbb{Z}_{\mathbf{Sm}}[X], \mathcal{E}[p])$. Since $\mathcal{E}^p = 0$ we have $H^p(X_{\mathbf{Sm}}, \tau^{>n}\mathcal{F}) = 0$ and the result follows.   □

Remark 2.18.1. Results 2.16–2.18 hold more generally for every X in $\mathbf{Sch}/k$. In particular, $H^p(X_{\mathbf{Sm}}, L/\ell(n)) \cong H^p_{\text{ét}}(X, \mu_\ell^{\otimes n})$ for every X in $\mathbf{Sch}/k$ and every $p \leq n$. This was proven by Suslin in [Sus03, §7], using alterations and the fact, proven in [SV96, §10], that étale cohomology agrees with cohomology for the h-topology on $\mathbf{Sch}/k$.

If $\mathcal{F}$ is a complex of presheaves, we may regard the usual Čech complex $\check{C}(\{X_i\}; \mathcal{F})$ as a double complex; by the Čech cohomology $H^* \check{C}(\{X_i\}; \mathcal{F})$ we mean the cohomology of the total complex.

Proposition 2.19.[3] *Suppose that W is a smooth semilocal scheme, and that $\{X_i\}$ are smooth closed subschemes of W with all intersections smooth. If $\mathcal{F}$ is a complex of sheaves with transfers on $\mathbf{Sm}/k$, having homotopy invariant cohomology, then the canonical map is an isomorphism:*

$$H^q((\cup X_i)_{\mathbf{Sm}}, \mathcal{F}) \to H^q \check{C}(\{X_i\}; \mathcal{F}).$$

Proof. If $\mathcal{F} \to \mathcal{E}$ is a fibrant replacement, $H^q((\cup X_i)_{\mathbf{Sm}}, \mathcal{F})$ is isomorphic to the cohomology of the total complex of $\check{C}(\{X_i\}, \mathcal{E})$, and the proposition asserts that $\check{C}(\{X_i\}, \mathcal{F}) \to \check{C}(\{X_i\}, \mathcal{E})$ is a quasi-isomorphism. By the comparison theorem for the spectral sequence of a double complex, it suffices to show that $\mathcal{F}(S) \to \mathcal{E}(S)$ is a quasi-isomorphism for each intersection S of the X_i.

The homotopy invariant presheaf $H^q(-, \mathcal{F})$ on $\mathbf{Sm}/k$ is a presheaf with transfers by [MVW, 13.4]. The associated sheaf $\mathcal{H}^q$ is also homotopy invariant by [MVW, 22.1–2], and has transfers by [MVW, 22.15]. If S is any smooth semilocal scheme then the hypercohomology spectral sequence for $H^*(S, \mathcal{F})$ degenerates: $E_2^{p,q} = H^p(S, \mathcal{H}^q) = 0$ for $p > 0$ by [MVW, 24.5], and $H^0(S, \mathcal{H}^q) = H^q(\mathcal{F}(S))$. Thus $E_2^{0,q} = H^q(\mathcal{F}(S)) \to H^q(S, \mathcal{F})$ is a quasi-isomorphism, as desired.   □

2.4 COHOMOLOGY WITH SUPPORTS

In this section we show that if one assumes BL(n − 1) then the cohomology with supports of $\mathbb{Z}/\ell^\nu(n)$ and $L/\ell^\nu(n)$ agree.

If Z is a closed subspace of a smooth scheme X, $\mathbb{Z}[X - Z]$ is a subsheaf of $\mathbb{Z}[X]$ and we set $\mathbb{Z}_Z[X] = \mathbb{Z}[X]/\mathbb{Z}[X - Z]$. The cohomology of $\mathcal{F}$ with supports on Z is $H^p_Z(X, \mathcal{F}) = \text{Hom}_{\mathbf{D}(Sm/k)}(\mathbb{Z}_Z[X], \mathcal{F}[p])$. If $Z \subset Y \subset X$ then there is an exact sequence of sheaves,

$$0 \to \mathbb{Z}_{Y-Z}[X - Z] \to \mathbb{Z}_Y[X] \to \mathbb{Z}_Z[X] \to 0, \qquad (2.20)$$

---

3. Proposition 2.19 is based on Proposition 1.11 of [Sus03].

and a long exact sequence for the cohomology of $\mathcal{F}$ with supports, natural in $\mathcal{F}$.

Let $\mathbf{D}(\mathbf{Cor}/k)$ denote the derived category of sheaves with transfers. Since forgetting transfers is an exact functor to sheaves, it induces a triangulated functor $\mathbf{D}(\mathbf{Cor}/k) \to \mathbf{D}(\mathbf{Sm}/k)$. Regarding $\mathbf{DM}_{\mathrm{nis}}^{\mathrm{eff}}$ as a full triangulated sub-category of $\mathbf{D}(\mathbf{Cor}/k)$ by [MVW, 14.11], every triangle in $\mathbf{DM}_{\mathrm{nis}}^{\mathrm{eff}}$ induces a triangle in $\mathbf{D}(\mathbf{Sm}/k)$. Here is an application of this technique.

Lemma 2.21. *Let (X, Z) be a smooth pair with codimension c, and suppose that $\mathcal{F}$ is a complex of sheaves with transfers, with homotopy invariant cohomology. Then there is a canonical isomorphism*

$$H^{p-c}(Z, \mathcal{F}_{-c}) \to H^p_Z(X, \mathcal{F}).$$

Proof. We may assume that $\mathcal{F}$ is bounded below, since replacing $\mathcal{F}$ by a truncation $\tau^{\leq N}\mathcal{F}$ for large N does not change the cohomology groups in question. There is a Gysin triangle $C_*\mathbb{Z}_{\mathrm{tr}}(X - Z) \to C_*\mathbb{Z}_{\mathrm{tr}}(X) \to C_*\mathbb{Z}_{\mathrm{tr}}(Z) \otimes_{\mathrm{tr}} \mathbb{L}^c$ in $\mathbf{DM}_{\mathrm{nis}}^{\mathrm{eff}}$; see [MVW, 15.15]. The maps $\mathbb{Z}[X] \to C_*\mathbb{Z}_{\mathrm{tr}}(X)$ allow us to compare the triangle defining $\mathbb{Z}_Z[X]$ to the induced Gysin triangle in $\mathbf{D} = \mathbf{D}(\mathbf{Sm}/k)$, yielding a canonical map $\mathbb{Z}_Z[X] \to C_*\mathbb{Z}_{\mathrm{tr}}(Z) \otimes \mathbb{L}^c$. Applying $\mathrm{Hom}(-, \mathcal{F}[p])$ yields a morphism of long exact sequences. Since

$$\mathrm{Hom}_{\mathbf{DM}_{\mathrm{nis}}^{\mathrm{eff}}}(C_*\mathbb{Z}_{\mathrm{tr}}(X), \mathcal{F}[p]) \xrightarrow{\simeq} \mathrm{Hom}_{\mathbf{D}}(\mathbb{Z}[X], \mathcal{F}[p]) = H^p(X, \mathcal{F})$$

is an isomorphism by [MVW, 13.4 and 13.5], and there is a similar isomorphism for $X - Z$, the 5-lemma shows that the canonical map yields an isomorphism

$$\mathrm{Hom}_{\mathbf{DM}_{\mathrm{nis}}^{\mathrm{eff}}}(C_*\mathbb{Z}_{\mathrm{tr}}(Z) \otimes \mathbb{L}^c, \mathcal{F}[p]) \xrightarrow{\simeq} \mathrm{Hom}_{\mathbf{D}}(\mathbb{Z}_Z(X), \mathcal{F}[p]) = H^p_Z(X, \mathcal{F}).$$

Replacing $\mathcal{F}$ by $\mathcal{F}_{-c}$, this isomorphism fits into a natural composition

$$H^{p-c}(Z, \mathcal{F}_{-c}) \cong \mathrm{Hom}_{\mathbf{DM}_{\mathrm{nis}}^{\mathrm{eff}}}(C_*\mathbb{Z}_{\mathrm{tr}}(Z), \underline{RHom}(\mathbb{L}^c, \mathcal{F})[p])$$

$$\cong \mathrm{Hom}_{\mathbf{DM}_{\mathrm{nis}}^{\mathrm{eff}}}(C_*\mathbb{Z}_{\mathrm{tr}}(Z) \otimes \mathbb{L}^c, \mathcal{F}[p]) \xrightarrow{\simeq} H^p_Z(X, \mathcal{F}).$$

The first map uses the isomorphism $\mathcal{F}_{-c} \cong \underline{RHom}(\mathbb{L}^c, \mathcal{F})[c]$ of Example 2.4(4), and the second map is the adjunction. $\qquad\square$

Example 2.22. Combining 2.21 with the formula for $\mathbb{Z}/\ell^\nu(n)_{-c}$ in Example 2.4(4), we have natural isomorphisms:

$$H^p_Z(X, \mathbb{Z}/\ell^\nu(n)) \cong \begin{cases} H^{p-2c}(Z, \mathbb{Z}/\ell^\nu(n-c)), & c \leq n; \\ 0, & c > n. \end{cases}$$

Combining 2.21 with Corollary 2.7, we have natural isomorphisms:

$$H_Z^p(X, L/\ell^\nu(n)) \cong \begin{cases} H^{p-2c}(Z, L/\ell^\nu(n-c)), & c \leq n; \\ 0, & c > n. \end{cases}$$

Example 2.23. Combining 2.21 with Example 2.4(3), we obtain the standard result that $H_Z^p(X_{\text{ét}}, \mathbb{Z}/\ell^\nu(n))$ is isomorphic to $H^{p-2c}(Z_{\text{ét}}, \mathbb{Z}/\ell^\nu(n-c))$ for $n \geq c$, and zero if $n < c$.

We now consider the map $\alpha_n : \mathbb{Z}/\ell^\nu(n) \to L/\ell^\nu(n)$ of (1.27).

Theorem 2.24. *Suppose that BL(n$-$1) holds. If (X, Z) is a smooth pair with* $\text{codim}_X Z > 0$ *then α_n induces isomorphisms on motivic cohomology with supports:*

$$H_Z^*(X, \mathbb{Z}/\ell^\nu(n)) \xrightarrow{\simeq} H_Z^*(X, L/\ell^\nu(n)).$$

Proof. By 2.22 and 2.23, we can identify the map in the theorem with the map $\alpha_{n-c} : H^{*-2c}(Z, \mathbb{Z}/\ell^\nu(n-c)) \to H^{*-2c}(Z, L/\ell^\nu(n-c))$. But α_{n-c} is an isomorphism, because BL(n$-$1) implies BL(n$-$c) by Theorem 2.9, Z is smooth, and therefore $\mathbb{Z}/\ell^\nu(n-c) \simeq L/\ell^\nu(n-c)$ by Lemma 1.29. □

Now assume that X is smooth over k, but Z is not necessarily smooth.

Corollary 2.25. *If X is smooth and BL(n$-$1) holds, then α_n induces an isomorphism for every closed subscheme Z with $\text{codim}_X Z > 0$ and every ν:*

$$H_Z^*(X, \mathbb{Z}/\ell^\nu(n)) \xrightarrow{\simeq} H_Z^*(X, L/\ell^\nu(n)).$$

Proof. Because ℓ is prime to the characteristic of k, a transfer argument 1.2 shows that we may assume that k is a perfect field. We may also assume that $Z = Z_{\text{red}}$. Thus if Z' denotes the singular locus of Z then $Z - Z'$ is smooth and not empty.

We proceed by induction on $\dim(Z)$. From (2.20) we see that we have a morphism of long exact sequences for coefficients $A = \mathbb{Z}/\ell^\nu(n)$ and $B = L/\ell^\nu(n)$:

$$
\begin{array}{ccccccccc}
\cdot & \longrightarrow & H_{Z'}^*(X, A) & \longrightarrow & H_Z^*(X, A) & \longrightarrow & H_{Z-Z'}^*(X - Z', A) & \longrightarrow & H_{Z'}^{*+1}(X, A) \\
& & \downarrow{\scriptstyle\cong} & & \downarrow & & \downarrow{\scriptstyle\cong} & & \downarrow{\scriptstyle\cong} \\
\cdot & \longrightarrow & H_{Z'}^*(X, B) & \longrightarrow & H_Z^*(X, B) & \longrightarrow & H_{Z-Z'}^*(X - Z', B) & \longrightarrow & H_{Z'}^{*+1}(X, B).
\end{array}
$$

The second and fifth verticals are isomorphisms by induction. The first and fourth verticals are isomorphisms by Theorem 2.24, because $(X-Z', Z-Z')$ is a smooth pair. The 5-lemma implies that the third vertical map is an isomorphism. □

We can extend Theorem 2.24 to singular X using the following definition.

Definition 2.26. If Z is a closed subscheme of any scheme X, $\mathbb{Z}_{\mathbf{Sm}}[X - Z]$ is a subsheaf of $\mathbb{Z}_{\mathbf{Sm}}[X]$. We set $\mathbb{Z}_Z[X] = \mathbb{Z}_{\mathbf{Sm}}[X]/\mathbb{Z}_{\mathbf{Sm}}[X - Z]$ and define $H_Z^p(X_{\mathbf{Sm}}, \mathcal{F})$ to be $\mathrm{Hom}_{\mathbf{D}(\mathbf{Sm}/k)}(\mathbb{Z}_Z[X], \mathcal{F}[p])$. If X is smooth, then $H_Z^p(X_{\mathbf{Sm}}, \mathcal{F})$ is the usual group $H_Z^p(X, \mathcal{F})$. By construction, and Definition 2.13, there is a long exact sequence

$$\xrightarrow{\partial} H_Z^n(X_{\mathbf{Sm}}, \mathcal{F}) \to H^n(X_{\mathbf{Sm}}, \mathcal{F}) \to H^n((X - Z)_{\mathbf{Sm}}, \mathcal{F}) \xrightarrow{\partial} H_Z^{n+1}(X_{\mathbf{Sm}}, \mathcal{F}) \to .$$

If $X = X_1 \cup X_2$ is the union of closed subschemes, and $Z_i = Z \cap X_i$, there is a long exact sequence $0 \to \mathbb{Z}_{Z_1 \cap Z_2}[X_1 \cap X_2] \to \mathbb{Z}_{Z_1}[X_1] \oplus \mathbb{Z}_{Z_2}[X_2] \to \mathbb{Z}_Z[X] \to 0$. Hence cohomology with supports satisfies Mayer–Vietoris for closed covers.

Recall that $\partial\Delta^m$ denotes $\mathrm{Spec}(k[t_0, \ldots, t_m]/(\prod t_i, 1 - \sum t_i))$, the union of all maximal faces $\partial_i(\Delta^{m-1})$ of the m-simplex. The (codimension p) *faces* are the intersections $F = \cap \partial_{i_r}\Delta$, $i_0 < \cdots < i_p$; each face is smooth. The *vertices* of $\partial\Delta^m$ are its codimension $m-1$ faces; they are the points where one t_i is 1 and all other t_j are zero. The next result is taken from [Sus03, 7.8] and [SV00a, 8.5].

Theorem 2.27. *Let Z be any closed subscheme of $\partial\Delta^m$ missing the vertices. If $BL(n-1)$ holds, then*

$$H_Z^*(\partial\Delta_{\mathbf{Sm}}^m, \mathbb{Z}/\ell^\nu(n)) \xrightarrow{\simeq} H_Z^*(\partial\Delta_{\mathbf{Sm}}^m, L/\ell^\nu(n)).$$

Proof. (Suslin) By Mayer–Vietoris, the cohomology with supports of the cover $\{\partial_i\Delta^m\}$ fits into Čech spectral sequences for $\mathcal{F} = \mathbb{Z}/\ell^\nu(n)$ and $\mathcal{F} = L/\ell^\nu(n)$,

$$E_1^{p,q}(\mathcal{F}) = \prod_{\substack{\text{faces } F \\ \mathrm{codim}(F) = p}} H_{Z \cap F}^q(F, \mathcal{F}) \Rightarrow H_Z^{p+q}(\partial\Delta_{\mathbf{Sm}}^m, \mathcal{F}),$$

and $E_*(\alpha_n)$ is a natural morphism between the spectral sequences. Because Z misses the vertices of each face, and the faces are smooth, Corollary 2.25 shows that it is an isomorphism on the E_1-page: $E_1^{p,q}(\mathbb{Z}/\ell^\nu(n)) \xrightarrow{\simeq} E_1^{p,q}(L/\ell^\nu(n))$. Hence $E_\infty(\alpha_n)$ is an isomorphism on the abutments, as asserted. $\square$

2.5 RATIONALLY CONTRACTIBLE PRESHEAVES

A presheaf $\mathcal{F}$ on $\mathbf{Sm}/k$ is called *contractible* if there is a presheaf morphism $s: \mathcal{F} \to C_1\mathcal{F} = \mathcal{F}(- \times \mathbb{A}^1)$ so that $i_0^* s = 0$ and $i_1^* s = \mathrm{id}_\mathcal{F}$, where $i_0, i_1 : X \to X \times \mathbb{A}^1$ send x to $(x, 0)$ and $(x, 1)$, respectively.

The standard example of a contractible presheaf is $\mathbb{Z}_{\mathbf{Sm}}[\mathbb{A}^m]/\mathbb{Z}_{\mathbf{Sm}}[0]$ (see Definition 2.13); s is induced by $s_X : \mathbb{Z}_{\mathbf{Sm}}[\mathbb{A}^m](X) \to \mathbb{Z}_{\mathbf{Sm}}[\mathbb{A}^m](X \times \mathbb{A}^1)$ sending $f : X \to \mathbb{A}^m$ to the morphism $sf : X \times \mathbb{A}^1 \to \mathbb{A}^m$, $(x, t) \mapsto tf(x)$. In fact, if v is any k-point of $\mathbb{A}_m$ then $\mathbb{Z}_{\mathbf{Sm}}[\mathbb{A}^m]/\mathbb{Z}_{\mathbf{Sm}}[v]$ is also contractible; one uses the linear translation $\tau_{sf} : (x, t) \mapsto tf(x) + v(1 - t)$.

In a similar fashion, $\mathbb{Z}_{\mathrm{tr}}(\mathbb{A}^m)/\mathbb{Z}_{\mathrm{tr}}(0)$ is contractible; the contraction s_X sends a finite correspondence Z (from X to $\mathbb{A}^m$) to the finite correspondence $\{(x, t, tv) : (x, v) \in Z\}$ from $X \times \mathbb{A}^1$ to $\mathbb{A}^m$. As before, the presheaf $\mathbb{Z}_{\mathrm{tr}}(\mathbb{A}^m)/\mathbb{Z}_{\mathrm{tr}}(v)$ is also contractible for any k-point v of $\mathbb{A}^m$; one uses τ_{sf}.

It is enough for our purposes to have contractions which are only defined "rationally," i.e., that there is a neighborhood U of $X \times \{0, 1\}$ in $X \times \mathbb{A}^1$ and maps $s_X : \mathcal{F}(X) \to \mathcal{F}(U)$ such that $i_0^* s = 0$ and $i_1^* s = \mathrm{id}_{\mathcal{F}}$. In order to make this functorial, we define $\tilde{C}_1 \mathcal{F}(X)$ to be the direct limit of the $\mathcal{F}(U)$, taken over all neighborhoods U of $X \times \{0, 1\}$ in $X \times \mathbb{A}^1$. This definition is natural in X, so $\tilde{C}_1 \mathcal{F}$ is a presheaf on $\mathbf{Sm}/k$. By construction, evaluation at $\{0, 1\}$ yields presheaf maps $i_0^*, i_1^* : \tilde{C}_1 \mathcal{F} \to \mathcal{F}$.

Definition 2.28. A presheaf $\mathcal{F}$ on $\mathbf{Sm}/k$ is called *rationally contractible* if there is a presheaf morphism $s : \mathcal{F} \to \tilde{C}_1 \mathcal{F}$ so that $i_0^* s = 0$ and $i_1^* s = \mathrm{id}_{\mathcal{F}}$.

Example 2.29. Any contractible presheaf is rationally contractible. In particular, $\mathbb{Z}_{\mathbf{Sm}}[\mathbb{A}^m]/\mathbb{Z}_{\mathbf{Sm}}[0]$ and $\mathbb{Z}_{\mathrm{tr}}(\mathbb{A}^m)/\mathbb{Z}_{\mathrm{tr}}(0)$ are rationally contractible, and so are $\mathbb{Z}_{\mathbf{Sm}}[\mathbb{A}^m]/\mathbb{Z}_{\mathbf{Sm}}[v]$ and $\mathbb{Z}_{\mathrm{tr}}(\mathbb{A}^m)/\mathbb{Z}_{\mathrm{tr}}(v)$ for any k-point v, by the above remarks.

If V is any open subscheme of $\mathbb{A}^m$ containing the point v, the presheaves $\mathbb{Z}_{\mathbf{Sm}}[V]/\mathbb{Z}_{\mathbf{Sm}}[v]$ and $\mathbb{Z}_{\mathrm{tr}}(V)/\mathbb{Z}_{\mathrm{tr}}(v)$ are also rationally contractible. Indeed, the composition if of a map (or correspondence) $X \xrightarrow{f} V$ with $V \xrightarrow{i} \mathbb{A}^m$ is a map (or correspondence), and $s_X(if) : X \times \mathbb{A}^1 \to \mathbb{A}^m$ sends some neighborhood U of $X \times \{0, 1\}$ into V; $f \mapsto s_X(if)|_U$ defines the required morphism $\mathcal{F} \to \tilde{C}_1 \mathcal{F}$ for these $\mathcal{F}$. (This example is based on [SV00a, 9.6].)

These remarks apply to $V = \mathbb{A}^1 - \{0\}$, V^m and $v_m = (1, \ldots, 1)$ to show that $\mathbb{Z}_{\mathrm{tr}}(\mathbb{G}_m) = \mathbb{Z}_{\mathrm{tr}}(V)/\mathbb{Z}_{\mathrm{tr}}(1)$ and $\mathbb{Z}_{\mathrm{tr}}(V^m)/\mathbb{Z}_{\mathrm{tr}}(v_m)$ are rationally contractible. Since summands of rationally contractible presheaves are also rationally contractible, they also apply to the smash product $\mathbb{Z}_{\mathrm{tr}}(\mathbb{G}_m^{\wedge m})$, defined in [MVW, 2.12–13] as a direct summand of $\mathbb{Z}_{\mathrm{tr}}(V^m)/\mathbb{Z}_{\mathrm{tr}}(v_m)$ complementary to the m coordinate inclusions of $\mathbb{Z}_{\mathrm{tr}}(V^{m-1})/\mathbb{Z}_{\mathrm{tr}}(v_{m-1})$. It follows that each $\mathbb{Z}_{\mathrm{tr}}(\mathbb{G}_m^{\wedge m})$ is rationally contractible.

We will need to evaluate rationally contractible sheaves on the following semilocal version of $\Delta^\bullet$. For each m, let Δ_{sl}^m denote the semilocal scheme of Δ^m at its vertices. The coface and codegeneracies of $\Delta^\bullet$ restrict to maps between the semilocal schemes, and make $\Delta_{\mathrm{sl}}^\bullet$ into a cosimplicial semilocal scheme.

Proposition 2.30. *([Sus03, 2.5]) If $\mathcal{F}$ is rationally contractible, the chain complex of abelian groups associated to $\mathcal{F}(\Delta_{\mathrm{sl}}^\bullet)$ is acyclic.*

Proof. Let S denote the semilocal scheme of $\Delta^m \times \Delta^1$ at the vertices $v \times 0$, $v \times 1$, where v runs over the vertices of Δ^m. Then $\mathcal{F}(S) = \mathrm{colim}\, \mathcal{F}(U)$ is $\tilde{C}_1 \mathcal{F}(\Delta_{\mathrm{sl}}^m)$ by construction so we have a map $s : \mathcal{F}(\Delta_{\mathrm{sl}}^m) \to \mathcal{F}(S)$. Mimicking the usual simplicial decomposition of $\Delta^m \times \Delta^1$, we have $m + 1$ morphisms $\psi_i : \Delta^{m+1} \to \Delta^m \times \Delta^1$;

ψ_i takes the initial $i+1$ vertices v to $v \times 0$ $(0 \leq v \leq i+1)$ and the remaining vertices v to $(v-1) \times 1$. Localizing, these maps induce morphisms $\Delta_{\mathrm{sl}}^{m+1} \to S$ and homomorphisms $\psi_i^* : \mathcal{F}(S) \to \mathcal{F}(\Delta_{\mathrm{sl}}^{m+1})$. Then the chain contraction σ of $\mathcal{F}(\Delta_{\mathrm{sl}}^\bullet)$ is given by the standard formula $\sigma(u) = \sum (-1)^i \psi_i^*(s(u))$. □

Corollary 2.31.[4] *If $\mathcal{F}$ is rationally contractible, the augmented Čech complex $\mathcal{F}(\Delta_{\mathrm{sl}}^m) \to \check{C}(\{\partial_i \Delta_{\mathrm{sl}}^m\}; \mathcal{F})$ is acyclic, except at $\mathcal{F}(\Delta_{\mathrm{sl}}^m)$.*

Of course, the homology at $\mathcal{F}(\Delta_{\mathrm{sl}}^m)$ is the kernel of $\mathcal{F}(\Delta_{\mathrm{sl}}^m) \to \prod \mathcal{F}(\partial_i \Delta_{\mathrm{sl}}^m)$.

Proof. Reindex the augmented complex so that $\mathcal{F}(\Delta_{\mathrm{sl}}^m)$ is in homological degree m, and $\prod \mathcal{F}(v)$ is in degree 0. It was proven in [FS02, 1.2] that this augmented complex is canonically quasi-isomorphic to the brutal truncation at homological level m of the chain complex associated to the simplicial abelian group $\mathcal{F}(\Delta_{\mathrm{sl}}^\bullet)$. As $\mathcal{F}(\Delta_{\mathrm{sl}}^\bullet)$ is acyclic by Proposition 2.30, its brutal truncation is acyclic except at the truncation degree m. □

Recall that $C_m \mathcal{F}$ denotes the presheaf $\mathcal{F}(- \times \Delta^m)$.

Lemma 2.32.[5] *If $\mathcal{F}$ is rationally contractible, so is $C_m \mathcal{F}$.*

Proof. We may assume by induction that $C_{m-1} \mathcal{F}$ is rationally contractible with section s^{m-1}. Using the face $\partial_0 : \Delta^{m-1} \to \Delta^m$ and degeneracy s_0, every element of $C_m \mathcal{F}(X)$ may be written uniquely as a sum $s_0^*(f) + z$ for an element f of $C_{m-1} \mathcal{F}(X)$ and an element z such that $\partial_0^*(z) = 0$. There is a similar decomposition of $\tilde{C}_1(C_m \mathcal{F})$, so we may view $h = s^{m-1}(f)$ as an element of $\tilde{C}_1(C_m \mathcal{F})(X)$ with $i_0^*(h) = 0$ and $i_1^*(h) = f$. Let $\mu : \mathbb{A}^m \times \mathbb{A}^1 \to \mathbb{A}^m$ denote the multiplication $(v, t) \mapsto tv$. Identifying Δ^m with $\mathbb{A}^m = \mathrm{Spec}(k[t_0, ..., t_{m-1}])$, the map

$$\mu^* : \mathcal{F}(X \times \Delta^m) \to \mathcal{F}(X \times \Delta^m \times \mathbb{A}^1) = C_1(C_m \mathcal{F})(X) \to \tilde{C}_1(C_m \mathcal{F})(X)$$

sends z to an element such that $\partial_0(\mu^* z) = 0$, $i_0^*(\mu^* z) = 0$ and $i_1^*(\mu^* z) = z$. We define $s^m : C_m \mathcal{F} \to \tilde{C}_1(C_m \mathcal{F})$ by $(s_0^* f + z) \mapsto s^{m-1}(f) + \mu^*(z)$. □

Theorem 2.33.[6] *If $\mathcal{F}$ is a rationally contractible presheaf, the Čech complex $\check{C}(\{\partial_i \Delta_{\mathrm{sl}}^m\}; C_* \mathcal{F})$ is acyclic in positive degrees.*

Proof. The Čech complex is the total complex of a bounded fourth-quadrant double complex whose $-q^{th}$ row is $\check{C}(\{\partial_i \Delta_{\mathrm{sl}}^m\}; C_q \mathcal{F})$, with $\prod_i C_q \mathcal{F}(\partial_i \Delta_{\mathrm{sl}}^m)$ in the $(0, -q)$ spot. Since each $C_q \mathcal{F}$ is rationally contractible by 2.32, the rows are acyclic in degrees $p > 0$ by 2.31. The result follows from the spectral sequence of a double complex, since $E_2^{p,q}$ is zero unless $p = 0$ and $q \leq 0$. □

---

4. Our 2.31 is taken from 2.6 of [Sus03].
5. Our 2.32 is taken from 2.4 of [Sus03].
6. Our 2.33 is taken from 2.7 of [Sus03].

Proposition 2.34. *If $\mathcal{F}$ is a rationally contractible sheaf with transfers, then*

$$H^p((\partial\Delta_{\mathrm{sl}}^m)_{\mathbf{Sm}}, C_*\mathcal{F}) = 0 \qquad \text{for } p > 0.$$

Proof. $C_*\mathcal{F}$ is a complex of sheaves with homotopy invariant cohomology, so by Proposition 2.19 the map $H^q((\partial\Delta_{\mathrm{sl}}^m)_{\mathbf{Sm}}, C_*\mathcal{F}) \to H^q \check{C}(\{\partial_i\Delta_{\mathrm{sl}}^m\}; C_*\mathcal{F})$ is an isomorphism. As the right side vanishes for $q > 0$ by 2.33, our statement follows. $\qquad\square$

Corollary 2.35. $H^p((\partial\Delta_{\mathrm{sl}}^m)_{\mathbf{Sm}}, \mathbb{Z}(n)) = H^p((\partial\Delta_{\mathrm{sl}}^m)_{\mathbf{Sm}}, \mathbb{Z}/\ell^\nu(n)) = 0$ *for $p > n$.*

Proof. $\mathbb{Z}_{\mathrm{tr}}(\mathbb{G}_m^{\wedge n})$ is rationally contractible by 2.29, and $C_*\mathbb{Z}_{\mathrm{tr}}(\mathbb{G}_m^{\wedge n}) = \mathbb{Z}(n)[n]$. By 2.34, $H^p((\partial\Delta_{\mathrm{sl}}^m)_{\mathbf{Sm}}, \mathbb{Z}(n)) = H^{p-n}((\partial\Delta_{\mathrm{sl}}^m)_{\mathbf{Sm}}, C_*\mathbb{Z}_{\mathrm{tr}}(\mathbb{G}_m^{\wedge n}))$ vanishes for $p > n$. This implies that the second group vanishes by the coefficient sequence. $\qquad\square$

2.6 BLOCH–KATO IMPLIES BEILINSON–LICHTENBAUM

Recall from 1.26 that $L/\ell(n)$ denotes $\tau^{\leq n}\mathbf{R}\pi_*\mathbb{Z}/\ell(n)_{\text{ét}}$ and that there is a natural morphism $\alpha_n : \mathbb{Z}/\ell(n) \to L/\ell(n)$. The condition BL(n) is that α_n is a quasi-isomorphism of sheaves on $\mathbf{Sm}/k$ for any field k containing $1/\ell$, i.e., that α_n induces isomorphisms $H^p(S, \mathbb{Z}/\ell(n)) \xrightarrow{\simeq} H^p(S, L/\ell(n)) = H^p_{\text{ét}}(S, \mu_\ell^{\otimes n})$ for every smooth local scheme S over k and every $p \leq n$.

One of the fundamental properties of a homotopy invariant presheaf with transfers F is that $F(V) \to F(S)$ is an injection for any dense open V of a smooth semilocal scheme S; see [MVW, 11.1]. This property plays a key role in our next result, which is based on Lemma 7.9 of [Sus03].

Lemma 2.36. *Assume that $H^n(k, \mathbb{Z}/\ell(n)) \to H^n_{\text{ét}}(k, L/\ell(n))$ is onto for all fields k with $1/\ell \in k$. Then $\alpha_n^* : H^n(S_{\mathbf{Sm}}, \mathbb{Z}/\ell(n)) \to H^n(S_{\mathbf{Sm}}, L/\ell(n))$ is a surjection for every semilocal scheme S which is a finite union of smooth semilocal schemes, all of whose finite intersections are smooth.*

Proof. Fix k and let C denote the presheaf cokernel of α_n^*. By assumption, $C(\mathrm{Spec}\,E) = 0$ for every field E over k. Since $H^n(-, \mathbb{Z}/\ell(n))$ and $H^n(-, L/\ell(n))$ are homotopy invariant presheaves with transfers on $\mathbf{Sm}/k$, so is the cokernel C of α_n^*. Hence $C(S) = 0$ for every smooth and semilocal S by [MVW, 11.1].

If S is not smooth, we use a trick due to Hoobler. The hypothesis on S implies that we can write S as $\mathrm{Spec}(R/I)$ for a smooth semilocal algebra R; let R_I^h denote the henselization of R along I. By Gabber's rigidity theorem, $H^n_{\text{ét}}(R_I^h, \mu_\ell^{\otimes n}) \cong H^n_{\text{ét}}(S, \mu_\ell^{\otimes n})$. Now R_I^h is the direct limit of étale extensions R' of R, each of which is semilocal, so the lemma follows from the diagram

$$\underrightarrow{\lim} H^n(R', \mathbb{Z}/\ell(n)) \xrightarrow{\cong} H^n(R_I^h, \mathbb{Z}/\ell(n)) \longrightarrow H^n(S_{\mathbf{Sm}}, \mathbb{Z}/\ell(n))$$

$$\text{onto} \downarrow \qquad\qquad\qquad\qquad\qquad\qquad\qquad \downarrow$$

$$\underrightarrow{\lim} H_{\text{ét}}^n(R', \mu_\ell^{\otimes n}) \xrightarrow{\cong} H_{\text{ét}}^n(R_I^h, \mu_\ell^{\otimes n}) \xrightarrow{\cong} H_{\text{ét}}^n(S, \mu_\ell^{\otimes n}),$$

using the observation that $H^n(S_{\mathbf{Sm}}, L/\ell(n)) \cong H_{\text{ét}}^n(S, \mu_\ell^{\otimes n})$ by Corollary 2.18.
$\square$

Remark 2.36.1. The proof of Lemma 2.36 goes through if the coefficients $\mathbb{Z}/\ell(n)$ are replaced by $\mathbb{Z}/\ell^\nu(n)$ or even $\mathbb{Q}/\mathbb{Z}_{(\ell)}(n)$.

Theorem 2.37. *Assume that $H^n(k, \mathbb{Z}/\ell(n)) \longrightarrow H_{\text{ét}}^n(k, \mu_\ell^{\otimes n})$ is onto for all fields k with $1/\ell \in k$. Then for every field k with $1/\ell \in k$:*

(a) BL(n) holds, i.e., $\alpha_n : \mathbb{Z}/\ell(n) \to L/\ell(n)$ is a quasi-isomorphism of complexes of sheaves on $\mathbf{Sm}/k_{\text{zar}}$.
(b) For all $p \le n$, we have an isomorphism:

$$H^p(\alpha_n) : H^p(k, \mathbb{Z}/\ell(n)) \to H^p(k, L/\ell(n)) \cong H_{\text{ét}}^p(k, \mu_\ell^{\otimes n}).$$

Proof. Since the hypothesis holds for $n-1$ by Corollary 2.12, we may use induction on n to prove that (a) and (b) hold. Recall that $\partial \Delta_{\text{sl}}^m$ is obtained from $\partial \Delta^m$ by removing Z, the union of all closed subschemes Z_α missing the vertices of $\partial \Delta^m$. Write $H_Z^*(\partial \Delta^m)$ for the direct limit of the $H_{Z_\alpha}^*(\partial \Delta_{\mathbf{Sm}}^m)$. By the definition of cohomology with supports (2.26), we have a commutative diagram, with $H^{p,n}(X)$ in the first row denoting $H^p(X_{\mathbf{Sm}}, \mathbb{Z}/\ell(n))$ and $H_L^{p,n}(X)$ in the second row denoting $H^p(X_{\mathbf{Sm}}, L/\ell(n))$.

$$
\begin{array}{ccccccccc}
H_Z^{n,n}(\partial \Delta^m) & \to & H^{n,n}(\partial \Delta^m) & \to & H^{n,n}(\partial \Delta_{\text{sl}}^m) & \longrightarrow & \cdot & \xrightarrow{\text{onto}} & H^{n+1,n}(\partial \Delta^m) \\
\cong \downarrow & & \downarrow & & \text{onto} \downarrow & & \cong \downarrow & & \downarrow \\
H_{Z,L}^{n,n}(\partial \Delta^m) & \to & H_L^{n,n}(\partial \Delta^m) & \to & H_L^{n,n}(\partial \Delta_{\text{sl}}^m) & \longrightarrow & \cdot & \longrightarrow & H_L^{n+1,n}(\partial \Delta^m).
\end{array}
$$

The first and fourth vertical maps are isomorphisms by Theorem 2.27, because BL(n − 1) holds by induction. The third vertical is a surjection by Lemma 2.36 applied to $S = \partial \Delta_{\text{sl}}^m$. It follows that the second vertical is a surjection. Taking $m = n + 1 - p$, it follows from (2.15) that $H^p(k, \mathbb{Z}/\ell(n)) \to H^p(k, L/\ell(n))$ is also a surjection.

By 2.35, we have $H^{n+1}((\partial \Delta_{\text{sl}}^m)_{\mathbf{Sm}}; \mathbb{Z}/\ell(n)) = 0$, so the final top horizontal map is onto. A diagram chase shows that the final vertical is an injection. Taking $m = n + 2 - p$, it follows from (2.15) that $H^p(k, \mathbb{Z}/\ell(n)) \to H^p(k, L/\ell(n))$ is also an injection. This establishes (b).

For (a), we note that: (i) the cohomology presheaves $H_{\text{zar}}^p(-, \mathbb{Z}/\ell(n))$, $H^p(-, \mathbf{R}\pi_* \mathbb{Z}/\ell(n))$ and hence $H_{\text{zar}}^p(-, L/\ell(n))$ are homotopy invariant by [MVW,

2.19] and [MVW, 9.24, 9.33], respectively; and (ii) α_n induces isomorphisms on cohomology $H^p(E, \mathbb{Z}/\ell(n)) \cong H^p(E, L/\ell(n))$ for every field E over k, by (b). Proposition 13.7 of [MVW] says that, because (i) and (ii) hold, α_n is a quasi-isomorphism. $\qquad\square$

2.7 CONDITION H90(n) IMPLIES BL(n)

In this section, we prove that H90(n) implies BL(n), and hence the conclusion of Theorems A and B, as promised in Theorem 1.8 and Lemma 1.30. Recall from Definition 1.5 that "H90(n) holds" means that $H^{n+1}_{\text{ét}}(k, \mathbb{Z}_{(\ell)}(n)) = 0$ for all fields k with $1/\ell \in k$.

By Lemma 1.6 there is an exact sequence

$$K_n^M(k) \otimes \mathbb{Q}/\mathbb{Z}_{(\ell)} \xrightarrow{\alpha_n} H^n_{\text{ét}}(k, \mathbb{Q}/\mathbb{Z}_{(\ell)}(n)) \to H^{n+1}_{\text{ét}}(k, \mathbb{Z}_{(\ell)}(n)) \to 0.$$

Thus $H^{n+1}_{\text{ét}}(k, \mathbb{Z}_{(\ell)}(n)) = 0$ holds if and only if the map $H^n(k, \mathbb{Q}/\mathbb{Z}_{(\ell)}(n)) \to H^n_{\text{ét}}(k, \mathbb{Q}/\mathbb{Z}_{(\ell)}(n))$ is onto.

Theorem 2.38. *Assume that H90(n) holds. Then BL(n) holds. In particular, for every field k with $1/\ell \in k$, and all $p \leq n$, the map $H^p(\alpha_n)$ is an isomorphism.*

$$H^p(\alpha_n) : H^p(k, \mathbb{Z}/\ell(n)) \to H^p(k, L/\ell(n)) \cong H^p_{\text{ét}}(k, \mu_\ell^{\otimes n})$$

Proof. The assertion that $H^p(\alpha_n)$ is an isomorphism for all fields implies BL(n), that α_n is a quasi-isomorphism of sheaves, by [MVW, 13.7], because $\mathbb{Z}/\ell(n)$ and $L/\ell(n)$ have homotopy invariant cohomology presheaves.

Write $L/\ell^\infty(n)$ for the truncation $\tau^{\leq n} \mathbf{R}\pi_* \mathbb{Q}/\mathbb{Z}_{(\ell)}(n)_{\text{ét}}$. Then for every semilocal scheme S which is a finite union of smooth semilocal schemes, the map $H^n(S, \mathbb{Q}/\mathbb{Z}_{(\ell)}(n)) \xrightarrow{\alpha_n^*} H^n(S, L/\ell^\infty(n))$ is a surjection by 2.36.1. Therefore the proof of Theorem 2.37(b) goes through with coefficients $\mathbb{Q}/\mathbb{Z}_{(\ell)}(n)$ to prove that

$$H^p(\alpha_n) : H^p(k, \mathbb{Q}/\mathbb{Z}_{(\ell)}(n)) \to H^p(k, L/\ell^\infty(n)) \qquad (2.38.1)$$

is an isomorphism for every field k with $1/\ell \in k$, and all $p \leq n$. The theorem now follows from the 5-lemma applied to the long exact cohomology sequence for the coefficients $0 \to \mathbb{Z}/\ell(n) \to \mathbb{Q}/\mathbb{Z}_{(\ell)}(n) \xrightarrow{\ell} \mathbb{Q}/\mathbb{Z}_{(\ell)}(n) \to 0$. $\qquad\square$

We conclude this section with a proof of Theorem 1.8, that H90(n) implies the Beilinson–Lichtenbaum conjecture for n. We will also need the following definition in Section 4.4.

Definition 2.39. Let $T(n)$ denote the truncation $\tau^{\leq n+1} \mathbf{R}\pi_* (\mathbb{Z}_{(\ell)}(n)_{\text{ét}})$ of the Nisnevich cochain complex representing étale motivic cohomology. We define

$K(n)$ to be the cone of the canonical map $\mathbb{Z}_{(\ell)}(n) \to T(n)$, so that

$$\mathbb{Z}_{(\ell)}(n) \to T(n) \to K(n) \to$$

is a triangle. From the remarks before 2.10, $K(n)$ and $T(n)$ are motivic complexes.

We have $H^{n+1}(Y_\bullet, T(n)) \cong H^{n+1}_{\text{ét}}(Y_\bullet, \mathbb{Z}_{(\ell)}(n))$ for every smooth simplicial $Y_\bullet$, because $H^p(Y_\bullet, \tau^{>n+1}C) = 0$ for $p \le n+1$ and any complex C.

Lemma 2.40. *If H90(n) holds, the Zariski sheaf associated to $H^{n+1}_{\text{ét}}(-, \mathbb{Z}_{(\ell)}(n))$ is zero.*

Proof. $H^{n+1}_{\text{ét}}(-, \mathbb{Z}_{(\ell)}(n))$ is a presheaf with transfers by [MVW, 6.21]. It is homotopy invariant by the 5-lemma, because the presheaves $H^p_{\text{ét}}(-, \mathbb{Q}/\mathbb{Z}_{(\ell)}(n))$ and $H^p_{\text{ét}}(-, \mathbb{Q}(n))$ are homotopy invariant for all p; see [MVW, 9.24, 9.33, 14.26]. By [MVW, 11.2], it suffices to show that $H^{n+1}_{\text{ét}}(E, \mathbb{Z}_{(\ell)}(n)) = 0$ for every field E; this is exactly the assumption that H90(n) holds. $\qquad\square$

Theorem 2.41.[7] *Assume that H90(n) holds. Then $\mathbb{Z}_{(\ell)}(n) \to T(n)$ is a quasi-isomorphism.*

Proof. It suffices to show that the cohomology sheaves are isomorphic, i.e., that $H^p(X, \mathbb{Z}_{(\ell)}(n)) \to H^p(X, T(n))$ is an isomorphism for every p and every smooth hensel local X; this is trivial for $p > n+1$. For $p = n+1$, it holds because both terms are zero by Lemma 2.40. For $p \le n$, we need to show that $H^p(X, \mathbb{Z}_{(\ell)}(n)) \to H^p(X, T(n)) \cong H^p_{\text{ét}}(X, \mathbb{Z}_{(\ell)}(n))$ is an isomorphism. The cohomology sequence

$$
\begin{array}{ccccccc}
H^p(X, \mathbb{Z}_{(\ell)}(n)) & \longrightarrow & H^p(X, \mathbb{Q}(n)) & \longrightarrow & H^p(X, \mathbb{Q}/\mathbb{Z}_{(\ell)}(n)) & \longrightarrow & H^{p+1}(X, \mathbb{Z}_{(\ell)}(n)) \\
\downarrow & & =\downarrow & & \downarrow{\scriptstyle BL(n)} & & \downarrow \\
H^p_{\text{ét}}(X, \mathbb{Z}_{(\ell)}(n)) & \longrightarrow & H^p_{\text{ét}}(X, \mathbb{Q}(n)) & \longrightarrow & H^p_{\text{ét}}(X, \mathbb{Q}/\mathbb{Z}_{(\ell)}(n)) & \longrightarrow & H^{p+1}_{\text{ét}}(X, \mathbb{Z}_{(\ell)}(n))
\end{array}
$$

shows that it suffices to show that $H^p(X, \mathbb{Q}/\mathbb{Z}_{(\ell)}(n)) \to H^p_{\text{ét}}(X, \mathbb{Q}/\mathbb{Z}_{(\ell)}(n))$ is an isomorphism for all $p \le n$. This is just equation (2.38.1). $\qquad\square$

Corollary 2.42.[8] *Assume that H90(n) holds. Then for any simplicial scheme X we have $H^{p,q}(X, \mathbb{Z}/\ell^\nu) \to H^p_{\text{ét}}(X, \mathbb{Z}/\ell^\nu(q))$ is an isomorphism for $p \le q \le n$.*

That is, property H90(n) implies Theorem B in the introduction. Corollary 2.42 is the converse of Theorem 1.7.

---

7. Theorem 2.41 is Theorem 6.6 of [Voe03a], using Cor. 1.31.
8. Cor. 2.42 is Cor. 6.9 of [Voe03a], using Cor. 1.31.

Proof. Recall from 1.26 and 2.39 that $L(n)$ and $T(n)$ denote the Nisnevich complexes $\tau^{\leq n}\mathbf{R}\pi_*\mathbb{Z}_\ell(n)_{\text{ét}}$ and $\tau^{\leq n+1}\mathbf{R}\pi_*\mathbb{Z}_\ell(n)_{\text{ét}}$, respectively. Tensoring the quasi-isomorphism $\mathbb{Z}_{(\ell)}(n) \xrightarrow{\sim} T(n)$ with $\mathbb{Z}/\ell^\nu$ yields a quasi-isomorphism

$$\mathbb{Z}/\ell^\nu(n) \xrightarrow{\sim} \tau^{\leq n}\big(T(n) \otimes^{\mathbf{L}} \mathbb{Z}/\ell^\nu\big) \xrightarrow{\sim} L(n) \otimes \mathbb{Z}/\ell^\nu = L/\ell^\nu(n). \qquad \square$$

2.8 HISTORICAL NOTES

In the early 1980s, S. Lichtenbaum [Lic84, §3] and A. Beilinson [Beĭ87, 5.10.D] formulated a set of conjectures describing the then-hypothetical complexes of sheaves $\mathbb{Z}(n)$ and properties they should enjoy. Among these properties are the assertions 1.5 and 1.28 that BL(n) and H90(n) hold.

The idea that the Beilinson–Lichtenbaum conjectures are closely related to the Bloch–Kato conjecture was worked out by Suslin and Voevodsky in July 1994, during a conference in Villa Madruzzo in Trento, Italy. The equivalence of BL(n) with the Bloch–Kato conjecture first appeared in preprint form in 1995; the published version [SV00a] is a greatly expanded version. This result, together with the fact that H90(n) implies the Bloch–Kato conjecture, was used in Voevodsky's 1996 preprint [Voe96] to prove Milnor's conjecture; a reworked version of this proof appeared in print as [Voe03a].

The original Suslin–Voevodsky proof that BL(n) was equivalent to the Bloch–Kato conjecture required resolution of singularities. Subsequent modifications due to Geisser–Levine [GL01, 1.1] and Suslin [Sus03] have allowed us to remove this assumption. A key role is played by the notion of a rationally contractible presheaf; although this briefly appeared in [SV00a, 9.5], our presentation follows Suslin's treatment in [Sus03]. Our Section 2.6 is based on Section 7 of [SV00a]. What appears in this chapter is an even further streamlined version, using the construction of $\mathcal{F}_{-1}$ (2.1) and the Cancellation Theorem (in the proof of Lemma 2.21).

Chapter Three

Hilbert 90 for K_n^M

IN THIS CHAPTER, we formulate a norm-trace relation for the Milnor K-theory and étale cohomology of a cyclic Galois extension, which we call *Hilbert 90 for K_n^M*. In Theorem 3.2, we prove that it follows from the condition BL(n).

In section 3.2, condition BL(n) is used to establish a related exact sequence in Galois cohomology (Theorem 3.6). Section 3.3 proves Theorem 3.11 (stated as Theorem 1.10 in chapter 1), establishing that condition BL(n − 1) implies the particular case of condition H90(n) for ℓ-special fields k such that $K_n^M(k)$ is ℓ-divisible. This case constitutes the first part of the inductive step in the proof of Theorem A; the remainder of this monograph explains how to reduce the general case to this particular one.

Section 3.4 uses Theorem 3.2 (for K_{n-1}^M) to construct nonzero cohomology elements $\delta \in H^{n,n-1}(\mathfrak{X}, \mathbb{Z}/\ell)$ and $\mu \in H^{2b+1,b}(\mathfrak{X}, \mathbb{Z}/\ell)$, where $b = d/(\ell - 1)$ and $\mathfrak{X}$ is the simplicial scheme associated to a Rost variety X of dimension $d = \ell^{n-1} - 1$ associated to an n-symbol. This will be used in section 5.4 to show that Rost motives exist.

3.1 HILBERT 90 FOR K_n^M

Let E/k be a cyclic Galois extension of degree ℓ and Galois group $G = \langle \sigma \rangle$. The classical Hilbert Theorem 90 says that $H^1(G, E^\times) = 0$, i.e., that the sequence $E^\times \xrightarrow{1-\sigma} E^\times \xrightarrow{N_{E/k}} k^\times$ is exact; see [Wei94, 6.4.8]. Here is the generalization from $K_1^M(k) = k^\times$ to $K_n^M(k)$.

Definition 3.1. We say that a field k satisfies *Hilbert 90 for K_n^M* if for every Galois extension E/k of degree ℓ and Galois group $G = \langle \sigma \rangle$, the sequence $K_n^M(E) \xrightarrow{1-\sigma} K_n^M(E) \xrightarrow{N_{E/k}} K_n^M(k)$ is exact.

This sequence is always exact modulo ℓ-torsion, and the cokernel of $N_{E/k}$ is ℓ-torsion. This is because (a) the usual transfer argument 1.2 implies that the map $K_n^M(k) \to K_n^M(E)^G$ is an isomorphism modulo ℓ-torsion, split by $N_{E/k}$, and (b) $A^G \to A_G$ is also an isomorphism modulo ℓ-torsion for any G-module A.

Recall from 1.28 that BL(n) holds if $\mathbb{Z}/\ell(n) \xrightarrow{\alpha_n} L/\ell(n)$ is a quasi-isomorphism for any field k containing $1/\ell$.

Theorem 3.2. *If BL(n) holds, then any field k of characteristic $\neq \ell$ satisfies Hilbert 90 for K_n^M. That is, for every Galois extension E/k of degree ℓ, with cyclic Galois group $G = \langle \sigma \rangle$, the following sequence is exact:*

$$K_n^M(E) \xrightarrow{\ 1-\sigma\ } K_n^M(E) \xrightarrow{\ N_{E/k}\ } K_n^M(k).$$

Proof. (Merkurjev) Since the sequence is exact up to ℓ-torsion, we may localize at ℓ. Since BL(n) implies that $K_n^M(k)_{(\ell)} \cong H_{\text{ét}}^n(k, \mathbb{Z}_{(\ell)})$ by Lemma 1.29(d), we are reduced to establishing exactness of the corresponding cohomology sequence.

Writing $\mathbb{Z}_{(\ell)}[G](n)$ for the induced representation $\mathbb{Z}[G] \otimes \mathbb{Z}_{(\ell)}(n)$, Shapiro's Lemma [Wei94, 6.3.2, 6.3.4] implies that $H_{\text{ét}}^n(E, \mathbb{Z}_{(\ell)}(n)) \cong H_{\text{ét}}^n(k, \mathbb{Z}_{(\ell)}[G](n))$, and that the composition with the augmentation map $H_{\text{ét}}^n(k, \mathbb{Z}_{(\ell)}[G](n)) \to H_{\text{ét}}^n(k, \mathbb{Z}_{(\ell)}(n))$ is the transfer map. Let $\mathcal{I}$ denote the augmentation ideal of the group ring $\mathbb{Z}[G]$; it is generated by $1 - \sigma$ and fits into an exact sequence $0 \to \mathbb{Z} \xrightarrow{N} \mathbb{Z}[G] \xrightarrow{1-\sigma} \mathcal{I} \to 0$. Tensoring with $\mathbb{Z}_{(\ell)}(n)$ and taking cohomology, we have an exact sequence

$$H_{\text{ét}}^n(k, \mathbb{Z}_{(\ell)}[G](n)) \xrightarrow{\ 1-\sigma\ } H_{\text{ét}}^n(k, \mathcal{I}_{(\ell)}(n)) \to H_{\text{ét}}^{n+1}(k, \mathbb{Z}_{(\ell)}(n)).$$

Since BL(n) holds, so does H90(n), by Lemma 1.30. That is, the term on the right vanishes, and therefore the left map is onto.

Tensoring $0 \to \mathcal{I} \to \mathbb{Z}[G] \to \mathbb{Z} \to 0$ with $\mathbb{Z}_{(\ell)}(n)$ and taking cohomology, we have the exact sequence forming the bottom row of the commutative diagram:

$$
\begin{array}{ccccc}
K_n^M(E)_{(\ell)} \cong H_{\text{ét}}^n(k, \mathbb{Z}_{(\ell)}[G](n)) & \xrightarrow{\ 1-\sigma\ } & K_n^M(E)_{(\ell)} & \xrightarrow{\ N_{E/k}\ } & K_n^M(k)_{(\ell)} \\
\text{onto} \downarrow {\scriptstyle 1-\sigma} & & \downarrow {\scriptstyle \cong} & & \downarrow {\scriptstyle \cong} \\
H_{\text{ét}}^n(k, \mathcal{I}_{(\ell)}(n)) & \longrightarrow & H_{\text{ét}}^n(k, \mathbb{Z}_{(\ell)}[G](n)) & \xrightarrow{\ \text{tr}\ } & H_{\text{ét}}^n(k, \mathbb{Z}_{(\ell)}(n)).
\end{array}
$$

The theorem now follows from a diagram chase. $\qquad\qquad\square$

Recall (1.9) that a field k is called ℓ-*special* if $1/\ell \in k$ and k has no finite field extensions of degree prime to ℓ. The next result will be used in Theorem 3.11.

Proposition 3.3. *([Voe03a, 5.6]) Suppose that k is ℓ-special and that every finite extension of k satisfies Hilbert 90 for K_{n-1}^M. Then for every degree ℓ field extension E/k such that $N_{E/k}: K_{n-1}^M(E) \to K_{n-1}^M(k)$ is onto, the Hilbert 90 sequence for K_n^M is also exact:*

$$K_n^M(E) \xrightarrow{\ 1-\sigma\ } K_n^M(E) \xrightarrow{\ N_{E/k}\ } K_n^M(k) \to 0.$$

Proof. Given $\underline{a} = \{a_2, \ldots, a_n\} \in K_{n-1}^M(k)$, choose $\underline{b} \in K_{n-1}^M(E)$ with $N_{E/k}(\underline{b}) = \underline{a}$. Since $N_{E/k}(\{a_1, \underline{b}\}) = \{a_1, N(\underline{b})\} = \{a_1, \ldots, a_n\}$, it follows that the norm $N_{E/k}$: $K_n^M(E) \to K_n^M(k)$ is onto.

Define $\phi(a_1, \underline{a})$ to be the chosen element $\{a_1, \underline{b}\}$ of $K_n^M(E)$. By Hilbert 90 for K_{n-1}^M, a different choice of $\underline{b}$ differs by an element $(1-\sigma)(c)$, and alters $\phi(a_1, \underline{a})$ by $(1-\sigma)\{a_1, c\}$. Thus ϕ and $N_{E/k}$ determine well-defined maps

$$(k^\times) \otimes K_{n-1}^M(k) \xrightarrow{\phi} K_n^M(E)/(1-\sigma) \xrightarrow{\bar{N}_{E/k}} K_n^M(k)$$

with $\bar{N}_{E/k}\phi(a_1 \otimes \underline{a}) = \{a_1, \underline{a}\}$. By Lemma 3.4, $K_n^M(E)$ is generated by symbols $\{a_1, \underline{b}\}$ with $a_1 \in k^\times$ and $\underline{b} \in K_{n-1}^M(E)$. Since $\{a_1, \underline{b}\} = \phi(a_1 \otimes N(\underline{b}))$, ϕ is onto.

We claim that ϕ factors through $K_n^M(k)$, which is the quotient of $k^\times \otimes K_{n-1}^M(k)$ by the subgroup J generated by symbols $\eta = a_1 \otimes \underline{a}$ with $a_1 = 1 - a_2$ and $\underline{a} = \{a_2, \ldots, a_n\}$. Since $\bar{N}_{E/k}\phi$ is the identity on $K_n^M(k)$, and $\bar{\phi}$ is a surjection, it will follow that $\bar{N}_{E/k}$ is an injection, as desired.

Fix η in J and set $\alpha = \sqrt[\ell]{a_1}$. If $\alpha \in E$ then $N_{E/k}(1-\alpha) = a_2$ and $\phi(\eta) = \{\alpha^\ell, 1-\alpha, a_3, \ldots\} = 0$. Otherwise, we pick $\underline{b} \in K_{n-1}^M(E)$ with $N_{E/k}(\underline{b}) = \underline{a}$ and note that the image $\underline{b}|_{E(\alpha)}$ of $\underline{b}$ in $K_{n-1}^M(E(\alpha))$ satisfies

$$N_{E(\alpha)/k(\alpha)}(\underline{b}|_{E(\alpha)}) = N_{E/k}(\underline{b})|_{k(\alpha)} = \underline{a}|_{k(\alpha)} = N_{E(\alpha)/k(\alpha)}(\{1-\alpha, a_3, \ldots\}).$$

Because the extension $E(\alpha)/k(\alpha)$ satisfies Hilbert 90 for K_{n-1}^M, we have $\underline{b}|_{E(\alpha)} = \{1-\alpha, a_3, \ldots\} + (1-\sigma)\underline{c}$ in $K_{n-1}^M(E(\alpha))$ for some $\underline{c}$. Thus $\phi(\eta) = \{a_1, \underline{b}\}$ equals

$$N_{E(\alpha)/E}(\{\alpha, \underline{b}\}) = N_{E(\alpha)/E}(\{\alpha, 1-\alpha, a_3 \ldots\}) + (1-\sigma)N_{E(\alpha)/E}\{\alpha, \underline{c}\},$$

which is zero in $K_n^M(E)/(1-\sigma)$, as claimed. □

Lemma 3.4. *If k is ℓ-special and $[E:k] = \ell$, then $K_n^M(E)$ is generated by symbols of the form $\{x, a_2 \ldots, a_n\}$ with $x \in E^\times$ and the $a_i \in k^\times$.*

Proof. (Tate) By induction, we may assume that $n = 2$. If $E = k(u)$ then every element of E is a polynomial in u of degree $< \ell$, and is a product of linear terms since k is ℓ-special. Terms $\{x, y\}$ in which one factor is in $k^\times$ have the desired form, so it suffices to consider the linear symbols $\{x, y\}$, where $x = u - a_1$ and $y = u - a_2$. Since $\{x, x\} = \{-1, x\}$ we may assume that $a = a_2 - a_1$ is nonzero. Since $1 = (a/x) + (y/x)$ we have

$$0 = \{(a/x), (y/x)\} = \{a, (y/x)\} - \{x, (y/x)\}$$
$$= \{a, y\} - \{a, x\} - \{x, y\} + \{-1, x\},$$

which shows that $\{x, y\}$ has the desired form. □

We conclude this section with the analogue of Hilbert 90 for K_n^M/ℓ. Suppose that k satisfies Hilbert 90 for K_n^M (3.1). Then we have an exact sequence

$$K_n^M(k)/\ell \oplus K_n^M(E)/\ell \xrightarrow{(\text{incl}_*,\, 1-\sigma)} K_n^M(E)/\ell \xrightarrow{N_{E/k}} K_n^M(k)/\ell.$$

To see this, consider the image $\bar{x} \in K_n^M(E)/\ell$ of some $x \in K_n^M(E)$. If $N_{E/k}(\bar{x}) = 0$ in $K_n^M(k)/\ell$, then $N_{E/k}(x) = \ell y$ for some $y \in K_n^M(k)$; then by Hilbert 90, $x - y = (1-\sigma)z$ for some $z \in K_n^M(E)$.

Provided that BL(n) holds, this yields an exact sequence in Galois cohomology:

$$H^n_{\text{ét}}(k, \mu_\ell^{\otimes n}) \longrightarrow H^n_{\text{ét}}(E, \mu_\ell^{\otimes n})/(1-\sigma) \xrightarrow{N} H^n_{\text{ét}}(k, \mu_\ell^{\otimes n}). \qquad (3.5)$$

3.2   A GALOIS COHOMOLOGY SEQUENCE

Recall that the nonzero elements of $H^1_{\text{ét}}(k, \mathbb{Z}/\ell) = \text{Hom}(\text{Gal}(\bar{k}/k), \mathbb{Z}/\ell)$ classify the Galois extensions E/k with $[E:k] = \ell$. If k contains the ℓth roots of unity then $E = k(\sqrt[\ell]{a})$ for some a and $[E] \in H^1_{\text{ét}}(k, \mathbb{Z}/\ell)$ corresponds to $[a]$ under the Kummer isomorphism $H^1_{\text{ét}}(k, \mathbb{Z}/\ell) \cong k^\times/k^{\times \ell}$.

Theorem 3.6.[1] *Assume that BL(n) holds, and that k contains the ℓth roots of unity. If E/k is a cyclic Galois field extension of degree ℓ then the cup product with $[E] \in H^1_{\text{ét}}(k, \mathbb{Z}/\ell)$ fits into an exact sequence*

$$H^n_{\text{ét}}(E, \mathbb{Z}/\ell) \xrightarrow{N} H^n_{\text{ét}}(k, \mathbb{Z}/\ell) \xrightarrow{\cup [E]} H^{n+1}_{\text{ét}}(k, \mathbb{Z}/\ell) \to H^{n+1}_{\text{ét}}(E, \mathbb{Z}/\ell).$$

Example 3.7. When $\ell = 2$, the exact sequence of Theorem 3.6 is part of the cohomology long exact sequence associated to $0 \to \mathbb{Z}/2 \to \mathbb{Z}/2[G] \to \mathbb{Z}/2 \to 0$:

$$H^n_{\text{ét}}(E, \mathbb{Z}/2) \xrightarrow{N} H^n_{\text{ét}}(k, \mathbb{Z}/2) \xrightarrow{\partial_2} H^{n+1}_{\text{ét}}(k, \mathbb{Z}/2) \to H^{n+1}_{\text{ét}}(E, \mathbb{Z}/2).$$

Indeed, an elementary cochain calculation shows that the map ∂_2 is the cup product with $[E]$. Thus we can and shall assume that $\ell > 2$.

When $\ell > 2$, the sequence of 3.6 does not extend to the left, even for $n = 1$.

The proof of Theorem 3.6 will be given at the end of this section. For the proof, it is convenient to introduce the following notation. Let $F_\ell = \mathbb{Z}/\ell[G]$, with $G = \text{Gal}(E/k)$ and hence $\text{Gal}(\bar{k}/k)$ acting via the coinduced structure: $F_\ell = \text{coind}_1^G(\mathbb{Z}/\ell)$. By Shapiro's Lemma, the Galois cohomology $H^*_{\text{ét}}(k, F_\ell)$ is $H^*_{\text{ét}}(E, \mathbb{Z}/\ell)$ and the map $N : H^*_{\text{ét}}(E, \mathbb{Z}/\ell) \to H^*_{\text{ét}}(k, \mathbb{Z}/\ell)$ is induced by the quotient map $F_\ell \to \mathbb{Z}/\ell$.

---

1. When k is ℓ-special, this is proven in [Voe03a, 5.2].

There is a G-module filtration $0 \subset \mathbb{Z}/\ell = F_1 \subset F_2 \subset \cdots \subset F_{\ell-1} \subset F_\ell$, where F_i is the kernel of $(1-\sigma)^i : F_\ell \to F_\ell$ and $F_i/F_{i-1} \cong \mathbb{Z}/\ell$. For $i > j$ there is a canonical surjection $F_i \to F_j$ sending c to $(1-\sigma)^{i-j}c$, with kernel F_{i-j}. For each $i \leq \ell$ we write γ_i for the exact sequence $\gamma_i : 0 \to F_{i-1} \to F_i \to \mathbb{Z}/\ell \to 0$. We will write η for the exact sequence $0 \to \mathbb{Z}/\ell \to F_\ell \xrightarrow{s} F_{\ell-1} \to 0$.

Since the cohomology $H^1(G, -)$ is defined to be $\mathrm{Ext}^1_{\mathbb{Z}[G]}(\mathbb{Z}/\ell, -)$, the G-module extensions γ_i define elements $[\gamma_i]$ of $H^1(G, F_{i-1})$. The restriction map $\mathrm{res} : H^1(G, F_{i-1}) \to H^1_{\text{ét}}(k, F_{i-1})$ sends $[\gamma_i]$ to an element $[\gamma_i]_{\text{ét}}$ of $H^1_{\text{ét}}(k, F_{\ell-1})$. Using $i : \mathbb{Z}/\ell = F_1 \hookrightarrow F_{\ell-1}$, we can compare $[\gamma_\ell]_{\text{ét}}$ to a multiple of $i_*[\gamma_2]_{\text{ét}}$.

Lemma 3.8. *For some $c \in \mathbb{Z}/\ell$, the element $[\gamma_\ell]_{\text{ét}} - i_*(c\,[\zeta]_{\text{ét}})$ is in the image of the canonical map $H^1_{\text{ét}}(k, F_\ell) \to H^1_{\text{ét}}(k, F_{\ell-1})$, induced by $s : F_\ell \to F_{\ell-1}$.*

Proof. The boundary map $\partial_\gamma : H^n(G, \mathbb{Z}/\ell) \to H^{n+1}(G, \mathbb{Z}/\ell)$ associated to γ_2 is the cup product with $\partial_\gamma(1) = [\gamma_2]$. Under the boundary map

$$\partial_\eta : H^1(G, F_{\ell-1}) \to H^2(G, \mathbb{Z}/\ell) \cong \mathrm{Ext}^2_G(\mathbb{Z}/\ell, \mathbb{Z}/\ell)$$

associated to the sequence $\eta : 0 \to \mathbb{Z}/\ell \to F_\ell \xrightarrow{s} F_{\ell-1} \to 0$, $[\gamma_\ell]$ maps to the class α of the extension

$$\alpha : \quad 0 \to \mathbb{Z}/\ell \to F_\ell \xrightarrow{\sigma-1} F_\ell \to \mathbb{Z}/\ell \to 0.$$

Since $H^*(G, \mathbb{Z}/\ell) \cong \mathbb{Z}/\ell[u,v]/(u^2)$ with Bockstein $\beta(u) = v$, and both $[\gamma_2]$ and α are nonzero, it follows that

$$\partial_\eta([\gamma_\ell]) = \alpha = c\beta([\gamma_2]) \quad \text{for a nonzero } c \in \mathbb{Z}/\ell. \tag{3.8a}$$

The morphism $\gamma_2 \to \eta$ of exact sequences yields the commutative diagram:

$$
\begin{array}{ccc}
H^1_{\text{ét}}(k, \mathbb{Z}/\ell) & \xrightarrow{\partial_\gamma} & H^2_{\text{ét}}(k, \mathbb{Z}/\ell) \\
\Big\downarrow{\scriptstyle i_*} & & \Big\| \\
H^1_{\text{ét}}(k, F_\ell) \xrightarrow{s_*} H^1_{\text{ét}}(k, F_{\ell-1}) & \xrightarrow{\partial_\eta} & H^2_{\text{ét}}(k, \mathbb{Z}/\ell).
\end{array}
\tag{3.8b}
$$

We see that for every $u \in H^1_{\text{ét}}(k, \mathbb{Z}/\ell)$,

$$\partial_\eta(i_* u) = \partial_\gamma(u) = [\gamma_2]_{\text{ét}} \cup u. \tag{3.8c}$$

Similarly, the choice of a root of unity ζ determines an isomorphism $\mathbb{Z}/\ell \cong \mu_\ell$, and the Bockstein $\beta : H^n_{\text{ét}}(k, \mathbb{Z}/\ell) \to H^{n+1}_{\text{ét}}(k, \mathbb{Z}/\ell)$ (the boundary map associated to $1 \to \mu_\ell \to \mu_{\ell^2} \to \mu_\ell \to 1$) is the cup product with $\beta(1) = [\zeta]_{\text{ét}}$. In particular,

$$\beta([\gamma_2]_{\text{ét}}) = [\gamma_2]_{\text{ét}} \cup [\zeta]_{\text{ét}}. \tag{3.8d}$$

Combining (3.8a) and (3.8d) yields $\partial_\eta([\gamma_\ell]_{\text{ét}}) = c[\gamma_2]_{\text{ét}} \cup [\zeta]_{\text{ét}}$. By (3.8c) with $u = [\zeta]_{\text{ét}}$, we have $\partial_\eta(c\, i_*[\zeta]_{\text{ét}}) = c[\gamma_2]_{\text{ét}} \cup [\zeta]_{\text{ét}}$. It follows from (3.8b) that $[\gamma_\ell]_{\text{ét}} - c\, i_*[\zeta]_{\text{ét}}$ comes from an element of $H^1_{\text{ét}}(k, F_\ell)$. □

Lemma 3.9. *Assume that $BL(n)$ holds, and k contains μ_ℓ. Then the canonical map $(i, s): \mathbb{Z}/\ell \oplus F_\ell \to F_{\ell-1}$, $(y, z) \mapsto y + (1 - \sigma)z$, induces a surjection*

$$H^n_{\text{ét}}(k, \mathbb{Z}/\ell) \oplus H^n_{\text{ét}}(k, F_\ell) \to H^n_{\text{ét}}(k, F_{\ell-1}).$$

Proof. By (3.5), we see that if $x \in H^n_{\text{ét}}(E, \mathbb{Z}/\ell) = H^n_{\text{ét}}(k, F_\ell)$ has $N(x) = 0$ then $x = y_E + (1 - \sigma)z$ for some $z \in H^n_{\text{ét}}(k, F_\ell)$ and some image y_E of a $y \in H^n_{\text{ét}}(k, \mathbb{Z}/\ell)$. Now the composite $F_\ell \to F_{\ell-1} \to F_\ell$ is the cup product with $1 - \sigma$, so from the exact cohomology sequence of γ_ℓ,

$$H^{n-1}_{\text{ét}}(k, \mathbb{Z}/\ell) \xrightarrow{\partial} H^n_{\text{ét}}(k, F_{\ell-1}) \to H^n_{\text{ét}}(k, F_\ell) \xrightarrow{N} H^n_{\text{ét}}(k, \mathbb{Z}/\ell),$$

we see that every element of $H^n_{\text{ét}}(k, F_{\ell-1})$ is equivalent modulo the image of $H^n_{\text{ét}}(k, \mathbb{Z}/\ell) \oplus H^n_{\text{ét}}(k, F_\ell)$ to an element $\partial(t)$ for some $t \in H^{n-1}(k, \mathbb{Z}/\ell)$. Since $\partial(t)$ is the cup product of t with the canonical element $[\gamma_\ell]_{\text{ét}}$ in the image of $H^0(k, \mathbb{Z}/\ell) \xrightarrow{\partial} H^1(k, F_{\ell-1})$, it suffices to observe that, according to Lemma 3.8, $[\gamma_\ell]_{\text{ét}}$ is a multiple of $i_*[\zeta]_{\text{ét}}$ plus the image of an element in $H^1_{\text{ét}}(k, F_\ell)$. □

Proposition 3.10.[2] *Suppose that $BL(n)$ holds, $\mu_\ell \subset k^\times$ and that $1 < i \le \ell$. Then the canonical maps $\mathbb{Z}/\ell \oplus F_i \to F_{i-1}$ induce a surjection*

$$H^n_{\text{ét}}(k, \mathbb{Z}/\ell) \oplus H^n_{\text{ét}}(k, F_i) \to H^n_{\text{ét}}(k, F_{i-1}).$$

Proof. The case $i = \ell$ is established in Lemma 3.9, so we may suppose that $i < \ell$. Consider the diagram with exact rows, induced by $\gamma_{i+1} \to \gamma_i$:

$$
\begin{array}{ccccc}
H^{n-1}_{\text{ét}}(k, \mathbb{Z}/\ell) & \longrightarrow & H^n_{\text{ét}}(k, F_i) & \longrightarrow & H^n(k, F_{i+1}) \\
\| & & \downarrow & \searrow{\scriptstyle 1-\sigma} & \\
H^{n-1}_{\text{ét}}(k, \mathbb{Z}/\ell) & \xrightarrow{\delta} & H^n_{\text{ét}}(k, F_{i-1}) \longrightarrow H^n_{\text{ét}}(k, F_i) & \xrightarrow{N} & H^n_{\text{ét}}(k, \mathbb{Z}/\ell). \\
& & \uparrow & & \\
& & H^n_{\text{ét}}(k, \mathbb{Z}/\ell) & &
\end{array}
$$

A diagram chase shows that the conclusion of Proposition 3.10 for $i < \ell$ is equivalent to the assertion that the following sequence is exact:

$$H^n_{\text{ét}}(k, \mathbb{Z}/\ell) \to H^n_{\text{ét}}(k, F_i)/(1 - \sigma) \xrightarrow{N} H^n_{\text{ét}}(k, \mathbb{Z}/\ell). \tag{3.10a}$$

---

2. Taken from [Voe03a, 5.4b].

We will prove that (3.10a) is exact by downward induction on i. It is exact when $i = \ell$, because it agrees with (3.5), which we saw is exact when BL(n) holds. Thus we may assume that $i < \ell$ and that (3.10a) is exact for $i + 1$, or equivalently, that Proposition 3.10 holds for $i + 1$.

Hence we can write any element of $H^n_{\text{ét}}(k, F_i)$ as $x_i + \bar{y}$ for $x \in H^n_{\text{ét}}(k, \mathbb{Z}/\ell)$ and $y \in H^n_{\text{ét}}(k, F_{i+1})$, where x_i and $\bar{y}$ represent the images of x and y in $H^n_{\text{ét}}(k, F_i)$. To check that (3.10a) is exact, we suppose that $N(x_i + \bar{y}) = 0$ in $H^n_{\text{ét}}(k, \mathbb{Z}/\ell)$. Because $i > 1$, the composition $\mathbb{Z}/\ell \to F_i \to \mathbb{Z}/\ell$ is zero; since N is induced by the natural map $F_i \to \mathbb{Z}/\ell$, we have $N(x_i) = 0$, so $y \in H^n_{\text{ét}}(k, F_{i+1})$ maps to zero in $H^n_{\text{ét}}(k, \mathbb{Z}/\ell)$. By exactness of (3.10a) for $i + 1$ we can write $y = (1 - \sigma)t + x'$ for x' in the image of $H^n_{\text{ét}}(k, \mathbb{Z}/\ell)$ and $t \in H^n_{\text{ét}}(k, F_{i+1})$. Since $\mathbb{Z}/\ell \to F_{i+1} \to F_i$ is zero, $\bar{y}$ is the image of $(1 - \sigma)t$ in $H^n_{\text{ét}}(k, F_i)$. Since the G-module map $F_{i+1} \to F_i$ induces a G-map on cohomology, we see that $\bar{y} = (1 - \sigma)\bar{t}$. This proves exactness of (3.10a) for i, which implies that Proposition 3.10 holds for i.   □

Proof of Theorem 3.6. Recall from Example 3.7 that we may assume that $\ell > 2$. Let η denote the exact sequence $0 \to \mathbb{Z}/\ell \to F_i \xrightarrow{1-\sigma} F_{i-1} \to 0$. The morphisms $\gamma_i \to \gamma_2 \to \eta$ yield a commutative diagram with exact rows for each $i \le \ell$:

$$
\begin{array}{ccccccccc}
H^n_{\text{ét}}(k, F_{i-1}) & \longrightarrow & H^n_{\text{ét}}(k, F_i) & \xrightarrow{N} & H^n_{\text{ét}}(k, \mathbb{Z}/\ell) \\
\downarrow & \text{3.10} & \downarrow & & \| \\
H^n_{\text{ét}}(k, \mathbb{Z}/\ell) & \longrightarrow & H^n_{\text{ét}}(k, F_2) & \xrightarrow{N} & H^n_{\text{ét}}(k, \mathbb{Z}/\ell) & \xrightarrow{\cup [E]} & H^{n+1}_{\text{ét}}(k, \mathbb{Z}/\ell) & \longrightarrow & H^{n+1}_{\text{ét}}(k, F_2) \\
& & \downarrow & \text{3.10} & \downarrow & & \| & & \downarrow \\
& & H^n_{\text{ét}}(k, F_i) & \longrightarrow & H^n_{\text{ét}}(k, F_{i-1}) & \xrightarrow{\delta} & H^{n+1}_{\text{ét}}(k, \mathbb{Z}/\ell) & \longrightarrow & H^{n+1}_{\text{ét}}(k, F_i).
\end{array}
$$

In the two squares marked "3.10," the lower right group is the sum of the images of the indicated maps, by Proposition 3.10. A chase on the above diagram shows that we have an exact sequence for each i:

$$
H^n_{\text{ét}}(k, F_i) \xrightarrow{N} H^n_{\text{ét}}(k, \mathbb{Z}/\ell) \xrightarrow{\cup [E]} H^{n+1}_{\text{ét}}(k, \mathbb{Z}/\ell) \to H^{n+1}_{\text{ét}}(k, F_i).
$$

Exactness of this sequence for $i = \ell$ is the desired conclusion of Theorem 3.6.   □

3.3   HILBERT 90 FOR ℓ-SPECIAL FIELDS

By Theorem 3.2, BL(n − 1) implies that every field of characteristic $\ne \ell$ satisfies Hilbert 90 for K^M_{n-1}. Our next theorem was used in section 1.2 (as Theorem 1.10) to prove that the norm residue map is an isomorphism.

Theorem 3.11.[3] *Suppose that k is ℓ-special, $K_n^M(k)/\ell = 0$ and $BL(n-1)$ holds.*

(a) $H_{\text{ét}}^n(k, \mathbb{Z}/\ell) = 0$, and
(b) $H_{\text{ét}}^{n+1}(k, \mathbb{Z}_{(\ell)}(n)) = 0$. In addition, for every finite field extension E/k:
(c) $K_{n-1}^M(E) \xrightarrow{N} K_{n-1}^M(k)$ is onto;
(d) $K_n^M(E)/\ell = 0$.

Proof. We first prove (c) and (d). Since k is ℓ-special, we may assume that $[E:k] = \ell$. In this case, $E = k(\sqrt[\ell]{a})$ for some $a \in k$. Consider the commutative diagram

$$
\begin{array}{ccccc}
K_{n-1}^M(E) & \xrightarrow{N} & K_{n-1}^M(k) & \xrightarrow{\cup[a]} & K_n^M(k)/\ell \\
{\scriptstyle\text{onto}}\downarrow & & {\scriptstyle\text{onto}}\downarrow & & \downarrow \\
H_{\text{ét}}^{n-1}(E, \mathbb{Z}/\ell) & \xrightarrow{N} & H_{\text{ét}}^{n-1}(k, \mathbb{Z}/\ell) & \xrightarrow{\cup[a]} & H_{\text{ét}}^n(k, \mathbb{Z}/\ell).
\end{array}
$$

Since $BL(n-1)$ holds, Theorem 3.6 implies that the bottom row is exact; the left two vertical maps are onto by Lemma 1.30. The upper right term vanishes by assumption, and the kernel $\ell K_{n-1}^M(k)$ of the middle vertical map is contained in the image of $K_{n-1}^M(E)$, so a diagram chase shows that (c) holds.

By (c), the hypotheses of Proposition 3.3 are satisfied, yielding a right exact sequence

$$
K_n^M(E) \xrightarrow{1-\sigma} K_n^M(E) \xrightarrow{N} K_n^M(k).
$$

Reducing it modulo ℓ shows that $(1-\sigma)$ is a surjection from $K_n^M(E)/\ell$ onto itself. But $(1-\sigma)^\ell$ is zero on any $\text{Gal}(E/k)$-module of exponent ℓ, so $K_n^M(E)/\ell$ must be zero. Thus (d) holds.

For (a), suppose that $x \in H_{\text{ét}}^n(k, \mathbb{Z}/\ell)$ is nonzero. Since x vanishes over the algebraic closure $\bar{k}$ of k, it vanishes over some finite field extension E; pick E minimal so that $x_E = 0$ in $H_{\text{ét}}^n(E, \mathbb{Z}/\ell)$. Because $\text{Gal}(E/k)$ is a finite ℓ-group, it has a subgroup H of order ℓ, corresponding to an intermediate subfield k' with $[E:k'] = \ell$. By the minimality of E, x remains nonzero in $H_{\text{ét}}^n(k', \mathbb{Z}/\ell)$.

Since $E = k'(\sqrt[\ell]{a})$ is a cyclic Galois extension of k', Theorem 3.6 yields the exact sequence

$$
H_{\text{ét}}^{n-1}(k', \mathbb{Z}/\ell) \xrightarrow{\cup[a]} H_{\text{ét}}^n(k', \mathbb{Z}/\ell) \to H_{\text{ét}}^n(E, \mathbb{Z}/\ell).
$$

Since $BL(n-1)$ implies that $K_{n-1}^M(k')/\ell \cong H_{\text{ét}}^{n-1}(k', \mathbb{Z}/\ell)$, and x vanishes in $H_{\text{ét}}^n(E, \mathbb{Z}/\ell)$, there is an element y of $K_{n-1}^M(k')$ such that the norm residue $K_n^M(k')/\ell \to H_{\text{ét}}^n(k', \mathbb{Z}/\ell)$ sends $\{y, a\}$ to x. Since $K_n^M(k')/\ell = 0$ by (d), it follows that $\{y, a\} = 0$, contradicting the assumption that $x \neq 0$ in $H_{\text{ét}}^n(k', \mathbb{Z}/\ell)$.

---

3. Parts (c) and (d) are taken from [Voe03a, 5.7 and 5.8].

For (b) we observed in Lemma 1.6 that $H_{\text{ét}}^{n+1}(k, \mathbb{Z}_{(\ell)}(n))$ is an ℓ-torsion group. The coefficient sequence for $\mathbb{Z}_{(\ell)}(n) \xrightarrow{\ell} \mathbb{Z}_{(\ell)}(n)$ shows that (a) implies (b).   □

3.4   COHOMOLOGY ELEMENTS

Using the cohomology operations Q_i, we will show that any nonzero element $\underline{a} = \{a_1, \ldots, a_n\}$ of $K_n^M(k)/\ell$ gives rise to nonzero elements $\delta \in H^{n,n-1}(\mathfrak{X}, \mathbb{Z}/\ell)$ and $\mu \in H^{2b+1,b}(\mathfrak{X}, \mathbb{Z})$, where $\mathfrak{X}$ is the simplicial scheme associated to a Rost variety X (see 1.24), and $b = (\ell^{n-1} - 1)/(\ell - 1)$.

We first show that the norm residue map is nonzero on symbols. To prove this, we will use the étale cohomology calculation contained in Theorem 3.6.

Lemma 3.12. *Suppose that $BL(n-1)$ holds, and that k satisfies Hilbert 90 for K_{n-1}^M. If $\{a_1, \ldots, a_n\}$ is a nonzero symbol in $K_n^M(k)/\ell$, its image in $H_{\text{ét}}^n(k, \mu_\ell^{\otimes n})$ is nonzero.*

Proof. By the standard transfer argument 1.2, we may assume k has no prime-to-ℓ extensions. For $E = k(\gamma)$, $\gamma = \sqrt[\ell]{a_n}$, we have a commutative diagram

$$
\begin{array}{ccccc}
K_{n-1}^M(E)/\ell & \xrightarrow{\text{norm}} & K_{n-1}^M(k)/\ell & \xrightarrow{\cup a_n} & K_n^M(k)/\ell \\
\cong \downarrow & & \cong \downarrow & & \downarrow \\
H_{\text{ét}}^{n-1}(E, \mathbb{Z}/\ell) & \xrightarrow{\text{norm}} & H_{\text{ét}}^{n-1}(k, \mathbb{Z}/\ell) & \xrightarrow{\cup [a_n]} & H_{\text{ét}}^n(k, \mathbb{Z}/\ell) \longrightarrow H_{\text{ét}}^n(E, \mathbb{Z}/\ell)
\end{array}
$$

in which the marked vertical maps are isomorphisms by $BL(n-1)$ and 1.29(d), and the bottom row is exact by Theorem 3.6. If $\{a_1, \ldots, a_{n-1}\} \cup \{a_n\} = \{a_1, \ldots, a_n\}$ vanishes in $H_{\text{ét}}^n(k, \mathbb{Z}/\ell)$ then $\{a_1, \ldots, a_{n-1}\}$ is the norm of some $s \in K_{n-1}^M(E)/\ell$. But then $\{a_1, \ldots, a_n\}$ is the norm of $\{s, a_n\} = \{s, \gamma\}^\ell = 0$ and hence is zero.   □

Lemma 3.13. *Assume that $BL(n-1)$ holds. If a smooth variety X splits a nonzero $\underline{a} \in K_n^M(k)/\ell$, then $\underline{a}$ lifts to a unique nonzero δ in $H^n(\mathfrak{X}, \mathbb{Z}/\ell(n-1))$.*

Proof. (See [Voe11, 6.5].) Set $A = \mathbb{Z}/\ell(n-1)$, so $A_{\text{ét}} \cong \mu_\ell^{\otimes n-1}$. By the standard transfer argument 1.2, we may assume that k contains μ_ℓ, so that $A_{\text{ét}} \cong \mu_\ell^{\otimes n}$ as well. Write C for the cone of $A \to \mathbf{R}\pi_*(A_{\text{ét}})$.

Now the hypercohomology spectral sequence for any smooth simplicial scheme $S_\bullet$ is $H_{\text{nis}}^p(S_\bullet, \mathcal{H}^q) \Rightarrow H_{\text{nis}}^{p+q}(S_\bullet, C)$, where $\mathcal{H}^q$ denotes the Nisnevich sheaf associated to $H_{\text{nis}}^q(-, C)$. By $BL(n-1)$ we have $C \cong \tau^{\geq n} \mathbf{R}\pi_*(A_{\text{ét}})$, so $\mathcal{H}^q$ is 0 for $q < n$, and the Nisnevich sheaf associated to $H_{\text{ét}}^q(-, C)$ if $q \geq n$. Hence the spectral sequence degenerates to yield $H_{\text{nis}}^{n-1}(S_\bullet, C) = 0$ and $H_{\text{nis}}^n(S_\bullet, C) \xrightarrow{\cong} H^0(S_\bullet, \mathcal{H}^n)$.

Thus for $S_\bullet = \mathfrak{X}$ we have the exact sequence forming the top row of the diagram:

$$\delta \longmapsto \underline{a} \longmapsto 0$$

$$0 = H_{\mathrm{nis}}^{n-1}(\mathfrak{X}, C) \longrightarrow H_{\mathrm{nis}}^n(\mathfrak{X}, A) \longrightarrow H_{\mathrm{\acute{e}t}}^n(\mathfrak{X}, A) \longrightarrow H_{\mathrm{nis}}^0(\mathfrak{X}, \mathcal{H}^n)$$

$$\cong \Big\uparrow 1.37 \qquad\qquad \Big\downarrow \text{into}$$

$$H_{\mathrm{\acute{e}t}}^n(k, A) \longrightarrow H_{\mathrm{\acute{e}t}}^n(k(X), A).$$

We claim that the right vertical is an injection. It is defined as the composition $H^0(\mathfrak{X}, \mathcal{H}^n) \to H^0(X, \mathcal{H}^n) \to H^0(k(X), \mathcal{H}^n) = H_{\mathrm{\acute{e}t}}^n(k(X), A)$. The first of these maps is an injection because for any simplicial scheme $S_\bullet$ and any sheaf $\mathcal{F}$, $H^0(S_\bullet, \mathcal{F})$ embeds in $H^0(S_0, \mathcal{F})$. Since the sheaf $\mathcal{F} = \mathcal{H}^n$ is a homotopy invariant Nisnevich sheaf with transfers by [MVW, 6.17 and 22.3], $H^0(X, \mathcal{H}^n)$ embeds in $H_{\mathrm{\acute{e}t}}^n(k(X), A)$ by [MVW, 11.1]. This establishes the claim.

The image of $\underline{a}$ in $H_{\mathrm{\acute{e}t}}^n(k, \mu_\ell^{\otimes n})$ is nonzero by Lemma 3.12 (and 3.2), and vanishes in $H_{\mathrm{\acute{e}t}}^n(k(X), \mu_\ell^{\otimes n})$ by hypothesis. It follows that $\underline{a}$ lifts to a nonzero δ in $H_{\mathrm{nis}}^n(\mathfrak{X}, A)$; δ is unique because $H_{\mathrm{nis}}^{n-1}(\mathfrak{X}, C) = 0$. $\qquad\square$

Lemma 3.14. *Suppose for some $i \geq 1$ that (a, b) and (c, d) in $\mathbb{Z}^2$ satisfy*

$$(a, b) + 2(2\ell^{i-1} - 1, \ell^{i-1} - 1) = (c, d) + (2\ell^i - 1, \ell^i - 1).$$

If (a, b) is in the plane region $\Omega = \{(x, y) : x - 1 \leq y < n\}$ then so is (c, d).

Proof. For $\ell = 2$ we have $(c - a, d - a) = (-1, -1)$. As a function of ℓ, $d - b$ is decreasing for $\ell \geq 2i/(i+1)$, and $c - a = 2(d - b) + 1$. $\qquad\square$

Recall from Definition 1.36 and Corollary 1.38 that the group $H^{p,q}(\mathfrak{X}, \mathbb{Z})$ has exponent ℓ when $p > q$, and hence injects into $H^{p,q}(\mathfrak{X}, \mathbb{Z}/\ell)$. In fact, $H^{p,q}(\mathfrak{X}, \mathbb{Z})$ is the kernel of the Bockstein $\beta : H^{p,q}(\mathfrak{X}, \mathbb{Z}/\ell) \to H^{p+1,q}(\mathfrak{X}, \mathbb{Z}/\ell)$. Now $\beta Q_i = -Q_i \beta$ by Lemmas 13.11 and 13.13, so each operation Q_i on $H^{p,q}(\mathfrak{X}, \mathbb{Z}/\ell)$ sends $H^{p,q}(\mathfrak{X}, \mathbb{Z})$ to the subgroup $H^{r,s}(\mathfrak{X}, \mathbb{Z})$ of $H^{r,s}(\mathfrak{X}, \mathbb{Z}/\ell)$, where $r = p + 2\ell^i - 1$ and $s = q + \ell^i - 1$.

Proposition 3.15. *Assume that $BL(n-1)$ holds, and X is a Rost variety. The cohomology operations $Q_0 = \beta$, $Q = Q_{n-2} \cdots Q_0$ and $Q_{n-1}Q$ are injections on $H^{n,n-1}(\mathfrak{X}, \mathbb{Z}/\ell)$.*

$$H^{n,n-1}(\mathfrak{X}, \mathbb{Z}/\ell) \xrightarrow{Q_{n-2} \cdots Q_1 Q_0} H^{2b+1,b}(\mathfrak{X}, \mathbb{Z}/\ell) \xrightarrow{Q_{n-1}} H^{2b\ell+2,b\ell}(\mathfrak{X}, \mathbb{Z}/\ell).$$

The maps $Q_{n-2} \cdots Q_1$ and $Q_{n-1} \cdots Q_1$ restrict to injections

$$H^{n+1,n-1}(\mathfrak{X}, \mathbb{Z}) \xrightarrow{Q_{n-2} \cdots Q_1} H^{2b+1,b}(\mathfrak{X}, \mathbb{Z}) \xrightarrow{Q_{n-1}} H^{2b\ell+2,b\ell}(\mathfrak{X}, \mathbb{Z}),$$

where $b = (\ell^{n-1} - 1)/(\ell - 1) = 1 + \ell + \cdots + \ell^{n-2}$.

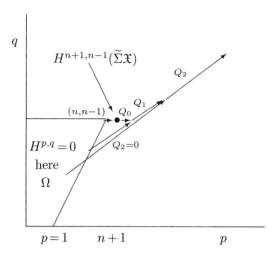

Figure 3.1: The composition $Q_2 Q_1 Q_0$ is an injection on $H^{n+1,n}(\widetilde{\Sigma}\mathfrak{X})$

Proof. We use the isomorphism $H^{p,q}(\mathfrak{X}, \mathbb{Z}/\ell) \cong H^{p+1,q}(\widetilde{\Sigma}\mathfrak{X}, \mathbb{Z}/\ell) = 0$ for $p > q$, where $\widetilde{\Sigma}\mathfrak{X}$ is the reduced suspension of $\mathfrak{X}$, pointed out in Definition 1.36. By 1.38, the integral groups have exponent ℓ, and each Q_i sends them to integral groups, so it suffices to show that each $Q_i \ldots Q_1$ is injective on the group $H^{n+1,n-1}(\mathfrak{X}, \mathbb{Z}/\ell) \cong H^{n+2,n-1}(\widetilde{\Sigma}\mathfrak{X}, \mathbb{Z}/\ell)$.

We proceed by induction on i; the case Q_0 was established in 1.40, so assume $i > 0$. Figure 3.1 shows the case $i = 2$. Because BL(n − 1) implies BL(q) for all $q < n$ by Theorem 2.9, Corollary 1.39 implies that

$$H^{p,q}\left(\widetilde{\Sigma}\mathfrak{X}, \mathbb{Z}/\ell\right) = 0 \text{ when } (p,q) \text{ is in the region } \Omega = \{(p,q) : p - 1 \le q < n\}.$$

Define (p,q) so that $Q_{i-1} \ldots Q_1$ takes $H^{n+2,n-1}(\widetilde{\Sigma}\mathfrak{X}, \mathbb{Z}/\ell)$ to $H^{p,q}(\widetilde{\Sigma}\mathfrak{X}, \mathbb{Z}/\ell)$. Since X is a Rost variety, there is a ν_i-variety X_i and a map $X_i \to X$. By Theorem 1.44 (proven in part III as Theorem 13.24) we have exact sequences

$$
\begin{aligned}
H^{a,b}(\widetilde{\Sigma}\mathfrak{X}, \mathbb{Z}/\ell) &\xrightarrow{Q_{i-1}} H^{*,*}(\widetilde{\Sigma}\mathfrak{X}, \mathbb{Z}/\ell) \xrightarrow{Q_{i-1}} H^{p,q}(\widetilde{\Sigma}\mathfrak{X}, \mathbb{Z}/\ell) \\
H^{c,d}(\widetilde{\Sigma}\mathfrak{X}, \mathbb{Z}/\ell) &\xrightarrow{Q_i} H^{p,q}(\widetilde{\Sigma}\mathfrak{X}, \mathbb{Z}/\ell) \xrightarrow{Q_i} H^{r,s}(\widetilde{\Sigma}\mathfrak{X}, \mathbb{Z}/\ell).
\end{aligned}
\tag{3.15a}
$$

By induction, (a,b) is in the region Ω; by Lemma 3.14, so is (c,d)—and therefore $H^{c,d}(\widetilde{\Sigma}\mathfrak{X}) = 0$. This implies that Q_i is injective on $H^{p,q}(\widetilde{\Sigma}\mathfrak{X}, \mathbb{Z}/\ell)$. $\qquad\square$

Recall from Lemma 3.13 that the symbol $\underline{a}$ determines a nonzero element δ in $H^{n,n-1}(\mathfrak{X}, \mathbb{Z}/\ell)$. Applying the cohomology operation $Q = Q_{n-2} \cdots Q_0$ of 3.15 yields an element μ of $H^{2b+1,b}(\mathfrak{X}, \mathbb{Z})$. We will use the element μ in chapter 5 to construct a Rost motive (defined in 4.11).

Corollary 3.16. *Both* $\mu = Q(\delta) \in H^{2b+1,b}(\mathfrak{X}, \mathbb{Z})$ *and* $Q_{n-1}(\mu)$ *are nonzero.*

Proof. The operations $Q : H^{n,n-1}(\mathfrak{X}, \mathbb{Z}/\ell) \to H^{2b+1,b}(\mathfrak{X}, \mathbb{Z})$ and $Q_{n-1}Q$ are injective by Proposition 3.15. $\qquad\square$

Proposition 3.17. *Assume that* $BL(n-1)$ *holds, and* X *is a Rost variety. Then* $Q_{n-1} \cdots Q_1$ *is an injection from* $H^{n+1,n}(\mathfrak{X}, \mathbb{Z})$ *to* $H^{2b\ell+2, b\ell+1}(\mathfrak{X}, \mathbb{Z})$.

Proof. This is similar to the proof of Proposition 3.15, using the vanishing of $H^{a,b}(\widetilde{\Sigma}\mathfrak{X}, \mathbb{Z}/\ell)$ in the region Ω granted by $BL(n-1)$ and Corollary 1.39. We omit the coefficients $\mathbb{Z}/\ell$ for clarity. For Q_1 we have the exact sequence

$$H^{a,b}(\widetilde{\Sigma}\mathfrak{X}) \xrightarrow{Q_1} H^{n+2,n}(\widetilde{\Sigma}\mathfrak{X}) \xrightarrow{Q_1} H^{p,q}(\widetilde{\Sigma}\mathfrak{X})$$

with $(a, b) = (n + 3 - 2\ell, n + 1 - \ell)$ in the region Ω. It follows that Q_1 is injective on $H^{n+1,n}(\mathfrak{X}) \cong H^{n+2,n}(\widetilde{\Sigma}\mathfrak{X})$. Inductively, if $Q_{i-1} \cdots Q_1$ takes $H^{n+2,n}(\widetilde{\Sigma}\mathfrak{X})$ to $H^{p,q}(\widetilde{\Sigma}\mathfrak{X})$ then from Lemma 3.14 applied to (a, b) and (c, d) in the exact sequences (3.15a), we see that $(c, d) \in \Omega$; this implies that $H^{c,d}(\widetilde{\Sigma}\mathfrak{X}) = 0$ and hence that Q_i is injective on $H^{p,q}(\widetilde{\Sigma}\mathfrak{X})$. It follows that $Q_i \cdots Q_1$ is injective on $H^{n+1,n}(\mathfrak{X}) \cong H^{n+2,n}(\widetilde{\Sigma}\mathfrak{X})$ for all $i = 1, \ldots, n-1$. $\qquad\square$

3.5 HISTORICAL NOTES

The Hilbert 90 condition for K_n^M takes its name from the fact that the case $n = 1$ is essentially Hilbert's original Theorem 90. For $n = 2$, it was proven in 1982 by Merkurjev–Suslin [MeS82]. For $n = 3$ (and $\ell = 2$), it was proven in 1990 by Merkurjev–Suslin in [MeS90]. The central role of the Hilbert 90 condition for K_n^M was recognized in the 1996 preprint [Voe96]; the material in this section is based upon sections 5 and 6 of the 2003 paper [Voe03a].

Chapter Four

Rost Motives and H90

IN THIS CHAPTER we introduce the notion of a Rost motive, which is a summand of the motive of a Rost variety X (see 4.11); the proof that Rost motives exist is deferred to chapter 5. The highlight of this chapter is Theorem 4.20: assuming that Rost motives exist and H90(n − 1) holds, then $H_{\text{ét}}^{n+1}(k, \mathbb{Z}(n))$ injects into $H_{\text{ét}}^{n+1}(k(X), \mathbb{Z}(n))$. By Theorem 1.12, this implies that H90(n) holds and—as we saw in Theorem 1.11—this will imply the Bloch–Kato conjecture for n.

While there may be many Rost varieties associated to a given symbol, there is essentially only one Rost motive (for a proof, see [KM13, Theorem 4.1]). The Rost motive captures the part of the cohomology of a Rost variety X which is essential for the proof of Theorem 4.20.

Since a Rost motive is a special kind of symmetric Chow motive, we begin by recalling what this means in section 4.1. The next section, 4.2, introduces the notion of $\mathfrak{X}$-duality. This duality plays an important role in the axioms defining Rost motives, which are introduced in section 4.3, as well as a role in the construction of the Rost motive in chapter 5. Finally, in section 4.4, we assume that Rost motives exist and prove the key Theorem 4.20, which we mentioned above.

4.1 CHOW MOTIVES

In this section, we briefly recall the construction of Grothendieck's (contravariant) category of *Chow motives* over k, following [Man68]. Recall that $CH^j(X)$ denotes the group of algebraic cycles of codimension j on a variety X, modulo rational equivalence.

We first form the category $\mathcal{M}$ of *Chow correspondences* (of degree 0). The objects of $\mathcal{M}$ are smooth projective varieties over k; morphisms from X to Y are elements of $CH^{\dim X}(X \times Y)$ and are referred to as correspondences. Composition of correspondences $f \in \text{Hom}(X_1, X_2)$ and $g \in \text{Hom}(X_2, X_3)$ is defined using pullback, intersection, and pushforward of cycles: if p_{ij} denotes the projection from $X_1 \times X_2 \times X_3$ onto $X_i \times X_j$ then $f \circ g = p_*^{13}(p_{12}^* f \cdot p_{23}^* g)$.

The category of smooth projective varieties over k (and morphisms of varieties) embeds *contravariantly* into $\mathcal{M}$, once we identify a morphism of varieties $f : Y \to X$ with the class of its graph Γ_f in $CH^{\dim X}(X \times Y) = \text{Hom}_{\mathcal{M}}(X, Y)$.

The category $\mathcal{M}$ is additive, with $\oplus$ being disjoint union, and has an internal tensor product: on objects, $X \otimes Y$ is the product $X \times Y$ of varieties X and Y.

One defect of the category $\mathcal{M}$ is that the composition of the structure map $X \to \mathrm{Spec}(k)$ and the inclusion of a k-point in X is a correspondence e on X which is *idempotent*, meaning that $e^2 = e$, yet there is no corresponding factor $e(X)$ of X in $\mathcal{M}$. For example, the endomorphisms of the projective line in $\mathcal{M}$ form the semisimple algebra $\mathrm{Hom}_{\mathcal{M}}(\mathbb{P}^1, \mathbb{P}^1) \cong \mathbb{Z} \times \mathbb{Z}$, yet $\mathbb{P}^1$ is irreducible. To fix this defect, we pass to the idempotent completion $\widehat{\mathcal{M}}$ of $\mathcal{M}$.

Given any category $\mathcal{C}$, the *idempotent completion* $\widehat{\mathcal{C}}$ is defined as follows. The objects of $\widehat{\mathcal{C}}$ are pairs (C, e) consisting of an object C of $\mathcal{C}$ and an idempotent endomorphism e; morphisms from (C, e) to (C', e') are just maps $\phi : C \to C'$ in $\mathcal{C}$ with $e'\phi = \phi e$. It is easy to see that $\mathcal{C}$ is a full subcategory of $\widehat{\mathcal{C}}$, and that every idempotent map in $\widehat{\mathcal{C}}$ factors as a projection and an inclusion. When $\mathcal{C}$ is additive, so is $\widehat{\mathcal{C}}$ and $(C, e) \times (C, 1 - e) = (C, 1) = C$; when $\mathcal{C}$ has an internal tensor product, so does $\widehat{\mathcal{C}}$.

Objects in the category $\widehat{\mathcal{M}}$ are called *effective Chow motives*. We define the *Lefschetz motive* $\mathbb{L}$ to be the unique factor of $\mathbb{P}^1$ in $\widehat{\mathcal{M}}$ so that e gives a direct sum decomposition $\mathbb{P}^1 = \mathrm{Spec}(k) \oplus \mathbb{L}$ in $\widehat{\mathcal{M}}$. In $\widehat{\mathcal{M}}$ we also have a decomposition $\mathbb{P}^n \cong \mathrm{Spec}(k) \oplus \mathbb{L} \oplus \mathbb{L}^2 \oplus \cdots \oplus \mathbb{L}^n$, where $\mathbb{L}^i$ denotes $\mathbb{L} \otimes \cdots \otimes \mathbb{L}$ (i times). This category contains the other graded pieces of the ring $CH^*(X)$, because $CH^i(X) \cong \mathrm{Hom}_{\widehat{\mathcal{M}}}(\mathbb{L}^i, X)$. There is a natural isomorphism

$$CH^{\dim X + i - j}(X \times Y) \cong \mathrm{Hom}_{\mathcal{M}}(X \otimes \mathbb{L}^i, Y \otimes \mathbb{L}^j).$$

Finally, Grothendieck's category *Chow* of Chow motives is obtained from the category $\widehat{\mathcal{M}}$ of effective Chow motives by formally inverting the Lefschetz motive $\mathbb{L}$ with respect to $\otimes$. That is, we add objects $M \otimes \mathbb{L}^i$ for negative values of i, for each effective motive M, so that every object of *Chow* has the form $(X, e) \otimes \mathbb{L}^i$ for some smooth projective variety X over k, some idempotent element e of $CH^{\dim X}(X, X)$, and some integer i. It is traditional to write (X, e, i) for $(X, e) \otimes \mathbb{L}^i$. Morphisms from (X, e, i) to (Y, f, j) are elements of the subgroup $f A e$ of the group $A = CH^{\dim(X) + i - j}(X \times Y)$.

Definition 4.1. Given any cycle ϕ in $X \times Y$, its *transpose* is the cycle ϕ^t in $Y \times X$ obtained by interchanging the two factors X and Y. The formula for composition of correspondences shows that $(\phi_1 \circ \phi_2)^t = \phi_2^t \circ \phi_1^t$.

The transpose M^t of an effective Chow motive (X, e) is defined to be (X, e^t). This makes sense because if e is an idempotent element of $CH^{\dim(X)}(X \times X)$, then so is its transpose e^t. We will say that a Chow motive (X, e) is *symmetric* if it is isomorphic to its transpose (X, e^t) by the canonical inclusion-projection map $(X, e) \to X \to (X, e^t)$.

The category *Chow* is a rigid tensor category, with dual $X^* = X \otimes \mathbb{L}^{-d_X}$ (where $d_X = \dim X$) and $\mathbb{L}^* = \mathbb{L}^{-1}$; $\underline{Hom}(X, Y) = X^* \otimes Y$ is the internal Hom.

The dual ϕ^* of a Chow morphism $\phi : X \to Y$ is the transpose cycle ϕ^t, regarded as a morphism $\phi^* : Y \otimes \mathbb{L}^{-d_Y} \to X \otimes \mathbb{L}^{-d_X}$ via the identification

$$\mathrm{Hom}_{Chow}(Y \otimes \mathbb{L}^{-d_Y}, X \otimes \mathbb{L}^{-d_X}) = CH^{\dim(X)}(Y \times X).$$

If ϕ is a Chow morphism from $X \otimes \mathbb{L}^i$ to $Y \otimes \mathbb{L}^j$, so that the codimension of ϕ is $\dim(X) + i - j$, then we shall commonly regard the transpose ϕ^t as a Chow morphism from $Y \otimes \mathbb{L}^{\delta-j}$ to $X \otimes \mathbb{L}^{-i}$ where $\delta = \dim(X) - \dim(Y)$. Formally, we have $\phi^t = \phi^* \otimes \mathbb{L}^{\dim(X)}$.

Example 4.2. As a special case, if $\dim(X) = d$ and $M = (X, e)$ is a Chow motive with projection $X \xrightarrow{p} M$, then the map $p^t : M^t \to X$ defining the transpose is the composition

$$M^t = M^* \otimes \mathbb{L}^d \xrightarrow{p^* \otimes \mathbb{L}^d} X^* \otimes \mathbb{L}^d \cong X.$$

By [MVW, 20.1], $\mathrm{Hom}_{Chow}(X, Y) \cong \mathrm{Hom}_{\mathbf{DM}}(M(Y), M(X))$, and this gives a fully faithful contravariant embedding of *Chow* into $\mathbf{DM}_{\mathrm{gm}}$. The Lefschetz motive $\mathbb{L}$ is identified with $\mathbb{L} = \mathbb{Z}(1)[2]$ under this embedding.

In the motivic category $\mathbf{DM}_{\mathrm{gm}}$, where $M(X)^* \otimes \mathbb{L}^{\dim(X)} \cong M(X)$, the dual of $\phi : M(Y) \to M(X)$ is the map $\phi^* : M(X)^* \to M(Y)^*$, and the above remarks show that $\phi^t = \phi^* \otimes \mathbb{L}^{\dim(X)}$ as a map from $M(X)$ to $M(Y) \otimes \mathbb{L}^\delta$.

4.2 $\mathfrak{X}$-DUALITY

Fix a scheme X and a commutative ring R. Forming the simplicial scheme $\mathfrak{X} = \check{C}(X)$ as in Definition 1.32, the remarks there show that there is a simplicial weak equivalence $\mathfrak{X} \times \mathfrak{X} \simeq \mathfrak{X}$. Hence $R_{\mathrm{tr}}(\mathfrak{X}) \otimes R_{\mathrm{tr}}(\mathfrak{X}) \cong R_{\mathrm{tr}}(\mathfrak{X})$ in the triangulated category $\mathbf{DM}$ of motives with coefficients in R; see [MVW, 14.1].

It is useful to introduce the notation ε for the motive of $R_{\mathrm{tr}}(\mathfrak{X})$, so we have $\varepsilon \otimes \varepsilon \cong \varepsilon$ in $\mathbf{DM}$. For any motive M, we write εM for the motive $R_{\mathrm{tr}}(\mathfrak{X}) \otimes M$, and let $\mathbf{DM}_{\mathrm{gm}}^\varepsilon$ denote the full triangulated subcategory of $\mathbf{DM}$ generated by the objects εM with M in $\mathbf{DM}_{\mathrm{gm}}$. The objects of $\mathbf{DM}_{\mathrm{gm}}^\varepsilon$ may be thought of as geometric motives over the simplicial scheme $\mathfrak{X}$, by Proposition 6.23. In this section we show that $\mathbf{DM}_{\mathrm{gm}}^\varepsilon$ is a rigid tensor category, and refer to its notion of duality as $\mathfrak{X}$-*duality*.

Example 4.3. We noted after Definition 1.32 that $\mathfrak{X} \times X \simeq X$, so $\varepsilon R_{\mathrm{tr}}(X) \cong R_{\mathrm{tr}}(X)$, and $\varepsilon M \cong M$ for Chow motives of the form $M = (X, e)$. More generally, whenever $\mathrm{Hom}(Y, X) \neq \emptyset$ we have $\mathfrak{X} \times Y \simeq Y$ and hence $\varepsilon R_{\mathrm{tr}}(Y) \cong R_{\mathrm{tr}}(Y)$.

If X has a k-rational point, $\mathfrak{X} \to \mathrm{Spec}(k)$ is a global weak equivalence and hence the augmentation $R_{\mathrm{tr}}(\mathfrak{X}) \to R = R_{\mathrm{tr}}(\mathrm{Spec}\ k)$ is an isomorphism in $\mathbf{DM}$; see Remark 1.32.1. In this case $\varepsilon \cong R$ in $\mathbf{DM}$, and $\mathbf{DM}_{\mathrm{gm}}^\varepsilon \cong \mathbf{DM}_{\mathrm{gm}}$.

Since $\varepsilon(\varepsilon M) \cong \varepsilon M$ for all M in $\mathbf{DM}$, the triangulated functor $M \mapsto \varepsilon M$ is idempotent up to isomorphism. Hence the full subcategories $\varepsilon \mathbf{DM}_{\mathrm{nis}}^{\mathrm{eff}}$, $\varepsilon \mathbf{DM}$ and $\mathbf{DM}_{\mathrm{gm}}^{\varepsilon}$ on the objects εM are all tensor triangulated subcategories.

Lemma 4.4. *For any effective motives L, N the natural map $\varepsilon N \to N$ induces isomorphisms $\mathrm{Hom}(\varepsilon L, \varepsilon N) \cong \mathrm{Hom}(\varepsilon L, N)$.*

Proof. Recall from 1.36 that there is a triangle $R_{\mathrm{tr}}(\mathfrak{X}) \to R \to R_{\mathrm{tr}}(\Sigma\mathfrak{X})$, where $\Sigma\mathfrak{X}$ is the suspension of $\mathfrak{X}$. Applying ε, we have $\varepsilon R_{\mathrm{tr}}(\Sigma\mathfrak{X}) = 0$. For every $f : \varepsilon L \to R_{\mathrm{tr}}(\Sigma\mathfrak{X}) \otimes N$, the commutative diagram

$$
\begin{array}{ccc}
\varepsilon(\varepsilon L) & \xrightarrow{\;\varepsilon(f)\;} & \varepsilon R_{\mathrm{tr}}(\Sigma\mathfrak{X}) \otimes N = 0 \\
\Big\downarrow{\cong} & & \Big\downarrow \\
\varepsilon L & \xrightarrow{\;\;f\;\;} & R_{\mathrm{tr}}(\Sigma\mathfrak{X}) \otimes N
\end{array}
$$

shows that $f = 0$. Hence $\mathrm{Hom}(\varepsilon L, R_{\mathrm{tr}}(\Sigma\mathfrak{X}) \otimes N) = 0$. Applying $\mathrm{Hom}(\varepsilon L, -)$ to the triangle $\varepsilon N \to N \to R_{\mathrm{tr}}(\Sigma\mathfrak{X}) \otimes N$ yields the result. $\qquad\square$

Corollary 4.5. *We have $\mathrm{End}_{\mathbf{DM}}(R_{\mathrm{tr}}(\mathfrak{X})) = R$. For all p, q*

$$\mathrm{Hom}_{\mathbf{DM}}(R_{\mathrm{tr}}(\mathfrak{X}), \varepsilon R(q)[p]) \cong H^{p,q}(\mathfrak{X}, R).$$

Proof. By Lemma 4.4, $\mathrm{End}(R_{\mathrm{tr}}(\mathfrak{X})) = \mathrm{Hom}(R_{\mathrm{tr}}(\mathfrak{X}), \varepsilon R) = \mathrm{Hom}(R_{\mathrm{tr}}(\mathfrak{X}), R)$, and $H^{0,0}(\mathfrak{X}, R) = R$ by Lemma 1.35. Similarly, $\mathrm{Hom}(R_{\mathrm{tr}}(\mathfrak{X}), \varepsilon R(q)[p]) = \mathrm{Hom}(R_{\mathrm{tr}}(\mathfrak{X}), R(q)[p]) = H^{p,q}(\mathfrak{X}, R)$. $\qquad\square$

For any geometric motive M, the functor $\varepsilon N \mapsto \varepsilon \, \underline{R\,\mathrm{Hom}}(M, \varepsilon N)$ from $\varepsilon \mathbf{DM}_{\mathrm{nis}}^{\mathrm{eff}}$ to itself is right adjoint to $\varepsilon L \mapsto \varepsilon L \otimes M \cong \varepsilon L \otimes \varepsilon M$; see [MVW, 14.12]. It follows that whenever $\varepsilon M \cong \varepsilon M'$ we have $\varepsilon \, \underline{R\,\mathrm{Hom}}(M, \varepsilon N) \cong \varepsilon \, \underline{R\,\mathrm{Hom}}(M', \varepsilon N)$, and that the object εM of $\mathbf{DM}_{\mathrm{gm}}^{\varepsilon}$ determines a well-defined functor $\underline{R\,\mathrm{Hom}}_{\varepsilon}(\varepsilon M, -)$ on $\varepsilon \mathbf{DM}_{\mathrm{nis}}^{\mathrm{eff}}$ sending εN to $\varepsilon \, \underline{R\,\mathrm{Hom}}(M, \varepsilon N)$.

Lemma 4.6. *For effective motives L, M, N with M geometric, the natural map $\varepsilon N \to N$ induces isomorphisms*

$$\mathrm{Hom}(\varepsilon L, \varepsilon \, \underline{R\,\mathrm{Hom}}(M, \varepsilon N)) \cong \mathrm{Hom}(\varepsilon L, \varepsilon \, \underline{R\,\mathrm{Hom}}(M, N)).$$

Thus $\varepsilon \, \underline{R\,\mathrm{Hom}}(M, \varepsilon N) \cong \varepsilon \, \underline{R\,\mathrm{Hom}}(M, N)$. If N is geometric this is in $\mathbf{DM}_{\mathrm{gm}}^{\varepsilon}$.

Proof. Replacing $\varepsilon L = \varepsilon \otimes L$ by $\varepsilon L \otimes M \cong \varepsilon \otimes (L \otimes M)$, Lemma 4.4 implies that the natural map from $\mathrm{Hom}(\varepsilon L \otimes M, \varepsilon N) \cong \mathrm{Hom}(\varepsilon L, \underline{R\,\mathrm{Hom}}(M, \varepsilon N))$ to $\mathrm{Hom}(\varepsilon L \otimes M, N) \cong \mathrm{Hom}(\varepsilon L, \underline{R\,\mathrm{Hom}}(M, N))$ is an isomorphism. The second assertion follows from this in a standard way, taking L to be $\underline{R\,\mathrm{Hom}}(M, \varepsilon N)$ and

$R \operatorname{Hom}(M, N)$. The final assertion follows from the observation in [MVW, 20.3] that $\underline{R \operatorname{Hom}}(M, N)$ is a geometric motive. □

If $M = R_{\mathrm{tr}}(Y)$ is the motive of a smooth Y and $d = \dim(Y)$ we define

$$(\varepsilon M)^{\dagger} = \underline{R \operatorname{Hom}}_{\varepsilon}(\varepsilon M, \varepsilon \mathbb{L}^d) \otimes \mathbb{L}^{-d} = \varepsilon \underline{R \operatorname{Hom}}(M, \varepsilon \mathbb{L}^d) \otimes \mathbb{L}^{-d}.$$

By Lemma 4.4 and [MVW, 20.6], this is isomorphic to $\varepsilon(M^*)$. We may now mimic the development of [MVW, §20] to prove the following proposition.

Proposition 4.7. *The tensor category* $\mathbf{DM}_{\mathrm{gm}}^{\varepsilon}$ *is rigid. More precisely:*

1. *For every M in* $\mathbf{DM}_{\mathrm{gm}}$, $(\varepsilon M)^{\dagger} \cong \varepsilon(M^*)$, *and hence* $(\varepsilon M)^{\dagger}$ *is in* $\mathbf{DM}_{\mathrm{gm}}^{\varepsilon}$; *in particular,* $\varepsilon^{\dagger} \cong \varepsilon$, *because* $R^* \cong R$.
2. $(-)^{\dagger}$ *extends to a contravariant triangulated endofunctor on* $\mathbf{DM}_{\mathrm{gm}}^{\varepsilon}$. *We call* $M^{\dagger}$ *the $\mathfrak{X}$-dual of M.*
3. *There is a natural isomorphism* $M \cong (M^{\dagger})^{\dagger}$ *in* $\mathbf{DM}_{\mathrm{gm}}^{\varepsilon}$.
4. *There are natural isomorphisms (for L, M, N in* $\mathbf{DM}_{\mathrm{gm}}^{\varepsilon}$)

$$\operatorname{Hom}(L \otimes M, N) \cong \operatorname{Hom}(L, M^{\dagger} \otimes N).$$

5. *The internal Hom functor on* $\mathbf{DM}_{\mathrm{gm}}^{\varepsilon}$ *is* $\underline{\operatorname{Hom}}_{\varepsilon}(M, N) = M^{\dagger} \otimes N$.

Not every object of $\mathbf{DM}_{\mathrm{gm}}^{\varepsilon}$ will be of the form εL for some geometric motive L, such as the motive A as the following example shows.

Example 4.8. Suppose that $A \to \varepsilon M \xrightarrow{\mu} \varepsilon N \to$ is a triangle in $\mathbf{DM}_{\mathrm{gm}}^{\varepsilon}$ with M, N geometric motives. By Proposition 4.7(2), there is an object $A^{\dagger}$ in $\mathbf{DM}_{\mathrm{gm}}^{\varepsilon}$ fitting into a triangle:

$$\varepsilon N^* \xrightarrow{\mu^{\dagger}} \varepsilon M^* \to A^{\dagger} \to \varepsilon N^*[1].$$

If M and N are invertible in $\mathbf{DM}_{\mathrm{gm}}$, it is easy to see that μ is self-dual in the sense that it equals the composite

$$\varepsilon M \cong \varepsilon M \otimes \varepsilon N^* \otimes \varepsilon N \xrightarrow{1_M \otimes \mu^{\dagger} \otimes 1_N} \varepsilon M \otimes \varepsilon M^* \otimes \varepsilon N \cong \varepsilon N.$$

It follows that A satisfies a twisted duality: $A \cong \varepsilon M \otimes A^{\dagger} \otimes \varepsilon N[-1]$. A special case, which will arise in chapter 5, is a triangle $A \to R_{\mathrm{tr}}(\mathfrak{X}) \xrightarrow{\mu} R_{\mathrm{tr}}(\mathfrak{X})$ $(q)[p] \to$ associated to an element μ of $H^{p,q}(\mathfrak{X})$. In this case we have $A \cong A^{\dagger}(q)[p-1]$.

When μ is not in the image of $\operatorname{Hom}_{\mathbf{DM}}(M, N)$, A cannot have the form εL for any geometric motive L. For example, suppose that $M = N = R = \operatorname{Spec}(k)$, so that $\varepsilon R = \varepsilon$; if E/k is a Galois field extension with group G and $X = \operatorname{Spec} E$ then

$\mathrm{Hom}_{\mathbf{DM}}(X, X) = R[G]$ by [MVW, 1.11], so we may take μ to be any element of $R[G]$ not in R.

Remark 4.9. (Tate objects) Following [Voe10b], we let DT^{ε} denote the smallest thick subcategory of $\mathbf{DM}_{\mathrm{nis}}^{\mathrm{eff}}$ generated by the objects $\varepsilon R(q)$, $q \geq 0$. This category has "slice filtrations"; $DT_{\geq n}^{\varepsilon}$ is the thick subcategory generated by the $\varepsilon R(q)$ with $q \geq n$, and $s_{\geq n}(M) = R \underline{\mathrm{Hom}}(\mathbb{L}^n, M) \otimes \mathbb{L}^n$ is a triangulated reflection functor from DT^{ε} to $DT_{\geq n}^{\varepsilon}$. This will be used in section 5.2. We refer the reader to section 6.4 for a fuller discussion of the slice filtration.[1]

Similarly, the thick subcategory $DT_{\leq n}^{\varepsilon}$ generated by the $\varepsilon R(q)$ with $q \leq n$ has a reflection functor $s_{\leq n}$, and the slice functor $s_n = s_{\geq n} s_{\leq n}$ takes values in $DT_n^{\varepsilon} = DT_{\geq n}^{\varepsilon} \cap DT_{\leq n}^{\varepsilon}$. Each DT_n^{ε} is equivalent to DT_0^{ε} by the functor $\otimes \mathbb{L}^n$, and DT_0^{ε} may be characterized as the abelian category of locally constant presheaves with transfers on $\mathbf{Sm}/X$.

Finally, the direct sum $\oplus s_n : DT^{\varepsilon} \to \oplus DT_n^{\varepsilon}$ is a conservative functor which is multiplicative in the sense that $s_n(M \otimes N) = \oplus_{p+q=n} s_p(M) \otimes s_q(N)$; this is proven in a more general context in Lemmas 6.17 and 6.18.

4.3   ROST MOTIVES

We now return to the category $\mathbf{DM}$. Fix a subring R of $\mathbb{Q}$ containing $1/(\ell-1)!$, such as the local ring $\mathbb{Z}_{(\ell)}$. Let X be a Rost variety, and form $\mathfrak{X}$ as in Definition 1.32 so that we have a canonical map $R_{\mathrm{tr}}(X) \to R_{\mathrm{tr}}(\mathfrak{X}) \xrightarrow{\epsilon} R$. Let $M = (X, e)$ be a Chow motive with coefficients in R (see section 4.1); M is a summand of $R_{\mathrm{tr}}(X)$, and we write y or y_M for the structure map $y : M \to R_{\mathrm{tr}}(X) \to R_{\mathrm{tr}}(\mathfrak{X})$. Note that the composition $\epsilon y : M \to R$ is the universal map, and that $\varepsilon M \cong M$ by 4.3.

By Proposition 4.7, $M^{\dagger} \cong M^{*}$ and the $\mathfrak{X}$-dual of y is $y^{\dagger} = R_{\mathrm{tr}}(\mathfrak{X}) \otimes (\epsilon y)^{*}$ (see 4.7). Recall from Example 4.2 that $M^t \cong M^* \otimes \mathbb{L}^d$, where $d = \dim X$. We define the *twisted dual* y^D of y to be the composite

$$R_{\mathrm{tr}}(\mathfrak{X}) \otimes \mathbb{L}^d \xrightarrow{y^{\dagger} \otimes 1_{\mathbb{L}^d}} M^{\dagger} \otimes \mathbb{L}^d \cong M^* \otimes \mathbb{L}^d \cong M^t. \qquad (4.10)$$

Definition 4.11. A *Rost motive* for $\underline{a}$ (with coefficients R) is a motive M, arising as a direct summand of the motive of a Rost variety X for $\underline{a}$, which satisfies:

(i) $M = (X, e)$ is a symmetric Chow motive;
(ii) the projection $X \to R$ factors $X \to (X, e) \to R$, i.e., is zero on $(X, 1-e)$;

---

1. The notation s_n, used here for the slice filtration, is not related to the notation $s_n(X)$ for the characteristic numbers of X in Definition 1.17 and chapter 8.

(iii) there is a motive D related to the structure map $y: M \to R_{\mathrm{tr}}(\mathfrak{X})$ and its twisted dual y^D by two distinguished triangles

$$D \otimes \mathbb{L}^b \to M \xrightarrow{y} R_{\mathrm{tr}}(\mathfrak{X}) \to , \qquad (4.11a)$$

$$R_{\mathrm{tr}}(\mathfrak{X}) \otimes \mathbb{L}^d \xrightarrow{y^D} M \to D \to . \qquad (4.11b)$$

(Recall that $b = d/(\ell - 1) = 1 + \ell + \cdots + \ell^{n-2}$.)

Remark 4.11.1. By Example 4.3, $\varepsilon M \cong M$ and $\varepsilon D \cong D$. By (4.11a), D must be isomorphic to $R_{\mathrm{tr}}(\mathfrak{X}) \otimes \widetilde{M} \otimes \mathbb{L}^{-b}$, where $\widetilde{M}$ is the geometric motive defined as the fiber of the structure map $M \to R$. Note that $D^\dagger = R_{\mathrm{tr}}(\mathfrak{X}) \otimes \widetilde{M}^* \otimes \mathbb{L}^b$.

Triangle (4.11b) is equivalent to the $\mathfrak{X}$-duality assertion that $D^\dagger \otimes \mathbb{L}^{d-b} \cong D$. To see this, observe that by Example 4.2 and 4.7(1), the $\mathfrak{X}$-dual of (4.11a) is

$$R_{\mathrm{tr}}(\mathfrak{X}) \otimes \mathbb{L}^d \xrightarrow{y^D} M \to D^\dagger \otimes \mathbb{L}^{d-b} \to .$$

Remark 4.11.2. Markus Rost has proposed a construction of M in [Ros06]. He shows that the element μ of Corollary 3.16 determines an equivalence class of symmetric Chow motives $M = (X, e)$ for any Rost variety X. We do not know if Rost's M satisfies the triangle condition (4.11b).

In chapter 5, we will prove the following theorem, with coefficient ring R either $\mathbb{Z}[1/(\ell-1)!]$ or $\mathbb{Z}_{(\ell)}$.

Theorem 4.12. *For every Rost variety X there is a Rost motive (X, e).*

Examples 4.13. (a) If $n = 2$, the Rost variety itself is a Rost motive: $M = X$. Since $b = 1$ we have $D = R_{\mathrm{tr}}(\mathfrak{X}) \otimes \widetilde{M}$. The condition (4.11b) is that $D^\dagger \otimes \mathbb{L} \cong D$, and is implicit in [MeS82].

(b) If $\ell = 2$, and X is a Pfister quadric, Rost showed in [Ros90] that there is a symmetric Chow motive (X, e) such that over any field F which has a point $x \in X(F)$ the motive (X, e) is given by the correspondence $X \times x + x \times X$. This is property 4.11(i); the proof that (X, e) also satisfies property 4.11(ii) is given in [Ros98c]. Property 4.11(iii) is proven in [Voe03a].

Lemma 4.14. *When $n \geq 2$, the structure map $H_{-1,-1}(M) \to H_{-1,-1}(k) \cong k^\times$ is injective for every Rost motive M.*

Proof. By Definition 1.24, the map $\overline{H}_{-1,-1}(X) \xrightarrow{N} k^\times$ is an injection for any Rost variety, and $X \to \mathfrak{X}$ induces $H_{-1,-1}(\mathfrak{X}) \cong \overline{H}_{-1,-1}(X)$ by Lemma 1.34. Thus it suffices to show that the structure map $y: H_{-1,-1}(M) \to H_{-1,-1}(\mathfrak{X})$ is an injection. By triangle (4.11a), we have an exact sequence

$$H_{-1,-1}(D \otimes \mathbb{L}^b) \to H_{-1,-1}(M) \to H_{-1,-1}(\mathfrak{X}),$$

so it suffices to observe that $H_{-1,-1}(D \otimes \mathbb{L}^b) = \mathrm{Hom}(R, D \otimes \mathbb{L}^b(1)[1])$ vanishes. This follows from triangle (4.11b) and the vanishing (granted by Lemma 1.33) of both $\mathrm{Hom}(R, M(b+1)[2b+1])$ and $\mathrm{Hom}(R, R_{\mathrm{tr}}(\mathfrak{X}) \otimes \mathbb{L}^{b+d}[1])$. $\qquad \square$

Lemma 4.15. *When* $n \geq 2$, $H^{2d+1,d+1}(D, R) = 0$.

Proof. Set $(p, q) = (2d+1, d+1)$, so that $H^{p-1,q}(\mathfrak{X} \otimes \mathbb{L}^d, R) \cong H^{0,1}(\mathfrak{X}, R)$. From (4.11b) we get an exact sequence

$$H^{0,1}(\mathfrak{X}, R) \to H^{p,q}(D, R) \to H^{p,q}(M, R) \xrightarrow{y^D} H^{p,q}(\mathfrak{X} \otimes \mathbb{L}^d, R).$$

The first group is zero by Lemma 1.35. Hence $H^{p,q}(D, R)$ is the kernel of y^D. Since $H^{p,q}(M, R) = \mathrm{Hom}(M^* \otimes \mathbb{L}^d, \mathbb{L}^d(1)[1]) \cong \mathrm{Hom}(M^*, R(1)[1])$, every element of $H^{p,q}(M, R)$ is $u^* \otimes \mathbb{L}^d$ for the dual u^* of a map $u : R(-1)[-1] \to M$. We may regard u as an element of $H_{-1,-1}(M) = \mathrm{Hom}(R, M(1)[1])$. Since the given structural map $H_{-1,-1}(M) \xrightarrow{y} H_{-1,-1}(\mathfrak{X}) \subseteq H_{-1,-1}(k)$ is an injection by 4.14, $yu \neq 0$ whenever $u^* \neq 0$. The $\mathfrak{X}$-dual of yu is the composition $u^* y^\dagger : R_{\mathrm{tr}}(\mathfrak{X}) \to M^* \to R_{\mathrm{tr}}(\mathfrak{X})(1)[1] \to R(1)[1]$. Tensoring with $\mathbb{L}^d$ yields a nonzero element of $H^{p,q}(\mathfrak{X} \otimes \mathbb{L}^d, R)$; by the definition of y^D it is represented by:

$$\mathfrak{X} \otimes \mathbb{L}^d \xrightarrow{y^D} M^t \xrightarrow{u^* \otimes \mathbb{L}^d} R(1)[1] \otimes \mathbb{L}^d = R(d+1)[2d+1].$$

Thus $(u^* \otimes \mathbb{L}^d) \circ y^D$ is nonzero whenever $u^* \neq 0$, as desired. $\qquad \square$

4.4 ROST MOTIVES IMPLY HILBERT 90

In this section, we will assume that a Rost variety X and a Rost motive $M = (X, e)$ exists for $\underline{a}$, and that BL(n − 1) holds. Given this, we will show that the map $H^{n+1}_{\text{ét}}(k, \mathbb{Z}_{(\ell)}(n)) \to H^{n+1}_{\text{ét}}(k(X), \mathbb{Z}_{(\ell)}(n))$ is an injection (see Theorem 4.20). For this, we need Propositions 3.15 and 3.17 about the Margolis homology of cohomology operations Q_i—which in turn depend upon Theorem 1.44, which will be established in part III.

We will assume in this section that $n \geq 2$. Recall that $b = (\ell^{n-1} - 1)/(\ell - 1)$.

Lemma 4.16. *The map* $H^{p-2b,q-b}(D, A) \cong H^{p,q}(D \otimes \mathbb{L}^b, A) \xrightarrow{s} H^{p+1,q}(\mathfrak{X}, A)$ *of triangle (4.11a) is an isomorphism if* $p > q + d$, *for any* $\mathbb{Z}_{(\ell)}$-*module* A.

Proof. The map s is from the cohomology sequence arising from triangle (4.11a):

$$H^{p,q}(M, A) \to H^{p,q}(D \otimes \mathbb{L}^b, A) \xrightarrow{s} H^{p+1,q}(\mathfrak{X}, A) \xrightarrow{y} H^{p+1,q}(M, A).$$

The second term is isomorphic to $H^{p-2b,q-b}(D, A)$ by Cancellation (see [Voe10a]). Whenever $p > q + d$, $H^{p,q}(X, A)$ is zero by the Vanishing Theorem [MVW, 3.6]. Since $H^{p,q}(M, A)$ is a summand of $H^{p,q}(X, A)$, it also vanishes in this range. □

Corollary 4.17. $H^{2b\ell+2,b\ell+1}(\mathfrak{X}, \mathbb{Z}_{(\ell)}) = 0$.

Proof. Since $b\ell = b + d$, $H^{2b\ell+2,b\ell+1}(\mathfrak{X}, \mathbb{Z}_{(\ell)}) \cong H^{2d+1,d+1}(D, \mathbb{Z}_{(\ell)})$ by Lemma 4.16. This group vanishes by Lemma 4.15. □

Proposition 4.18. *Assume that $BL(n-1)$ holds. Then $H^{n+1,n}(\mathfrak{X}, \mathbb{Z}_{(\ell)}) = 0$.*

Proof. By Proposition 3.17, which requires $BL(n-1)$, the cohomology operation $Q_{n-1} \cdots Q_1$ is an injection from $H^{n+1,n}(\mathfrak{X}, \mathbb{Z}_{(\ell)})$ into $H^{2b\ell+2,b\ell+1}(\mathfrak{X}, \mathbb{Z}_{(\ell)})$. The latter group is zero by Corollary 4.17. □

If F is any homotopy invariant presheaf with transfers, the contraction F_{-1} was defined in 2.1 so that $F(Y \times (\mathbb{A}^1 - \{0\})) = F(Y) \oplus F_{-1}(Y)$ for all Y.

Recall from Definition 2.39 that $T(n)$ denotes $\tau^{\leq n+1} \mathbf{R}\pi_*(\mathbb{Z}_{(\ell)}(n)_{\text{ét}})$ and $K(n)$ denotes the cone of the canonical map $\mathbb{Z}_{(\ell)}(n) \to T(n)$.

Lemma 4.19. *If $BL(n-1)$ holds, then for every smooth Y and every integer q:*

$$H^q(Y \times (\mathbb{A}^1 - \{0\}), K(n)) \cong H^q(Y, K(n)), \quad \text{and} \quad H^q(Y \otimes \mathbb{L}, K(n)) = 0.$$

Proof. (See [Voe03a, 6.12].) We have to show that $H^*(Y, K(n))_{-1} = 0$; see (2.2.1). By the definition of $K(n)$, this is equivalent to showing that the map $H^*(Y, \mathbb{Z}_{(\ell)}(n))_{-1} \to H^*(Y, T(n))_{-1}$ is an isomorphism. By Lemma 2.2, $H^*(-, \mathcal{F})_{-1} \cong H^*(-, \mathcal{F}_{-1})$ for $\mathcal{F}$ either $\mathbb{Z}_{(\ell)}(n)$ or $T(n)$, and $\mathbb{Z}_{(\ell)}(n)_{-1} \cong \mathbb{Z}_{(\ell)}(n-1)[-1]$ by Example 2.4(4). By functoriality,

$$\pi^* \mathbb{Z}_{(\ell)}(n)_{-1} \cong \pi^* \mathbb{Z}_{(\ell)}(n-1)[-1].$$

Applying $\tau^{\leq n+1} \mathbf{R}\pi_*$ yields $T(n)_{-1} \cong T(n-1)[-1]$. We are reduced to seeing that the map $\mathbb{Z}_{(\ell)}(n-1) \to T(n-1)$ is a quasi-isomorphism. By [MVW, 13.7], it suffices to show it is a quasi-isomorphism at every field E over k, i.e., that

$$H^p(E, \mathbb{Z}_{(\ell)}(n-1)) \longrightarrow H^p(E, T(n-1)) = \begin{cases} H^p_{\text{ét}}(E, \mathbb{Z}_{(\ell)}(n-1)), & p \leq n; \\ 0, & p > n \end{cases}$$

is an isomorphism for all p. This is trivial for $p > n$. The assumption that $BL(n-1)$ implies the case $p = n$ (that $H90(n-1)$ holds) by Lemma 1.30, and the case $p < n$, by Lemma 1.29(c). □

Remark 4.19.1. It follows from Lemma 4.19 that $H^*(Y, K(n))$ is a birational invariant. Voevodsky proves this in [Voe03a, Lemma 6.13].

Theorem 4.20. *Assume that X is a Rost variety over k and that a Rost motive (X, e) exists. If $BL(n-1)$ holds, then $H^{n+1}_{\text{ét}}(k, \mathbb{Z}_{(\ell)}(n)) \to H^{n+1}_{\text{ét}}(k(X), \mathbb{Z}_{(\ell)}(n))$ is an injection.*

Proof. Set $E = k(X)$, and recall that $K(n)$ is the cone of $\mathbb{Z}_{(\ell)}(n) \to T(n)$. Since $X \to \text{Spec}(k)$ factors through $\mathfrak{X}$, $\text{Spec}(E) \to \text{Spec}(k)$ factors through a map $\text{Spec}(E) \to \mathfrak{X}$. This map induces a commutative diagram with exact rows, in which the lower left group $H^{n+1}(E, \mathbb{Z}_{(\ell)}(n))$ vanishes by [MVW, 3.6]:

$$
\begin{array}{ccccc}
0 \overset{4.18}{=} H^{n+1}_{\text{nis}}(\mathfrak{X}, \mathbb{Z}_{(\ell)}(n)) & \longrightarrow & H^{n+1}_{\text{nis}}(\mathfrak{X}, T(n)) & \longrightarrow & H^{n+1}_{\text{nis}}(\mathfrak{X}, K(n)) \\
\downarrow & & \downarrow & & \downarrow \\
0 = H^{n+1}_{\text{nis}}(E, \mathbb{Z}_{(\ell)}(n)) & \longrightarrow & H^{n+1}_{\text{nis}}(E, T(n)) & \longrightarrow & H^{n+1}_{\text{nis}}(E, K(n)).
\end{array}
$$

Since $H^{n+1}(E, T(n)) \cong H^{n+1}_{\text{ét}}(E, \mathbb{Z}_{(\ell)}(n))$ and (using Lemma 1.37)

$$
H^{n+1}(\mathfrak{X}, T(n)) \cong H^{n+1}_{\text{ét}}(\mathfrak{X}, \mathbb{Z}_{(\ell)}(n)) \cong H^{n+1}_{\text{ét}}(k, \mathbb{Z}_{(\ell)}(n)),
$$

we may identify the map $H^{n+1}_{\text{ét}}(k, \mathbb{Z}_{(\ell)}(n)) \to H^{n+1}_{\text{ét}}(k(X), \mathbb{Z}_{(\ell)}(n))$ with the middle vertical arrow. To prove that it is an injection, we will show that the right vertical is an injection. The right vertical factors as

$$
H^{n+1}(\mathfrak{X}, K(n)) \overset{y^*}{\longrightarrow} H^{n+1}(M, K(n)) \subset H^{n+1}(X, K(n)) \overset{\cong}{\longrightarrow} H^{n+1}(E, K(n)).
$$

In this factorization, $H^{n+1}(M, K(n))$ is a summand of $H^{n+1}(X, K(n))$, and we argue that $H^q(X, K(n)) \to H^q(E, K(n))$ is an isomorphism for all q. By [MVW, 13.8], the presheaves $H^q = H^q(-, K(n))$ are homotopy invariant, so by [MVW, 22.1, 22.15] the associated Zariski sheaves $\mathcal{H}^q$ are homotopy invariant presheaves with transfers. By Lemma 4.19, we have $\mathcal{H}^q_{-1} = 0$. Then [MVW, 24.11] says that each $\mathcal{H}^q$ is a constant Zariski sheaf on X. Therefore $H^q(X, K(n)) = H^0(X, \mathcal{H}^q)$ equals $H^q(E, K(n))$ for all integers q.

It suffices to show that the map $y^* : H^{n+1}(\mathfrak{X}, K(n)) \to H^{n+1}(M, K(n))$, induced by $y \colon M \to \mathfrak{X}$, is an injection. By (4.11a) it suffices to show that $H^n(D \otimes \mathbb{L}^b, K(n))$ vanishes. By (4.11b) we have an exact sequence

$$
\begin{aligned}
H^{n-1}(\mathfrak{X} &\otimes \mathbb{L}^{b+d}, K(n)) \\
&\to H^n(D \otimes \mathbb{L}^b, K(n)) \to H^n(M \otimes \mathbb{L}^b, K(n)).
\end{aligned} \tag{4.20.1}
$$

We are reduced to showing the outside terms vanish in (4.20.1). By Lemma 4.19, $H^n(X \otimes \mathbb{L}^b, K(n)) = 0$; as M is a summand of X by 4.11(i), $H^n(M \otimes \mathbb{L}^b, K(n)) = 0$ in (4.20.1). Lemma 4.19 also implies that (for any $m > 1$)

the spectral sequence $E^1_{pq} = H^q(X^p \otimes \mathbb{L}^m, K(n)) \Rightarrow H^{p+1}(\mathfrak{X} \otimes \mathbb{L}^m, K(n))$ colla-pses, yielding $H^{n-1}(\mathfrak{X} \otimes \mathbb{L}^m, K(n)) = 0$. Thus the left term in (4.20.1) also vanishes. $\qquad\qquad\qquad\qquad\qquad\qquad\qquad\qquad\qquad\qquad\qquad\qquad\qquad\qquad\square$

4.5   HISTORICAL NOTES

The proof of the Norm Residue Isomorphism Theorem for $\ell = 2$ (the Milnor Conjecture) invoked a theorem of Rost that the Pfister quadric $Q_{\underline{a}}$ had a direct summand $M = (Q_{\underline{a}}, e)$ with maps $\psi^* : \mathbb{L}^d \to M$ related to the structure map $\psi_* : M \to \mathbb{Z}$ (see [Voe03a, 4.3]). Voevodsky showed in [Voe03a, 4.4] that this data fits into a triangle

$$M(\mathfrak{X}) \otimes \mathbb{L}^d \to M \to M(\mathfrak{X}) \to .$$

This is the triangle (4.11a) with $D = M(\mathfrak{X})$; triangle (4.11b) holds by $\mathfrak{X}$-duality 4.7. Note that M is a twisted form of $\mathbb{Z} \oplus \mathbb{L}^d$, because the triangle splits with $\mathfrak{X}_E \cong \mathbb{Z}_E$ over any extension field E for which $Q_a(E) \neq \emptyset$; see [Voe03a, 4.3].

This material was further developed in Voevodsky's 2003 preprint [Voe03b]. Rost observed in [Ros06] that the existence of these triangles could be used to give another proof of [Voe03b, Lemma 6.13] (see [Voe11, (5.3–5)] for the published version). This led to the axiomatic approach to the notion of a Rost motive, developed in [Wei08]. Most of the material in this chapter is taken from [Wei08].

Chapter Five

Existence of Rost Motives

IN THIS CHAPTER, we fix a Rost variety X for a sequence $\underline{a}$ of n units in a field k with $1/\ell \in k$. Under the inductive assumption that $\mathrm{BL}(n-1)$ holds, we will construct a Rost motive $M = (X, e)$ with coefficients $\mathbb{Z}_{(\ell)}$; see Theorem 5.18. Later on, in chapter 11, we will show that a Rost variety always exists when k is ℓ-special of characteristic 0. As we saw in Theorem 4.20, the existence of X and M suffices to prove Theorem 1.11, which in turn proves Theorem 1.12 and hence Theorems A and B, the main theorems of this book.

Recall that $b = d/(\ell - 1) = 1 + \ell + \cdots + \ell^{n-2}$ and that $\mathfrak{X}$ denotes the simplicial Čech variety (or 0-coskeleton) associated to X in Definition 1.32. According to Definition 4.11, a motive M is a *Rost motive* associated to X if there is a map $\lambda : R_{\mathrm{tr}}(X) \to M$ such that the following three axioms are satisfied:

(i) λ expresses M as a symmetric Chow motive of the form (X, e).
(ii) The projection $R_{\mathrm{tr}}(X) \to R$ factors through λ.
(iii) There is a motive D, related to the structure map $M \xrightarrow{y} R_{\mathrm{tr}}(\mathfrak{X})$ and the twisted dual y^D of (4.10) by two distinguished triangles:

$$ D \otimes \mathbb{L}^b \to M \xrightarrow{y} R_{\mathrm{tr}}(\mathfrak{X}) \to ; \qquad R_{\mathrm{tr}}(\mathfrak{X}) \otimes \mathbb{L}^d \xrightarrow{y^D} M \to D \to . $$

Given any cohomology class z of $H^{2b+1,b}(\mathfrak{X}, \mathbb{Z}_{(\ell)})$, it is not hard to construct a motive M satisfying axioms (ii) and (iii); see Propositions 5.7 and 5.9. In order to establish axiom (i), making M a Rost motive, we need to assume that z is *suitable* in the sense of Definition 5.13. As pointed out in Example 5.13.1, the element $\mu = Q(\delta)$ of Corollary 3.16 is suitable in this sense. The assumption that $\mathrm{BL}(n-1)$ holds is used to show that δ exists and that μ is nonzero. This proves that Rost motives exist whenever Rost varieties exist.

The proof that axiom (i) holds requires one result which we have postponed to chapter 6: the proof that the function ϕ^V defined in 5.10 agrees with the cohomology operation βP^b (Proposition 5.16 uses Corollary 6.34). This uses a recognition criterion for the cohomology operation βP^b which we have placed in part III (in Theorem 15.38) because its proof requires advanced motivic homotopy techniques. Some other facts about the cohomology operations Q_i which are used in section 5.4 are given in chapter 13 of part III.

5.1   A CANDIDATE FOR THE ROST MOTIVE

Let R be any ring, X any smooth scheme of dimension $d = \ell^{n-1} - 1$, and let $\mathfrak{X}$ denote the simplicial scheme associated to X in Definition 1.32.

We may interpret any $z \in H^{2b+1,b}(\mathfrak{X}, R)$ as a morphism $R_{\mathrm{tr}}(\mathfrak{X}) \to \mathbb{L}^b[1]$ in $\mathbf{DM}^{\mathrm{eff}}_{\mathrm{nis}}(k, R)$. Since $\mathrm{Hom}(R_{\mathrm{tr}}(\mathfrak{X}), N) \cong \mathrm{Hom}(R_{\mathrm{tr}}(\mathfrak{X}), \varepsilon N)$ by Lemma 4.4, z lifts to a morphism $R_{\mathrm{tr}}(\mathfrak{X}) \to R_{\mathrm{tr}}(\mathfrak{X}) \otimes \mathbb{L}^b[1]$. Up to a non-canonical isomorphism, the morphism z determines a motive A and maps x, y in $\mathbf{DM}^{\varepsilon}_{\mathrm{gm}}$, fitting into a triangle

$$R_{\mathrm{tr}}(\mathfrak{X}) \otimes \mathbb{L}^b \xrightarrow{x} A \xrightarrow{y} R_{\mathrm{tr}}(\mathfrak{X}) \xrightarrow{z} R_{\mathrm{tr}}(\mathfrak{X}) \otimes \mathbb{L}^b[1]. \tag{5.1}$$

By Example 4.8, we have $A \cong A^{\dagger} \otimes \mathbb{L}^b$. Setting $y^D = y^{\dagger} \otimes \mathbb{L}^b$, the tensor product of the $\mathfrak{X}$-dual of (5.1) with $\mathbb{L}^b$ is the triangle:

$$R_{\mathrm{tr}}(\mathfrak{X}) \otimes \mathbb{L}^b \xrightarrow{y^D} A^{\dagger} \otimes \mathbb{L}^b \xrightarrow{x^{\dagger} \otimes \mathbb{L}^b} R_{\mathrm{tr}}(\mathfrak{X}) \xrightarrow{z^{\dagger}[1]} R_{\mathrm{tr}}(\mathfrak{X}) \otimes \mathbb{L}^b. \tag{5.2}$$

Lemma 5.3. *There is a map* $\lambda_1 : R_{\mathrm{tr}}(X) \to A$ *such that the structure map* $\iota : R_{\mathrm{tr}}(X) \to R_{\mathrm{tr}}(\mathfrak{X})$ *factors in* **DM** *as:*

$$R_{\mathrm{tr}}(X) \xrightarrow{\lambda_1} A \xrightarrow{y} R_{\mathrm{tr}}(\mathfrak{X}).$$

Proof. The group $\mathrm{Hom}(R_{\mathrm{tr}}(X), \mathbb{L}^b[1]) \cong H^{2b+1,b}(X, R)$ is zero by the Vanishing Theorem [MVW, 19.3]. Applying $\mathrm{Hom}_{\mathbf{DM}}(R_{\mathrm{tr}}(X), -)$ to (5.1) yields the exact sequence

$$\mathrm{Hom}(R_{\mathrm{tr}}(X), A) \xrightarrow{y} \mathrm{Hom}(R_{\mathrm{tr}}(X), R_{\mathrm{tr}}(\mathfrak{X})) \xrightarrow{z} \mathrm{Hom}(R_{\mathrm{tr}}(X), \mathbb{L}^b[1]) = 0.$$

Hence $\iota \in \mathrm{Hom}(R_{\mathrm{tr}}(X), R_{\mathrm{tr}}(\mathfrak{X}))$ is $y \circ \lambda_1$ for some $\lambda_1 \in \mathrm{Hom}(R_{\mathrm{tr}}(X), A)$. $\quad\square$

We now present our candidates for Rost motives.

Example 5.4. When $\ell = 2$ and X is a Rost variety for $\underline{a}$, $M = A$ is a Rost motive for $\underline{a}$ (with $R = \mathbb{Z}_{(2)}$), where λ is the map $\lambda_1 : R_{\mathrm{tr}}(X) \to A$ of Lemma 5.3. To see this, note that axiom (ii) is satisfied by Lemma 5.3, and axiom (iii) is satisfied by the triangles (5.1) and (5.2), since $b = d$. The hard part to verify is axiom (i), that λ_1 expresses A as a symmetric Chow motive (X, e). This will follow from Theorem 5.18, where we prove that the composition $A \xrightarrow{\lambda_1^D} R_{\mathrm{tr}}(X) \xrightarrow{\lambda_1} A$ is an isomorphism. Here $\lambda_1^D = \lambda_1^{\dagger} \otimes \mathbb{L}^d$ is the twisted dual of λ_1 in the sense of (4.10).

For $\ell > 2$, when $b \neq d$, we will show that the $(\ell-1)^{st}$ symmetric power of A, $S^{\ell-1}(A)$, is a Rost motive for $\underline{a}$ with coefficients $R = \mathbb{Z}_{(\ell)}$.

Recall that the symmetric power $S^i(N)$ of a motive N is defined whenever $i!$ is invertible in the coefficient ring R. Indeed, the group ring $R[\Sigma_i]$ of the symmetric group Σ_i contains the symmetrizing idempotent $e_i = \sum_{\sigma \in \Sigma_i} \sigma / i!$ and we define $S^i(N)$ to be the summand $e_i(N^{\otimes i})$ of $N^{\otimes i}$, where $R[\Sigma_i]$ acts on $N^{\otimes i}$ in the evident way.

Definition 5.5. Given a nonzero $z \in H^{2b+1,b}(\mathfrak{X}, \mathbb{Z}_{(\ell)})$, we define $M = S^{\ell-1}(A)$ and $D = S^{\ell-2}(A)$, where $A = A(z)$ is given by (5.1).

The goal of the rest of this chapter is to prove that M is a Rost motive for suitable z. Axioms (ii) and (iii) of 4.11 are verified for $\ell > 2$ in the next section (Propositions 5.7 and 5.9), and axiom (i) is verified in section 5.4.

5.2 AXIOMS (ii) AND (iii)

Fix a ring R containing $1/(\ell-1)!$ and a smooth scheme X. In this section, we show that the motive $M = S^{\ell-1}(A)$ of Definition 5.5 satisfies axioms (ii) and (iii) of 4.11. It is useful to work in slightly greater generality in this section, so we fix an element $z \in H^{2p+1,q}(\mathfrak{X}, R)$, where $p, q \geq 0$ are arbitrary integers. Axioms (ii) and (iii) will follow from Propositions 5.9 and 5.7 in the special case $p = q = b$.

For axiom (iii), recall that $\mathfrak{X}$ denotes the simplicial scheme associated to X, as defined in 1.32. Setting $T = R(q)[2p]$, we may interpret an element $z \in H^{2p+1,q}(\mathfrak{X}, R)$ as a morphism $R_{\mathrm{tr}}(\mathfrak{X}) \to T[1]$ in $\mathbf{DM}_{\mathrm{nis}}^{\mathrm{eff}}$. Since $R_{\mathrm{tr}}(\mathfrak{X}) \simeq R_{\mathrm{tr}}(\mathfrak{X}) \otimes R_{\mathrm{tr}}(\mathfrak{X})$, it follows from Lemma 4.4 that z lifts to a morphism $R_{\mathrm{tr}}(\mathfrak{X}) \to R_{\mathrm{tr}}(\mathfrak{X}) \otimes T[1]$. As in (5.1), z determines a motive A (well-defined up to non-canonical isomorphism) and maps x, y which fit into a triangle

$$R_{\mathrm{tr}}(\mathfrak{X}) \otimes T \xrightarrow{x} A \xrightarrow{y} R_{\mathrm{tr}}(\mathfrak{X}) \xrightarrow{z} R_{\mathrm{tr}}(\mathfrak{X}) \otimes T[1]. \qquad (5.6)$$

Remark 5.6.1. From equation (5.6), we can read off the slice filtration on A, as described in Remark 4.9. There are only two layers: the top slice, $s_0(A) = R_{\mathrm{tr}}(\mathfrak{X})$, and the bottom slice, $s_q(A) = R_{\mathrm{tr}}(\mathfrak{X}) \otimes T$.

As in Definition 5.5, our hypothesis that $(\ell-1)!$ is invertible in R implies that $S^i A = e_i(A^{\otimes i})$ is defined for $i < \ell$. The inclusion $\Sigma_{i-1} \subset \Sigma_i$ allows us to talk about the idempotent $e_{i-1} \otimes 1$ of $R[\Sigma_i]$, and $(e_{i-1} \otimes 1) A^{\otimes i} = S^{i-1}(A) \otimes A$. Since $e_i(e_{i-1} \otimes 1) = e_i$, there is a corestriction map $S^{i-1}(A) \otimes A \to S^i(A)$. There is also a transfer map $\mathrm{tr}\colon S^i(A) \to S^{i-1}(A) \otimes A$, induced by the endomorphism

$$a_1 \otimes \cdots \otimes a_i \longmapsto \sum_j (\cdots \otimes \hat{a}_j \otimes \cdots) \otimes a_j$$

of $A^{\otimes i}$. Now $S^i(A) \cong R_{\mathrm{tr}}(\mathfrak{X}) \otimes S^i(A)$. Composing tr with $1 \otimes y$ and $1 \otimes x$ with corestriction yields maps

$$u = (1 \otimes y) \circ \mathrm{tr}\colon S^i(A) \to S^{i-1}(A) \text{ and } v = \mathrm{cores} \circ (1 \otimes x)\colon S^{i-1}(A) \otimes T \to S^i(A).$$

Since the ith symmetric power of $T = R(q)[2p]$ is T^i (see [MVW, 15.7]), we have $S^i(R_{\mathrm{tr}}(\mathfrak{X}) \otimes T) \cong R_{\mathrm{tr}}(\mathfrak{X}) \otimes S^i(T) \cong R_{\mathrm{tr}}(\mathfrak{X}) \otimes T^i$. Thus $S^i x$ maps $R_{\mathrm{tr}}(\mathfrak{X}) \otimes T^i$ to $S^i A$.

Proposition 5.7. *If $0 < i < \ell$ and $1/(\ell-1)! \in R$, then there exist unique morphisms r and s so that $(S^i x, u, r)$ and $(v, S^i y, s)$ are distinguished triangles:*

(a) $R_{\mathrm{tr}}(\mathfrak{X}) \otimes T^i \xrightarrow{S^i x} S^i(A) \xrightarrow{u} S^{i-1}(A) \xrightarrow{r} R_{\mathrm{tr}}(\mathfrak{X}) \otimes T^i[1]$.

(b) $S^{i-1}(A) \otimes T \xrightarrow{v} S^i(A) \xrightarrow{S^i y} R_{\mathrm{tr}}(\mathfrak{X}) \xrightarrow{s} S^{i-1}(A) \otimes T[1]$.

Moreover, u identifies $S^{i-1}(A)$ with the slice quotient $s_{<qi}(S^i A)$, v identifies $S^{i-1}(A) \otimes T$ with the slice filtration $s_{>0}(S^i A)$, and $S^i y$ identifies $R_{\mathrm{tr}}(\mathfrak{X})$ with $s_0(S^i A)$.

Proof. (Cf. [Voe11, 3.1].) By multiplicativity (4.9), the slice $s_n(A^{\otimes i})$ of $A^{\otimes i}$ is the sum of the $s_{n_1}(A) \otimes \cdots \otimes s_{n_i}(A)$ with $\sum n_i = n$. Since the symmetric group acts on everything here, the slice $s_n(S^i A)$ is the symmetric part of this, viz., $R(qj)[2pj]$ if $n = qj$ for $0 \le j \le i$ and zero otherwise. The same is true for the slices of $S^{i-1} A$, except that $s_{qi} S^{i-1} A = 0$. Since $R_{\mathrm{tr}}(\mathfrak{X}) \otimes T^i$ is concentrated in slice filtration qi, we see that $S^i x$ is an isomorphism onto $s_{qi}(S^i A)$.

By inspection of the formula for the transfer, and hence for the map u, we see that the morphism $s_{qj}(u)$ is multiplication by $i - j$; this is an isomorphism for $j < i$. The existence and uniqueness of r now follows from Lemma 5.8.

Similarly, $S^i y$ is an isomorphism of $s_0(A)$ onto $R_{\mathrm{tr}}(\mathfrak{X})$, and $s_{qj}(v)$ is an isomorphism for $j > 0$ because by multiplicativity (4.9):

$$s_{qj}\left(S^{i-1}(A) \otimes R(q)[2p]\right) = s_{q(j-1)}(S^{i-1} A) \otimes R(q)[2p].$$

The existence and uniqueness of s now follows from Lemma 5.8. □

The following result is due to Voevodsky [Voe10b, Lem. 5.18].

Lemma 5.8. *Let $A \xrightarrow{a} B \xrightarrow{b} C$ be a sequence in DT^ε and $n \ge 0$ such that $A \in DT^\varepsilon_{\ge n}$ and $C \in DT^\varepsilon_{<n}$. If $s_i(a)$ is an isomorphism for all $i \ge n$ and $s_i(b)$ is an isomorphism for all $i < n$, then there is a unique morphism $c : C \to A[1]$ such that $A \xrightarrow{a} B \xrightarrow{b} C \xrightarrow{c} A[1]$ is a distinguished triangle, and identifies A and C with $s_{\ge n} B$ and $s_{<n} B$, respectively.*

Proof. Choose a distinguished triangle $A \xrightarrow{a} B \to C' \to A[1]$. The composition ba is zero for weight reasons. Hence b factors through a morphism $\phi : C' \to C$. The hypotheses imply that $s_i(\phi)$ is an isomorphism for all i; as the slice functor is conservative (Remark 4.9), this implies that ϕ is an isomorphism. Letting c be the composite of ϕ^{-1} and $C' \to A[1]$ yields the triangle (a, b, c).

If (a, b, c') is a second triangle on (A, B, C) then there is an $f : C \to C$ so that $(1, 1, f)$ is a morphism from the first to the second triangle. Since $s_{<n}(b)$ is

an isomorphism, it follows that $f = s_{<n}(f)$ is the identity of C and hence that $c = c'$. The final assertion follows since $s_{\geq n} C = 0$ and $s_{<n} A = 0$ by conservativity of the slice functors (see 4.9). □

Now set $p = q = b$. For any z, the distinguished triangles of axiom (iii) exist with $D = S^{\ell-2} A$, by Proposition 5.7. Next, we show that $M = S^{\ell-1} A$ also satisfies axiom (ii).

Proposition 5.9. *If $1/(\ell-1)! \in R$, there are maps $\lambda : R_{\mathrm{tr}}(X) \to S^{\ell-1}(A)$ such that the inclusion $\iota : R_{\mathrm{tr}}(X) \to R_{\mathrm{tr}}(\mathfrak{X})$ factors in* **DM** *as:*

$$R_{\mathrm{tr}}(X) \xrightarrow{\lambda} S^{\ell-1}(A) \xrightarrow{S^{\ell-1}y} R_{\mathrm{tr}}(\mathfrak{X}).$$

In fact[1], ι factors as $R_{\mathrm{tr}}(X) \xrightarrow{\lambda} S^{\ell-1}(A) \xrightarrow{u} \cdots \xrightarrow{u} S^2(A) \xrightarrow{u} A \xrightarrow{y} R_{\mathrm{tr}}(\mathfrak{X})$.

Proof. We saw in Lemma 5.3 that ι factors through a map $\lambda_1 : R_{\mathrm{tr}}(X) \to A$. By induction on $i < \ell$, ι factors through a map $\lambda_{i-1} : R_{\mathrm{tr}}(X) \to S^{i-1}(A)$. Applying $\mathrm{Hom}_{\mathbf{DM}}(R_{\mathrm{tr}}(X), -)$ to the triangle 5.7(a) yields exact sequences

$$\mathrm{Hom}(R_{\mathrm{tr}}(X), S^i(A)) \xrightarrow{u} \mathrm{Hom}(R_{\mathrm{tr}}(X), S^{i-1}(A)) \xrightarrow{r} \mathrm{Hom}(R_{\mathrm{tr}}(X), \mathfrak{X} \otimes T^i[1]).$$

The group on the right is $H^{2pi+1,qi}(X, \mathbb{Z})$, which is zero for $p = q$ (see [MVW, 19.3]), so λ_{i-1} lifts to a map $\lambda_i : R_{\mathrm{tr}}(X) \to S^i(A)$ with $\lambda_{i-1} = u\lambda_i$. By the construction of u, $yu^i = S^i y : S^i(A) \to R_{\mathrm{tr}}(\mathfrak{X})$. □

We end this section by defining a function ϕ^V from $H^{2p+1,q}(\mathfrak{X}, R)$ to $H^{2p\ell+2,q\ell}(\mathfrak{X}, R)$ for every $p, q \geq 0$; we will use ϕ^V in section 5.4 when $p = q = b = d/(\ell-1)$. Later, in chapter 6, we will extend ϕ^V to a cohomology operation $H^{2p+1,q}(-, R) \to H^{2p\ell+2,q\ell}(-, R)$.

Definition 5.10. Suppose that $1/(\ell-1)! \in R$. Then the function

$$\phi^V : H^{2p+1,q}(\mathfrak{X}, R) \to H^{2p\ell+2,q\ell}(\mathfrak{X}, R)$$

is defined as follows. Given $z \in H^{2p+1,q}(\mathfrak{X}, R)$, form A as in (5.6), and r, s as in Proposition 5.7. The composition of s and $r \otimes 1$ yields a map (for $i = \ell-1$)

$$\phi^V(z) : R_{\mathrm{tr}}(\mathfrak{X}) \xrightarrow{s} S^{\ell-2}(A) \otimes T[1] \xrightarrow{r \otimes 1} R_{\mathrm{tr}}(\mathfrak{X}) \otimes T^\ell[2] \to T^\ell[2],$$

i.e., an element of $H^{2p\ell+2,q\ell}(\mathfrak{X}, R)$.

Remark 5.10.1. Since the construction of A and (5.6) is natural in X and z, so are the triangles in Proposition 5.7. It follows that if $f : Y \to X$ is a morphism of

---

1. Compare with (5.6) and 5.11 of [Voe11].

smooth schemes then $f^*(\phi^V(z)) = \phi^V(f^*z)$. This shows that the ϕ^V of Definition 5.10 is natural in the pair (X, z). The proof that it is the restriction of a natural transformation, i.e., a cohomology operation in the sense of Definition 13.1, is given in Proposition 6.28 and Remark 6.27.1 of chapter 6.

5.3   End(M) IS A LOCAL RING

When $R = \mathbb{Z}_{(\ell)}$ and X is smooth, any $z \in H^{2p+1,q}(\mathfrak{X}, R)$ determines a motive A by the triangle (5.6). In this section we show that if z is nonzero modulo ℓ then the endomorphism rings $\text{End}(S^i A)$ are local rings for $i < \ell$. We will need this in the proof of Theorem 5.18.

Lemma 5.11. *If $R = \mathbb{Z}_{(\ell)}$ and z is nonzero modulo ℓ, then $\text{End}(A)$ is a local ring, and its maximal ideal is the kernel of $\text{End}(A) \xrightarrow{s_0} \mathbb{Z}_{(\ell)} \to \mathbb{Z}/\ell$.*

Proof. Recall that there are natural isomorphisms $\text{End}(R_{\text{tr}}(\mathfrak{X})) \cong \mathbb{Z}_{(\ell)}$ and $\text{End}(R_{\text{tr}}(\mathfrak{X}) \otimes T) \cong \mathbb{Z}_{(\ell)}$ by Corollary 4.5, where $T = R(q)[2p]$. Hence any endomorphism h of A determines numbers $c_0 = s_0(h) \in \mathbb{Z}_{(\ell)}$ and $c_1 = s_q(h)$ in $\text{End}(R_{\text{tr}}(\mathfrak{X}) \otimes T) \cong \mathbb{Z}_{(\ell)}$ and a morphism of triangles (where, by abuse of notation, we write $\mathfrak{X}$ for $R_{\text{tr}}(\mathfrak{X})$):

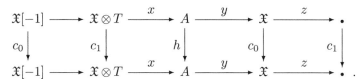

Since z is nonzero modulo ℓ, and $zc_0 = c_1 z$, we see that $c_0 \equiv c_1 \pmod{\ell}$. Let J be the kernel of $\text{End}(A) \to \mathbb{Z}/\ell$; if $h \in 1 + J$ then $c_0 \equiv c_1 \equiv 1 \pmod{\ell}$, and h is an isomorphism by the 5-lemma. It follows that $\text{End}(A)$ is a local ring.   □

Recall from Proposition 5.7 that the top slice of $S^i A$ is $s_0(S^i A) = R_{\text{tr}}(\mathfrak{X})$. Since s_0 is a functor, this gives a natural ring homomorphism $s_0 : \text{End}(S^i A) \to \text{End}(R_{\text{tr}}(\mathfrak{X})) = \mathbb{Z}_{(\ell)}$.

Proposition 5.12. *When $R = \mathbb{Z}_{(\ell)}$, z is nonzero modulo ℓ and $1 \leq i < \ell$, the ring $\text{End}(S^i A)$ is a local ring; its maximal ideal is the kernel of the natural surjection $\text{End}(S^i A) \xrightarrow{s_0} \mathbb{Z}_{(\ell)} \to \mathbb{Z}/\ell$.*
    Thus a map $h \in \text{End}(S^i A)$ is an isomorphism if and only if $c = s_0(h) \in \mathbb{Z}_{(\ell)}$ is not congruent to zero modulo ℓ.[2]

Proof. Again, we use the slice filtration (see 4.9) and induction on i, the case $i = 1$ being Lemma 5.11. By Proposition 5.7(a), the endomorphism h induces an endomorphism $h' = s_{<qi}(h)$ on $S^{i-1} A$, and $\text{End}(S^i A) \to \mathbb{Z}_{(\ell)}$ factors through

---

2. Compare with [Voe11, 15.10].

$\text{End}(S^{i-1}A) \to \mathbb{Z}_{(\ell)}$. By Proposition 5.7(b), h induces an endomorphism $s_{>0}h$ on $s_{>0}(S^i A)$, which is identified with $S^{i-1}(A) \otimes T$ via v. Moreover, the isomorphism $\text{End}(S^{i-1}A \otimes T) \cong \text{End}(S^{i-1}A)$ identifies $s_{>0}h$ with an endomorphism h'' of $S^{i-1}A$ and the slice $s_q : \text{End}(S^{i-1}A \otimes T) \to \text{End}(T) \cong \mathbb{Z}_{(\ell)}$ of h with the augmentation $s_0(h'')$ of h''.

Therefore if $s_0(h) \equiv 1 \pmod{\ell}$ the map h' is an isomorphism by induction, and $s_{qj}(h)$ is an isomorphism for every $j < i$. In particular, the augmentation $s_q(h)$ and hence the augmentation of h'' is an isomorphism. Since $\text{End}(S^{i-1}A)$ is a local ring by induction, it follows that the slices $s_{qj}(h)$ are isomorphisms for every $j > 1$. Because the slice functor is conservative (Remark 4.9), h is an isomorphism. Since every h of augmentation 1 is an isomorphism, $\text{End}(S^i A)$ is a local ring. $\qquad\square$

Remark 5.12.1. As in the proof of Lemma 5.11, it is easy to show by induction that the slices of an endomorphism h have degrees which are all congruent modulo ℓ.

5.4 EXISTENCE OF A ROST MOTIVE

In this section, we will show that if z is chosen suitably in the sense of Definition 5.13 and $R = \mathbb{Z}_{(\ell)}$, then $\lambda : R_{\text{tr}}(X) \to M$ splits in such a way that $M = S^{\ell-1}(A)$ is a symmetric Chow motive of the form (X, e). This verifies the final axiom 4.11(i), proving that M is a Rost motive whenever X is a Rost variety.

In this section, X is a smooth projective variety of dimension $d = \ell^{n-1} - 1$, $b = (\ell^{n-1} - 1)/(\ell - 1)$, and $\mathfrak{X}$ is the 0-coskeleton of X (see 1.32). We will use the motivic cohomology operations Q_i on $H^{*,*}(-, \mathbb{Z}/\ell)$ described in 1.43. One result, Corollary 6.34, is postponed to chapter 6 in order to not interrupt the flow of this section.

Definition 5.13. We say that $z \in H^{2b+1,b}(\mathfrak{X}, \mathbb{Z}_{(\ell)})$ is *suitable* if its mod-ℓ reduction $\bar{z}$ satisfies $Q_{n-1}(\bar{z}) \neq 0$ and $Q_i(\bar{z}) = 0$ for $0 \leq i \leq n - 2$.

An element of the form $z = \tilde{\beta} Q_1 \cdots Q_{n-2}(\delta)$ is suitable if and only if $Q_{n-1}(\bar{z}) \neq 0$, where $\tilde{\beta}$ is the integral Bockstein 1.42(3), since its mod-ℓ reduction is $\bar{z} = Q_0 \cdots Q_{n-2}(\delta)$. The fact that $Q_i(\bar{z}) = 0$ for $i \leq n - 2$ follows from the fact that the Q_i anticommute and $Q_i^2 = 0$; see [Mil58, 4a] (which is recorded in Lemma 13.11 below).

Example 5.13.1. Suppose that $\text{BL}(n - 1)$ holds. If X splits $\underline{a}$ and δ is the lift of $\underline{a}$ given by Lemma 3.13, then the element $\mu = \tilde{\beta} Q_1 \cdots Q_{n-2}(\delta)$ is suitable, because $Q_{n-1}(\mu) \neq 0$ by Corollary 3.16.

The mod-ℓ reduction $\bar{z}$ of any suitable z in $H^{2b+1,b}(\mathfrak{X}, \mathbb{Z}_{(\ell)})$ satisfies the hypotheses of the following lemma.

Lemma 5.14. *Let* $\bar{z} \in H^{2b+1,b}(\mathfrak{X}, \mathbb{Z}/\ell)$ *be such that* $Q_i(\bar{z}) = 0$ *for* $0 \leq i \leq n - 2$. *Then* $\beta P^b(\bar{z}) = (-1)^{n-1} Q_{n-1}(\bar{z})$.

Proof. Recall that $b = (\ell^{n-1} - 1)/(\ell - 1)$. If $\ell \neq 2$ then by 13.12 we have

$$\beta P^b(\tilde{z}) = \sum_{i=0}^{n-1} (-1)^i P^{b-(\ell^i-1)/(\ell-1)} Q_i(\tilde{z}).$$

If $\ell = 2$ we have the same formula, by induction applied to Corollary 13.14. By hypothesis, all terms on the right vanish except the term for $i = n-1$, which is $(-1)^{n-1} Q_{n-1}(\tilde{z})$ because $P^0 x = x$ by Axiom 13.6.1. $\qquad\square$

Our next proposition uses the fundamental class τ of X. To define it, recall that by motivic duality ([MVW, 20.11]) we have $R_{\mathrm{tr}}(X) \cong R_{\mathrm{tr}}(X)^* \otimes \mathbb{L}^d$. Hence

$$\mathrm{Hom}(\mathbb{L}^d, R_{\mathrm{tr}}(X)) \cong \mathrm{Hom}(\mathbb{L}^d, R_{\mathrm{tr}}(X)^* \otimes \mathbb{L}^d) \cong \mathrm{Hom}(R_{\mathrm{tr}}(X), R) \cong H^0(X, R).$$

Note that motivic duality holds when $\mathrm{char}(k) = p$ and $1/p \in R$, even if k does not admit resolution of singularities, by [Kel13, 5.5.14].

Definition 5.15. The fundamental class $\tau : \mathbb{L}^d \to R_{\mathrm{tr}}(X)$ is defined to be the map corresponding to $1 \in H^0(X, R)$. Since $R_{\mathrm{tr}}(\mathfrak{X}) \otimes R_{\mathrm{tr}}(X) \cong R_{\mathrm{tr}}(X)$, the class τ determines a map from $R_{\mathrm{tr}}(\mathfrak{X}) \otimes \mathbb{L}^d$ to $R_{\mathrm{tr}}(X)$.

We now use Corollary 6.34, that $\phi^V = \beta P^b$ when $R = \mathbb{Z}/\ell$.

Proposition 5.16.[3] *Let $z \in H^{2b+1,b}(\mathfrak{X}, R)$, $R = \mathbb{Z}/\ell$, be such that $Q_{n-1}(z) \neq 0$ and $Q_i(z) = 0$ for $0 \leq i \leq n-2$. If $s_d(X) \not\equiv 0 \pmod{\ell^2}$, the composition*

$$R_{\mathrm{tr}}(\mathfrak{X}) \otimes \mathbb{L}^d \xrightarrow{\tau} R_{\mathrm{tr}}(X) \xrightarrow{\lambda} S^{\ell-1}(A)$$

is nonzero, where λ is given by Proposition 5.9.

Proof. By Proposition 5.9, the structure map of $R_{\mathrm{tr}}(X)$ factors as $S^{\ell-1} y \circ \lambda$. Since $s_d(X) \not\equiv 0 \pmod{\ell^2}$, the motivic degree theorem 13.25 states that $\lambda\tau : \mathfrak{X} \otimes \mathbb{L}^d \to S^{\ell-1}(A)$ is nonzero, provided there is a nonzero $\alpha \in H^{b\ell+2,b\ell}(\mathfrak{X}, \mathbb{Z}/\ell)$ such that $Q_{n-1}(\alpha) = 0$ and $(S^{\ell-1}y)^*(\alpha) = 0$ in $H^{b\ell+2,b\ell}(S^{\ell-1}(A), \mathbb{Z}/\ell)$.

By assumption, the element $\alpha = (-1)^{n-1} Q_{n-1}(z)$ of $H^{b\ell+2,b\ell}(\mathfrak{X}, \mathbb{Z}/\ell)$ is nonzero, and $Q_{n-1}(\alpha) = 0$ because $Q_{n-1}^2 = 0$. We are reduced to showing that $(S^{\ell-1}y)^*(\alpha) = 0$.

By Lemma 5.14, $\alpha = \beta P^b(z)$; by Corollary 6.34, $\phi^V(z) = \beta P^b(z)$. By the definition of ϕ^V in 5.10, $(S^{\ell-1}y)^*\phi^V(z)$ is the composite $(r \otimes 1) \circ s \circ S^{\ell-1}y$, and is zero because $s \circ S^{\ell-1}y = 0$ by 5.7(b). $\qquad\square$

Regarding $z : R_{\mathrm{tr}}(\mathfrak{X}) \to R_{\mathrm{tr}}(\mathfrak{X}) \otimes \mathbb{L}^b[1]$ as a map between Tate objects, it is self-dual (by Example 4.8, $z^D = z^\dagger \otimes \mathbb{L}^b$ under the identification of $R_{\mathrm{tr}}(\mathfrak{X})$ with

---

3. Compare to Proposition 5.12 of [Voe11].

$R_{\mathrm{tr}}(\mathfrak{X})^{\dagger})$. It follows that $A \cong A^{\dagger} \otimes \mathbb{L}^{b}$. Since $S^{i}(M^{\dagger}) \cong (S^{i}M)^{\dagger}$ for every M we also have $S^{i}(A) \cong S^{i}(A)^{\dagger} \otimes \mathbb{L}^{bi}$. (See Proposition 4.7(2).)

Definition 5.17. Assume that $1/(\ell-1)! \in R$. For the map λ of Proposition 5.9, we write λ^{D} for the dual map

$$\lambda^{D}: S^{\ell-1}(A) \cong S^{\ell-1}(A)^{\dagger} \otimes \mathbb{L}^{d} \xrightarrow{\lambda^{\dagger} \otimes \mathbb{L}^{d}} R_{\mathrm{tr}}(X)^{\dagger} \otimes \mathbb{L}^{d} \cong R_{\mathrm{tr}}(X).$$

Theorem 5.18. Set $R = \mathbb{Z}_{(\ell)}$. Suppose that $z \in H^{2b+1,b}(\mathfrak{X}, \mathbb{Z}_{(\ell)})$ is suitable in the sense of Definition 5.13. If $s_{d}(X) \not\equiv 0 \pmod{\ell^{2}}$, then: the composition $\lambda \circ \lambda^{D}$ is an isomorphism on $S^{\ell-1}(A)$; there is a unit $c \in R$ so that $\lambda \tau = c \cdot S^{\ell-1}x$; and the following diagram commutes:

$$
\begin{array}{ccc}
S^{\ell-1}(A) & \xrightarrow[\simeq]{\lambda \circ \lambda^{D}} & S^{\ell-1}(A) \\
{\scriptstyle S^{\ell-1}y} \downarrow & & \downarrow {\scriptstyle S^{\ell-1}y} \\
R_{\mathrm{tr}}(\mathfrak{X}) & \xrightarrow[\simeq]{c} & R_{\mathrm{tr}}(\mathfrak{X}).
\end{array}
$$

In particular, $S^{\ell-1}(A)$ is a direct summand of $R_{\mathrm{tr}}(X)$, i.e., a Chow motive.

Proof. By Proposition 5.7, $S^{\ell-1}y$ is the natural projection onto the top slice, $R_{\mathrm{tr}}(\mathfrak{X})$. We saw in Corollary 4.5 that $\mathrm{End}(R_{\mathrm{tr}}(\mathfrak{X})) = R$, so the diagram always commutes when $c \in R$ is the top slice of $\lambda \circ \lambda^{D}$. By Proposition 5.12, $\mathrm{End}(M)$ is a local ring and the top slice $\mathrm{End}(M) \to R$ is a local homomorphism. Thus c is a unit of R if and only if $u = \lambda \circ \lambda^{D}$ is an isomorphism, in which case $e = \lambda^{D} \circ u^{-1} \circ \lambda$ is an idempotent element of $\mathrm{End}(R_{\mathrm{tr}}(X))$ and $S^{\ell-1}(A) = eR_{\mathrm{tr}}(X)$ is a direct summand of $R_{\mathrm{tr}}(X)$. Therefore it suffices to prove that $c \not\equiv 0 \pmod{\ell}$.

Let us abbreviate $R_{\mathrm{tr}}(\mathfrak{X})$ by $\mathfrak{X}$. By 4.5 and 4.7, $\mathrm{Hom}(\mathfrak{X} \otimes \mathbb{L}^{r}, \mathfrak{X}) = 0$ for all $r > 0$. Using the exact sequences of 5.7(b),

$$\mathrm{Hom}(\mathfrak{X} \otimes \mathbb{L}^{r}, S^{i-1}A \otimes \mathbb{L}^{b}) \xrightarrow{v} \mathrm{Hom}(\mathfrak{X} \otimes \mathbb{L}^{r}, S^{i}A) \xrightarrow{S^{i}y} \mathrm{Hom}(\mathfrak{X} \otimes \mathbb{L}^{r}, \mathfrak{X}) = 0,$$

it follows by induction on $i \geq 0$, $i < \ell$, that $\mathrm{Hom}(\mathfrak{X} \otimes \mathbb{L}^{r}, S^{i}A) = 0$ for $r > qi$. From triangle 5.7(a) we have an exact sequence

$$
\mathrm{Hom}(\mathfrak{X} \otimes \mathbb{L}^{d}, \mathfrak{X} \otimes \mathbb{L}^{d}) \xrightarrow{S^{\ell-1}x} \mathrm{Hom}(\mathfrak{X} \otimes \mathbb{L}^{d}, S^{\ell-1}A) \xrightarrow{u} \mathrm{Hom}(\mathfrak{X} \otimes \mathbb{L}^{d}, S^{\ell-2}A) = 0.
$$
$$\exists\, C \qquad \mapsto \qquad \lambda\tau \not\equiv 0 \pmod{\ell}.$$

Since $d > (\ell-2)b$, the last term is zero. Hence the composite $\lambda\tau$ lifts to an element C of $R \cong \mathrm{Hom}(\mathfrak{X} \otimes \mathbb{L}^{d}, \mathfrak{X} \otimes \mathbb{L}^{d})$. Since $\lambda\tau \not\equiv 0 \pmod{\ell}$ by 5.16, $C \not\equiv 0 \pmod{\ell}$. That is, $(S^{\ell-1}x) \circ C = \lambda\tau$.

Dualizing $\lambda\tau = (S^{\ell-1}x)C$ yields the left square in the following diagram, since $S^{\ell-1}y$ is dual to $S^{\ell-1}x$ and ι is dual to $\tau\colon \mathfrak{X}\otimes\mathbb{L}^d \to R_{\mathrm{tr}}(X)$, so $\iota\circ\lambda^D$ is dual to $\lambda\tau$.

$$
\begin{array}{ccc}
S^{\ell-1}(A) & \xrightarrow{\ \lambda^D\ } R_{\mathrm{tr}}(X) \xrightarrow{\ \lambda\ } & S^{\ell-1}(A) \\
{\scriptstyle S^{\ell-1}y}\big\downarrow & \quad\iota\big\downarrow \quad {}^{\textstyle S^{\ell-1}y} & \\
\mathfrak{X} & \xrightarrow{\quad C\quad} \mathfrak{X}. &
\end{array}
$$

The right triangle commutes by Proposition 5.9. Thus C is the top slice of $\lambda\circ\lambda^D$, i.e., $C = c$. $\qquad\square$

Corollary 5.19. *If $z \in H^{2b+1,b}(\mathfrak{X}, \mathbb{Z}_{(\ell)})$ is suitable and $s_d(X) \not\equiv 0 \pmod{\ell^2}$, then $M = S^{\ell-1}(A)$ is a symmetric Chow motive.*

Proof. Setting $e = c^{-1}\lambda^D\circ\lambda$, Theorem 5.18 shows that $S^{\ell-1}(A) \cong (X, e)$ is a Chow motive. By Example 4.2, its transpose (X, e^t) is defined by

$$\lambda^t = \lambda^* \otimes \mathbb{L}^d : S^{\ell-1}(A) \cong S^{\ell-1}(A)^* \otimes \mathbb{L}^d \to X.$$

Via $\mathrm{Hom}(S^{\ell-1}(A), R_{\mathrm{tr}}(\mathfrak{X})) \cong \mathrm{Hom}(S^{\ell-1}(A), R_{\mathrm{tr}}(X))$, we may identify λ^D and λ^t. Since $e = c^{-1}\lambda^t\circ\lambda$, we have $e = e^t$. That is, $S^{\ell-1}(A) = (X, e)$ is a symmetric Chow motive in the sense of Definition 4.1. $\qquad\square$

Corollary 5.20. *Suppose that $R = \mathbb{Z}_{(\ell)}$, and X is a Rost variety for $\underline{a}$. When $z = \mu$, the maps λ and λ^t make the direct summand $M = S^{\ell-1}(A)$ of $R_{\mathrm{tr}}(X)$ into a symmetric Chow motive, i.e., the following composition is an isomorphism:*

$$M \cong M^* \otimes \mathbb{L}^d \xrightarrow{\ \lambda^t\ } R_{\mathrm{tr}}(X)^* \otimes \mathbb{L}^d \cong R_{\mathrm{tr}}(X) \xrightarrow{\ \lambda\ } M.$$

Hence M is a Rost motive for $\underline{a}$.

Proof. By Example 5.13.1, μ satisfies the hypothesis of Theorem 5.18. The corollary is just a restatement of Theorem 5.18 in the form of property 4.11(i). Since we have already seen that properties 4.11(ii-iii) hold for M, by 5.7 and 5.9, M is a Rost motive for $\underline{a}$, as claimed. $\qquad\square$

5.5 HISTORICAL NOTES

When $\ell = 2$, the Pfister quadric X associated to $\underline{a}$ is a Rost variety, and it was Rost who constructed a symmetric Chow motive $M = (X, e)$ in this case, around 1990; see [Ros90] and [Ros98c]. This is the eponymous Rost motive.

Voevodsky's 1996 proof of the Milnor Conjecture (the case $\ell = 2$) centered around the observation that Rost's motive fits into the motivic triangle (5.1) in [Voe03a, Thm. 4.4].

Voevodsky's 2003 preprint [Voe03b], announced in 1998, contained the construction of the Rost motive as presented in this chapter, as well as our next chapter which proves that the ϕ^V we define in 5.10 is a cohomology operation. We will prove in part III that ϕ^V equals βP^b; the original proof in the 2003 preprint [Voe03b] had a gap which was patched in 2007, and finally published in [Voe11]. We have followed the presentation in [Wei09].

Chapter Six

Motives over S

THE PURPOSE OF this chapter is to show that the operation ϕ^V of Definition 5.10 extends to a cohomology operation over k, and that it satisfies the recognition criterion of Theorem 15.38, so that ϕ^V must be βP^b (see Corollary 6.34).

Our construction of the cohomology operation utilizes the machinery of motives over a simplicial noetherian scheme, which is presented in sections 6.1–6.3, and the slice filtration in section 6.4.

This chapter is independent of chapter 5, but does use the duality material of chapter 4 (section 4.2) and the symmetric powers $S^i(N)$ of a motive N, defined just before 5.5.

6.1 MOTIVES OVER A SCHEME

In this section, we summarize the basic theory of motives over a scheme S, following [MVW] and [SV00b]. Other approaches to this theory are given in [Ayo07] and [CD13].

If S is a noetherian scheme, there is a tensor triangulated category $\mathbf{DM}_{\mathrm{nis}}^{\mathrm{eff}}(S)$. It may be constructed by replacing $\mathrm{Spec}(k)$ by S, using the finite correspondences described in appendix 1A of [MVW], and also in [SV00b].

In more detail, we begin with the category $\mathbf{Sm}/S$ of smooth schemes over S. An *elementary correspondence* W from X to Y is an irreducible closed subset of $X \times_S Y$ whose associated integral scheme is finite and surjective over a component of X. Elements of the free abelian group generated by the elementary correspondences from X to Y are called *cycles*; a cycle is called a *finite correspondence* if it is a "universally integral relative cycle." (See loc. cit.) If S is regular, every cycle is a finite correspondence; see [SV00b, 3.3.15, 3.4.8].

The set $\mathbf{Cor}_S(X, Y)$ of all finite correspondences from X to Y is a subgroup of the group of all cycles. These sets form the morphisms of an additive category $\mathbf{Cor}_S$ whose objects are the smooth schemes over S; composition of finite correspondences is explained in [MVW, 1A.11].

Presheaves with transfers over S (with coefficients in a fixed ring R) are just additive contravariant functors from $\mathbf{Cor}_S$ to R-modules. They form the objects of an abelian category $\mathbf{PST}(S)$, morphisms being natural transformations. If Y is a smooth scheme over S, we write $R_{\mathrm{tr}}(Y/S)$ for the associated representable presheaf with transfers. As in Lecture 8 of [MVW], the $R_{\mathrm{tr}}(Y/S)$ are projective objects (and every projective is a direct summand of a direct sum of representables) so we can use resolutions to construct a right exact tensor product $\otimes^{\mathrm{tr}}$ on $\mathbf{PST}(S)$ with unit $R_{\mathrm{tr}}(S/S)$ determined by the formula

$$R_{\mathrm{tr}}(X/S) \otimes^{\mathrm{tr}} R_{\mathrm{tr}}(Y/S) = R_{\mathrm{tr}}(X \times_S Y/S),$$

as well as a total tensor product $\otimes^{\mathrm{tr}}_{\mathbb{L}}$ on the derived category $\mathbf{D}^-(\mathbf{PST}(S))$ making it into a tensor triangulated category.

Sheafifying for the Nisnevich topology is compatible with $\otimes^{\mathrm{tr}}_{\mathbb{L}}$, and makes the associated derived category $\mathbf{D}^-_{\mathrm{nis}}$ of Nisnevich sheaves with transfers over S into a tensor triangulated category, as in [MVW, 14.2]. Form the smallest thick subcategory $\mathcal{E}$ of $\mathbf{D}^-$, closed under direct sums and containing the cone of every projection $R_{\mathrm{tr}}(X \times \mathbb{A}^1) \to R_{\mathrm{tr}}(X)$; maps whose cone is in $\mathcal{E}$ are called $\mathbb{A}^1$-*weak equivalences*. The triangulated category $\mathbf{DM}^{\mathrm{eff}}_{\mathrm{nis}}(S)$ is defined to be the localization of $\mathbf{D}^-_{\mathrm{nis}}$ at the class of $\mathbb{A}^1$-weak equivalences, or equivalently, the Verdier quotient $\mathbf{D}^-_{\mathrm{nis}}/\mathcal{E}$. By the argument of [MVW, 9.6], $\mathbf{DM}^{\mathrm{eff}}_{\mathrm{nis}}(S)$ is a tensor triangulated category.

If $f : S \to T$ is a morphism of schemes, the pullback $\mathbf{Sm}/T \to \mathbf{Sm}/S$ induces a pullback $\mathbf{Cor}_T \to \mathbf{Cor}_S$ and therefore adjoint functors

$$\mathbf{PST}(T) \overset{f^*}{\underset{f_*}{\rightleftarrows}} \mathbf{PST}(S). \tag{6.1}$$

By inspection, $f_* F(U) = F(U \times_T S)$ is exact and induces a functor between the derived categories. The inverse image functor f^* is right exact and (strongly) monoidal in the sense that $f^*(F \otimes^{\mathrm{tr}} G) \cong f^*(F) \otimes^{\mathrm{tr}} f^*(G)$. If Y is in $\mathbf{Sm}/T$ then $f^* R_{\mathrm{tr}}(Y/T) \cong R_{\mathrm{tr}}(Y \times_T S/S)$ because for each X in $\mathbf{Sm}/S$ we have $\mathbf{Cor}_S (X, Y \times_T S) \cong \mathbf{Cor}_T(X, Y)$. Using projective resolutions, f^* has a total left derived functor $\mathbf{L}f^*$ which is left adjoint to f_*.

Remark 6.2. If $f : S \to T$ is smooth, then $f^* : \mathbf{PST}(T) \to \mathbf{PST}(S)$ is exact, with $(f^*G)(X) = G(X)$. This is because we may regard any X in $\mathbf{Sm}/S$ as smooth over T. It follows that f^* has a left adjoint $f_\#$, satisfying $f_\# R_{\mathrm{tr}}(X/S) = R_{\mathrm{tr}}(X/T)$. This formula determines the right exact functor $f_\#$ because every object has a projective resolution.

6.2 MOTIVES OVER A SIMPLICIAL SCHEME

If $S_\bullet$ is a simplicial noetherian scheme, there is also a tensor triangulated category $\mathbf{DM}^{\mathrm{eff}}_{\mathrm{nis}}(S_\bullet)$ of motives over $S_\bullet$. In order to construct it, we make a few judicious modifications to the template of section 6.1; further details may be found in [Voe10b]. This theory will be used in order to construct the cohomology operations ϕ_i and ϕ^V in sections 6.6 and 6.7.

We first form the category $\mathbf{Sm}/S_\bullet$. The class of objects of $\mathbf{Sm}/S_\bullet$ is the disjoint union of the objects of the $\mathbf{Sm}/S_i$; a morphism from $X \to S_i$ to $Y \to S_j$ is a compatible pair of morphisms $(X \to Y, S_i \to S_j)$. Thus $\mathbf{Sm}/S_\bullet$ is fibered over the simplicial category. We will not need to introduce a category $\mathbf{Cor}_{S_\bullet}$.

Definition 6.3. A *presheaf with transfers* over $S_\bullet$ (with coefficients in R) is a collection $\{F_i\}_{i \geq 0}$ of presheaves with transfers F_i over S_i together with bonding maps $\alpha^*(F_j) \to F_i$ in $\mathbf{PST}(S_j)$ for every simplicial map $\alpha: [j] \to [i]$ (where $\alpha: S_i \to S_j$ is the simplicial structure map), natural in α and such that α^* is the identity map when α is.

We write $\mathbf{R}$ (or $\mathbf{R}_{S_\bullet}$) for the distinguished presheaf with transfers with $F_i = R_{\mathrm{tr}}(S_i/S_i)$. For Y in $\mathbf{Sm}/S_i$, we write $R_{\mathrm{tr}}(Y, i)$ for the representable presheaf with transfers over $S_\bullet$ corresponding to Y, i.e., $X \mapsto \mathrm{Hom}_{\mathbf{Sm}/S_\bullet}(X, Y)$.

As before, $\mathbf{PST}(S_\bullet)$ is an abelian category with enough projectives, projective objects being the same thing as summands of direct sums of representable presheaves with transfers. It is also a symmetric monoidal category with the termwise tensor product $\{F_i\} \otimes^{\mathrm{tr}} \{G_i\} = \{F_i \otimes^{\mathrm{tr}}_{S_i} G_i\}$ of section 6.1; the unit is $\mathbf{R}$.

We will say that a presheaf with transfers $\{F_i\}$ is *termwise projective* if each F_i is a projective object in $\mathbf{PST}(S_i)$. Clearly $\mathbf{R}$ is termwise projective, and each representable object $R_{\mathrm{tr}}(Y, i)$ is termwise projective because

$$R_{\mathrm{tr}}(Y, i)_j = \oplus_{\varphi: [i] \to [j]} R_{\mathrm{tr}}(S_j \times_\varphi Y).$$

Lemma 6.4. *If $F \xrightarrow{\sim} F'$ is a quasi-isomorphism of complexes of termwise projectives in $\mathbf{PST}(S_\bullet)$, then the map $F \otimes^{\mathrm{tr}} G \to F' \otimes^{\mathrm{tr}} G$ is a quasi-isomorphism for every G in $\mathbf{PST}(S_\bullet)$.*

Proof. By hypothesis, each $F_i \xrightarrow{\sim} F'_i$ is a quasi-isomorphism of projective objects in $\mathbf{PST}(S_i)$, so $F_i \otimes^{\mathrm{tr}}_{S_i} G_i \simeq F'_i \otimes^{\mathrm{tr}}_{S_i} G_i$. $\qquad\square$

Again following [MVW, Lect. 8], $\mathbf{D}^-(\mathbf{PST}(S_\bullet))$ is a tensor triangulated category whose total tensor product $\otimes^{\mathrm{tr}}_{\mathbb{L}}$ is formed using projective resolutions, or even termwise projective resolutions. In particular, $\mathbf{R}$ is the unit object for $\otimes^{\mathrm{tr}}_{\mathbb{L}}$. Sheafifying makes the triangulated category $\mathbf{D}^-_{\mathrm{nis}}$ of Nisnevich sheaves with transfers into a tensor triangulated category.

The unit $\mathbf{R}$ is not representable in general. To form a projective resolution of $\mathbf{R}$, we begin with the canonical map $R_{\mathrm{tr}}(S_0, 0) \xrightarrow{\epsilon} \mathbf{R}$. Consider the simplicial object in $\mathbf{PST}(S_\bullet)$ which is $R_{\mathrm{tr}}(S_i, i)$ in degree i, and let Λ_* denote the associated normalized chain complex: $\cdots \to R_{\mathrm{tr}}(S_1, 1) \to R_{\mathrm{tr}}(S_0, 0) \to 0$.

Lemma 6.5. $\Lambda_* \xrightarrow{\epsilon} \mathbf{R}$ *is a projective resolution.*

Proof. By construction, each Λ_i is projective. Evaluating Λ_* at $S_i \in \mathbf{Sm}/S_i$ yields the normalized chain complex associated to the simplicial abelian group

$R[\Delta^i]$, and evaluating at X in $\mathbf{Sm}/S_i$ yields the complex associated to $R[\Delta^i] \otimes_R H^0(X, R)$. Since the R-module augmentation $R[\Delta^i] \xrightarrow{\epsilon} R$ is a simplicial equivalence, $\Lambda_*(X) \xrightarrow{\epsilon} \mathbf{R}(X)$ is a quasi-isomorphism for all X. This shows that $\Lambda_* \to \mathbf{R}$ is a projective resolution. $\square$

Let $\mathcal{E}(S_\bullet)$ denote the smallest thick subcategory of $\mathbf{D}_{\text{nis}}^-$, closed under direct sums and containing every $R_{\text{tr}}(X \times \mathbb{A}^1, i) \to R_{\text{tr}}(X, i)$; maps whose cone is in $\mathcal{E}$ are called $\mathbb{A}^1$-*weak equivalences*.

Definition 6.6. ([Voe10b, 4.2]) The triangulated category $\mathbf{DM}_{\text{nis}}^{\text{eff}}(S_\bullet)$ is defined to be the localization of $\mathbf{D}_{\text{nis}}^-$ at the class of $\mathbb{A}^1$-weak equivalences, i.e., as the Verdier quotient $\mathbf{D}_{\text{nis}}^-/\mathcal{E}$. Again by [MVW, 9.6], it is a tensor triangulated category.

Functoriality in $S_\bullet$ follows the pattern of (6.1). If $f : S_\bullet \to T_\bullet$ is a morphism of simplicial schemes, the direct image $f_* : \mathbf{PST}(S_\bullet) \to \mathbf{PST}(T_\bullet)$ is defined by $(f_* F)(U/T_i) = F(U \times_{T_i} S_i)$ and is exact, so it passes to a functor $f_* : \mathbf{D}^-(\mathbf{PST}(S_\bullet)) \to \mathbf{D}^-(\mathbf{PST}(T_\bullet))$. There is also an inverse image functor f^* fitting into an adjunction like (6.1); f^* is monoidal because its components are, by the discussion after (6.1). If Y is in $\mathbf{Sm}/T_i$, then $f^* R_{\text{tr}}(Y, i) \cong R_{\text{tr}}(Y \times_{T_i} S_i)$. In particular, $f^*(\mathbf{R}_{T_\bullet}) \cong \mathbf{R}_{S_\bullet}$. As in the non-simplicial case, there is a total left derived functor $\mathbf{L}f^* : \mathbf{D}^-(\mathbf{PST}(T_\bullet)) \to \mathbf{D}^-(\mathbf{PST}(S_\bullet))$ which is adjoint to f_*, defined using projective resolutions. Since f^* preserves projectives, we have canonical isomorphisms in $\mathbf{D}^-(\mathbf{PST}(S_\bullet))$ of the form

$$\mathbf{L}f^*(K \otimes_{\mathbb{L}}^{\text{tr}} L) \xrightarrow{\sim} \mathbf{L}f^*(K) \otimes_{\mathbb{L}}^{\text{tr}} \mathbf{L}f^*(L). \tag{6.7}$$

6.3 MOTIVES OVER A SMOOTH SIMPLICIAL SCHEME

In this section we assume that $S_\bullet$ is a smooth simplicial scheme over a perfect field k, with maps $c_i : S_i \to \text{Spec}(k)$ forming the structure map $c : S_\bullet \to \text{Spec}(k)$. If M is a presheaf with transfers over k, the pullback $c^*(M)$ is $\{c_i^*(M)\}$, and $c^*(R) = \mathbf{R}$. By Remark 6.2, c^* is exact and has a left adjoint $c_\# : \mathbf{PST}(S_\bullet) \to \mathbf{PST}(k)$.

Because c^* is exact, $c_\#$ is determined by the formula $c_\# R_{\text{tr}}(Y, i) = R_{\text{tr}}(Y)$. Since $c_\#$ commutes with direct sums, it takes $\mathcal{E}(S_\bullet)$ to $\mathcal{E}(\text{Spec } k)$. Therefore its total left derived functor $\mathbf{L}c_\#$ (defined via projective resolutions) induces a functor $\mathbf{L}c_\# : \mathbf{DM}_{\text{nis}}^{\text{eff}}(S_\bullet) \to \mathbf{DM}_{\text{nis}}^{\text{eff}}(k)$, left adjoint to $c^* = \mathbf{L}c^*$.

Lemma 6.8. *If $S_\bullet$ is a smooth simplicial scheme over k, then*

$$\mathbf{L}c_\#(c^* F) \cong R_{\text{tr}}(S_\bullet) \otimes_{\mathbb{L}}^{\text{tr}} F$$

for all F in $\mathbf{DM}_{\text{nis}}^{\text{eff}}(k)$. In particular, $\mathbf{L}c_\#(\mathbf{R}) \cong R_{\text{tr}}(S_\bullet)$.

Proof. It suffices to consider the case when $F = R_{\mathrm{tr}}(X)$ for X in $\mathbf{Sm}/k$. By Lemma 6.4, the tensor product of c^*F with the quasi-isomorphism $\Lambda_* \xrightarrow{\sim} \mathbf{R}$ of 6.5 is a quasi-isomorphism $\Lambda_* \otimes c^*F \xrightarrow{\sim} c^*F$. It is a projective resolution because each $R_{\mathrm{tr}}(S_i, i) \otimes c^*F \cong R_{\mathrm{tr}}(S_i \times X, i)$ is a projective object. It follows that $\mathbf{L}c_\#(c^*F) = c_\#[\Lambda_* \otimes c^*F]$ is the chain complex $R_{\mathrm{tr}}(S_\bullet) \otimes^{\mathrm{tr}} R_{\mathrm{tr}}(X)$ sending i to $c_\# R_{\mathrm{tr}}(S_i \times X, i) = R_{\mathrm{tr}}(S_i \times X)$, with the evident differentials. $\qquad \square$

Corollary 6.9. *For all F, G in $\mathbf{DM}_{\mathrm{nis}}^{\mathrm{eff}}(k)$:*

$$\mathrm{Hom}_{\mathbf{DM}(S_\bullet)}(c^*F, c^*G) \cong \mathrm{Hom}_{\mathbf{DM}(k)}(R_{\mathrm{tr}}(S_\bullet) \otimes_{\mathbb{L}}^{\mathrm{tr}} F, G).$$

Proof. Both sides are $\mathrm{Hom}_{\mathbf{DM}(k)}(\mathbf{L}c_\#(c^*F), G)$, by adjunction and 6.8. $\qquad \square$

For each p and $q \geq 0$, we write $\mathbf{R}(q)[p]$ for $c^*(R(q)[p])$ and call them the *Tate objects*. Their role in motivic cohomology is illustrated by our next proposition.

Proposition 6.10. *If $S_\bullet$ is a smooth simplicial scheme over k, then for all p, q:*

$$\mathrm{Hom}_{\mathbf{DM}(S_\bullet)}(\mathbf{R}, \mathbf{R}(q)[p]) \cong H^{p,q}(S_\bullet, R).$$

Proof. Take $F = R$ and $G = R(q)[p]$ in Corollary 6.9 and apply $H^{p,q}(S_\bullet, R) = \mathrm{Hom}_{\mathbf{DM}(k)}(R_{\mathrm{tr}}(S_\bullet), R(q)[p])$. $\qquad \square$

Definition 6.11. Let $\mathbf{DM}_{\mathrm{gm}}^{\mathrm{eff}} = \mathbf{DM}_{\mathrm{gm}}^{\mathrm{eff}}(S_\bullet)$ denote the triangulated subcategory of $\mathbf{DM}_{\mathrm{nis}}^{\mathrm{eff}}(S_\bullet)$ generated by the objects c^*M with M in $\mathbf{DM}_{\mathrm{gm}}^{\mathrm{eff}}(k)$. The formula (6.7) shows that $\mathbf{DM}_{\mathrm{gm}}^{\mathrm{eff}}$ is a tensor subcategory of $\mathbf{DM}_{\mathrm{nis}}^{\mathrm{eff}}(S_\bullet)$.

The goal of the rest of this section is to provide the tensor category $\mathbf{DM}_{\mathrm{gm}}^{\mathrm{eff}}$ with an internal Hom, i.e., a right adjoint <u>Hom</u> to the tensor which is bi-distributive for $\otimes_{\mathbb{L}}^{\mathrm{tr}}$ (Lemma 6.13), and then extend this to a rigid tensor category $\mathbf{DM}_{\mathrm{gm}}(S_\bullet)$, i.e., one with an internal Hom, equipped with a dual such that $A \to (A^*)^*$ is an isomorphism for every A (Theorem 6.14).

For K in $\mathbf{DM}_{\mathrm{gm}}^{\mathrm{eff}}$, write $K(q)$ for $K \otimes_{\mathbb{L}}^{\mathrm{tr}} \mathbf{R}(q)$.

Theorem 6.12 (Cancellation). *For all K and L in $\mathbf{DM}_{\mathrm{gm}}^{\mathrm{eff}}(S_\bullet)$, the canonical map is an isomorphism:*

$$\mathrm{Hom}_{\mathbf{DM}(S_\bullet)}(K, L) \xrightarrow{\simeq} \mathrm{Hom}_{\mathbf{DM}(S_\bullet)}(K(1), L(1)).$$

Proof. It suffices to consider $K = c^*M$, $L = c^*N$, as these generate $\mathbf{DM}_{\mathrm{gm}}^{\mathrm{eff}}$. By Corollary 6.9, the map in question is identified with

$$\mathrm{Hom}_{\mathbf{DM}(k)}(R_{\mathrm{tr}}(S_\bullet) \otimes M, N) \longrightarrow \mathrm{Hom}_{\mathbf{DM}(k)}(R_{\mathrm{tr}}(S_\bullet) \otimes M(1), N(1)),$$

which is an isomorphism by [Voe10a]. $\qquad \square$

We may form the category $\mathbf{DM}_{\mathrm{gm}}(S_{\bullet})$ by formally inverting the Tate twist functor $K \mapsto K(1)$ in $\mathbf{DM}_{\mathrm{gm}}^{\mathrm{eff}}$; every object K has the form $M(-r)$ for an effective M. By Cancellation 6.12, $\mathbf{DM}_{\mathrm{gm}}^{\mathrm{eff}}$ embeds as a full tensor subcategory of $\mathbf{DM}_{\mathrm{gm}}(S_{\bullet})$, and $\mathrm{Hom}_{\mathbf{DM}(S_{\bullet})}(K, L) \xrightarrow{\sim} \mathrm{Hom}_{\mathbf{DM}(S_{\bullet})}(K(r), L(r))$ for all r.

Recall from [MVW, 14.12] that if M is an effective geometric motive over k then $-\otimes_{\mathbb{L}}^{\mathrm{tr}} M$ has a right adjoint $\underline{RHom}_k(M, -)$ on $\mathbf{DM}_{\mathrm{nis}}^{\mathrm{eff}}(k)$. If N is another effective geometric motive then $\underline{RHom}_k(M, N)$ is also effective geometric.

Lemma 6.13. *For any objects $c^* M$ in $\mathbf{DM}_{\mathrm{gm}}^{\mathrm{eff}}(S_{\bullet})$ and $c^* N$ in $\mathbf{DM}_{\mathrm{nis}}^{\mathrm{eff}}(S_{\bullet})$, $\underline{\mathrm{Hom}}(M, N) = c^* \underline{RHom}_k(M, N)$ is an internal Hom object in $\mathbf{DM}_{\mathrm{nis}}^{\mathrm{eff}}(S_{\bullet})$ from $c^* M$ to $c^* N$, and is independent (up to canonical isomorphism) of the choice of M and N.*

It follows that $\mathbf{DM}_{\mathrm{gm}}^{\mathrm{eff}}(S_{\bullet})$ has an internal Hom functor, and that the object $\underline{RHom}_{S_{\bullet}}(\mathbf{R}(n), K)$ exists for all K in $\mathbf{DM}_{\mathrm{gm}}^{\mathrm{eff}}(S_{\bullet})$.

Proof. Writing $\otimes$ for $\otimes_{\mathbb{L}}^{\mathrm{tr}}$, Corollary 6.9 yields

$$\mathrm{Hom}_{\mathbf{DM}(S_{\bullet})}(c^* L \otimes c^* M, c^* N) \cong \mathrm{Hom}_{\mathbf{DM}(k)}(R_{\mathrm{tr}}(S_{\bullet}) \otimes L \otimes M, N)$$

$$\cong \mathrm{Hom}_{\mathbf{DM}(k)}(R_{\mathrm{tr}}(S_{\bullet}) \otimes L, \underline{RHom}_k(M, N))$$

$$\cong \mathrm{Hom}_{\mathbf{DM}(S_{\bullet})}(c^* L, c^* \underline{RHom}_k(M, N)).$$

This shows that $c^* \underline{RHom}_k(M, N)$ is an internal Hom object; the canonical map $c^* \underline{RHom}_k(M, N) \otimes c^* M \xrightarrow{e} c^* N$ is obtained by taking $L = \underline{RHom}_k(M, N)$.

If $c^* M \cong c^* M'$ and $c^* N \cong c^* N'$ then taking $L = \underline{RHom}(M', N')$ yields a canonical map $c^* \underline{RHom}(M', N') \to c^* \underline{RHom}(M, N)$; the usual argument shows that it is an isomorphism. Thus $-\otimes c^* M$ has a right adjoint $c^* N \mapsto c^* \underline{RHom}_k (M, N)$ in $\mathbf{DM}_{\mathrm{nis}}^{\mathrm{eff}}(S_{\bullet})$. $\square$

Recall [MVW, 20.4] that the dual $M^* = \underline{RHom}_k(M, R)$ of a geometric motive over k is again a geometric motive and that $(M^*)^* \cong M$. We may now borrow the construction in [MVW, 20.6]: any geometric motive over $S_{\bullet}$ has the form $K = M(-r)$ for an effective geometric motive M and we want $M(-r)^*$ to be $M^*(r)$. Thus we define the dual to be $\underline{RHom}_k(K(r), \mathbf{R}(i))(r-i)$ for large i, an object which is independent of i and r by Cancellation 6.12. Since

$$(M(-r)^*)^* \cong M^*(r)^* \cong M^{**}(-r) \cong M(-r),$$

this shows that the dual satisfies $K \cong (K^*)^*$.

We may now copy the rest of the development in [MVW, §20] to prove that M^* is a geometric motive over $S_{\bullet}$, $M \cong M^{**}$ and $\underline{\mathrm{Hom}}(M, N) \cong M^* \otimes N$. This proves:

Theorem 6.14. *The tensor category* $\mathbf{DM}_{\mathrm{gm}}(S_\bullet)$ *is rigid, with internal Hom* $\underline{\mathrm{Hom}}(K, L) \cong K^* \otimes L$.

6.4   THE SLICE FILTRATION

In this section we fix a smooth $S_\bullet$ and write $\mathbf{DM}_{\mathrm{gm}}^{\mathrm{eff}}$ for $\mathbf{DM}_{\mathrm{gm}}^{\mathrm{eff}}(S_\bullet)$. For $n \geq 0$, let $\mathbf{DM}_{\mathrm{gm}}(n)$ denote the full subcategory of $\mathbf{DM}_{\mathrm{gm}}^{\mathrm{eff}}$ on the objects of the form $M(n)$ with M in $\mathbf{DM}_{\mathrm{gm}}^{\mathrm{eff}}$. The *slice filtration* of $\mathbf{DM}_{\mathrm{gm}}^{\mathrm{eff}}$ is the sequence of subcategories

$$\cdots \subset \mathbf{DM}_{\mathrm{gm}}(n+1) \subset \mathbf{DM}_{\mathrm{gm}}(n) \subset \cdots \subset \mathbf{DM}_{\mathrm{gm}}(0) = \mathbf{DM}_{\mathrm{gm}}^{\mathrm{eff}}.$$

Lemma 6.15. $\mathbf{DM}_{\mathrm{gm}}(n)$ *is a coreflective subcategory of* $\mathbf{DM}_{\mathrm{gm}}^{\mathrm{eff}}$; *the functor*

$$s_{\geq n}M = \underline{RHom}(\mathbf{R}(n), M)(n)$$

is right adjoint for the inclusion $\mathbf{DM}_{\mathrm{gm}}(n) \subset \mathbf{DM}_{\mathrm{gm}}^{\mathrm{eff}}$.

Proof. It is clear that $s_{\geq n}$ is a triangulated functor from $\mathbf{DM}_{\mathrm{gm}}^{\mathrm{eff}}$ to $\mathbf{DM}_{\mathrm{gm}}(n)$. For M, N in $\mathbf{DM}_{\mathrm{gm}}^{\mathrm{eff}}$,

$$\begin{aligned}
\mathrm{Hom}(M(n), N) &= \mathrm{Hom}(M \otimes^{\mathrm{tr}} \mathbf{R}(n), N) \\
&\cong \mathrm{Hom}(M, \underline{RHom}(\mathbf{R}(n), N)) \\
&\xrightarrow{\cong} \mathrm{Hom}(M(n), \underline{RHom}(\mathbf{R}(n), N)(n)) \\
&= \mathrm{Hom}(M(n), s_{\geq n}N).
\end{aligned}$$

Thus $s_{\geq n}$ is right adjoint to the inclusion. By the Cancellation Theorem 6.12, $s_{\geq n}(M(n)) \xrightarrow{\cong} \underline{RHom}(R, M)(n) \cong M(n)$ for all M, so $s_{\geq n}$ is a coreflection functor. $\square$

Definition 6.16. We let $s_{<n}M$ denote the mapping cone of the adjunction counit $\varepsilon : s_{\geq n}M \to M$. It is unique up to canonical isomorphism, and fits into a distinguished triangle:

$$s_{\geq n}M \xrightarrow{\varepsilon} M \to s_{<n}M \to .$$

We let s_nM denote the mapping cone for the adjunction $\varepsilon : s_{\geq n+1}M \to s_{\geq n}M$ associated to $\mathbf{DM}_{\mathrm{gm}}(n+1) \subset \mathbf{DM}_{\mathrm{gm}}(n)$. It is also unique up to canonical isomorphism, and fits into a distinguished triangle:

$$s_{\geq n+1}M \xrightarrow{\varepsilon} s_{\geq n}M \to s_nM \to .$$

The s_n are called the *slice functors* over $S_\bullet$.

Remark 6.16.1. The slice filtration extends to an (exhaustive) slice filtration on $\mathbf{DM}_{\mathrm{gm}}(S_{\bullet})$. The existence of this extension is straightforward from the observation that, by Lemma 6.15, there is a well-defined coreflection $s_{\geq 0} : \mathbf{DM}_{\mathrm{gm}}(S_{\bullet}) \to \mathbf{DM}_{\mathrm{gm}}^{\mathrm{eff}}(S_{\bullet})$.

We write $DT_{\geq 0} = DT_{\geq 0}(S_{\bullet})$ for the thick subcategory of $\mathbf{DM}_{\mathrm{nis}}^{\mathrm{eff}}(S_{\bullet})$ generated by the Tate objects $\mathbf{R}(q)[p]$ with $q \geq 0$.

Lemma 6.17. *The slice functor is conservative on $DT_{\geq 0}$. That is, an object A is zero if and only if every $s_n(A)$ is zero.*

Proof. ([Voe10b, 5.14]) If $s_n(A) = 0$ then $s_{\geq n+1}(A) \cong s_{\geq n}(A)$. Since $s_{\geq 0}(A) = A$, we see by induction on $n \geq 0$ that if every $s_i(A) = 0$ then $s_{\geq n}(A) \cong A$ for all n. On the other hand, an induction on the number of triangles needed to define A in terms of Tate objects shows that some $s_{\geq n}(A)$ is zero. The lemma follows. $\square$

For all $i, j \geq 0$ the tensor product on $\mathbf{DM}_{\mathrm{gm}}^{\mathrm{eff}}$ restricts to a natural transformation $\eta_{i,j} : s_{\geq i}(M) \otimes s_{\geq j}(N) \to s_{\geq i+j}(M \otimes N)$, defined as the right adjoint of the map $s_{\geq i}(M) \otimes s_{\geq j}(N) \to (M \otimes N)$. As these are compatible, they induce natural transformations $\theta_{i,j} : s_i(M) \otimes s_j(N) \to s_{i+j}(M \otimes N)$.

Lemma 6.18. *The slice functor is multiplicative on $DT_{\geq 0}$ in the sense that the $\theta_{i,j}$ induce an isomorphism*

$$s_n(M \otimes N) = \bigoplus_{p+q=n} s_p(M) \otimes s_q(N).$$

Proof. If $M = \mathbf{R}(q)[p]$ and $N = \mathbf{R}(q')[p']$, $\theta_{q,q'}$ is the canonical isomorphism $\mathbf{R}(q)[p] \otimes \mathbf{R}(q')[p'] \cong \mathbf{R}(q+q')[p+p']$, and all other $\theta_{i,j}$ are the isomorphism $0 \cong 0$. By the 5-lemma, the subcategory of (M, N) in $DT_{\geq 0} \times DT_{\geq 0}$ for which Lemma 6.18 holds is closed under extensions. Since it contains the generators, and is closed under shifts and summands, it must be all of $DT_{\geq 0} \times DT_{\geq 0}$. $\square$

Lemma 6.19. *Let $A \xrightarrow{a} B \xrightarrow{b} C$ be a sequence in $DT_{\geq 0}$, and $n \geq 0$ such that $A \in DT_{\geq n}$ and $C \in DT_{<n}$. If $s_i(a)$ is an isomorphism for all $i \geq n$ and $s_i(b)$ is an isomorphism for all $i < n$, then there is a unique morphism $c : C \to A[1]$ such that $A \xrightarrow{a} B \xrightarrow{b} C \xrightarrow{c} A[1]$ is a distinguished triangle, which identifies A and C with $s_{\geq n}B$ and $s_{<n}B$, respectively.*

The proof that we gave for Lemma 5.8 goes through verbatim to prove 6.19.

Proof. Choose a distinguished triangle $A \xrightarrow{a} B \to C' \to A[1]$. The composition ba is zero for weight reasons. Hence b factors through a morphism $\phi : C' \to C$. The hypotheses imply that $s_i(\phi)$ is an isomorphism for all i; as the slice functor is conservative (Lemma 6.17), this implies that ϕ is an isomorphism. Letting c be the composite of ϕ^{-1} and $C' \to A[1]$ yields the triangle (a, b, c).

If (a, b, c') is a second triangle on (A, B, C) then there is an $f : C \to C$ so that $(1, 1, f)$ is a morphism from the first to the second triangle. Since $s_{<n}(b)$ is an isomorphism, it follows that $f = s_{<n}(f)$ is the identity and hence that $c = c'$. The final assertion follows since $s_{\geq n} C = 0$ and $s_{<n} A = 0$ by 6.17. $\qquad\square$

6.5 EMBEDDED SCHEMES

In section 4.2 we considered the simplicial scheme $\mathfrak{X}$ of Definition 1.32 associated to a smooth scheme X over k, and the subcategory $\mathbf{DM}^{\varepsilon}_{\mathrm{gm}}$ of $\mathbf{DM}^{\mathrm{eff}}_{\mathrm{nis}}(k)$ generated by the objects $R_{\mathrm{tr}}(\mathfrak{X}) \otimes M$ for M in $\mathbf{DM}_{\mathrm{gm}}(k)$. In this section we generalize from $\mathfrak{X}$ to embedded schemes; see Proposition 6.23. Our exposition is based on section 6 of [Voe10b].

Definition 6.20. ([Voe10b, 6.1]) A smooth simplicial scheme $S_{\bullet}$ over k is called an *embedded scheme* if the natural map $c \otimes 1 : R_{\mathrm{tr}}(S_{\bullet}) \otimes R_{\mathrm{tr}}(S_{\bullet}) \xrightarrow{\simeq} R_{\mathrm{tr}}(S_{\bullet})$ is an isomorphism. The most important example of an embedded simplicial scheme is the simplicial scheme $\mathfrak{X}$ of 1.32 associated to a smooth scheme X over k.

Remark 6.20.1. The term "embedded" is suggested by the following observation. Let us say that a morphism of schemes $\pi : S \to T$ is an *embedding* if the projections $S \times_T S \to S$ are isomorphisms, or equivalently, if it is topologically an injection, and $\mathcal{O}_{T, \pi(s)} \to \mathcal{O}_{S, s}$ is a surjection for every $s \in S$. Then S (considered as a constant simplicial scheme) is an embedded scheme over T.

In Proposition 6.23 we will see that $\mathbf{DM}^{\varepsilon}_{\mathrm{gm}}$ is equivalent to the full subcategory $\mathbf{DM}^{\mathrm{eff}}_{\mathrm{gm}}(\mathfrak{X})$ of $\mathbf{DM}(\mathfrak{X})$ introduced in section 6.3 above.

Recall from section 6.3 that the structure map $c : S_{\bullet} \to \mathrm{Spec}(k)$ induces an adjunction $(\mathbf{L}c_{\#}, c^*)$ between $\mathbf{DM}^{\mathrm{eff}}_{\mathrm{nis}}(S_{\bullet})$ and $\mathbf{DM}^{\mathrm{eff}}_{\mathrm{nis}}(k)$. The unit and counit of the adjunction are the natural transformations $\eta_N : N \to c^* \mathbf{L}c_{\#} N$ in $\mathbf{DM}^{\mathrm{eff}}_{\mathrm{nis}}(S_{\bullet})$ and $\varepsilon_M : \mathbf{L}c_{\#} c^* M \to M$ in $\mathbf{DM}^{\mathrm{eff}}_{\mathrm{nis}}(k)$.

Lemma 6.21. *The natural transformation $c^* \varepsilon_M : c^* \mathbf{L}c_{\#}(c^* M) \to c^* M$ is an isomorphism when $S_{\bullet}$ is an embedded scheme.*

Proof. As with any adjunction, the composite $(c^* \varepsilon) \eta_{c^* M}$ is the identity on $c^* M$, so it is sufficient to show that the other composite $\eta_{c^* M} \circ c^* \varepsilon_M$ is the identity map on $c^* \mathbf{L}c_{\#}(c^* M)$. The latter two maps are adjoint to the maps

$$\mathbf{L}c_{\#} c^* \varepsilon_M : (\mathbf{L}c_{\#} c^*)^2 \to \mathbf{L}c_{\#} c^*, \quad \text{and} \quad \varepsilon_{\mathbf{L}c_{\#} c^*} : (\mathbf{L}c_{\#} c^*)^2 \to \mathbf{L}c_{\#} c^*,$$

respectively. Using the identification $\mathbf{L}c_{\#}(c^* M) \cong R_{\mathrm{tr}}(S_{\bullet}) \otimes^{\mathrm{tr}}_{\mathbb{L}} M$ of Lemma 6.8, these two maps may be identified with the two projections

$$M \otimes R_{\mathrm{tr}}(S_{\bullet}) \otimes R_{\mathrm{tr}}(S_{\bullet}) \cong M \otimes R_{\mathrm{tr}}(S_{\bullet} \times S_{\bullet}) \rightrightarrows M \otimes R_{\mathrm{tr}}(S_{\bullet}).$$

Because $S_{\bullet}$ is embedded, these two projections are isomorphisms and have the diagonal as their common inverse, so they must be equal. Since $\mathbf{L}c_{\#}c^{*}\varepsilon_{M} = \varepsilon_{\mathbf{L}c_{\#}c^{*}}$, their adjoints are equal: $1 = \eta_{c^{*}M} \circ c^{*}\varepsilon_{M}$. Hence $c^{*}\varepsilon_{M}$ is an isomorphism, as required. $\qquad\square$

Corollary 6.22. *For any objects* M, N *of* $\mathbf{DM}_{\mathrm{nis}}^{\mathrm{eff}}(k)$, *there is a natural bijection*

$$\mathrm{Hom}_{k}(\mathbf{L}c_{\#}c^{*}M, \mathbf{L}c_{\#}c^{*}N) \cong \mathrm{Hom}_{S_{\bullet}}(c^{*}M, c^{*}\mathbf{L}c_{\#}c^{*}N) \xrightarrow{c^{*}\varepsilon} \mathrm{Hom}_{S_{\bullet}}(c^{*}M, c^{*}N).$$

Proof. The first map is the adjunction isomorphism for the adjoint pair $(\mathbf{L}c_{\#}, c^{*})$, and the second map is an isomorphism by Lemma 6.21. $\qquad\square$

Let $\mathbf{DM}_{S_{\bullet}}^{\mathrm{eff}}$ denote the full triangulated subcategory of $\mathbf{DM}_{\mathrm{nis}}^{\mathrm{eff}}(S_{\bullet})$ generated by objects of the form $c^{*}(M)$ for M in $\mathbf{DM}_{\mathrm{nis}}^{\mathrm{eff}}(k)$. It contains the subcategory $\mathbf{DM}_{\mathrm{gm}}^{\mathrm{eff}}(S_{\bullet})$, defined in 6.11, which is generated by objects of the form $c^{*}(M)$ for M in $\mathbf{DM}_{\mathrm{gm}}^{\mathrm{eff}}(k)$.

Proposition 6.23. *The functor* $\mathbf{L}c_{\#} : \mathbf{DM}_{\mathrm{nis}}^{\mathrm{eff}}(S_{\bullet}) \to \mathbf{DM}_{\mathrm{nis}}^{\mathrm{eff}}(k)$ *induces an equivalence between* $\mathbf{DM}_{S_{\bullet}}^{\mathrm{eff}}$ *and the full subcategory of* $\mathbf{DM}_{\mathrm{nis}}^{\mathrm{eff}}(k)$ *consisting of objects* M *such that* $c \otimes M : R_{\mathrm{tr}}(S_{\bullet}) \otimes M \xrightarrow{\simeq} M$ *is an isomorphism.*

The subcategory $\mathbf{DM}_{\mathrm{gm}}^{\mathrm{eff}}(S_{\bullet})$ *is equivalent to the full subcategory of* $\mathbf{DM}_{\mathrm{nis}}^{\mathrm{eff}}(k)$ *consisting of objects* $R_{\mathrm{tr}}(S_{\bullet}) \otimes M$ *with* M *in* $\mathbf{DM}_{\mathrm{gm}}^{\mathrm{eff}}(k)$. *In particular,* $\mathbf{DM}_{\mathrm{gm}}^{\mathrm{eff}}(\mathfrak{X})$ *is equivalent to* $\mathbf{DM}_{\mathrm{gm}}^{\varepsilon}$.

Proof. Corollary 6.22 implies that $\mathbf{L}c_{\#}$ induces an isomorphism on Hom-sets, so it is a full embedding on $\mathbf{DM}_{S_{\bullet}}^{\mathrm{eff}}$. By Lemma 6.8, the image consists of objects $\mathbf{L}c_{\#}(c^{*}M) \cong R_{\mathrm{tr}}(S_{\bullet}) \otimes M$. Since $S_{\bullet}$ is an embedded scheme we have $R_{\mathrm{tr}}(S_{\bullet}) \otimes M \cong R_{\mathrm{tr}}(S_{\bullet}) \otimes R_{\mathrm{tr}}(S_{\bullet}) \otimes M$, so $N \cong \mathbf{L}c_{\#}(c^{*}M)$ if and only if $R_{\mathrm{tr}}(S_{\bullet}) \otimes N \xrightarrow{\simeq} N$ is an isomorphism. $\qquad\square$

The equivalence of Proposition 6.23 also preserves tensor products.

Lemma 6.24. *If* $S_{\bullet}$ *is embedded, the functor* $\mathbf{L}c_{\#}$ *preserves tensor products in the sense that there is a canonical isomorphism for* M, N *in* $\mathbf{DM}_{S_{\bullet}}^{\mathrm{eff}}$.

$$\mathbf{L}c_{\#}(M \otimes N) \xrightarrow{\simeq} \mathbf{L}c_{\#}(M) \otimes \mathbf{L}c_{\#}(N).$$

Proof. Since c^{*} commutes with tensor products by (6.7), we have a canonical map

$$M \otimes N \xrightarrow{\eta \otimes \eta} c^{*}\mathbf{L}c_{\#}(M) \otimes c^{*}\mathbf{L}c_{\#}(N) \cong c^{*}\left(\mathbf{L}c_{\#}(M) \otimes \mathbf{L}c_{\#}(N)\right).$$

The adjoint map is a map $\mathbf{L}c_\#(M \otimes N) \longrightarrow \mathbf{L}c_\#(M) \otimes \mathbf{L}c_\#(N)$ defined for all M, N in $\mathbf{DM}_{\mathrm{nis}}^{\mathrm{eff}}(S_\bullet)$.

Let $\mathcal{E}$ denote the subcategory of $\mathbf{DM}_{S_\bullet} \times \mathbf{DM}_{S_\bullet}$ consisting of pairs (M, N) for which this adjoint map is an isomorphism. Since the map is bi-triangulated, $\mathcal{E}$ is a localizing subcategory. Therefore it suffices to show that $\mathcal{E}$ contains pairs $(M, N) = (c^* R_{\mathrm{tr}}(X), c^* R_{\mathrm{tr}}(Y))$ for X and Y smooth over k. Since $\mathbf{L}c_\# M \cong R_{\mathrm{tr}}(S_\bullet) \otimes R_{\mathrm{tr}}(X)$ and $\mathbf{L}c_\# N \cong R_{\mathrm{tr}}(S_\bullet) \otimes R_{\mathrm{tr}}(Y)$ by Lemma 6.8, the map of the lemma coincides with the map

$$\mathbf{L}c_\#(M \otimes N) \cong R_{\mathrm{tr}}(S_\bullet) \otimes R_{\mathrm{tr}}(X) \otimes R_{\mathrm{tr}}(Y)$$
$$\rightarrow R_{\mathrm{tr}}(S_\bullet) \otimes R_{\mathrm{tr}}(S_\bullet) \otimes R_{\mathrm{tr}}(X) \otimes R_{\mathrm{tr}}(Y)$$
$$\cong R_{\mathrm{tr}}(S_\bullet) \otimes R_{\mathrm{tr}}(X) \otimes R_{\mathrm{tr}}(S_\bullet) \otimes R_{\mathrm{tr}}(Y) = \mathbf{L}c_\#(M) \otimes \mathbf{L}c_\#(N)$$

induced by the diagonal $S_\bullet \to S_\bullet \times S_\bullet$. Since $S_\bullet$ is embedded, this map is an isomorphism, as required. $\qquad\qquad\qquad\qquad\qquad\qquad\qquad\qquad\qquad\qquad\square$

6.6   THE OPERATIONS ϕ_i

We are now able to construct the cohomology operations $\phi_i = \phi_i^{p,q}$. We first define $\phi_i(z)$ for any $z \in H^{2p+1,q}(S_\bullet, R)$, where $S_\bullet$ is any smooth simplicial scheme over k, R contains $1/(\ell - 1)!$, and $p, q \geq 0$ are arbitrary.

It is convenient to work in $\mathbf{DM}_{\mathrm{gm}}(S_\bullet)$ and set $T = \mathbf{R}(q)[2p]$. By Proposition 6.10 we may interpret z as a morphism $\mathbf{R} \to \mathbf{R}(q)[2p + 1] = T[1]$ in $\mathbf{DM}_{\mathrm{gm}}(S_\bullet)$. Define A, x and y in $\mathbf{DM}_{\mathrm{gm}}(S_\bullet)$ by the triangle

$$T \xrightarrow{\ x\ } A \xrightarrow{\ y\ } \mathbf{R} \xrightarrow{\ z\ } T[1]. \tag{6.25}$$

Recall that the motive $S^i(A)$ is defined for $i < \ell$ as the symmetric part $e_i(A^{\otimes i})$ of the $R[\Sigma_i]$-module $A^{\otimes i}$, where $e_i = \sum_{\sigma \in \Sigma_i} \sigma/i!$. Since $\Sigma_{i-1} \subset \Sigma_i$, we also have $e_{i-1} \in R[\Sigma_i]$ and $S^{i-1}(A) \otimes A = e_{i-1}(A^{\otimes i})$. Because $e_i e_{i-1} = e_i$, we get a corestriction map cores: $S^{i-1}(A) \otimes A \to S^i(A)$. There is also a transfer map tr: $S^i(A) \to S^{i-1}(A) \otimes A$, induced by the endomorphism

$$a_1 \otimes \cdots \otimes a_i \longmapsto \Sigma(\cdots \otimes \hat{a}_j \otimes \cdots) \otimes a_j$$

of $A^{\otimes i}$. Composing tr with $1 \otimes y$ yields a map $u\colon S^i(A) \to S^{i-1}(A)$; composing $1 \otimes x$ with corestriction yields a map $v\colon S^{i-1}(A) \otimes T \to S^i(A)$.

The following result generalizes Proposition 5.7 from $\mathfrak{X}$ to $S_\bullet$, and has almost the same proof. It is taken from [Voe11, 3.1].

Proposition 6.26. *If $0 < i < \ell$ and $1/(\ell - 1)! \in R$, then there exist unique morphisms r and s so that $(S^i x, u, r)$ and $(v, S^i y, s)$ are distinguished triangles:*

(a) $T^i \xrightarrow{S^i x} S^i(A) \xrightarrow{u} S^{i-1}(A) \xrightarrow{r} T^i[1]$.

(b) $S^{i-1}(A) \otimes T \xrightarrow{v} S^i(A) \xrightarrow{S^i y} \mathbf{R} \xrightarrow{s} S^{i-1}(A) \otimes T[1]$.

Moreover, u identifies T^i and $S^{i-1}(A)$ with $s_{qi}(A)$ and $s_{<qi}(S^i A)$, respectively, and v identifies $S^{i-1}(A) \otimes T$ and $\mathbf{R}$ with $s_{>0}(S^i A)$ and $s_0(S^i A)$, respectively.

Proof. By multiplicativity (Lemma 6.18), the slice $s_n(A^{\otimes i})$ of $A^{\otimes i}$ is the sum of the $s_{n_1}(A) \otimes \cdots \otimes s_{n_i}(A)$ with $\sum n_j = n$. Since the symmetric group acts on everything here, the slice $s_n(S^i A)$ is the symmetric part of this, viz., $\mathbf{R}(qj)[2pj]$ if $n = qj$ for $0 \le j \le i$ and zero otherwise.

By inspection, $S^i x$ induces an isomorphism $T^i \cong s_{qi}(S^i A)$, and each morphism $s_{qj}(u)$ is multiplication by $i - j$, which is an isomorphism for $0 \le j < i$. The existence and uniqueness of r now follows from Lemma 6.19.

Similarly, y is the top slice transformation, as $s_0(A) \cong \mathbf{R}$. Thus $S^i y$ induces $s_0(S^i A) \cong \mathbf{R}$, and $s_{qj}(v)$ is an isomorphism for $j > 0$ because by multiplicativity (6.18):

$$s_{qj}\left[S^{i-1}(A) \otimes T\right] = s_{q(j-1)}(S^{i-1} A) \otimes T.$$

The existence and uniqueness of s now follows from Lemma 6.19. □

Definition 6.27. Let $S_\bullet$ be a smooth simplicial scheme. The functions

$$\phi_i : H^{2p+1,q}(S_\bullet, R) \to H^{2p(i+1)+2, q(i+1)}(S_\bullet, R)$$

are defined for each $i = 1, \ldots, \ell - 1$ as follows. Given z, the composition of the maps s and $r \otimes 1$ of Proposition 6.26 yields a map

$$\phi_i(z) : \mathbf{R} \xrightarrow{s} S^{i-1}(A) \otimes T[1] \xrightarrow{r \otimes 1} T^i[1] \otimes T[1] \cong T^{i+1}[2]$$
$$= \mathbf{R}(q(i+1))[2p(i+1) + 2],$$

i.e., an element of $H^{2p(i+1)+2, q(i+1)}(S_\bullet, R)$.

Remark 6.27.1. When $S_\bullet$ is an embedded scheme, the above construction of $\phi_i(z)$ can be performed in the subcategory $\mathbf{DM}^{\text{eff}}_{S_\bullet}$ of $\mathbf{DM}^{\text{eff}}_{\text{nis}}(k)$, by Proposition 6.23 and Lemma 6.24. This was carried out in section 5.2 for the simplicial Čech variety $\mathfrak{X}$ associated to a smooth X in Definition 1.32. (Compare Propositions 5.7 and 6.26.) Thus Definitions 5.10 and 6.27 of $\phi^V(z) = \phi_{\ell-1}$ agree for z in $H^{2p+1,q}(\mathfrak{X}, R)$.

Proposition 6.28. *The functions* $H^{2p+1,q}(S_\bullet, R) \xrightarrow{\phi_i} H^{2p(i+1)+2, q(i+1)}(S_\bullet, R)$ *assemble to form a cohomology operation* $\phi_i = \phi_i^{p,q}$ *for each* p, q *and* i, $0 < i < \ell$.

Proof. Given a map $f : S'_\bullet \to S_\bullet$ and an element z, triangle (6.25) induces a triangle $f^*(T) \to f^*(A) \to f^*(\mathbf{R}) \xrightarrow{f^* z} f^*(T)[1]$. By functoriality in $S_\bullet$, f^* maps

R and T for $S_\bullet$ to **R** and T for $S'_\bullet$. Thus the construction of (6.25) and A is natural in z. Since f^* is strongly monoidal, we see that u, v, r and s are natural, and we also have $f^*(\phi_i(z)) = \phi_i(f^*(z))$. $\qquad\qquad\qquad\qquad\qquad\qquad\qquad\square$

Example 6.28.1. When $i = 1$, it is immediate from the definition that $r = s = z$ and hence $\phi_1(z) = z^2$. This is zero unless $\ell = 2$, because z has odd degree.

Another case when it is easy to describe ϕ_i is when $p = q = 0$ and $R = \mathbb{Z}/\ell$. In this case each $\phi_i = \phi_i^{0,0}$ goes from $H^{1,0}(-, \mathbb{Z}/\ell)$ to $H^{2,0}(-, \mathbb{Z}/\ell)$ and we have:

Proposition 6.29. *The operations* $\phi_i : H^{1,0}(-, \mathbb{Z}/\ell) \to H^{2,0}(-, \mathbb{Z}/\ell)$ *are zero for* $i < \ell - 1$, *and the Bockstein* β *for* $i = \ell - 1$.

To set up the proof, let G denote the group $\mathbb{Z}/\ell$, and let R denote the ring $\mathbb{Z}/\ell$. Recall that in topology the classical Eilenberg–Mac Lane space $BG = K(G,1)$ represents cohomology in the sense that $H^1_{\text{top}}(X, R) = [X, BG]$. There is a canonical element $\alpha_1 \in H^1_{\text{top}}(BG, R)$, corresponding to the identity of BG, and the Yoneda Lemma yields a 1–1 correspondence between cohomology operations $\psi : H^1_{\text{top}}(-, R) \to H^p_{\text{top}}(-, R)$ and elements of $H^p_{\text{top}}(BG, R)$, given by $\psi \mapsto \psi(\alpha_1)$.

In motivic cohomology, we have the simplicial classifying space $B_\bullet(G)$, which is the simplicial set BG regarded as a simplicial scheme; it is a disjoint union of ℓ^{i+1} copies of $\text{Spec}(k)$ in simplicial degree i, and the simplicial structure comes from the group structure of G. Then $H^{1,0}(-, \mathbb{Z}/\ell)$ is represented by $B_\bullet(G)$ in the motivic homotopy category; see [MV99, p. 130]. Again by the Yoneda Lemma, there is a 1–1 correspondence between motivic cohomology operations $\psi : H^{1,0}(-, \mathbb{Z}/\ell) \to H^{p,q}(-, \mathbb{Z}/\ell)$ and elements of $H^{p,q}(B_\bullet G, \mathbb{Z}/\ell)$, given by $\psi \mapsto \psi(\alpha_1)$, where α_1 is the canonical element of $H^{1,0}(B_\bullet G, \mathbb{Z}/\ell)$.

Proof. To compare $\phi_{\ell-1}(\alpha_1)$ and $\beta(\alpha_1)$, we use Yoneda extensions and the identification in Lemma 13.3 of $H^{p,0}(B_\bullet G, \mathbb{Z}/\ell)$ with

$$H^p_{\text{top}}(BG, \mathbb{Z}/\ell) = \text{Ext}^p_{\mathbb{Z}[G]}(\mathbb{Z}, \mathbb{Z}/\ell) \cong \text{Ext}^p_{\mathbb{Z}/\ell[G]}(\mathbb{Z}/\ell, \mathbb{Z}/\ell).$$

Recall from section 3.2 that the regular representation has an $\mathbb{Z}/\ell[G]$-module filtration $\mathbb{Z}/\ell = F_1 \subset F_2 \subset \cdots F_\ell = \mathbb{Z}/\ell[G]$, where $\dim_{\mathbb{Z}/\ell} F_i = i$. The canonical element α_1 corresponds to the extension $0 \to \mathbb{Z}/\ell \to F_2 \to \mathbb{Z}/\ell \to 0$.

To compute $\phi_i(\alpha_1)$, we represent α_1 by the distinguished triangle

$$\mathbb{Z}/\ell \to F_2 \to \mathbb{Z}/\ell \xrightarrow{\alpha_0} \mathbb{Z}/\ell[1]$$

in $\mathbf{D}(\mathbb{Z}/\ell[G])$. The symmetric power $S^i(F_2)$ is isomorphic to F_{i+1}, and $S^{\ell-1}(F_2)$ is the regular representation $\mathbb{Z}/\ell[G]$ of G over $\mathbb{Z}/\ell$. Hence the triangles (6.26) (a,b) are

$$\mathbb{Z}/\ell \to F_{i+1} \xrightarrow{u} F_i \xrightarrow{r} \mathbb{Z}/\ell[1] \quad \text{and} \quad F_i \xrightarrow{v} F_{i+1} \to \mathbb{Z}/\ell \xrightarrow{s} F_i[1];$$

thus r and s represent the $\mathbb{Z}/\ell[G]$-module extensions $0 \to \mathbb{Z}/\ell \to F_{i+1} \to F_i \to 0$ and $0 \to F_i \to F_{i+1} \to \mathbb{Z}/\ell \to 0$, respectively. Since $v \circ u : F_{i+1} \to F_{i+1}$ is multiplication by $(\sigma - 1)$, where σ is a chosen generator of G, the composition of these extensions is the $\mathbb{Z}/\ell[G]$-module Yoneda extension representing $\phi_i(\alpha_1)$. In particular, $\phi_{\ell-1}(\alpha_1)$ is represented by the Yoneda extension

$$0 \to \mathbb{Z}/\ell \to \mathbb{Z}/\ell[G] \xrightarrow{\sigma-1} \mathbb{Z}/\ell[G] \xrightarrow{\epsilon} \mathbb{Z}/\ell \to 0.$$

If $i < \ell - 1$, the following diagram shows that $\phi_i(\alpha_1) = 0$:

$$
\begin{array}{ccccccc}
\mathbb{Z}/\ell & \longrightarrow & \mathbb{Z}/\ell[G] & \xrightarrow{\sigma-1} & \mathbb{Z}/\ell[G] & \xrightarrow{\epsilon} & \mathbb{Z}/\ell \\
\downarrow{\scriptstyle 0} & & \downarrow & & \downarrow & & \| \\
\mathbb{Z}/\ell & \longrightarrow & F_{i+1} & \xrightarrow{\sigma-1} & F_{i+1} & \xrightarrow{\epsilon} & \mathbb{Z}/\ell.
\end{array}
$$

To compare $\phi_{\ell-1}(\alpha_1)$ to $\beta(\alpha_1)$, we need to use the natural isomorphism $\mathrm{Ext}^*_{\mathbb{Z}/\ell[G]}(\mathbb{Z}/\ell, \mathbb{Z}/\ell) \cong \mathrm{Ext}^*_{\mathbb{Z}[G]}(\mathbb{Z}, \mathbb{Z}/\ell)$ (see [Wei94, Ex. 6.2]). As an extension of $\mathbb{Z}[G]$-modules, α_1 is represented by the extension

$$0 \to \mathbb{Z}/\ell \xrightarrow{\sigma-1} \mathbb{Z}[G]/I^2 \to \mathbb{Z} \to 0,$$

where I is the augmentation ideal and $\mathbb{Z}/\ell \cong I/I^2$. Since the integral Bockstein $\tilde\beta$ is represented by the extension $0 \to \mathbb{Z} \xrightarrow{\ell} \mathbb{Z} \to \mathbb{Z}/\ell \to 0$, the composition $\tilde\beta(\alpha_1) \in \mathrm{Ext}^2_{\mathbb{Z}[G]}(\mathbb{Z}, \mathbb{Z})$ is represented by the $\mathbb{Z}[G]$-module Yoneda extension in the bottom row (and hence the top row) of the following commutative diagram.

$$
\begin{array}{ccccccc}
\mathbb{Z} & \xrightarrow{N} & \mathbb{Z}[G] & \xrightarrow{\sigma-1} & \mathbb{Z}[G] & \xrightarrow{\epsilon} & \mathbb{Z} \\
\downarrow & & \downarrow{\scriptstyle \epsilon} & & \downarrow & & \| \\
\mathbb{Z} & \xrightarrow{\ell} & \mathbb{Z} & \xrightarrow{\sigma-1} & \mathbb{Z}[G]/I^2 & \xrightarrow{\epsilon} & \mathbb{Z}
\end{array}
$$

It follows that $\beta(\alpha_1)$ is represented by the $\mathbb{Z}[G]$-module Yoneda extension

$$0 \to \mathbb{Z}/\ell \xrightarrow{N} \mathbb{Z}[G]/\ell N \cdot \mathbb{Z} \xrightarrow{\sigma-1} \mathbb{Z}[G] \xrightarrow{\epsilon} \mathbb{Z} \to 0.$$

Under the natural isomorphism $\mathrm{Ext}^2_{\mathbb{Z}/\ell[G]}(\mathbb{Z}/\ell, \mathbb{Z}/\ell) \cong \mathrm{Ext}^2_{\mathbb{Z}[G]}(\mathbb{Z}, \mathbb{Z}/\ell)$, this corresponds to the Yoneda extension given above for $\phi_{\ell-1}(\alpha_1)$. $\qquad\square$

6.7   THE OPERATION ϕ^V

Consider the map $\phi^V = \phi_{\ell-1}$ from $H^{2p+1,q}(S_\bullet, R)$ to $H^{2p\ell+2,q\ell}(S_\bullet, R)$ constructed in the last section (Definition 6.27). In this section, we show that ϕ^V agrees with the Steenrod operation βP^b when $p = q = b$. For notational simplicity, we shall write R for $\mathbb{Z}/\ell$ in this section. Recall that the presheaf $\mathbf{R}$ is defined in 6.3, and that $0 < i < \ell$.

Theorem 6.30. *Given morphisms* $\mathbf{R} \xrightarrow{\gamma} \mathbf{R}(r)[2s]$ *and* $\mathbf{R} \xrightarrow{\sigma} \mathbf{R}(p)[2q+1]$, *for* $p, r > 0$ *and* $q, s \geq 0$, *we have*

$$\phi_i(\gamma\sigma) = \gamma^{i+1}\phi_i(\sigma).$$

Proof. ([Voe11, 3.4]) For simplicity of notation we will write T_0 for $\mathbf{R}(r)[2s]$ and T_1 for $\mathbf{R}(p)[2q+1]$. Let M_γ and M_σ be the fibers of γ and σ, i.e., the objects defined (up to an isomorphism) by the distinguished triangles

$$T_0[-1] \to M_\gamma \to \mathbf{R} \xrightarrow{\gamma} T_0, \qquad T_1[-1] \xrightarrow{x_\sigma} M_\sigma \xrightarrow{y_\sigma} \mathbf{R} \xrightarrow{\sigma} T_1.$$

Step 1: Let α denote $\mathbf{R} \cong \mathbf{R} \otimes \mathbf{R} \xrightarrow{\gamma \otimes \sigma} T_0 \otimes T_1$, and M_α the fiber of α. We can represent α in two ways as a composition: $(\gamma \otimes \mathrm{id}_{T_1}) \circ \sigma$ and $(\mathrm{id}_{T_0} \otimes \sigma) \circ \gamma$. The octahedral axiom applied to these compositions yields two distinguished triangles:

$$M_\sigma \xrightarrow{f} M_\alpha \longrightarrow M_\gamma \otimes T_1 \to M_\sigma[1],$$
$$M_\gamma \longrightarrow M_\alpha \xrightarrow{g} T_0 \otimes M_\sigma \to M_\gamma[1],$$

such that $y_\alpha f = y_\sigma$ and $\gamma\, y_\alpha = (T_0 \otimes y_\sigma)g$. Thus the vertical arrows in diagram (6.30a) form morphisms of triangles.

$$\begin{array}{ccccccc}
T_1[-1] & \xrightarrow{x_\sigma} & M_\sigma & \xrightarrow{y_\sigma} & \mathbf{R} & \xrightarrow{\sigma} & T_1 \\
{\scriptstyle\gamma \otimes T_1[-1]}\downarrow & & {\scriptstyle f}\downarrow & & \| & & \downarrow{\scriptstyle\gamma \otimes T_1} \\
T_0 \otimes T_1[-1] & \xrightarrow{x_\alpha} & M_\alpha & \xrightarrow{y_\alpha} & \mathbf{R} & \xrightarrow{\alpha} & T_0 \otimes T_1 \\
\| & & {\scriptstyle g}\downarrow & & {\scriptstyle\gamma}\downarrow & & \| \\
T_0 \otimes T_1[-1] & \longrightarrow & T_0 \otimes M_\sigma & \xrightarrow{T_0 \otimes y_\sigma} & T_0 & \xrightarrow{T_0 \otimes \sigma} & T_0 \otimes T_1
\end{array} \qquad (6.30\text{a})$$

Because $(\gamma \otimes M_\sigma, \gamma, \gamma \otimes T_1)$ is also a morphism of triangles between the top and bottom rows of (6.30a), the difference between gf and $\gamma \otimes M_\sigma$ is the image of a map $d: M_\sigma \to T_0 \otimes T_1[-1]$. We may modify f by the image of d under $T_0 \otimes T_1[-1] \xrightarrow{x_\alpha} M_\alpha$ to assume that $g \circ f = \gamma \otimes M_\sigma$.

Applying the slice functor s_0 to the top middle square of (6.30a), we see that $s_0(M_\sigma) \cong s_0(M_\alpha) \cong \mathbf{R}$ and $s_0(f) = 1$. Applying the bottom slice functor s_{p+r} to the lower left square in (6.30a), and using Proposition 6.18, we see that $s_{p+r}(g) = 1$.

Step 2: Consider the following commutative diagram, in which the left square is $S^{i-1}(f)[1]$ tensored with the upper left square in (6.30a) and the right square commutes by naturality of the corestriction map $S^{i-1}(A) \otimes A \to S^i(A)$ with respect to f (see 6.26(b)).

$$
\begin{array}{ccccc}
S^{i-1}(M_\sigma) \otimes T_1 & \xrightarrow{1 \otimes x_\sigma[1]} & S^{i-1}(M_\sigma) \otimes M_\sigma[1] & \xrightarrow{\text{cores}} & S^i(M_\sigma)[1] \\
\Big\downarrow{\scriptstyle S^i f \otimes \gamma \otimes 1} & & \Big\downarrow{\scriptstyle S^{i-1}f \otimes f} & & \Big\downarrow{\scriptstyle S^i f[1]} \\
S^{i-1}(M_\alpha) \otimes T_0 \otimes T_1 & \xrightarrow{1 \otimes x_\alpha[1]} & S^{i-1}(M_\alpha) \otimes M_\alpha[1] & \xrightarrow{\text{cores}} & S^i(M_\alpha)[1]
\end{array}
$$

The horizontal composites in this diagram are the maps $v[1] = \text{cores}(1 \otimes x[1])$ of the triangles 6.26(b) for σ and α. It follows that there is a morphism $c : \mathbf{R} \to \mathbf{R}$ completing the large outer square above into a morphism between the triangles of 6.26(b) for σ and α.

$$
\begin{array}{ccccccc}
S^i(M_\sigma) & \xrightarrow{S^i y_\sigma} & \mathbf{R} & \xrightarrow{s_\sigma} & S^{i-1}(M_\sigma) \otimes T_1 & \xrightarrow{v[1]} & S^i(M_\sigma)[1] \\
\Big\downarrow{\scriptstyle S^i(f)} & & \exists c \Big\downarrow (=1) & & \Big\downarrow{\scriptstyle S^{i-1}f \otimes \gamma \otimes T_1} & & \Big\downarrow{\scriptstyle S^i(f)[1]} \\
S^i(M_\alpha) & \xrightarrow{S^i y_\alpha} & \mathbf{R} & \xrightarrow{s_\alpha} & S^{i-1}(M_\alpha) \otimes T_0 \otimes T_1 & \xrightarrow{v[1]} & S^i(M_\alpha)[1]
\end{array}
$$

Applying the slice functor s_0 to the left square, we conclude from 6.26(a) that $c = s_0(S^i f) = 1$. In particular,

$$s_\alpha = \left(S^{i-1} f \otimes \gamma \otimes T_1\right) \circ s_\sigma. \tag{6.30b}$$

Step 3: Similarly, we have a commutative diagram in which the left square commutes by naturality of the transfer, and the right square is just $S^{i-1}g$ tensored with the bottom middle square in (6.30a):

$$
\begin{array}{ccccccc}
S^i(M_\alpha) & \xrightarrow{\text{tr}} & S^{i-1}(M_\alpha) \otimes M_\alpha & \xrightarrow{1 \otimes y_\alpha} & S^{i-1}(M_\alpha) \\
\Big\downarrow{\scriptstyle S^i g} & & \Big\downarrow{\scriptstyle S^{i-1}g \otimes g} & & \Big\downarrow{\scriptstyle S^{i-1}g \otimes \gamma} \\
T_0^i \otimes S^i(M_\sigma) & \xrightarrow{\text{tr}} & (T_0^{i-1} \otimes S^{i-1}(M_\sigma)) \otimes (T_0 \otimes M_\sigma) & \xrightarrow{1 \otimes y_\sigma} & T_0^i \otimes S^{i-1}(M_\sigma).
\end{array}
$$

The horizontal composites in this diagram are the maps u_α and $1 \otimes u_\sigma$ of the triangles 6.26(a) for α and σ. It follows that there is a morphism $c : T_0^i \otimes T_1^i[1] \to T_0^i \otimes T_1^i[1]$ completing the large outer square above into a morphism of triangles, from the triangle 6.26(a) for α to the tensor product of T_0^i with the triangle 6.26(a) for σ:

$$
\begin{array}{ccccccc}
S^i(M_\alpha) & \xrightarrow{\;u_\alpha\;} & S^{i-1}(M_\alpha) & \xrightarrow{\;r_\alpha\;} & T_0^i \otimes T_1^i[1] & \longrightarrow & S^i(M_\alpha)[1] \\
\downarrow{\scriptstyle S^i(g)} & & \downarrow{\scriptstyle S^{i-1}(g)\otimes\gamma} & \exists c \downarrow{\scriptstyle (=1)} & & & \downarrow{\scriptstyle S^i(g)[1]} \\
T_0^i \otimes S^i(M_\sigma) & \xrightarrow{\;1\otimes u_\sigma\;} & T_0^i \otimes S^{i-1}(M_\sigma) & \xrightarrow{\;1\otimes r_\sigma\;} & T_0^i \otimes T_1^i[1] & \longrightarrow & T_0^i \otimes S^i(M_\sigma)[1].
\end{array}
$$

Applying the bottom slice functor, we conclude from the commutativity of the right square and $s_{ip+ir}(T_0^i \otimes A) = s_{ip}(A)[2si]$ that $c = s_{ip+ir}(S^i g) = 1$. Thus the middle square yields

$$
r_\alpha = \left(T_0^i \otimes r_\sigma \right) \circ \left(S^{i-1} g \otimes \gamma \right). \tag{6.30c}
$$

Step 4: By Definition 6.27, $\phi_i(\sigma) = (r_\sigma \otimes T_1) \circ s_\sigma$, where r_σ and s_σ are given in 6.26, and $\phi_i(\alpha) = (r_\alpha \otimes T) \circ s_\alpha$, where s_α and r_α are given in (6.30b) and (6.30c) and $T = T_0 \otimes T_1$. Thus $\phi_i(\alpha)$ is the composition

$$
(1 \otimes r_\sigma \otimes T)(S^{i-1} g \otimes \gamma \otimes T_1) \circ (S^{i-1} f \otimes \gamma \otimes T_1) \circ s_\sigma
$$
$$
= (1 \otimes r_\sigma \otimes T) \circ (\rho \otimes T_1) \circ s_\sigma,
$$

where ρ is the composite

$$
S^{i-1}(M_\sigma) \xrightarrow{\;S^{i-1} f \otimes \gamma\;} S^{i-1}(M_\alpha) \otimes T_0 \xrightarrow{\;S^{i-1} g \otimes \gamma\;} S^{i-1}(M_\sigma) \otimes T_0^{i+1}.
$$

Because $g \circ f = \gamma \otimes M_\sigma$ we have

$$
S^{i-1}(g) \circ S^{i-1}(f) = S^{i-1}(g \circ f) = S^{i-1} M_\sigma \otimes S^{i-1}(\gamma) = S^{i-1} M_\sigma \otimes \gamma^{i-1}.
$$

Thus $\rho = S^{i-1} M_\sigma \otimes \gamma^{i+1}$. Combining ρ, r_σ and s_σ, we get the desired identity $\phi_i(\alpha) = \gamma^{i+1} \phi_i(\sigma)$. $\qquad\square$

Recall from [Voe03a, 2.4] that if S_s^1 is the simplicial circle and $\Sigma S_\bullet$ denotes the simplicial suspension $S_\bullet \wedge S_s^1$ of a pointed simplicial space $S_\bullet$ then there is a suspension isomorphism $\Sigma: \widetilde{H}^{2p,q}(S_\bullet, R) \xrightarrow{\;\simeq\;} \widetilde{H}^{2p+1,q}(\Sigma S_\bullet, R)$ sending γ to $\gamma \cup \sigma_s$, where $\sigma_s \in H^{1,0}(S_s^1, R) \cong H^*_{\mathrm{top}}(S^1, R)$ is the fundamental class of S^1.

Proposition 6.31. *The operations $\phi_i = \phi_i^{p,q}$ satisfy:*

(a) $\phi_i(cz) = c^{i+1} \phi_i(z)$ for $c \in \mathbb{Z}$;
(b) $\phi_i(\Sigma \gamma) = 0$ in $\widetilde{H}^{2p(i+1)+2, q(i+1)}(\Sigma S_\bullet, R)$ for all $\gamma \in \widetilde{H}^{2p,q}(S_\bullet, R)$.

Proof. Part (a) is the case $c = \gamma$ of Theorem 6.30. Part (b) is the case where $z = \sigma_s \in H^{1,0}(S_s^1, R)$, because by Theorem 6.30 we have

$$\phi_i(\sigma_s \cup \gamma) = \phi_i(\sigma_s) \cup \gamma^{i+1},$$

and this is zero because $\phi_i(\sigma_s)$ is an element of $H^{2,0}(S_s^1) = 0$. $\square$

The group $H^{*,0}(S_\bullet, R)$ is just the topological cohomology $H^*_{\text{top}}(\pi_0(S_\bullet), R)$; see Lemma 13.3. Under this identification, we show that $\phi_{\ell-1}$ is classical.

Corollary 6.32. $\phi_{\ell-1} : H^{2p+1,0}(-, R) \to H^{2p\ell+2,0}(-, R)$ *is the classical Steenrod operation* βP^p_{top} *(which is* Sq^{2p+1}_{top} *if* $\ell = 2$*).*

Proof. Since $\phi_{\ell-1}$ vanishes on suspensions, it must be $c\beta P^p_{\text{top}}$ for some $c \in R$ by Example 13.4.1. To see that $\phi_{\ell-1} = \beta P^p_{\text{top}}$, it suffices to find one element x such that $\phi_{\ell-1}(x)$ is nonzero and equal to $\beta P^p_{\text{top}}(x)$.

For this, we recall from the proof of Proposition 6.29 that there is a smooth simplicial scheme $B_\bullet(G)$ and an element α_1 of $H^{1,0}(B_\bullet(G), R)$ with $\gamma = \phi_{\ell-1}(\alpha_1) = \beta(\alpha_1)$ nonzero. For $x = \gamma^p \cdot \alpha_1$, Theorem 6.30 yields

$$\phi_{\ell-1}(x) = \gamma^{p\ell} \cdot \beta(\alpha_1) = \gamma^{p\ell+1}.$$

By [Ste62, V.5.3], this is a nonzero element of $H^*_{\text{top}}(BG, R)$. On the other hand, if $\ell = 2$ then $\gamma = \alpha_1^2$ and $x = \alpha_1^{2p+1}$, so $Sq^{2p+1}_{\text{top}}(x) = \alpha_1^{4p+2} = \gamma^{2p+1}$, while if $\ell \neq 2$ the Cartan formula yields $P^p_{\text{top}}(x) = P^p_{\text{top}}(\gamma^p)\alpha_1 = \gamma^{p\ell}\alpha_1$ and hence $\beta P^p_{\text{top}}(x) = \gamma^{p\ell}\beta(\alpha_1) = \gamma^{p\ell+1}$. $\square$

We now establish a motivic analogue of the classical assertion in Corollary 6.32, using a similar proof. For this we need the following theorem, which will be proven in chapter 15, as Theorem 15.38.

Theorem 6.33. *Let* $\phi \colon H^{2n+1,n}(-, \mathbb{Z}) \to H^{2n\ell+2,n\ell}(-, \mathbb{Z}/\ell)$ *be a cohomology operation such that for all* X *and all* $x \in H^{2n+1,n}(X, \mathbb{Z})$*:*

1. $\phi(bx) = b\phi(x)$ *for* $b \in \mathbb{Z}$*;*
2. *if* $x = \Sigma y$ *for* $y \in H^{2n,n}(X, \mathbb{Z})$ *then* $\phi(x) = 0$*.*

Then ϕ *is a multiple of the mod-ℓ reduction of* βP^n*.*

Corollary 6.34. *When* $R = \mathbb{Z}/\ell$ *and* $p = q = b$*, the operation* $\phi^V = \phi_{\ell-1}$ *is the motivic cohomology operation*

$$\beta P^b : H^{2b+1,b}(-, \mathbb{Z}/\ell) \to H^{2b\ell+2,b\ell}(-, \mathbb{Z}/\ell).$$

Proof. By Propositions 6.28 and 6.31, ϕ^V satisfies the hypotheses of Theorem 6.33, which says that in this case ϕ^V is a $\mathbb{Z}/\ell$-multiple of βP^b.

To determine which multiple of βP^b, consider the canonical line element $u \in H^{2,1}(\mathbb{P}^N, R)$ for $N > b\ell$. Then $x = u^b \alpha_1$ is an element of $H^{2b+1,b}(\mathbb{P}^N \times B_\bullet(G))$ and by combining Propositions 6.31(a) and 6.29 we obtain $\phi^V(x) = u^{b\ell}\beta(\alpha_1)$. By the projective bundle formula [MVW, 15.12], multiplication by u^s is an injection from $H^{i,0}(B_\bullet(G), R)$ into $H^{2s+i,s}(B_\bullet(G) \times \mathbb{P}^N, R)$ for all $s \leq N$. It follows that $\phi^V(x) \neq 0$.

When $\ell = 2$, the only $\mathbb{Z}/2$-multiples of Sq^{2b+1} are Sq^{2b+1} and 0. Since $\phi^V \neq 0$, we have $\phi^V = Sq^{2b+1}$. Now suppose that $\ell \neq 2$. To compute $\beta P^b(x)$, we refer to the properties of the motivic operations P^i, listed in Axioms 13.6. Because $P^b(u^b) = u^{b\ell}$, the Cartan formula yields

$$\beta P^b(x) = \beta\left(u^{b\ell}\alpha_1\right) = u^{b\ell}\beta(\alpha_1).$$

Since $\beta P^b(x) = \phi^V(x)$, we must have $\phi^V = \beta P^b$. $\qquad\square$

6.8 HISTORICAL NOTES

Sections 6.1–6.5 are taken from the 2010 paper [Voe10b], based upon a preprint written in 2003. The idea of working with motives over S originated much earlier, in the 1994 preprint of [SV00b]. Sections 6.6–6.7 are taken from the 2011 paper [Voe11]; this material first appeared in the 2003 preprint [Voe03b].

Cohomology operations in motivic cohomology have their origins in the 1996 preprint [Voe96], and appeared in [Voe03c] and [Voe03a]. The construction of the cohomology operations ϕ_i first appeared in the 2003 preprint [Voe03b], as did the identification of ϕ^V with βP^b up to a constant c (the fact that $c = 1$ is taken from [Wei09]). The power of simplicial schemes to mimic simplicial sets in manipulating cohomology operations is amply illustrated by the constructions in section 6.6.

Chapter Seven

The Motivic Group $H^{BM}_{-1,-1}$

IN THIS SHORT chapter, we develop some more of the properties of the Borel–Moore homology groups $H^{BM}_{-1,-1}(X)$ which we shall need in part II.

The Borel–Moore homology group $H^{BM}_{-1,-1}(X) = H^{BM}_{-1,-1}(X,\mathbb{Z})$ is defined as $\operatorname{Hom}_{\mathbf{DM}}(\mathbb{Z}, M^c(X)(1)[1])$, and was briefly discussed in section 1.3. It is contravariant in X for finite flat maps [MVW, 16.1] [SV00b, 3.6], and has a functorial pushforward for proper maps. If X is smooth and proper (in characteristic 0), $H^{BM}_{-1,-1}(X)$ agrees with $H^{2d+1,d+1}(X,\mathbb{Z})$ [MVW, 16.24], and has a nice presentation (Proposition 1.19), which we will explore in section 7.1.

If k is a field, we saw in Example 1.21 that $H^{BM}_{-1,-1}(\operatorname{Spec} k) \cong k^\times$. Thus if X is proper over k there is a pushforward $N : H^{BM}_{-1,-1}(X) \to H^{BM}_{-1,-1}(k) \cong k^\times$; it factors through the quotient $\overline{H}_{-1,-1}(X)$ of $H^{BM}_{-1,-1}(X)$ by the difference of the two projections from $H^{BM}_{-1,-1}(X \times X)$.

The main result in this chapter is Proposition 7.7: if X is a norm variety for $\underline{a}$ and k is ℓ-special then the image of $H^{BM}_{-1,-1}(X) \to k^\times$ is the group of units b such that $\underline{a} \cup b$ vanishes in $K^M_{n+1}(k)/\ell$. We will see later, in the Norm Principle 11.1, that each of these units is a norm from a degree ℓ extension of k. This will be used in Theorem 10.17 to prove that norm varieties exist, and again in Theorem 11.2 to prove that norm varieties are Rost varieties.

7.1 PROPERTIES OF $\overline{H}_{-1,-1}$

If x is a closed point on X then the proper map $x \to X$ induces a map $k(x)^\times \to H^{BM}_{-1,-1}(X)$; we write $[x,\alpha]$ for the image of $\alpha \in k(x)^\times$ in $H^{BM}_{-1,-1}(X)$. As noted in Example 1.21, $N : H^{BM}_{-1,-1}(X) \to k^\times$ sends $[x,\alpha]$ to the norm of α, $N_{k(x)/k}(\alpha)$. When X is smooth, we saw in Proposition 1.19 that $H^{BM}_{-1,-1}(X)$ is generated by the symbols $[x,\alpha]$. This is true even when X is singular:

Lemma 7.1. *For any reduced scheme X over a perfect field k, $H^{BM}_{-1,-1}(X)$ is generated by symbols $[x,\alpha]$, where x is a closed point of X and $\alpha \in k(x)^\times$.*

Proof. We proceed by induction on $\dim(X)$, the case $\dim(X) = 0$ being given by Proposition 1.19. Let Z be the singular locus of X and U the complement.

Then we have an exact localization sequence

$$H^{BM}_{-1,-1}(Z) \longrightarrow H^{BM}_{-1,-1}(X) \longrightarrow H^{BM}_{-1,-1}(U)$$

by [Voe00b, 4.1.5] or [MVW, 16.20] when char$(k) = 0$, and by [Kel13, 5.5.5] when char$(k) > 0$. Since U is smooth and dim$(Z) < $ dim(X), the left and right groups are generated by $[x, \alpha]$ with x a closed point in Z (resp., U) and $\alpha \in k(x)^\times$. As these may be regarded as elements of $H^{BM}_{-1,-1}(X)$, the result follows. $\qquad\square$

Here are some elementary facts about the reduced group $\overline{H}_{-1,-1}(X)$, which first appeared in [SJ06, 1.5–7].

If E is a finite field extension of $k(x)$ for some closed point $x \in X$ then the proper map Spec$(E) \to x \to X$ induces a map

$$E^\times = H^{BM}_{-1,-1}(\mathrm{Spec}\ E) \to H^{BM}_{-1,-1}(x) \to H^{BM}_{-1,-1}(X).$$

By inspection, the composite sends $\alpha \in E^\times$ to the class of $[x, N_{E/k(x)}\alpha]$.

More generally, if $f : Y \to X$ is proper and $y = \mathrm{Spec}(E)$ is a closed point of Y, then the pushforward f_* sends $[y, \alpha]$ to $[f(y), N_{E/k(x)}\alpha]$, $x = f(y)$.

To illustrate the advantage of passing to $\overline{H}_{-1,-1}$, consider a cyclic field extension E/k. Then $\overline{H}_{-1,-1}(\mathrm{Spec}\ E)$ is the quotient of $H^{BM}_{-1,-1}(\mathrm{Spec}\ E) = E^\times$ by the relation $x \sim \sigma(x)$, where σ generates Gal(E/k), and $\overline{H}_{-1,-1}(\mathrm{Spec}\ E)$ injects into $k^\times$ because Hilbert's Theorem 90 gives an exact sequence

$$0 \to \overline{H}_{-1,-1}(\mathrm{Spec}\ E) \xrightarrow{N} k^\times \to \mathrm{Br}(E/k) \to 0.$$

If $f : Y \to X$ is a finite flat map of degree d, there is a natural pullback $f^* : H^{BM}_{-1,-1}(X) \to H^{BM}_{-1,-1}(Y)$, and an induced map $\overline{H}_{-1,-1}(X) \to \overline{H}_{-1,-1}(Y)$. The composition $f_* f^*$ is multiplication by d. This follows for example from the fact that $M^c(X) \xrightarrow{f^*} M^c(Y) \xrightarrow{f_*} M^c(X)$ is multiplication by d.

Given a closed point Spec$(E) \to X \times X$, the projections to X send it to points $x_i = \mathrm{Spec}(E_i)$ of X, for intermediate subfields E_1, E_2 such that $E = E_1 E_2$; every closed point of $X \times X$ has this form for suitable $x_1, x_2 \in X$. It follows that $\overline{H}_{-1,-1}(X)$ is the quotient of $H^{BM}_{-1,-1}(X)$ by the relations that $[x_1, N_{E/E_1}\alpha] \sim [x_2, N_{E/E_2}\alpha]$ for $\alpha \in E^\times$.

Lemma 7.2. *Suppose that $x_1, x_2 \in X$ are closed points, and $\alpha \in k(x_1)^\times$.*

1. If $\sigma : k(x_1) \xrightarrow{\sim} k(x_2)$ then $[x_1, \alpha] \sim [x_2, \sigma(\alpha)]$ in $\overline{H}_{-1,-1}(X)$.
2. If there is a field embedding $k(x_2) \hookrightarrow k(x_1)$, then in $\overline{H}_{-1,-1}(X)$ we have:

$$[x_1, \alpha] = [x_2, N_{k(x_1)/k(x_2)}\alpha].$$

3. If $X(k) \neq \emptyset$ and X is proper over k, then $N : \overline{H}_{-1,-1}(X) \cong k^\times$.

4. If X has a closed point x with $[k(x):k]=m$, and X is proper over k, then the kernel and cokernel of $N : \overline{H}_{-1,-1}(X) \to k^{\times}$ have exponent m.

Proof. The first two parts follow from the above remarks by taking the pushforward of $\alpha \in k(x_1)^{\times}$ along the proper map $\operatorname{Spec} k(x_1) \to (x_1, x_2) \to X \times X$. If X has a k-point x_2, then (2) implies that every element of $\overline{H}_{-1,-1}(X)$ is equivalent to one of the form $[x_2, \alpha]$; since $\overline{H}_{-1,-1}(x_2) \to \overline{H}_{-1,-1}(X) \to k^{\times}$ is an isomorphism, we obtain part (3). Finally, in the situation of part (4) the cokernel of N is a quotient of the cokernel of $N_{k(x)/k} : k(x)^{\times} \to k^{\times}$, which has exponent m, and the kernel of N is contained in the kernel of the finite flat pullback $f^* : \overline{H}_{-1,-1}(X) \to \overline{H}_{-1,-1}(X_{k(x)}) = k(x)^{\times}$, which has exponent m because $f_* f^* = m$. $\qquad\square$

Lemma 7.3. *Let $f : Y \to X$ be a finite flat morphism over k of degree prime to ℓ. If F is an ℓ-special field over k, $Y(F) \to X(F)$ is onto.*

Proof. Suppose that $\operatorname{Spec}(F) \to X$ is an F-point with image $x \in X$. Since the degree of $Y \times_X \operatorname{Spec}(F) \to \operatorname{Spec}(F)$ is $\deg(f)$ it is prime to ℓ. Since F is ℓ-special, the map splits, yielding an F-point of Y. $\qquad\square$

Lemma 7.4. *Let $f : Y \to X$ be a finite flat morphism of degree prime to ℓ over an ℓ-special field k. Then $f_* : \overline{H}^{BM}_{-1,-1}(Y) \to \overline{H}^{BM}_{-1,-1}(X)$ is onto, and $f^* f_*$ is multiplication by $\deg(f)$.*
If X and Y are proper over k, then f_* is an isomorphism.

Proof. For each closed $x \in X$, $k(x)$ is ℓ-special, so there is a $y \in Y$ with $f(y) = x$ and $k(x) = k(y)$ by Lemma 7.3. Since we have $f_*([y, \alpha]) = [x, \alpha]$ for all $\alpha \in k(x)^{\times}$, the map f_* is onto. To see that $f^* f_* = \deg(f)$, consider a generator $[y, \beta]$ of $H^{BM}_{-1,-1}(Y)$ and set $x = f(y)$, $F = k(x)$. By Lemma 7.3, there is a $y' \in Y$ with $f(y) = f(y')$ and $k(y') = F$. By Lemma 7.2(2), $[y, \beta] = [y', N_{k(y)/F}\beta]$. Replacing y by y', we may assume that $k(y) = F$.

Suppose that $f^{-1}(x)$ consists of points $y = y_1, \ldots, y_r$ counted with multiplicities, so $\sum [k(y_i) : F] = \deg(f)$. Then in $\overline{H}^{BM}_{-1,-1}(Y)$, again by 7.2(2),

$$f^* f_*([y, \beta]) = f^*([x, \beta]) = \sum [y_i, \beta] = [y, \prod N_{k(y_i)/F}(\beta)].$$

Since $\prod N_{k(y_i)/F}(\beta) = \prod \beta^{[k(y_i):F]} = \beta^{\deg(f)}$, $f^* f_*([y, \beta]) = \deg(f)[y, \beta]$.

Finally, if X and Y are proper over k then the pushforwards $N_{X/k}$ and $N_{Y/k}$ are defined, and $N_{Y/k} = N_{X/k} \circ f_*$. By Lemma 7.2(4) and the fact that k is ℓ-special, the kernel of $N_{Y/k}$ and hence the kernel of f_* has exponent a power of ℓ. Hence the kernel of f_* is zero, since it also has exponent $\deg(f) = f^* f_*$, which is prime to ℓ. $\qquad\square$

Theorem 7.5. *Let $f : W \to X$ be a dominant morphism of projective varieties over the ℓ-special field k of characteristic 0. Suppose that $\dim(W) = \dim(X)$ and the degree of f is prime to ℓ. Then:*

(a) For every ℓ-special field F over k, the image of $W(F) \to X(F)$ contains every F-point of X lying in the smooth locus of X.

(b) The map $f_* : H^{BM}_{-1,-1}(W) \to H^{BM}_{-1,-1}(X)$ is a surjection.

(c) If W and X are smooth projective then $f_* : \overline{H}_{-1,-1}(W) \to \overline{H}_{-1,-1}(X)$ is an isomorphism.

Proof. (a) We first note an easy case: if X_1 is the blow-up of X along a smooth center, then every F-point in the smooth locus of X is in the image of $X_1(F) \to X(F)$ by construction. By induction, if $X_n \to X$ is obtained by a sequence of n blow-ups along smooth centers, then the image of $X_n(F) \to X(F)$ contains the smooth points of $X(F)$.

In the general case, the Raynaud–Gruson platification theorem [RG71] says that there is a blow-up $b : X' \to X$ so that the proper pullback $f' : W' \to X'$ is flat. Because f is proper, so is f'; since f is generically finite, it follows that f' is finite. Since $\deg(f') = \deg(f)$ is prime to ℓ, $W'(F) \to X'(F)$ is onto by Lemma 7.3. Thus it suffices to show that if $x : \mathrm{Spec}(F) \to X$ is in the smooth locus of X then x is in the image of $X'(F) \to X(F)$.

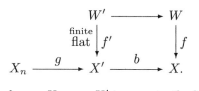

Consider the rational map $X \dashrightarrow X'$ inverse to the birational map $b : X' \to X$. To eliminate the indeterminacy of this map, we can find a smooth X_n, a map $\pi : X_n \to X$ obtained by a sequence of blow-ups along smooth centers, and a morphism $g : X_n \to X'$ extending the rational map. Moreover, $\pi = bg$. By the easy case noted above, $x = \pi(x_n)$ for some F-point $x_n \in X_n(F)$. Hence the F-point $x' = g(x_n)$ satisfies $b(x') = \pi(x_n) = x$. This proves part (a).

(b) Let $j : U \hookrightarrow X$ be a dense open subset, smooth over k, such that f is finite and flat over U. Since every closed point in X can be connected by a chain of curves to a closed point in U, $H^{BM}_{-1,-1}(X)$ is generated by elements $[x, \alpha]$ with $x \in U$. Similarly, $H^{BM}_{-1,-1}(W)$ is generated by $[w, \beta]$ with $w \in f^{-1}(U)$. The surjectivity of f_* is now immediate: by (a), any $x \in U$ can be lifted to a point w of W with $k(x) = k(w)$, and $f_*([w, \alpha]) = [x, \alpha]$ for each $\alpha \in k(x)^\times$.

(c) To show that f_* is injective, we use the fact that the pullback $f^* : H^{BM}_{-1,-1}(X) \to H^{BM}_{-1,-1}(W)$ is defined. This is because the map $H^{2d+1,d+1}(X) \cong H_{-1,-1}(X) \to H^{BM}_{-1,-1}(X)$, induced by $M(X) \to M^c(X)$, is an isomorphism when X is a smooth projective variety (see [MVW, 16.24] or [FV00, p. 186]):

$$H^{BM}_{-1,-1}(X) \xleftarrow{\;\cong\;} H^{2d+1,d+1}(X) \xrightarrow{\;f^*\;} H^{2d+1,d+1}(W) \xrightarrow{\;\cong\;} H^{BM}_{-1,-1}(W).$$

If w is in $f^{-1}(U)$, with $x = f(w)$, the calculation in Lemma 7.4 shows that $f^* f_*([w, \beta]) = (\deg f)[w, \beta]$ in $\overline{H}^{BM}_{-1,-1}(W)$. Hence if $z = \sum n_i [w_i, \beta_i]$ satisfies $f_*(z) = 0$ then $(\deg f)z = 0$ in $\overline{H}^{BM}_{-1,-1}(W)$. Since k is ℓ-special, X has a closed point x with $[k(x) : k] = \ell^\nu$ for some $\nu > 0$. By Lemma 7.2(4), $\ker(f_*)$ is also annihilated by ℓ^ν. Since $\ell^\nu z = (\deg f)z = 0$, we have $z = 0$. This shows that f_* is injective. $\qquad\qquad\square$

We conclude this section by relating the pushforward $H^{BM}_{-1,-1}(X) \to k^\times$ to the fundamental class $\mathbb{L}^d \xrightarrow{\tau} R_{\mathrm{tr}}(X)$, where X is smooth and proper of dimension d and $\mathbb{L} = R(1)[2]$ as usual. Recall from Definition 5.15 that τ is the map dual to the structure map $R_{\mathrm{tr}}(X) \xrightarrow{\pi} R$. Therefore composing with τ yields a motivic duality map from $H^{n,i}(X, R) = \mathrm{Hom}(R_{\mathrm{tr}}(X), R(i)[n])$ to $\mathrm{Hom}(\mathbb{L}^d, R(i)[n])$, which is $H^{n-2d, i-d}(\mathrm{Spec}\, k, R)$ when $n \geq 2d$ and $i \geq d$.

Lemma 7.6. *If $char(k) = 0$ and $X \xrightarrow{\pi} \mathrm{Spec}(k)$ is smooth and proper of dimension d then, for all p and q, the motivic duality maps for X and k fit into a commutative diagram:*

$$
\begin{array}{ccc}
H^{2d+p, d+q}(X, R) & \xrightarrow{\tau} & H^{p,q}(k, R) \\
\Big\downarrow{\cong} & & \Big\downarrow{\cong} \\
H_{-p,-q}(X, R) & \xrightarrow{\pi_*} & H_{-p,-q}(k, R).
\end{array}
$$

In particular, the fundamental class $\tau : H^{2d+1, d+1}(X) \to H^{1,1}(k) \cong k^\times$ is identified by motivic duality with $N : H_{-1,-1}(X) \to k^\times$.

Proof. Recall that $H^{2d+p, d+q}(X, R) \cong \mathrm{Hom}(R_{\mathrm{tr}}(X), \mathbb{L}^d(q)[p])$. By motivic duality [MVW, 16.24], an element f of this group corresponds to the element of $H_{-p,-q}(X, R) \cong \mathrm{Hom}(R(-q)[-p], R_{\mathrm{tr}}(X))$ represented by the dual map

$$
R(-q)[-p] \cong \mathbb{L}^d \otimes \mathbb{L}^d(q)[p]^* \xrightarrow{1 \otimes f^*} \mathbb{L}^d \otimes R_{\mathrm{tr}}(X)^* \cong R_{\mathrm{tr}}(X).
$$

On the other hand, the element $f \circ \tau$ of $H^{p,q}(k, R) = \mathrm{Hom}(\mathbb{L}^d, \mathbb{L}^d(q)[p])$ corresponds to the element $1 \otimes \tau^* f^*$ of $H_{-p,-q}(k, R)$ represented by the dual map

$$
R(-q)[-p] \cong \mathbb{L}^d \otimes \mathbb{L}^d(q)[p]^* \xrightarrow{1 \otimes f^*} \mathbb{L}^d \otimes R_{\mathrm{tr}}(X)^* \xrightarrow{1 \otimes \tau^*} \mathbb{L}^d \otimes \mathbb{L}^{-d} \cong R.
$$

Since $1 \otimes \tau^*$ is the canonical projection $R_{\mathrm{tr}}(X) \xrightarrow{\pi} R$, $1 \otimes \tau^* f^*$ is the image of $1 \otimes f^*$ under π^*. $\qquad\qquad\square$

7.2   THE CASE OF NORM VARIETIES

Recall from Definition 1.13 that a *norm variety* for a symbol $\underline{a}$ in $K_n^M(k)/\ell$ is a smooth projective variety X of dimension $\ell^{n-1}-1$ such that $\underline{a}$ vanishes in $K_n^M(k(X))/\ell$, and which is ℓ-*generic* in the sense that any splitting field F of $\underline{a}$ has a finite extension E of degree prime to ℓ with $X(E) \neq \emptyset$.

In this section, we assume that there is a norm variety X for the symbol $\underline{a} = \{a_1, \ldots, a_n\}$, and show that the image of the pushforward map $N : H_{-1,-1}(X) \to k^\times$ is the group of units β of k such that $\{a_1, \ldots, a_n, \beta\}$ vanishes in $K_{n+1}^M(k)/\ell$. This will be used in section 10.4 to construct a norm variety for $\{a_1, \ldots, a_n, a_{n+1}\}$, by induction on n.

Proposition 7.7. *Let* $\underline{a} = \{a_1, \ldots, a_n\} \in K_n^M(k)/\ell$ *be a nontrivial symbol. Assume that $BL(n-1)$ holds, that k is ℓ-special, and that X is a norm variety for $\underline{a}$. Then the following sequence is exact:*

$$H_{-1,-1}^{BM}(X) \xrightarrow{\ N\ } k^\times \xrightarrow{\ \underline{a} \cup (-)\ } K_{n+1}^M(k)/\ell.$$

Proof. We first show that the composition is trivial. It suffices to consider the image of a generator $[x, \beta]$. Since X is a smooth splitting variety for $\underline{a}$, $\underline{a}$ vanishes in $K_n^M(k(X))/\ell$ and also in $K_n^M(k(x))/\ell$ (by specialization; see Remark 1.13.1). By the projection formula, $\{\underline{a}, N(\beta)\} = N_{k(x)/k}\{\underline{a}, \beta\} = 0$.

Since $F = k(\sqrt[\ell]{a_1})$ splits $\underline{a}$, X is an ℓ-generic splitting variety, and F is ℓ-special, we conclude that X has an F-point x. Since $N([x, b]) = b^\ell$ for $b \in k$, the image of N contains $k^{\times \ell}$. Since $H_{-1,-1}(X, \mathbb{Z})/\ell \cong H_{-1,-1}(X, \mathbb{Z}/\ell)$, this shows that it suffices to show that the following sequence is exact:

$$H_{-1,-1}(X, \mathbb{Z}/\ell) \to k^\times/k^{\times \ell} \to H_{\text{ét}}^{n+1}(k, \mu_\ell^{\otimes n+1}).$$

We claim that this sequence forms the top row of a large commutative diagram:

$$
\begin{array}{ccccc}
H_{-1,-1}(X, \mathbb{Z}/\ell) & \xrightarrow{\ N\ } & k^\times/k^{\times \ell} & \xrightarrow{\ \underline{a}\cup\ } & H_{\text{ét}}^{n+1}(k, \mu_\ell^{\otimes n+1}) \\
\Big\uparrow{\scriptstyle \cong} & & \Big\uparrow{\scriptstyle \cong} & & \Big\uparrow{\scriptstyle \text{injection}} \\
H^{2d+1,d+1}(X, \mathbb{Z}/\ell) & \xrightarrow{\ \tau\ } & H^{1,1}(k, \mathbb{Z}/\ell) & \xrightarrow{\ \delta\cup\ } & H^{n+1,n}(\mathfrak{X}, \mathbb{Z}/\ell) \\
\Big\uparrow{\scriptstyle \lambda} & & \Big\downarrow{\scriptstyle \cong} & {\scriptstyle Q_{n-1}\mu\cup} & \Big\downarrow{\scriptstyle Q_{n-1}\cdots Q_0} \\
H^{2d+1,d+1}(M, \mathbb{Z}/\ell) & \xrightarrow{\ c\cdot S^{\ell-1}x\ } & H^{1,1}(\mathfrak{X}, \mathbb{Z}/\ell) & \xrightarrow{\ \pm s^* \circ r^*\ } & H^{2b\ell+3, b\ell+1}(\mathfrak{X}, \mathbb{Z}/\ell).
\end{array}
$$

We first describe the top half of this diagram. The map labelled τ is composition with the fundamental class $\tau : \mathbb{L}^d \to R_{\mathrm{tr}}(X)$, defined in 5.15 and dual to the structure map $\pi : R_{\mathrm{tr}}(X) \to R$; the upper left square commutes by Lemma 7.6.

Recall that $d = \ell^{n-1} - 1$ and $b = d/(\ell - 1)$. Since BL(n − 1) holds, Lemma 3.13 states that $\underline{a} \in H^n_{\text{ét}}(k, \mu_\ell^{\otimes n})$ is the image of an element δ under the injection $H^{n,n-1}(\mathfrak{X}, \mathbb{Z}/\ell) \hookrightarrow H^n_{\text{ét}}(k, \mu_\ell^{\otimes n})$. Thus the upper right square commutes, and the upper right vertical is an injection. (The middle vertical isomorphisms $H^{1,1}(\mathfrak{X}, \mathbb{Z}/\ell) \cong H^{1,1}(k, \mathbb{Z}/\ell) \cong k^\times / k^{\times \ell}$ come from Lemma 1.35.)

In section 5.2 we used the element $\mu = Q_{n-2} \cdots Q_0(\delta)$ to construct a motive A and its symmetric powers $D = S^{\ell-2}(A)$ and $M = S^{\ell-1}(A)$, fitting into the triangles defining a Rost motive (see Proposition 5.7). Part of the cohomology exact sequence of the first of these triangles is

$$H^{2d+1,d+1}(M, \mathbb{Z}/\ell) \xrightarrow{S^{\ell-1}x} H^{1,1}(\mathfrak{X}, \mathbb{Z}/\ell) \xrightarrow{r^*} H^{2d+2,d+1}(D, \mathbb{Z}/\ell).$$

We also have an isomorphism $s^* : H^{2d+2,d+1}(D, \mathbb{Z}/\ell) \to H^{2b\ell+3,b\ell+1}(\mathfrak{X}, \mathbb{Z}/\ell)$, by Lemma 4.16. The map $\lambda : R_{\mathrm{tr}}(X) \to M$ is defined in Proposition 5.9, and the lower left square commutes (for some unit c) by Theorem 5.18. This also establishes exactness of the bottom row in the large diagram.

It remains to show that the triangles in the lower right square commute. Given this, the exactness of the bottom row formally implies that the top row is exact, proving the proposition.

Because the operations Q_i are $K^M_*(k)$-linear by 13.15 we have

$$Q_{n-1} \cdots Q_0(\delta \cup \beta) = Q_{n-1} \cdots Q_0(\delta) \cup \beta = Q_{n-1}(\mu) \cup \beta$$

for every unit $\beta \in k^\times$. Thus the upper right triangle in the lower right square commutes. Note that the elements μ of $H^{2b+1,b}(\mathfrak{X}, \mathbb{Z})$ and $Q_{n-1}(\mu)$ of the group $H^{2b\ell+2,b\ell+1}(\mathfrak{X}, \mathbb{Z})$ are both nonzero by Corollary 3.16.

Referring to Definition 5.10, we see that the composition $s^* \circ r^*$ in the above diagram is multiplication by the element $\phi^V(\mu)$ of $H^{2b\ell+2,b\ell}(\mathfrak{X}, \mathbb{Z}/\ell)$. We showed in Corollary 6.34 (which uses Theorem 15.38) that ϕ^V agrees with βP^b. In addition, since μ is annihilated by the Q_i with $i \leq n - 2$ we have $\beta P^b(\mu) = (-1)^{n-1} Q_{n-1}(\mu)$ by Lemma 5.14. This shows that the bottom triangle in the lower right square also commutes. This completes the proof. $\square$

7.3 HISTORICAL NOTES

The group we refer to as $H_{-1,-1}(X)$ first surfaced in the early 1970s as the K-cohomology group $H^d(X, \mathcal{K}_{d+1})$ ($d = \dim(X)$), computed via the Gersten resolution of the sheaf $\mathcal{K}_{d+1}$. For curves, the groups $H^1(X, \mathcal{K}_2)$ were an important tool in understanding $K_1(X)$, especially in the work of Bloch. However, the K-cohomology groups $H^d(X, \mathcal{K}_{d+1})$ for $d > 1$ did not attract much attention until Rost's 1988 paper [Ros88] where $A_0(X, K_1)$ was defined and $N : A_0(X, K_1)$ $\to k^\times$ was shown to be injective for certain quadric hypersurfaces X. Rost's 1996 paper [Ros96] considered $A_0(X, K_1)$ in the context of cycle modules. This thread was picked up by Déglise in [Dég03], who connected it with motives (and homotopy invariant sheaves with transfers).

The observation that this group was also the Borel–Moore motivic homology group $H_{-1,-1}(X)$ was established in [SJ06, 1.1], and most of section 7.1 is taken from [SJ06]. Proposition 7.7, which is due to Voevodsky, appeared first for $\ell = 2$ as Theorem 2.1 in the 1996 preprint version of [OVV07], and later (for all ℓ) in the appendix to [SJ06].

Part II

The goal of part II is to show that Rost varieties exist. Recall that Rost varieties for a symbol $\underline{a} = \{a_1, ..., a_n\}$ were defined in section 1.3; a Rost variety X must have dimension $d = \ell^{n-1} - 1$, its function field must split the symbol, it must satisfy some conditions about characteristic classes, and a certain homology group $\bar{H}_{-1,-1}(X)$ must inject into the group $k^\times$ of units of the field.

The strategy is to define a related notion, that of a *norm variety* for $\underline{a}$, show that norm varieties exist, and then show that norm varieties are Rost varieties. The existence of norm varieties will be established in chapter 10, using Rost's Chain Lemma 9.1; the fact that they are Rost varieties will be given in chapter 11.

We remind the reader that we are proceeding by induction on n, so that we shall assume that $\mathrm{BL}(n-1)$ holds, and use the material from part I which depends upon it. We will also invoke some results about cohomology operations from part III.

To streamline the presentation, we begin with some preliminary chapters. In chapter 8, we use algebraic cobordism to establish some degree formulas (due to Rost) that we shall need. In chapter 9, we shall establish the Chain Lemma.

Chapter Eight

Degree Formulas

LET δ BE a function from a class of smooth projective varieties over a field k to some abelian group. A *degree formula* for δ is a formula relating $\delta(X)$, $\delta(Y)$, and $\deg(f)$ for any generically finite map $f : Y \to X$ in this class. The formula is usually $\delta(Y) = \deg(f)\delta(X)$.

In this chapter, we recall three degree formulas (8.7, 8.9, and 8.12) involving the algebraic cobordism ring $\Omega_*(k)$ over a field k of characteristic 0, in a form due to Levine and Morel [LM07].

These are used in Theorem 10.12 and Proposition 10.14 to prove that any norm variety over k is a ν_{n-1}-variety. In particular, this is the case for the norm variety we construct in Theorem 11.24.

Using a standard result (8.16) for the complex bordism ring MU_*, which uses a gluing argument of equivariant bordism theory, we establish Rost's *DN Theorem* (8.18) for degrees, and define the invariant $\eta(X/S)$ of a pseudo-Galois cover, which are used in the proof of the Norm Principle 11.27 (the initialism "DN" is for "Degree" and "Norm Principle").

8.1 ALGEBRAIC COBORDISM

If k is a field, the algebraic cobordism ring $\Omega_*(k)$ is a positively graded $\mathbb{Z}$-algebra. Although its definition in [LM07, 2.4.10] is a bit lengthy, we shall only need a few facts about it. These facts are summarized in this section.

The ring $\Omega_*(k)$ is generated as a group by the classes $[X] \in \Omega_{\dim X}(k)$ of smooth projective k-varieties X. The unit of the ring is $1 = [\operatorname{Spec} k]$ and the product is determined by $[X][Y] = [X \times Y]$. As a functor, $\Omega_*(k)$ is covariant in k; if $k \subset k'$, $\Omega_*(k) \to \Omega_*(k')$ sends $[X]$ to $[X \times_k \operatorname{Spec} k']$; by convention, if X is a disjoint union of varieties X_i we write $[X]$ for $\sum[X_i]$.

The Lazard ring $\mathbb{L}_*$ is the polynomial ring $\mathbb{Z}[c_1, c_2, \ldots]$. We grade it by placing c_n in degree n; it is the coefficient ring of the universal formal group law. By definition [LM07, 2.4.10], $\Omega_*(k)$ has a canonical formal group law, so there is a canonical graded ring homomorphism $\mathbb{L}_* \to \Omega_*(k)$.

When $k = \mathbb{C}$, there is a canonical map $\Omega_n(\mathbb{C}) \to MU_{2n}$ sending the class of a smooth projective X to the class of the complex manifold $X(\mathbb{C})$; see [LM07, 4.3.1]. By Quillen's theorem [Ada74, II.8], the cobordism ring MU_{2*} has a formal group law, and the universal map $\mathbb{L}_* \to MU_{2*}$ is an isomorphism. This isomorphism factors through the map $\Omega_n(\mathbb{C}) \to MU_{2n}$, which is also an isomorphism; see [LM07, 4.3.7].

Now suppose that k is any field of characteristic 0. Then the canonical maps $\mathbb{L}_n \to \Omega_n(k)$ are also isomorphisms by [LM07, 4.3.7]. By universality of $\mathbb{L}_*$, we have graded ring isomorphisms:

$$
\begin{array}{ccc}
& \mathbb{L}_* & \xrightarrow{\cong} & MU_{2*} \\
{\cong} \nearrow & {\cong} \downarrow & & \uparrow {\cong} \\
\Omega_*(k) & \xleftarrow{\cong} \Omega_*(\mathbb{Q}) & \xrightarrow{\cong} & \Omega_*(\mathbb{C}).
\end{array}
\tag{8.1}
$$

Example 8.1.1. $\Omega_0(k) \cong \mathbb{Z}$, and if k'/k is a finite separable field extension of degree e then $[\mathrm{Spec}(k')] = e$ in $\Omega_0(k)$. If $\mathrm{char}(k) = 0$, this is immediate from the fact that $\mathrm{Spec}(k' \otimes_k \bar{k}) \cong \sqcup_1^e \mathrm{Spec}(\bar{k})$, together with the observation that the isomorphism $\Omega_0(k) \xrightarrow{\cong} \Omega_0(\bar{k})$ of (8.1) takes $[\mathrm{Spec}(k')]$ to $[\mathrm{Spec}(k' \otimes_k \bar{k})] = e[\mathrm{Spec}(\bar{k})] = e$. If $\mathrm{char}(k) > 0$, the isomorphism $\Omega_0(k) \cong \mathbb{Z}$ is established in [LM07, 2.3.4], using geometric relations for $\Omega_*(k)$ that we have not mentioned.

The homology theory Ω_*

Except for section 8.2, we shall not need any other facts about $\Omega_*(k)$. For that result, we shall need to know that $\Omega_*(k)$ extends to a functor Ω_* which assigns a graded $\Omega_*(k)$-module $\Omega_*(X)$ to every k-variety X, and that each group $\Omega_n(X)$ is generated by classes $[Y \xrightarrow{f} X]$ where Y is a smooth n-dimensional variety over k and f is projective; see [LM07, 2.5.11]. For each proper map $p : X' \to X$, there is a pushforward map p_* sending $[Y \xrightarrow{f} X']$ to $[Y \xrightarrow{pf} X]$; p_* is not a ring homomorphism. In fact, Ω_* is an "oriented Borel–Moore functor" in the sense of [LM07, 2.1.2, 2.4.10], but we shall not need the full strength of this definition.

Example 8.2. The pushforwards $\mathbb{Z} = \Omega_0(k(x)) \to \Omega_0(X)$ associated to closed points $x \in X$ induce a map from the group of zero-cycles on X onto $\Omega_0(X)$. By [LM07, 4.5.1], it induces an isomorphism $CH_0(X) \cong \Omega_0(X)$.

For each open $j : U \subset X$, the restriction $[f] \mapsto [f|_U]$ defines a pullback map j^*, making Ω_* into a presheaf of $\Omega_*(k)$-modules on each X. Moreover, if $\mathrm{char}(k) = 0$ then for each open $j : U \hookrightarrow X$ with closed complement $i : Z \to X$ the sequence

$$
\Omega_*(Z) \xrightarrow{i_*} \Omega_*(X) \xrightarrow{j^*} \Omega_*(U) \to 0
\tag{8.3}
$$

is exact; see [LM07, 3.2.7].

If X is a d-dimensional k-variety, and $F = k(X)$, the pullback maps sending $[Y \to X]$ to $[Y_F]$ induce a natural homomorphism $\Omega_{n+d}(X) \to \Omega_n(F)$. With this as motivation, we write $\Omega_n(k(X)/k)$ for the stalk of the presheaf Ω_{n+d} at the generic point of X, i.e., the colimit of the groups $\Omega_{d+n}(U)$, taken over the poset of all open U in X. This is a birational invariant of X, and hence depends only on the field $F = k(X)$. The pullback maps sending $[Y \to U]$ to $[Y_F]$ induce a natural factorization $\Omega_{n+d}(X) \to \Omega_n(F/k) \to \Omega_n(F)$, and $\Omega_n(F/k) \to \Omega_n(F)$ is an isomorphism if $\mathrm{char}(k) = 0$; see [LM07, 4.4.2].

Definition 8.4 (Degree map). ([LM07, 4.4.4]) If $\mathrm{char}(k) = 0$ and $\dim(X) = d$, the degree homomorphism is the natural composition

$$\deg_X : \Omega_{n+d}(X) \to \Omega_n(k(X)/k) \cong \Omega_n(k(X)) \cong \Omega_n(k).$$

It is not hard to see that $\deg_X([X]) = 1$ in $\Omega_0(k)$ and that if $\alpha \in \Omega_*(X)$ has $\deg_X(\alpha) = 0$ then there is an open $U \subset X$ such that α vanishes in $\Omega_*(U)$.

Example 8.4.1. If $Y \overset{f}{\to} X$ is a map of finite degree, then $\deg_X([Y \overset{f}{\to} X]) = \deg(f)$ in $\Omega_0(k(X)) \cong \mathbb{Z}$. This follows from Example 8.1.1 because the fiber $Y_{k(X)}$ over the generic point of X is $\mathrm{Spec}\, k(Y)$, and $[k(Y) : k(X)] = \deg(f)$.

8.2   THE GENERAL DEGREE FORMULA

In this section, we prove the General Degree Formula 8.7. Our formulation is taken from [LM07, Theorem 4.4.15]. We begin with a useful calculation, modified from [LM07, 4.4.7].

Theorem 8.5. *Suppose that X is a smooth k-variety, with $\mathrm{char}(k) = 0$. Then $\Omega_*(X)$ is generated as a graded $\Omega_*(k)$-module by $[X]$ and the subgroups $\Omega_i(X)$ with $i < \dim X$.*

Proof. We proceed by induction on $d = \dim(X)$, the case $d = 0$ being the observation that for each finite field extension $k \subset k'$ the ring isomorphism $\Omega_*(k) \overset{\cong}{\longrightarrow} \Omega(k')$ of (8.1) sends $[\mathrm{Spec}\, k]$ to $[\mathrm{Spec}\, k']$.

Consider a class $[Y \overset{f}{\to} X]$ in $\Omega_{n+d}(X)$ with $n \geq 0$, and set $\lambda = \deg_X([Y \overset{f}{\to} X])$. Then λ is an element of $\Omega_n(k)$, and $\alpha = [Y \overset{f}{\to} X] - \lambda \cdot [X]$ is in $\Omega_{n+d}(X)$. By construction, $\deg_X(\alpha) = 0$. Thus there is a dense open U of X such that α vanishes in $\Omega_{n+d}(U)$. We see from (8.3) that $\alpha = i_*\beta$ for some $\beta \in \Omega_*(Z)$, $Z = X - U$. By induction, $\beta = \sum c_i[Z_i' \to Z]$ where the $c_i \in \Omega_*(k)$ and the Z_i' are smooth over k of dimension at most $\dim Z < \dim(X)$. Applying the pushforward, $i_*\beta$ is in the $\Omega_*(k)$-submodule of $\Omega_*(X)$ generated by the $\Omega_i(X)$ with $i < \dim X$, and $[Y \overset{f}{\to} X] = \lambda \cdot [X] + i_*\beta$.   □

Corollary 8.5.1. *Suppose that X is any k-variety, with $\mathrm{char}(k) = 0$, and that $\tilde{X} \overset{p}{\to} X$ is a resolution of singularities. Then $\Omega_*(X)$ is generated as a graded $\Omega_*(k)$-module by $[\tilde{X} \overset{p}{\to} X]$ and the subgroups $\Omega_i(X)$ with $i < \dim X$.*

Proof. There is a dense open U of X over which p is an isomorphism. Let $Z \subset X$ and $\tilde{Z} \subset \tilde{X}$ denote the respective closed complements of U. The corollary follows immediately from the following diagram, whose rows are exact by (8.3), and Theorem 8.5.

$$
\begin{array}{ccccccc}
\Omega_*(\tilde{Z}) & \longrightarrow & \Omega_*(\tilde{X}) & \longrightarrow & \Omega_*(U) & \longrightarrow & 0 \\
\downarrow{\scriptstyle p_*} & & \downarrow{\scriptstyle p_*} & & \| & & \\
\Omega_*(Z) & \longrightarrow & \Omega_*(X) & \longrightarrow & \Omega_*(U) & \longrightarrow & 0
\end{array}
\qquad \square
$$

Definition 8.6. For each projective X, let $\mathcal{I}(X)$ denote the ideal of $\Omega_*(k)$ generated by the classes $[Z]$ of smooth projective varieties Z such that $\dim(Z) < \dim(X)$ and there is a k-morphism $Z \to X$.

That is, $\mathcal{I}(X)$ is the subgroup of $\Omega_*(k)$ generated by the classes $[Z \times W] = [Z] \cdot [W]$, where W and Z are smooth projective varieties, $\dim(Z) < \dim(X)$, and there is a k-morphism $Z \to X$.

Remark 8.6.1. Applying the pushforward $p_* : \Omega_*(X) \to \Omega_*(k)$ to Theorem 8.5, we see that for smooth projective X the ideal $p_* \Omega_*(X)$ of $\Omega_*(k)$ is generated by the ideal $\mathcal{I}(X)$ of Definition 8.6 and the element $[X]$.

Theorem 8.7 (General Degree Formula). *Let $f : Y \to X$ be a morphism of smooth projective k-varieties, with $\dim(X) = \dim(Y)$ and $char(k) = 0$. Then*

$$
[Y] - \deg(f)[X] \in \mathcal{I}(X).
$$

Proof. First suppose that $Y \xrightarrow{f} X$ is not dominant, so that $\deg(f) = 0$ and $f(Y) = Z$ for some subvariety Z of X. By Corollary 8.5.1, $[Y \to Z]$ is in the $\Omega_*(k)$-linear span of the $\Omega_i(Z)$ for $i \le \dim(Z)$. As $Z \subset X$, $[Y] \in \Omega_*(k)$ is in $\mathcal{I}(X)$. Since $[Y] - 0[X]$ is in $\mathcal{I}(X)$, the formula holds in this case.

Now suppose that $Y \xrightarrow{f} X$ is dominant, hence of finite degree. In this case, the degree map $\Omega_d(X) \to \Omega_0(k(X)) \cong \mathbb{Z}$ sends $[Y \xrightarrow{f} X]$ to $\deg(f)$, by Example 8.4.1. Therefore the element $\alpha = [Y \xrightarrow{f} X] - \deg(f)[X]$ of $\Omega_d(X)$ vanishes under the degree map $\Omega_d(X) \to \Omega_0(k(X))$. Thus (as in the proof of Theorem 8.5) there is a dense open $U \subset X$ such that α vanishes in $\Omega_d(U)$. By (8.3), $\alpha = i_* \beta$ for some $\beta \in \Omega_*(Z)$, $Z = X \setminus U$. Hence the pushforward $p_* \alpha$ equals the pushforward $(pi)_* \beta$ under $\Omega_*(Z) \to \Omega_*(k)$. By Theorem 8.5, $(pi)_* \beta$ is in the ideal $\mathcal{I}(X)$, so we are done. $\square$

Corollary 8.8. *Let $f : X' \to X$ be a morphism of smooth projective k-varieties of the same dimension, with $char(k) = 0$. If the degree of f is prime to ℓ then $\mathcal{I}(X')_{(\ell)} = \mathcal{I}(X)_{(\ell)}$ as ideals of $\Omega_*(k) \otimes \mathbb{Z}_{(\ell)}$.*

Proof. By definition $\mathcal{I}(X') \subseteq \mathcal{I}(X)$. We will show by induction on $\dim(Y)$ that if $Y \to X$ is a projective k-morphism with Y smooth and $\dim(Y) < \dim(X)$ then $[Y] \in \mathcal{I}(X')_{(\ell)}$. If $\dim(Y) = 0$, i.e., Y is a closed point x of X, then $[x]$ is an element of $\Omega_0(k) \cap \mathcal{I}(X)$. Since $\ell \nmid \deg(f)$, there is a closed point x' of X' over x with $e = [k(x') : k(x)]$ prime to ℓ, and $[x'] \in \mathcal{I}(X')$. Since $[x'] = e[x]$ by Example 8.1.1, we have $[x] = [x']/e$ in $\Omega_0(k)_{(\ell)} = \mathbb{Z}_{(\ell)}$.

Choose an irreducible component Y' of $Y \times_X X'$ such that $e = [k(Y') : k(Y)]$ is prime to ℓ, and let $\tilde{Y}$ be a resolution of singularities of Y'. By Theorem 8.7, $[\tilde{Y}] - e[Y]$ is in $\mathcal{I}(Y)$. Since $[\tilde{Y}] \in \mathcal{I}(X')$ and $\mathcal{I}(Y) \subseteq \mathcal{I}(X')_{(\ell)}$ by the inductive hypothesis, we see that $e[Y]$ and hence $[Y]$ is in $\mathcal{I}(X')_{(\ell)}$. $\qquad\square$

8.3 OTHER DEGREE FORMULAS

Other important degree formulas arise by applying suitable homomorphisms from $\Omega_d(k)$ to abelian groups. We enumerate a few, proving those we shall need.

The characteristic number $s_d(X)$ of a smooth d-dimensional projective variety X provides an illustration of this method; it is used in Propositions 10.14 and 10.15 as part of the proof that norm varieties are Rost varieties.

Recall from section 1.3 that the number $s_d(X)$ is defined to be the degree of the characteristic class $s_d(T_X) \in CH^d(X)$. It is shown in [LM07, 4.4.19] that $X \mapsto s_d(X)$ is an algebraic cobordism invariant, i.e., it factors through a homomorphism $s_d : \Omega_d(k) \to \mathbb{Z}$ satisfying $s_d([X]) = s_d(X)$.

Theorem 8.9 (Rost's Degree Formula for s_d). *Let $f : W \to X$ be a morphism of smooth projective varieties of dimension $d = \ell^\nu - 1$, over a field of characteristic 0. Suppose that the degree of every 0-cycle on X is divisible by ℓ. Then*

$$s_d(W) \equiv \deg(f) s_d(X) \pmod{\ell^2}.$$

Proof. (See [Ros02], [LM07, Thm. 4.4.23].) Consider the element $a = [W] - \deg(f)[X]$ in $\Omega_d(k)$. Since s_d is additive, $s_d(a) = s_d(W) - \deg(f) s_d(X)$. Since $a \in \mathcal{I}(X)$ by Theorem 8.7, it suffices to show that s_d vanishes modulo ℓ^2 on $\mathcal{I}_d(X)$, the degree d part of $\mathcal{I}(X)$. By Definition 8.6, $\mathcal{I}_d(X)$ is generated as a group by the classes $[Y \times Z]$, where $Z \to X$ is a map with $\dim(Z) < d$ and $\dim(Y) + \dim(Z) = d$.

It is a general fact that when $\dim(Y), \dim(Z) > 0$ we have $s_{\dim Y + \dim Z}(Y \times Z) = 0$. For stably complex manifolds this is [MS74, 16.3 and 16.5]; for algebraic varieties this is [LM07, (4.8), p. 132]. In the case at hand, when $\dim(Z) < d$, we have $s_d(Y \times Z) = 0$ unless $\dim(Z) = 0$. When $\dim(Z) = 0$, i.e., Z is a 0-cycle, we have $\ell | \deg(Z)$ by assumption. In this case, $\dim(Y) = d = \ell^{\nu-1} - 1$, and we saw in section 1.3 that $\ell | s_d(Y)$. Hence $\ell^2 | \deg(Z) s_d(Y) = s_d(Y \times Z)$, as desired. $\qquad\square$

Remark 8.9.1. The characteristic numbers s_d in [MS74] and in this book are written as S_d in [LM07]; the class called $s_d(X)$ in [LM07] is our class $s_d(X)/\ell$.

A second class of examples, including the Levine–Morel "higher degree formula" (Theorem 8.12), comes from graded ring homomorphisms

$$\psi = \psi_{\ell,n} : \mathbb{L}_* \to \mathbb{F}_\ell[v]$$

with $v = v_n$ in degree $d = \ell^n - 1$, associated to ℓ-typical formal group laws of height n. Note that $\psi(x) = 0$ for $x \in \Omega_i(k)$ unless $d \mid i$, and that ψ takes $\Omega_{rd}(k) \cong \mathbb{L}_{rd}$ to the 1-dimensional vector space generated by v^r.

Construction 8.10. Here is a sketch of the construction of ψ; useful references for this material are appendix A2 of Ravenel's book [Rav86] and [Lan73]. There is a universal ℓ-typical formal group law, which we fix; it is represented by a ring homomorphism from $\mathbb{L}_*$ to $\mathbb{Z}_{(\ell)}[v_1, v_2, ...]$, with v_i in degree $\ell^i - 1$. More restrictively, we may consider ℓ-typical formal group laws of height n; a universal such law is represented by the quotient homomorphism $\psi = \psi_{\ell,n} : \mathbb{L}_* \to \mathbb{F}_\ell[v_n]$, obtained by going modulo ℓ and killing all the v_i for $i \neq n$.

The following definition of the numbers $t_{d,r}(X)$ is taken from [LM07, 4.4].

Definition 8.11. Fix a graded ring homomorphism $\psi : \mathbb{L}_* \to \mathbb{F}_\ell[v]$ corresponding to an ℓ-typical formal group law of height n, as above. For $r > 0$, the homomorphism

$$t_{d,r} : \Omega_{rd}(k) \cong \mathbb{L}_{rd} \to \mathbb{F}_\ell$$

is defined by $\psi(x) = t_{d,r}(x)v^r$ for $x \in \Omega_{rd}(k)$. That is, $t_{d,r}(x)$ is the coefficient of v_n^r in $\psi(x)$. If X is a smooth projective variety over k, of dimension rd, then X determines a class $[X]$ in $\Omega_{rd}(k)$, and we write $t_{d,r}(X)$ for $t_{d,r}([X])$.

Note that a different choice of ψ corresponds to an automorphism of $\mathbb{F}_\ell[v]$, so the number $t_{d,r}$ depends upon the choice of formal group law above, but only up to a unit of $\mathbb{F}_\ell$.

Here is the Levine–Morel "higher degree formula" (cf. [LM07, Thm. 4.4.23]). We fix an n and set $d = \ell^n - 1$, $\psi = \psi_{\ell,n}$.

Theorem 8.12 (Higher Degree Formula)*. Let X be a smooth projective variety of dimension rd over a field k of characteristic 0 which admits a sequence of surjective morphisms with $X^{(i)}$ smooth and $\dim X^{(i)} = d \cdot i$:*

$$X = X^{(r)} \to X^{(r-1)} \to \cdots \to X^{(0)} = \mathrm{Spec}(k).$$

Suppose moreover that ℓ divides the degree of every zero-cycle on each of the $X^{(i)} \times_{X^{(i-1)}} k(X^{(i-1)})$. Then $\psi(\mathcal{I}(X)) = 0$.

In particular, if W is also smooth projective of dimension rd then for every morphism $f : W \to X$ we have $\psi([W]) = \deg(f) \cdot \psi([X])$, i.e.,

$$t_{d,r}(W) = \deg(f) \cdot t_{d,r}(X).$$

Proof. Since $[W] - \deg(f)[X]$ is in $\mathcal{I}(X)$ by Theorem 8.7, it suffices to show that $\psi(\mathcal{I}(X)) = 0$. We will show that $\psi(\mathcal{I}(X^{(i)})) = 0$ by induction on i.

We first show that $\psi(\mathcal{I}(X^{(1)})) = 0$. By assumption, if z is a closed point of $X^{(1)}$ then the degree of z is divisible by ℓ; since $[z] = [k(z) : k]$ in $\Omega_0(k)$ by Example 8.4.1, we have $\psi([z]) = 0$. Next, note that if d does not divide $\dim(Z)$, then $\psi([Z]) = 0$ as $\mathbb{F}_\ell[v]$ is zero in degree $\dim(Z)$. Since $\dim X^{(1)} = d$, $\psi([Z]) = 0$ for every generator $[Z]$ of $\mathcal{I}(X^{(1)})$.

In the general case, consider a generator $[Z]$ of the ideal $\mathcal{I}(X^{(i)})$, defined by a k-morphism $f : Z \to X^{(i)}$ with $\dim(Z) < di$. If $d \nmid \dim(Z)$ then $\psi([Z]) = 0$ for degree reasons. If $\dim(Z) < d(i-1)$ then $[Z] \in \mathcal{I}(X^{(i-1)})$ via the obvious composition $f_i : Z \to X^{(i-1)}$, and $\psi([Z]) = 0$ by induction. Thus we may assume that $\dim(Z) = d(i-1) = \dim(X^{(i-1)})$. By the General Degree Formula 8.7, $[Z] - \deg(f_i)[X^{(i-1)}]$ is in $\mathcal{I}(X^{(i-1)})$. Since $\psi(\mathcal{I}(X^{(i-1)})) = 0$, we get $\psi([Z]) = \deg(f_i)\psi([X^{(i-1)}])$. Thus in order to show that $\psi([Z]) = 0$, it suffices to show that $\deg(f_i) \equiv 0 \pmod{\ell}$.

If f_i is not dominant, then $\deg(f_i) = 0$ by definition. On the other hand, if f_i is dominant, then the generic point of Z maps to a closed point η of $X^{(i)} \times_{X^{(i-1)}} k(X^{(i-1)})$. By the 0-cycles hypothesis of this theorem, ℓ divides $\deg(\eta) = \deg(f_i)$. $\qquad\square$

Example 8.12.1. Here is a typical application. Suppose that $t_{d,r}(W) \neq 0$ in $\mathbb{F}_\ell$. Then the degree of any dominant map $f : W \to X$ must be prime to ℓ.

Here are some properties of the numbers $t_{d,r}$ that we shall need below, in Corollary 8.17 and Theorem 8.18. Recall from section 1.3 (or [MS74, 16.6/16.E] and [Sto68, p. 130]) that if $d = \ell^n - 1$ then ℓ divides the characteristic number $s_d(X)$, so that $s_d(X)/\ell$ is an integer.

Lemma 8.13. *Let X/k be a smooth projective variety of dimension rd, and let $k \subseteq \mathbb{C}$ be an embedding.*

1. *There is a $u \in \mathbb{F}_\ell^\times$ such that $t_{d,1}(x) \equiv u\, s_d(x)/\ell \pmod{\ell}$ for all $x \in \Omega_d(k)$.*
2. *If $X = \prod_{i=1}^r X_i$ and $\dim(X_i) = d$, then $t_{d,r}(X) = \prod_{i=1}^r t_{d,1}(X_i)$.*
3. *$t_{d,r}(X)$ depends only on the class of the manifold $X(\mathbb{C})$ in the complex bordism ring MU_*.*

Proof. Part (1) is [LM07, Proposition 4.4.22(3)]. Part (2) is immediate from the graded multiplicative structure on $\Omega_*(k)$: since $[X] = \prod[X_i]$, $\psi([X]) = \prod \psi([X_i]) = \prod t_{d,1}(X_i)v$. Finally, part (3) is a consequence of the fact that the natural homomorphism $\Omega_*(k) \to MU_{2*}$ is an isomorphism (since both rings are canonically isomorphic to the Lazard ring $\mathbb{L}_*$). $\qquad\square$

8.4   AN EQUIVARIANT DEGREE FORMULA

In this section we consider G-varieties, where $G = \mu_\ell^n$, and use the Higher Degree Formula 8.12 to prove the DN Theorem 8.18. This theorem shows that (for a certain class of G-varieties X) the degrees of maps $W \to X$ must be prime to ℓ for a class of varieties W. At its heart, it follows the pattern of Example 8.12.1.

We will need the following standard consequence of equivariant bordism theory of complex manifolds, which is usually attributed to Conner and Floyd. Let $\psi : \mathbb{L}_* \to \mathbb{F}_\ell[v_n]$ be as in Definition 8.11, with $v = v_n$ in degree $d = \ell^n - 1$.

Lemma 8.14. *Suppose that the abelian ℓ-group $G = (\mathbb{Z}/\ell)^n$ acts without fixed points on a stably complex manifold M, preserving the stably complex structure. Then $\psi([M]) = 0$ in $\mathbb{F}_\ell[v_n]$.*

Proof. Let $I(n)$ denote the ideal in MU_* generated by $\{\ell, [M_1], \ldots, [M_{n-1}]\}$, where each M_i is a complex manifold with $\dim_{\mathbb{C}}(M_i) = \ell^i - 1$ which generates $MU_{2\ell^i - 2}$ modulo decomposable elements (a "Milnor manifold"). By [tD70] and [Flo71], $I(n)$ is the ideal of bordism classes represented by smooth G-manifolds without fixed points. Since M has no fixed points, $[M]$ is in $I(n)$.

Since ℓ is the only generator of $I(n)$ whose dimension is a multiple of $d = \ell^n - 1$, the map $\psi : MU_{2*} \to \mathbb{F}_\ell[v_n]$ is zero on every generator and hence on $I(n)$. It follows that $\psi([M]) = 0$. $\qquad\square$

Definition 8.15. For any group G, we say that two stably complex G-manifolds are *G-fixed point equivalent* if $\mathrm{Fix}_G X$ and $\mathrm{Fix}_G Y$ are 0-dimensional, and there is a bijection $\mathrm{Fix}_G X \to \mathrm{Fix}_G Y$ under which the families of tangent spaces at the fixed points are isomorphic as G-representations.

We say that two G-varieties X and Y over k are *G-fixed point equivalent* if $\mathrm{Fix}_G X$ and $\mathrm{Fix}_G Y$ are 0-dimensional, lie in the smooth locus of X and Y, and there is a separable extension K of k and a bijection $\mathrm{Fix}_G X_K \to \mathrm{Fix}_G Y_K$ under which the families of tangent spaces at the fixed points are isomorphic as G-representations over K.

Theorem 8.16. *Let G be $(\mathbb{Z}/\ell)^n$ and let X and Y be compact complex G-manifolds which are G-fixed point equivalent. Then $\psi([X]) = \psi([Y])$ in $\mathbb{F}_\ell[v_n]$, i.e., $t_{d,r}(X) = t_{d,r}(Y)$ for all $r > 0$.*

$$M = X \cup -Y$$

Proof. Remove equivariantly isomorphic small balls about the fixed points of X and Y, and let $M = X \cup -Y$ denote the result of joining the rest of X

and Y, using the opposite (stably complex) orientation on Y, as in the figure. This construction is possible because X and Y are G-fixed point equivalent. Then M has a canonical stably complex structure, G acts on M with no fixed points, and $[X] - [Y] = [M]$ in MU_*; see [Sto68, II]. By Lemma 8.14, $\psi([X]) - \psi([Y]) = \psi([M]) = 0$. By Definition 8.11, the final assertion follows automatically. $\qquad\square$

Theorem 8.16 applies to complex G-varieties, via realization, and even to G-varieties over a field k with an embedding into $\mathbb{C}$. By Lemma 8.13(1), we conclude:

Corollary 8.17. *For d-dimensional X and Y as in Theorem 8.16, we have $s_d(X) \equiv 0 \pmod{\ell^2}$ iff $s_d(Y) \equiv 0 \pmod{\ell^2}$.*

We can now establish the "DN" degree theorem. It will be used in chapter 11 to verify the Norm Principle.

To motivate it, suppose we are given varieties X and Y, a finite field extension $k(Y) \subset F$ and an F-point of X, $\mathrm{Spec}(F) \to X$. Up to birational equivalence, this data determines a smooth variety W with $k(W) = F$ together with a dominant map $g : W \to Y$ and a map $f : W \to X$ compatible with the data in the sense of the following diagram:

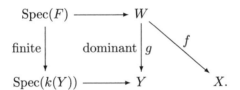

Indeed, the data determines a smooth variety W' up to birational equivalence together with a dominant rational map $Y \dashleftarrow W'$ and a rational map $W' \dashrightarrow X$; we may replace W' by a blow-up W to eliminate the points of indeterminacy and obtain the desired morphisms $f : W \to X$, $g : W \to Y$, with $\deg(g) = [F : k(Y)]$.

Theorem 8.18 (DN Theorem). *Let $u_1, ..., u_r$ be r symbols in $K^M_{n+1}(k)/\ell$ and let $X_1, ..., X_r$ be smooth projective G-varieties, each of dimension $d = \ell^n - 1$ such that for each i:*

1. *$k(X_i)$ splits u_i;*
2. *u_i is nonzero over $k(X_1 \times \cdots \times X_{i-1})$; and*
3. *$s_d(X_i) \not\equiv 0 \pmod{\ell^2}$.*

Set $X = \prod_1^r X_i$ and let Y be a smooth projective G-variety which is G-fixed point equivalent to the disjoint union of m copies of X, where $\ell \nmid m$.

If $g : W \to Y$ is a dominant map of degree prime to ℓ, then any map $f : W \to X$ is dominant and of degree prime to ℓ.

Proof. We will apply the Higher Degree Formula 8.12 to $W \xrightarrow{f} X$ and the sequence of projections between the $X^{(i)} = \prod_{j=1}^{i} X_j$, $1 \le i \le r$. We must first check that the zero-cycle hypothesis of Theorem 8.12 is satisfied. For a fixed i, it is convenient to set $F_i = k(X_1 \times \cdots \times X_{i-1})$ and $X' = X^{(i)} \times_{X^{(i-1)}} F_i$. The hypotheses (1–2) of the DN Theorem say that the symbol u_i is nonzero over F_i but splits over the generic point of X'; by specialization (see Remark 1.13.1), u_i splits over all closed points of X'. This implies that ℓ divides the degree of any closed point η of X', by the following transfer argument 1.2: the composition $K_*^M(k)/\ell \to K_*^M(k(\eta))/\ell \to K_*^M(k)/\ell$ is multiplication by $\deg(\eta) = [k(\eta) : k]$; since it sends the nonzero element u_i to 0, $\ell \mid \deg(\eta)$. Hence Theorem 8.12 applies and we have $\psi([W]) = \deg(f)\,\psi([X])$.

Now $\psi(\mathcal{I}(X)) = 0$ by Theorem 8.12, and $\mathcal{I}(W) \subseteq \mathcal{I}(X)$ since $W \to \mathrm{Spec}(k)$ factors through X, so $\psi(\mathcal{I}(W)) = 0$. We now localize at ℓ; since $\deg(g)$ is prime to ℓ, Corollary 8.8 implies that $\mathcal{I}(Y)_{(\ell)} = \mathcal{I}(W)_{(\ell)}$ as ideals of the localization $\Omega_*(k)_{(\ell)}$. Since the map $\psi : \Omega_*(k) \to \mathbb{F}_\ell[v]$ factors through $\Omega_*(k)_{(\ell)}$, ψ is zero on $\mathcal{I}(Y)_{(\ell)}$.

Applying the General Degree Formula 8.7 to g, we conclude that $\psi([W]) = \deg(g)\,\psi([Y])$, so that $\deg(f)\,\psi([X]) = \psi([W]) = \deg(g)\,\psi([Y])$, i.e.,

$$\deg(f)\, t_{d,r}(X) = \deg(g)\, t_{d,r}(Y).$$

On the other hand, since Y is G-fixed point equivalent to m copies of X, we have $m \cdot t_{d,r}(X) = t_{d,r}(Y)$ by Theorem 8.16. Condition (3) of the DN Theorem and Lemma 8.13(1,2) imply that $t_{d,1}(X_i) \ne 0$ for all i and hence that $t_{d,r}(X) \ne 0$. It follows that $\deg(f) = m\deg(g) \ne 0$ in $\mathbb{Z}/\ell$, as required. $\qquad\square$

8.5   THE η-INVARIANT

In this section we define an invariant $\eta(X/S)$ in $\mathbb{Z}/\ell$ for pseudo-Galois covers (morphisms which are generically Galois), and provide a degree formula for $\eta(X/S)$ which will be needed in chapter 11. Although η can be formulated in terms of an invariant in $\Omega_0(k) \cong \mathbb{Z}$, it is simpler to define it directly. We assume in this section that k is a field containing $1/\ell$ and μ_ℓ.

Definition 8.19. Let $p : X \to S$ be a finite surjective map of varieties over k. We say that p is a *pseudo-Galois* cover if $k(X)/k(S)$ is a Galois field extension with group $G = \mathrm{Gal}(k(X)/k(S))$, and the action of G on the field $k(X)$ extends to an action of G on X. Note that p is étale over an open subvariety U of S.

Examples 8.20. (1) A *Galois covering* is a finite étale map $X \to S$ of varieties such that $k(X)/k(S)$ is a Galois field extension and the action of $G = \mathrm{Gal}(k(X)/k(S))$ on $k(X)$ extends to an action of G on X. Equivalently, $X \times_S X$ is isomorphic to the disjoint union $X \times G = \sqcup_{g \in G} X$ of copies of X. (See [SGA71, V(2.8)].)

(2) If $\mu_\ell \subset k$, $G = \mu_\ell$ acts on $\mathbb{A}^1 = \mathrm{Spec}(k[t])$; $\zeta \in \mu_\ell$ sends t to ζt. Then $\mathbb{A}^1$ is pseudo-Galois over $S = \mathrm{Spec}(k[t^\ell])$ but is not étale over the origin.

(3) If a finite group G acts on a variety X, the geometric quotient X/G has function field $k(X)^G$, and $X \to X/G$ is pseudo-Galois. If $X \to S$ is any pseudo-Galois cover with group G, the canonical map $X/G \to S$ is birational. If in addition S is normal then $X/G \cong S$, by Zariski's Main Theorem.

We will now assume that $X \xrightarrow{p} S$ is a pseudo-Galois cover with group $G = \mathbb{Z}/\ell$ and that $\mu_\ell \subset k$. Choosing an isomorphism $G \xrightarrow{\simeq} \mu_\ell$, the Galois group G acts on $\mathbb{A}^1$ (as in Example 8.20(2)), and we can form $\mathbb{A}^1 \times_G X$. If $U \subset S$ is the (open) locus where X is étale over S, then $L = \mathbb{A}^1 \times_G (p^{-1}U)$ is a line bundle over U. If in addition U is smooth, we may form the divisor class $c_1(L)$ in $CH^1(U)$ and the zero-cycle $z = c_1(L)^{\dim(U)}$ in $CH_0(U)$.

Lemma 8.21. *If U is a smooth open subvariety of a projective variety S, and every closed point s in the complement of U has $[k(s):k] \equiv 0 \pmod{\ell}$, the degree $CH_0(S) \to \mathbb{Z}$ induces a map $\deg_U : CH_0(U) \to \mathbb{Z}/\ell$.*

Suppose that $f : S' \to S$ is a projective morphism with $\dim(S') = \dim(S)$, and $U' \subseteq f^{-1}(U) \subset S'$ is such that every point s' in the complement $S' \setminus U'$ of U' has $[k(s'):k] \equiv 0 \pmod{\ell}$. If U is smooth and $f_U : U' \to U$ then for every 0-cycle z on U,

$$\deg_{U'}(f_U^* z) = (\deg f) \deg_U(z) \qquad in \ \mathbb{Z}/\ell.$$

Proof. There is an exact sequence $CH_0(S \setminus U) \to CH_0(S) \to CH_0(U) \to 0$. Since $\deg : CH_0(S) \to \mathbb{Z}$ sends $[s]$ to $[k(s):k]$, the image of $CH_0(S \setminus U)$ is contained in $\ell\mathbb{Z}$, whence the first assertion.

For the second assertion, note that we may replace U' by $f^{-1}(U)$ since ℓ divides the degree of every point in $f^{-1}(U) \setminus U'$. In this case, the restriction $f_U : U' \to U$ is proper, and the projection formula yields

$$\deg_{U'}(f_U^* z) = \deg_U(f_* f_U^* z) = (\deg f) \deg_U(z). \qquad \square$$

Definition 8.22. Suppose that $p : X \to S$ is a pseudo-Galois cover with group $G = \mathbb{Z}/\ell$, with S projective. We say that $\eta(X/S)$ *is defined* if there is a smooth open $U \subset S$ such that p is étale over U, and ℓ divides $[k(s):k]$ for every closed $s \in S \setminus U$.

In this case, we define the invariant $\eta(X/S)$ in $\mathbb{Z}/\ell$ to be the degree of $z = c_1(L)^{\dim(U)}$ in $\mathbb{Z}/\ell$, i.e., $\eta(X/S) = \deg(z)$. This is well defined in $\mathbb{Z}/\ell$ and independent of the choice of U by the case $S' = S$ of Lemma 8.21.

Remark 8.22.1. We could have defined $\eta(X/S)$ in terms of algebraic cobordism. Suppose given a pseudo-Galois cover $X \to S$ with $d = \dim(X)$. Then $[U]$ is an element of $\Omega_d(U)$, and the standard cycle $[U, L, \ldots, L]$ is an element of $\Omega_0(U)$, in the sense of [LM07, 2.1.6, 2.4.10].

If $\operatorname{char}(k) = 0$, we have $CH_0(S) \cong \Omega_0(S)$ by Example 8.1.1. This extends to a morphism $\theta : \Omega_*(U) \to CH_*(U)$ of oriented Borel–Moore homology theories [LM07, 4.5.1]. By [LM07, 2.1.8], $\theta([U, L, \dots, L]) = c_1(L)^d$. Now take the degree.

The next lemma shows that $\eta(X/S)$ is essentially a birational invariant.

Lemma 8.23. *Suppose that X/S is a pseudo-Galois cover with group $G = \mathbb{Z}/\ell$. If $\bar{f} : S' \to S$ is a birational morphism, and X' is the normalization of $X \times_S S'$ in the field $k(X)$, then X'/S' is pseudo-Galois.*

Suppose in addition that S and S' are projective, S' is smooth, and $\eta(X/S)$ is defined (8.22). Then $\eta(X'/S')$ is defined, and equals $\eta(X/S)$.

Proof. The group G acts on X' because it is the normalization of the G-variety $X \times_S S'$, so $X' \to S'$ is pseudo-Galois and $\bar{f} : S' \to S$ lifts to a map $f : X' \to X$.

To see that $\eta(X'/S')$ is defined, let U be a smooth open in S such that $[k(s) : k] \equiv 0 \pmod{\ell}$ for all $s \in S \setminus U$, and set $U' = \bar{f}^{-1}(U)$. Since S' is smooth, so is U'. If $s' \in S' \setminus U'$ then $s = \bar{f}(s') \notin U$ and hence $[k(s') : k] \equiv 0 \pmod{\ell}$. Consider the open subscheme $V = X \times_S U$ of X; since $V \to U$ is étale, the pullback $V' = V \times_U U'$ is étale over U' and hence smooth. Since X' is normal, it follows that the open subscheme $X' \times_{S'} U'$ of X' is isomorphic to V' and hence that $\eta(X'/S')$ is defined.

Finally, the equality $\eta(X/S) = \eta(X'/S')$ is a special case of the calculation in Lemma 8.24. Note that $X'/G \to S'$ is an isomorphism by Example 8.20(3), since S' is smooth. $\qquad\square$

Lemma 8.24. *Suppose that $X \xrightarrow{p} S$ is pseudo-Galois with group G, and that $f : X' \to X$ is a G-equivariant morphism of projective varieties of dimension d. Set $S' = X'/G$. If $\eta(X/S)$ and $\eta(X'/S')$ are both defined then*

$$\eta(X'/S') = \deg(f)\eta(X/S).$$

Proof. Since f is equivariant, it induces a morphism $\bar{f} : X'/G \to X/G \to S$. Let $U \subset S$ be such that X is étale over U, and $[k(s) : k] \equiv 0 \pmod{\ell}$ for every $s \in S \setminus U$. Setting $U' = \bar{f}^{-1}(U)$, we have the G-equivariant diagram:

$$
\begin{array}{ccc}
p'^{-1}(U') & \xrightarrow{\ f\ } & p^{-1}(U) \\
\downarrow{\scriptstyle p'} & & \downarrow{\scriptstyle p} \\
U' & \xrightarrow{\ \bar{f}\ } & U.
\end{array}
$$

Since the vertical maps are Galois covers, it is easy to check that this is a pullback diagram. Thus the line bundle L' on U' is the pullback $\bar{f}^* L$ of the line bundle L on U, and $\bar{f}^* : CH_0(U) \to CH_0(U')$ sends $z = c_1(L)^d$ to $z' = c_1(L')^d$.

Since $\deg : CH_0(S') \to \mathbb{Z}$ factors through $\bar{f}_*$, Lemma 8.21 yields

$$\eta(X'/S') = \deg_{U'}(z') = \deg_U(\bar{f}_* \bar{f}^* z) = \deg(\bar{f}) \deg_U(z) = \deg(\bar{f}) \eta(X/S). \quad \square$$

Corollary 8.24.1. *Let X/S be a pseudo-Galois cover with group $G = \mathbb{Z}/\ell$, and set $S' = X/G$. Then $\eta(X/S) = \eta(X/S')$.*

Proof. The cover X/S' is pseudo-Galois by Example 8.20(3). Let $U \subset S$ be an open associated to $\eta(X/S)$. Since $X|_U \to U$ is a Galois cover, $X/G \to S$ is an isomorphism over U. It follows immediately that $\eta(X/S')$ is defined, so Lemma 8.24 applies (with $f : X \to X$ the identity). $\qquad\qquad\qquad\qquad\qquad\quad \square$

Theorem 8.25. *Suppose that $\mathrm{char}(k) = 0$ and that X/S and X'/S' are two pseudo-Galois covers with group $G = \mathbb{Z}/\ell$. Suppose in addition that X and X' are projective of the same dimension, and that $\eta(X/S)$ and $\eta(X'/S')$ are defined. Then for any G-equivariant rational map $f : X' \dashrightarrow X$,*

$$\eta(X'/S') = \deg(f) \cdot \eta(X/S).$$

Proof. By definition, f induces a G-map from a G-invariant open U' in X' to X. Hence f induces a unique map $U'/G \to X/G$, i.e., a rational map $X'/G \dashrightarrow X/G$ and hence (by Example 8.20(3)) a unique rational map $\bar{f} : S' \dashrightarrow S$ compatible with f. Clearly, $\deg(\bar{f}) = \deg(f)$.

There is a birational map $\widetilde{S} \to S'$ with $\widetilde{S}$ smooth projective, which eliminates the points of indeterminacy of the rational map $\bar{f}$ in the sense that $\bar{f}$ extends to the morphism $\widetilde{S} \to S$. Let $\widetilde{X}$ denote the normalization of $X' \times_{S'} \widetilde{S}$ in $k(X')$; by Lemma 8.23, $\eta(\widetilde{X}/\widetilde{S})$ is defined and equals $\eta(X'/S')$. By Lemma 8.24, $\eta(\widetilde{X}/\widetilde{S}) = \deg(f) \cdot \eta(X/S)$, and the result follows. $\qquad\qquad\qquad\qquad\qquad\quad \square$

8.6   HISTORICAL NOTES

The notion of framed bordism was introduced by Pontryagin in the 1938 paper [Pon38], in order to study homotopy groups. Bordism for real (and oriented) manifolds was introduced by René Thom [Tho54] in 1954. It was promoted to a generalized cohomology theory (cobordism) by Atiyah [Ati61a] in 1961. The extension to stably complex structures (with coefficient ring MU_*) was introduced by Milnor and Novikov (see [Mil60]) and made into a generalized cohomology theory in [Las63]. The development up to 1968 is summarized in [Sto68]; Quillen's identification of MU_* with the Lazard ring $\mathbb{L}_*$ followed soon after, and all of the cobordism-theoretic versions of the results in this chapter were known by 1975 (see [Ada74]). This includes Lemma 8.13, Lemma 8.14, and Theorem 8.16 (birational invariance of $s_d(X)/\ell$).

The applications to zero-cycles on varieties in this chapter first appeared in Rost's 1998 preprint [Ros98b], where the passage to stably complex manifolds and the complex cobordism ring MU_* played an important role. With the appearance of the algebraic cobordism ring $\Omega_*(k)$ in 2001 (published as the book [LM07] in 2007), it became possible to recast many of these results in a cleaner algebraic way. The proofs in [LM07] only use resolution of singularities (and weak factorization), which holds in characteristic 0.

The definition of a pseudo-Galois covering originated in the 1996 paper [SV96] in connection with qfh sheaves. The invariant $\eta(X/S)$ of Definition 8.22 is due to Rost, and first appeared in [SJ06].

This chapter is based upon the lectures of Markus Rost at the Institute for Advanced Study in spring 2005, and [SJ06]. Section 8.5 is based on section 3 of [SJ06]. The name "DN Theorem" (Degree Theorem for the Norm Principle) is due to Rost.

Chapter Nine

Rost's Chain Lemma

IN THIS CHAPTER we state and prove Rost's Chain Lemma (Theorem 9.1). The proof (due to Markus Rost) does not use the inductive assumption of part I that $\mathrm{BL}(n-1)$ holds. The Chain Lemma is used in chapter 10 to construct norm varieties.

Throughout this chapter, ℓ is a fixed prime, and k is a field containing $1/\ell$ and all ℓ^{th} roots of unity. We fix an integer $n \geq 2$ and an n-tuple $(a_1, ..., a_n)$ of units in k, such that the symbol $\underline{a} = \{a_1, ..., a_n\}$ is nontrivial in the Milnor K-group $K_n^M(k)/\ell$.

Here is the statement of the Chain Lemma, which we prove in section 9.6; the special case $n=2$ is proven in section 9.2. The notion of an ℓ-form on a locally free sheaf over S is introduced in section 9.1; Definition 9.4 shows how ℓ-forms may be used to define elements of $K_n^M(k(S))/\ell$.

Theorem 9.1 (Rost's Chain Lemma). *Let $\underline{a} = \{a_1, ..., a_n\}$ be a nontrivial symbol in $K_n^M(k)/\ell$, where k is a field containing $1/\ell$. Then there exists a smooth projective cellular variety S/k and a collection of invertible sheaves $J = J_1, J_1', \ldots, J_{n-1}, J_{n-1}'$ on S equipped with nonzero ℓ-forms*

$$\gamma = \gamma_1, \gamma_1', \ldots, \gamma_{n-1}, \gamma_{n-1}'$$

satisfying the following conditions:

1. $\dim S = \ell(\ell^{n-1} - 1) = \ell^n - \ell$;
2. $\{a_1, \ldots, a_n\} = \{a_1, \ldots, a_{n-2}, \gamma_{n-1}, \gamma_{n-1}'\} \in K_n^M(k(S))/\ell,$
   $\{a_1, \ldots, a_{i-1}, \gamma_i\} = \{a_1, \ldots, a_{i-2}, \gamma_{i-1}, \gamma_{i-1}'\} \in K_i^M(k(S))/\ell$ *for* $2 \leq i < n;$
   in particular, $\{a_1, \ldots, a_n\} = \{\gamma, \gamma_1', \ldots, \gamma_{n-1}'\} \in K_n^M(k(S))/\ell;$
3. $\gamma \notin \Gamma(S, J)^{\otimes(-\ell)};$
4. *for any i and any $s \in V(\gamma_i) \cup V(\gamma_i')$, the field $k(s)$ splits $\{a_1, \ldots, a_n\};$*
5. *for any i and any $s \in V(\gamma_i) \cup V(\gamma_i')$, ℓ divides $[k(s):k];$*
6. *the degree of $c_1(J)^{\dim S}$ is relatively prime to ℓ.*

Note that parts (3) and (5) are immediate from parts (2) and (4).

9.1   FORMS ON VECTOR BUNDLES

We begin with a review of some well-known facts about ℓ-forms.

If V is a vector space over a field k, an ℓ-*form* on V is a symmetric ℓ-linear function on V, i.e., a k-linear map $\phi : \mathrm{Sym}^{\ell}(V) \to k$. It determines an ℓ-*ary form*, i.e., a function $\varphi : V \to k$ satisfying $\varphi(\lambda v) = \lambda^{\ell} \varphi(v)$, by $\varphi(v) = \phi(v, v, \dots, v)$. If $\ell!$ is invertible in k, ℓ-forms are in 1–1 correspondence with ℓ-ary forms.

When $\dim(V) = 1$, there is little difference between ℓ-forms and ℓ-ary forms. If $V = k$ then every ℓ-form ϕ may be written as $\phi(\lambda_1, \dots, \lambda_{\ell}) = a \prod \lambda_i$ for some $a \in k$, and ϕ is determined by its associated ℓ-ary form, $\varphi(\lambda) = a\lambda^{\ell}$. If $\dim(V) = 1$, the choice of an isomorphism $f : V \to k$ determines a form $f^* \varphi$ on V. Up to isometry, nonzero 1-dimensional ℓ-forms are in 1–1 correspondence with elements of $k^{\times}/k^{\times \ell}$, because the isomorphism $v \mapsto sf(v)$ determines the form $s^{\ell} f^* \varphi$. Therefore an n-tuple of forms φ_i on a 1-dimensional V determines a well-defined element of $K_n^M(k)/\ell$ which we write as $\{\varphi_1, \dots, \varphi_n\}$.

Definition 9.2. Given an ℓ-ary form φ on a 1-dimensional vector space V, the *Kummer algebra* $A = A_{\varphi}(V)$ is the quotient of the symmetric algebra $\mathrm{Sym}^*(V)$ by the relation $u^{\ell} = \varphi(u)$ for $u \in V$. If u is a nonzero element of V and $a = \varphi(u) \in k$ then $A \cong k[u]/(u^{\ell} - a)$. If $a \notin k^{\times \ell}$ then A is the field $k(\sqrt[\ell]{a})$; if $a \in k^{\ell} - \{0\}$ then $A \cong \prod k$.

Of course the notion of an ℓ-form on a projective module over a commutative ring makes sense, but it is a special case of ℓ-forms on locally free modules (algebraic vector bundles) which we now define.

Definition 9.3. If $\mathcal{E}$ is a locally free $\mathcal{O}_X$-module over a scheme X then an ℓ-form on $\mathcal{E}$ is a symmetric ℓ-linear function on $\mathcal{E}$, i.e., an $\mathcal{O}_X$-linear map $\phi : \mathrm{Sym}^{\ell}(\mathcal{E}) \to \mathcal{O}_X$. If $\mathcal{E}$ is invertible, we will sometimes identify the ℓ-form with the diagonal ℓ-ary form $\varphi = \phi \circ \Delta : \mathcal{E} \to \mathcal{O}_X$; locally, if v is a section generating $\mathcal{E}$ then the form is $\varphi(tv) = a\, t^{\ell}$, where $a = \varphi(v)$.

Remark 9.3.1. The geometric vector bundle over a scheme X whose sheaf of sections is $\mathcal{E}$ is $\mathbb{V} = \mathbf{Spec}(\mathrm{Sym}^*(\mathcal{E}\tilde{\ })) $, where $\mathcal{E}\tilde{\ }$ is the dual $\mathcal{O}_X$-module of $\mathcal{E}$. We will sometimes describe ℓ-forms in terms of $\mathbb{V}$.

The projective space bundle associated to $\mathcal{E}$ is $\pi : \mathbb{P}(\mathcal{E}) = \mathbf{Proj}(S^*) \to X$, $S^* = \mathrm{Sym}^*(\mathcal{E}\tilde{\ })$. The tautological sheaf $\mathcal{O}(-1)$ on $\mathbb{P}(\mathcal{E})$ is the sheaf of sections of the line bundle $\mathbb{L} = \mathbf{Spec}(\mathrm{Sym}^* \mathcal{O}(1))$. The multiplication $S^* \otimes \mathcal{E}\tilde{\ } \to S^*(1)$ in the symmetric algebra induces a surjection of locally free sheaves $\pi^*(\mathcal{E}\tilde{\ }) \to \mathcal{O}(1)$ and hence an injection $\mathcal{O}(-1) \to \pi^*(\mathcal{E})$; this yields a canonical morphism $\mathbb{L} \to \pi^*(\mathbb{V})$ of the associated geometric vector bundles.

We will use the following notational shorthand. For a scheme Z, a point z on some Z-scheme and a locally free sheaf $\mathcal{E}$ on Z we write $\mathcal{E}|_z$ for the fiber of $\mathcal{E}$ at z, i.e., the $k(z)$-vector space $z^*(\mathcal{E})$ for $z \to Z$. If φ is an ℓ-ary form on an invertible sheaf L, $0 \neq u \in L|_z$ and $a = \varphi|_z(u)$, then $\varphi|_z : (L|_z) \to k(z)$ is the ℓ-ary form $\varphi|_z(tu) = at^\ell$. By $V(\varphi)$ we mean $\{z \in Z : \varphi|_z = 0\}$.

The notation $\{\gamma, \ldots, \gamma'_{n-1}\}$ in the Chain Lemma 9.1 is a special case of the notation in the following definition.

Definition 9.4. Given invertible sheaves $H_1, \ldots, H_n$ on X, ℓ-forms α_i on H_i, and a point $x \in X$ at which each form $\alpha_i|_x$ is nonzero, we write $\{\alpha_1, \ldots, \alpha_n\}|_x$ for the element $\{\alpha_1|_x, \ldots, \alpha_n|_x\}$ of $K_n^M(k(x))/\ell$ described before 9.2: if u_i is a generator of $H_i|_x$ and $\alpha_i|_x(u_i) = a_i$ then $\{\alpha_1, \ldots, \alpha_n\}|_x = \{a_1, \ldots, a_n\}$.

We record the following useful consequence of this construction. Recall that if $(R, \mathfrak{m})$ is a regular local ring with quotient field F, any regular sequence $r_1, \ldots$ generating $\mathfrak{m}$ determines a specialization map $K_*^M(F) \to K_*^M(R/\mathfrak{m})$; if $a_1, \ldots$ are units of R, this specialization sends $\{a_1, \ldots\}$ to $\{\bar{a}_1, \ldots\}$. (See Remark 1.13.1.)

Lemma 9.5. *Suppose that the ℓ-forms α_i are all nonzero at the generic point η of a smooth X. The set U of points x in X on which each $\alpha_i|_x \neq 0$ is open, and for every $x \in U$, the symbol $\{\alpha_1|_x, \ldots, \alpha_n|_x\}$ in $K_n^M(k(x))/\ell$ is obtained by specialization from the symbol in $K_n^M(k(X))/\ell$.*

Definition 9.6. Any ℓ-form $\psi : \mathrm{Sym}^\ell(\mathcal{E}) \to \mathcal{O}_X$ on $\mathcal{E}$ induces a canonical ℓ-form ϵ on the tautological sheaf $\mathcal{O}(-1)$ on $\mathbb{P}(\mathcal{E})$:

$$\epsilon : \mathcal{O}(-\ell) = \mathrm{Sym}^\ell(\mathcal{O}(-1)) \to \mathrm{Sym}^\ell(\pi^*\mathcal{E}) = \pi^*\mathrm{Sym}^\ell(\mathcal{E}) \xrightarrow{\psi} \pi^*\mathcal{O}_X = \mathcal{O}_{\mathbb{P}(\mathcal{E})}.$$

Example 9.7. Given an invertible sheaf L on X, and an ℓ-ary form φ on L, the sheaf $V = \mathcal{O} \oplus L$ has the ℓ-ary form $\psi(t, u) = t^\ell - \varphi(u)$. Then $\mathbb{P}(V) \to X$ is a $\mathbb{P}^1$-bundle, and its tautological sheaf $\mathcal{O}(-1)$ has the ℓ-form ϵ described in 9.6.

Over a point in $\mathbb{P}(V)$ of the form $\infty = (0 : u)$, the ℓ-form on $\mathcal{O}(-1)|_\infty$ is $\epsilon(0, \lambda u) = -\lambda^\ell \varphi(u)$. If $q = (1 : u)$ is any other point on $\mathbb{P}(V)$ then the 1-dimensional subspace $\mathcal{O}(-1)|_q$ of the vector space $V|_q$ is generated by $v = (1, u)$ and the ℓ-form $\epsilon|_q$ on $\mathcal{O}(-1)|_q$ is determined by $\epsilon(v) = \psi(1, u) = 1 - \varphi(u)$ in the sense that $\epsilon(\lambda v) = \lambda^\ell(1 - \varphi(u))$.

One application of these ideas is the formation of the sheaf of Kummer algebras associated to an ℓ-form.

Definition 9.8. If L is an invertible sheaf on X, equipped with an ℓ-ary form φ, the *Kummer algebra* $\mathcal{A}_\varphi(L)$ is the sheaf $\mathcal{A}(L) = \bigoplus_{i=0}^{\ell-1} L^{\otimes i}$ regarded as a sheaf of algebras in the following way: locally, if u is a section generating L then $\mathcal{A}(L) \cong \mathcal{O}[u]/(u^\ell - \varphi(u))$. If $x \in X$ and $a = \varphi|_x(u)$ is nonzero then the $k(x)$-algebra $\mathcal{A}|_x$ is the Kummer algebra $k(x)(\sqrt[\ell]{a})$, defined in 9.2; it is a field if $a \notin k(x)^{\times \ell}$ and $\prod k(x)$ otherwise.

Since the norm on $\mathcal{A}_\varphi(L)$ is given by a homogeneous polynomial of degree ℓ, we may regard the norm as a map from $\mathrm{Sym}^\ell \mathcal{A}_\varphi(L)$ to $\mathcal{O}$.

Using the well-known isomorphism $\text{Sym}^n(\mathbb{A}^1) \cong \mathbb{A}^n$, it is easy to check that if L is an invertible sheaf on X then the $(\ell-1)$st symmetric power of the projective line bundle $\mathbb{P}(\mathcal{O} \oplus L)$ over X is $\text{Sym}_X^{\ell-1}\mathbb{P}(\mathcal{O} \oplus L) = \mathbb{P}(\mathcal{A}(L))$, where $\mathcal{A}(L) = \bigoplus_{i=0}^{\ell-1} L^{\otimes i}$. The canonical ℓ-form ϵ on the tautological sheaf $\mathcal{O}(-1)$ on the projective bundle $\mathbb{P} = \mathbb{P}(\mathcal{A}(L))$, given in 9.6, is the natural ℓ-form:

$$\mathcal{O}(-1)^{\otimes \ell} \to \text{Sym}^\ell \pi^* \mathcal{A}(L) \xrightarrow{N} \mathcal{O}_{\mathbb{P}},$$

where $\pi : \mathbb{P} \to X$ is the structure map and the canonical inclusion of $\mathcal{O}(-1)$ into $\pi^*(\mathcal{A}(L)) = \oplus_0^{\ell-1} \pi^* L^{\otimes i}$ induces the first map.

Recall from 9.6 and 9.8 that ϕ is an ℓ-form on L, $\psi = (1, -\phi)$ is an ℓ-form on $\mathcal{O} \oplus L$, and ϵ is the canonical ℓ-form on $\mathcal{O}(-1)$ induced from ψ.

Lemma 9.9. *Suppose that $x \in X$ has $\phi|_x \neq 0$ and that $0 \neq u \in L|_x$. Then $\epsilon|_{(0:u)} \neq 0$. Moreover, $\phi(u) \in k(x)^{\times \ell}$ iff there is a point $[l] \in \mathbb{P}(\mathcal{O} \oplus L)$ over x so that $\epsilon|_{[l]} = 0$.*

Proof. Let $w = (t, su)$ be a point of $\mathcal{O}(-1)|_x$ over $[l] = (t : su) \in \mathbb{P}(\mathcal{O} \oplus L)|_x$. If $t = 0$ then $\ell = (0 : u)$ and $\epsilon(w) = -s^\ell \phi(u)$, which is nonzero for $s \neq 0$. If $t \neq 0$ then $\epsilon|_{[l]}$ is determined by the scalar $\epsilon(w) = \psi(t, su) = t^\ell - s^\ell \phi(u)$. Thus $\epsilon|_{[l]} = 0$ iff $\phi(u) = (t/s)^\ell$. $\qquad\qquad\square$

Remark 9.9.1. Here is an alternative proof, using the Kummer algebra $K = k(x)(a)$, $a = \sqrt[\ell]{\phi(u)}$. Since $\epsilon(w) = \psi(t, su)$ is the norm of the nonzero element $t - sa$ in K, the norm $\epsilon(w)$ is zero iff the Kummer algebra is split, i.e., $\phi(u) = a^\ell \in k(x)^{\times \ell}$.

9.2 THE CHAIN LEMMA WHEN $N = 2$

In this section, we prove the Chain Lemma 9.1 for a symbol $\{a_1, a_2\}$. To do this, we construct certain iterated projective bundles Y_r together with invertible sheaves and ℓ-forms on them; the variety S in the Chain Lemma will be Y_ℓ.

We begin with a generic construction, which starts with a pair L_0, L_{-1} of invertible sheaves on a variety $Y_0 = Y_{-1}$ and produces a tower of varieties Y_r, equipped with distinguished invertible sheaves L_r. Each Y_r is a product of $\ell - 1$ projective line bundles over Y_{r-1}, so Y_r has relative dimension $r(\ell - 1)$ over Y_0.

$$Y_\ell \to \cdots \to Y_r \xrightarrow{f_r} Y_{r-1} \xrightarrow{f_{r-1}} Y_{r-2} \to \cdots \to Y_1 \xrightarrow{f_1} Y_0 = Y_{-1}.$$

Definition 9.10. Given a morphism $f_{r-1} : Y_{r-1} \to Y_{r-2}$ and invertible sheaves L_{r-1} on Y_{r-1}, L_{r-2} on Y_{r-2}, we form the projective line bundle $\mathbb{P}(\mathcal{O} \oplus L_{r-1})$ over Y_{r-1} and its tautological sheaf $\mathcal{O}(-1)$. We define Y_r to be the product $\prod_1^{\ell-1} \mathbb{P}(\mathcal{O} \oplus L_{r-1})$ over Y_{r-1}. Writing f_r for the projection $Y_r \to Y_{r-1}$, and

$\mathcal{O}_{Y_r}(-1,\ldots,-1)$ for the external product $\mathcal{O}(-1)\boxtimes\cdots\boxtimes\mathcal{O}(-1)$ on Y_r, we define the invertible sheaf L_r on Y_r to be $L_r=(f_r\circ f_{r-1})^*(L_{r-2})\otimes\mathcal{O}_{Y_r}(-1,\ldots,-1)$.

Example 9.11 (k-tower). The k-*tower* is the tower obtained when we start with $Y_0=\mathrm{Spec}(k)$, using the trivial invertible sheaves L_{-1}, L_0. Note that $Y_1=\prod\mathbb{P}^1$ and $L_1=\mathcal{O}_{Y_1}(-1,\ldots,-1)$, while Y_2 is a product of projective line bundles over $\prod\mathbb{P}^1$, and $L_2=\mathcal{O}_{Y_2}(-1,\ldots,-1)$.

Remark 9.11.1. In the Chain Lemma (Theorem 9.1) for $n=2$ we have $S=Y_\ell$ in the k-tower, and the invertible sheaves are $J=J_1=L_\ell$, $J_1'=f_\ell^*(L_{\ell-1})$.

Before defining the ℓ-forms γ_1 and γ_1' in 9.16, we quickly establish 9.15; this verifies part (6) of Theorem 9.1, that the degree of $c_1(L_\ell)^{\ell^2-\ell}$ is prime to ℓ.

If L is an invertible sheaf over Y, and $\lambda=c_1(L)$, the Chow ring of $\mathbb{P}=\mathbb{P}(\mathcal{O}\oplus L)$ is $CH^*(\mathbb{P})=CH^*(Y)[z]/(z^2+\lambda z)$, where $z=c_1(\mathcal{O}(-1))$. If $\pi:\mathbb{P}\to Y$ then $\pi_*(z)=-1$ in $CH^*(Y)$. Applying this observation to the construction of Y_r out of $Y=Y_{r-1}$ with $\lambda_{r-1}=c_1(L_{r-1})$, we have

$$CH^*(Y_r)=CH^*(Y_{r-1})[z_{r,1},\ldots,z_{r,\ell-1}]/(\{z_{r,j}^2+\lambda_{r-1}z_{r,j}\mid j=1,\ldots,\ell-1\}),$$

where $z_{r,j}$ is the first Chern class of the jth tautological sheaf $\mathcal{O}(-1)$, and the inclusion of $CH^*(Y_{r-1})$ as a subring of $CH^*(Y_r)$ is via the pullback of cycles. In $CH^*(Y_r)$, $\lambda_r=c_1(L_r)$ equals $\lambda_{r-2}+\sum z_{r,i}$. By induction on r, this yields the following result:

Lemma 9.12. $CH^*(Y_r)$ *is a free* $CH^*(Y_0)$-*module. A basis consists of the monomials* $\prod z_{i,j}^{e_{i,j}}$ *for* $e_{i,j}\in\{0,1\}$, $0<i\le r$ *and* $0<j<\ell$. *As a graded algebra,* $CH^*(Y_r)/\ell\cong CH^*(Y_0)/\ell\otimes_{R_0}R_r$, *where* $R_0=\mathbb{F}_\ell[\lambda_0,\lambda_{-1}]$ *and*

$$R_r=\mathbb{F}_\ell[\lambda_{-1},\lambda_0,\ldots,\lambda_r,z_{1,1},\ldots,z_{r,\ell-1}]/I_r,$$

$$I_r=\left(\{z_{i,j}^2+\lambda_{i-1}z_{i,j}\mid 1\le i\le r,\ 0<j<\ell\},\ \left\{\lambda_i-\lambda_{i-2}-\sum_{j=1}^{\ell-1}z_{i,j}\mid 1\le i\le r\right\}\right).$$

Definition 9.13. For $i=1,\ldots,\ell$, set $z_i=\sum_{j=1}^{\ell-1}z_{i,j}$ and $\zeta_i=\prod z_{i,j}$. It follows from Lemma 9.12 that $\lambda_i=\lambda_{i-2}+z_i$, $z_{i,j}^{k+1}=(-1)^k\lambda_{i-1}^k z_{i,j}$ and

$$z_i^\ell=\sum_j z_{i,j}^\ell=\sum_j z_{i,j}\lambda_{i-1}^{\ell-1}=z_i\lambda_{i-1}^{\ell-1}$$

in the ring R_i, and hence in $CH^*(Y_i)/\ell$.

By Lemma 9.12, if $1\le r\le\ell$ then multiplication by $\prod\zeta_i\in CH^{r(\ell-1)}(Y_r)$ is an isomorphism $CH_0(Y_0)/\ell\xrightarrow{\sim}CH_0(Y_r)/\ell$. If $Y_0=\mathrm{Spec}(k)$ then $CH_0(Y_r)/\ell$ is isomorphic to $\mathbb{F}_\ell$, and generated by $\prod\zeta_i$.

Lemma 9.14. *If $y \in CH_0(Y_0)$, the degree of $y \cdot \zeta_1 \cdots \zeta_r$ is $(-1)^{r(\ell-1)} \deg(y)$.*

Proof. The degree on Y_r is the composition of the $(f_i)_*$ with the degree map on Y_0. The projection formula implies that $(f_r)_*(\zeta_r) = (-1)^{\ell-1}$, and

$$(f_r)_*(y \cdot \zeta_1 \cdots \zeta_r) = (y \cdot \zeta_1 \cdots \zeta_{r-1}) \cdot (f_r)_*(\zeta_r) = (-1)^{\ell-1} y \cdot \zeta_1 \cdots \zeta_{r-1}.$$

By induction on r, $(f_1 \circ \cdots \circ f_r)_*(y \cdot \zeta_1 \cdots \zeta_r)$ equals $(-1)^{r(\ell-1)} y$. The result follows. $\square$

Proposition 9.15. *For every 0-cycle y on Y_0 and $1 \leq r \leq \ell$, $\lambda_r = c_1(L_r)$ satisfies*

$$y \, \lambda_r^{r(\ell-1)} \equiv y \, \zeta_1 \cdots \zeta_r \text{ in } CH_0(Y_r)/\ell, \text{ and } \deg(y\lambda_r^{r(\ell-1)}) \equiv \deg(y) \pmod{\ell}.$$

For the k-tower 9.11 (with $y=1$), we have $\deg(\lambda_\ell^{\ell^2-\ell}) \equiv 1 \pmod{\ell}$.

Proof. If $r=1$ this follows from $y\lambda_{-1} = y\lambda_0 = 0$ in $CH^*(Y_0)$: $\lambda_1 = z_1 + \lambda_{-1}$ and $y \cdot \zeta_1 \equiv y \, \lambda_1^{\ell-1}$. For $r \geq 2$, we have $\lambda_r = z_r + \lambda_{r-2}$ and $z_r^\ell = z_r \lambda_{r-1}^{\ell-1}$ by 9.13. Because $\ell - r \geq 0$, we have

$$\lambda_r^{r(\ell-1)} = (z_r + \lambda_{r-2})^{\ell(r-1)+(\ell-r)} \equiv (z_r^\ell + \lambda_{r-2}^\ell)^{r-1} \cdot (z_r + \lambda_{r-2})^{\ell-r} \quad \mathrm{mod}\ \ell$$

$$= (z_r\lambda_{r-1}^{\ell-1} + \lambda_{r-2}^\ell)^{r-1}(z_r + \lambda_{r-2})^{\ell-r} \equiv \zeta_r \, \lambda_{r-1}^{(r-1)(\ell-1)} + T \quad \mathrm{mod}\ \ell,$$

where $T \in CH^*(Y_{r-1})[z_r]$ is a homogeneous polynomial of total degree less than $\ell-1$ in z_r.

By 9.12, the coefficients of yT are elements of $CH^*(Y_{r-1})$ of degree $> \dim(Y_{r-1})$, so yT must be zero. Then by the inductive hypothesis,

$$y \, \lambda_r^{r(\ell-1)} \equiv (-1)^{r-2} y \, \zeta_r \lambda_{r-1}^{(r-1)(\ell-1)} \equiv (-1)^{r-1} y \, \zeta_r \cdot (\zeta_1 \cdots \zeta_{r-1})$$

in $CH^*(Y_r)/\ell$, as claimed. The degree assertion follows from Lemma 9.14. $\square$

The ℓ-forms for $n=2$

We now turn to the ℓ-forms in the Chain Lemma 9.1, using the k-tower 9.11. We will inductively equip the invertible sheaves $\mathcal{O}_{Y_r}(-1, \ldots, -1)$ and L_r of 9.11 with ℓ-forms Ψ_r and φ_r; the γ_1 and γ_1' of the Chain Lemma 9.1 will be φ_ℓ and $\varphi_{\ell-1}$.

When $r=0$, we equip the trivial invertible sheaves L_{-1}, L_0 on $Y_0 = \mathrm{Spec}(k)$ with the ℓ-forms $\varphi_{-1}(t) = a_2 t^\ell$ and $\varphi_0(t) = a_1 t^\ell$. The ℓ-form φ_{r-1} on L_{r-1} induces an ℓ-form $\psi(t, u) = t^\ell - \varphi_{r-1}(u)$ on $\mathcal{O} \oplus L_{r-1}$ and an ℓ-form $\epsilon(y)$ on the tautological sheaf $\mathcal{O}(-1)$, as in Example 9.7. As observed in Example 9.7, at the point $q = (1:x)$ of $\mathbb{P}(\mathcal{O} \oplus L_{r-1})$ the subspace $\mathcal{O}(-1)|_q$ is generated by $y = (1, x)$ and we have $\epsilon(y) = \psi(1, x) = 1 - \varphi_{r-1}(x)$.

Definition 9.16. The ℓ-form Ψ_r on $\mathcal{O}_{Y_r}(-1,\ldots,-1)$ is the product form $\prod \epsilon$:

$$\Psi_r(y_1 \boxtimes \cdots \boxtimes y_{\ell-1}) = \prod \epsilon(y_i).$$

The ℓ-form φ_r on $L_r = (f_{r-1} \circ f_r)^*(L_{r-2}) \otimes \mathcal{O}_{Y_r}(-1,\ldots,-1)$ is defined to be

$$\varphi_r = (f_{r-1} \circ f_r)^*(\varphi_{r-2}) \otimes \Psi_r.$$

Proposition 9.17. *Let* $x = (x_1,\ldots,x_{\ell-1})$ *be a point of* Y_r *with residue field* $E = k(x)$. *For* $-1 \leq i \leq r$, *choose generators* u_i *and* v_i *for the 1-dimensional* E-*vector spaces* $L_i|_x$ *and* $\mathcal{O}_{Y_i}(-1,\ldots,-1)|_x$, *respectively, in such a way that* $u_i = u_{i-2} \otimes v_i$.

1. *If* $\varphi_i|_x = 0$ *for some* $1 \leq i \leq r$ *then* $\{a_1, a_2\}_E = 0$ *in* $K_2(E)/\ell$.
2. *If* $\varphi_i|_x \neq 0$ *for all* i, $1 \leq i \leq r$, *then*

$$\{a_1, a_2\}_E = (-1)^r \{\varphi_{r-1}(u_{r-1}), \varphi_r(u_r)\} \text{ in } K_2(E)/\ell.$$

Proof. By induction on r. Both parts are obvious if $r = 0$. To prove the first part, we may assume that $\varphi_i|_x \neq 0$ for $1 \leq i \leq r-1$, but $\varphi_r|_x = 0$. We have $u_r = u_{r-2} \otimes v_r$ and by the definition of φ_r, we conclude that

$$0 = \varphi_r(u_r) = \varphi_{r-2}(u_{r-2})\Psi_r(v_r),$$

whence $\Psi_r(v_r) = 0$. Now the element $v_r \neq 0$ is a tensor product of sections w_j and $\Psi_r(v_r) = \prod \epsilon(w_j)$, so $\epsilon(w_j) = 0$ for a nonzero section w_j of $\mathcal{O}(-1)|_{x_j}$. By Lemma 9.9, $\varphi_{r-1}(u_{r-1})$ is an ℓth power in E. Consequently, $\{\varphi_{r-2}(u_{r-2}), \varphi_{r-1}(u_{r-1})\}_E = 0$ in $K_2(E)/\ell$. This symbol equals $\pm\{a_1, a_2\}_E$ in $K_2(E)/\ell$, by (2) and induction. This finishes the proof of the first assertion.

For the second claim, we can assume by induction that

$$\{a_1, a_2\}_E = \pm\{\varphi_{r-2}(u_{r-2}), \varphi_{r-1}(u_{r-1})\}_E.$$

Now $\varphi_r(u_r) = \varphi_{r-2}(u_{r-2})\Psi_r(v_r)$. Letting $K = k(x)(\alpha)$, $\alpha = \sqrt[\ell]{\varphi_{r-1}(u_{r-1})}$, we have $N_{K/k(x)}(t - s\alpha) = t^\ell - s^\ell\varphi_{r-1}(u_{r-1}) = \psi(t, su_{r-1})$, and hence

$$\Psi_r(v_r) = \prod \epsilon(w_j) = \prod \psi(t_i, s_i u_{r-1}) = N_{K/k(x)}(v')$$

for some $v' \in K$. But $\{\varphi_{r-1}(u_{r-1}), N_{K/k(x)}(v')\} = 0$ by Lemma 9.18. We conclude that

$$\{\varphi_{r-2}(u_{r-2}), \varphi_{r-1}(u_{r-1})\}_E \equiv -\{\varphi_{r-1}(u_{r-1}), \varphi_r(u_r)\}_E \quad \text{mod } \ell;$$

this concludes the proof of the second assertion. $\qquad\square$

Lemma 9.18. *For any field* k, *any* $a \in k^\times$, *and any* b *in* $K_a = k[\sqrt[\ell]{a}]$, *the symbol* $\{a, N_{K_a/k}(b)\}$ *is trivial in* $K_2(k)/\ell$.

Proof. Because $\{a, b\} = \ell\{\sqrt[\ell]{a}, b\}$ vanishes in $K_2(K_a)/\ell$, we have $\{a, N(b)\} = N\{a, b\} = \ell N(\{\sqrt[\ell]{a}, b\}) = 0$. $\qquad\qquad\qquad\qquad\qquad\qquad\qquad\qquad\qquad\qquad\qquad\square$

Proof of the Chain Lemma 9.1 for $n = 2$. We verify the conditions for the variety $S = Y_\ell$ in the k-tower 9.11; the invertible sheaves $J = J_1 = L_\ell$, $J'_1 = f^*_\ell(L_{\ell-1})$; and the ℓ-forms γ_1 and γ'_1 in 9.1, which are the forms φ_ℓ and $\varphi_{\ell-1}$ of 9.16. Part (1) of Theorem 9.1 is immediate from the construction of $S = Y_\ell$; parts (2) and (4) were proven in Proposition 9.17; parts (3) and (5) follow from (2) and (4); and part (6) is Proposition 9.15 with $y = 1$. $\qquad\qquad\square$

9.3   THE SYMBOL CHAIN

In this section, we describe the pattern of the Chain Lemma in all weights. Using downward induction on n, we will construct a tower

$$S = S_1 \longrightarrow S_2 \longrightarrow \cdots \longrightarrow S_{n-2} \longrightarrow S_{n-1} \longrightarrow S_n = \mathrm{Spec}(k), \qquad (9.19)$$

along with sheaves J_i, J'_i on S_i carrying ℓ-forms γ_i, γ'_i satisfying the conditions of the Chain Lemma 9.1. The relative dimension of S_{i-1} over S_i will be $\ell^i - \ell^{i-1}$, so that $S = S_1$ will have dimension $\ell^n - \ell$, as required by Theorem 9.1(1), and each function field $k(S_{i-1})$ will be purely transcendental over $k(S_i)$.

The construction will use a family of functions Φ_r and Ψ_r To define them, we start with the function $\Phi_0(t) = t^\ell$ and a sequence $a_1, a_2, \ldots, a_n$ of units of k. For $r \geq 1$, we inductively define functions Φ_r in ℓ^r variables and Ψ_r in $\ell^r - \ell^{r-1}$ variables, taking values in k, and prove (in 9.23) that $\{a_1, ..., a_r, \Phi_r(\mathbf{x})\} \equiv 0$ (mod ℓ). Note that Φ_r and Ψ_r depend only upon the units $a_1, ..., a_r$. We write $\mathbf{x}_i$ for a sequence of ℓ^r variables x_{ij} (where $j = (j_1, \ldots, j_r)$ and $0 \leq j_t < \ell$), so that $\Phi_r(\mathbf{x}_i)$ is defined for $i = 0, ..., \ell-1$, and we inductively define

$$\Psi_{r+1}(\mathbf{x}_1, ..., \mathbf{x}_{\ell-1}) = \prod_{i=1}^{\ell-1} [1 - a_{r+1}\Phi_r(\mathbf{x}_i)], \qquad (9.20)$$

$$\Phi_{r+1}(\mathbf{x}_0, ..., \mathbf{x}_{\ell-1}) = \Phi_r(\mathbf{x}_0)\Psi_{r+1}(\mathbf{x}_1, ..., \mathbf{x}_{\ell-1}). \qquad (9.21)$$

We say that two rational functions are *birationally equivalent* if they can be transformed into one another by an automorphism (over the base field k) of the field of rational functions.

Example 9.22. $\Psi_1(x_1, ..., x_{\ell-1})$ is $\prod(1 - a_1 x_i^\ell)$ and $\Phi_1(x_0, ..., x_{\ell-1})$ is $x_0^\ell \prod (1 - a_1 x_i^\ell)$, the norm of the element $x_0 \prod(1 - a_1 x_i)$ in the Kummer extension $k(\mathbf{x})(\alpha_1)$, $\alpha_1 = \sqrt[\ell]{a_1}$. Thus Φ_1 is birationally equivalent to symmetrizing in the x_i, followed by the norm from $k[\sqrt[\ell]{a_1}]$ to k.

More generally, $\Psi_r(\mathbf{x}_1, ..., \mathbf{x}_{\ell-1})$ is the product of norms of elements in Kummer extensions $k(\mathbf{x}_1, ..., \mathbf{x}_{\ell-1})(\sqrt[\ell]{b_i})$ of $k(\mathbf{x}_1, ..., \mathbf{x}_{\ell-1})$.

It is useful to interpret the map Φ_1 geometrically. Given a field extension $k(\alpha)$ of degree ℓ over k, let $\mathbb{A}^{k(\alpha)}$ denote the variety, isomorphic to $\mathbb{A}^\ell$, whose F-points $(x_0, ..., x_{\ell-1})$ correspond to elements $\sum x_i \alpha^i$ of $F(\alpha)$. In fact, $\mathbb{A}^{k(\alpha)}$ is the Weil restriction $R_{k(\alpha)/k}\mathbb{A}^1$ of the affine line over $k(\alpha)$; see section 11.3. Corresponding to the norm map $k(\alpha) \to k$, there is a morphism $N: \mathbb{A}^{k(\alpha)} \to \mathbb{A}^1$. Similarly, the function $k^\ell \to k(\alpha)$ defined by

$$(x_0, t_1, ..., t_{\ell-1}) \mapsto x_0(1 - t_1\alpha + t_2\alpha^2 - \cdots \pm t_{\ell-1}\alpha^{\ell-1})$$

induces a birational isomorphism $\mathbb{A}^\ell \xrightarrow{m} \mathbb{A}^{k(\alpha)}$. Finally, let $\mathbb{A}^{\ell-1} \xrightarrow{q} \mathbb{A}^{\ell-1}/\Sigma_{\ell-1} \cong \mathbb{A}^{\ell-1}$ be the symmetrizing map sending $(x_1, ...)$ to the elementary symmetric functions $(s_1, ...)$. Then the following diagram commutes:

$$\mathbb{A}^\ell = \mathbb{A}^1 \times \mathbb{A}^{\ell-1} \xrightarrow{1 \times q} \mathbb{A}^1 \times \mathbb{A}^{\ell-1} \xrightarrow[\text{birat.}]{m} \mathbb{A}^{k(\alpha)} \cong \mathbb{A}^\ell$$

with diagonal map Φ_1 and the right vertical map N to $\mathbb{A}^1$.

Remark 9.22.1. If $\ell = 2$, $\Phi_1(x_0, x_1) = x_0^2(1 - a_1 x_1^2)$ is birationally equivalent to the norm form $u^2 - a_1 v^2$ for $k(\sqrt{a_1})/k$, and $\Phi_2 = \Phi_1(\mathbf{x}_0)[1 - a_2\Phi_1(\mathbf{x}_1)]$ is birationally equivalent to the norm form $\langle\langle a_1, a_2 \rangle\rangle = (u^2 - a_1 v^2)[1 - a_2(w^2 - a_1 t^2)]$ for the quaternionic algebra $A_{-1}(a_1, a_2)$.

More generally, Φ_n is birationally equivalent to the Pfister form

$$\langle\langle a_1, ..., a_r \rangle\rangle = \langle\langle a_1, ..., a_{r-1} \rangle\rangle \perp a_n \langle\langle a_1, ..., a_{r-1} \rangle\rangle$$

and Ψ_r is equivalent to the restriction of the Pfister form to the subspace defined by the equations $\mathbf{x}_0 = (1, ..., 1)$.

Remark 9.22.2. (Rost [Ros99]) Suppose that $\ell = 3$, and that k contains a cube root of unity, ζ. Then Φ_2 is birationally equivalent to (symmetrizing, followed by) the reduced norm of the algebra $A_\zeta(a_1, a_2)$ and Φ_3 is equivalent to the norm form of the exceptional Jordan algebra $J(a_1, a_2, a_3)$. When $r = 4$, Rost showed that the set of nonzero values of Φ_4 is a subgroup of $k^\times$.

For the next lemma, it is useful to introduce the function field F_r over k in the ℓ^r variables $x_{j_1, ..., j_r}$, $0 \leq j_t < \ell$. Note that F_r is isomorphic to the tensor product of ℓ copies of F_{r-1}.

Lemma 9.23. $\{a_1, ..., a_r, \Phi_r(\mathbf{x})\} = \{a_1, ..., a_r, \Psi_r(\mathbf{x})\} = 0 \in K^M_{r+1}(F_r)/\ell$.
If $b \in k$ is a nonzero value of Φ_r, then $\{a_1, ..., a_r, b\} = 0 \in K^M_{r+1}(k)/\ell$.

Proof. By Lemma 9.18, $\{a_r, \Psi_r(\mathbf{x})\} = 0$ because $\Psi_r(\mathbf{x})$ is a product of norms of elements of $k(\mathbf{x})(\alpha_r)$ by Example 9.22. If $r = 1$ then $\{a_1, \Phi_1(\mathbf{x})\} = \{a_1, x_0^\ell\} \equiv 0$ as well. The result for F_r follows by induction, using (9.21):

$$\{a_1, ..., a_{r+1}, \Phi_{r+1}(\mathbf{x})\} = \{a_1, ..., a_{r+1}, \Phi_r(\mathbf{x}_0)\}\{a_1, ..., a_{r+1}, \Psi_{r+1}(\mathbf{x})\} = 0.$$

The result for b follows from the first assertion, and specialization from F_r to k, using the regular local ring at the point $\mathbf{c}$ where $\Phi_r(\mathbf{c}) = b$. $\qquad\square$

The Chain Lemma is based upon the observation that certain manipulations (or "moves") of Milnor symbols do not change the class in $K_n^M(k)/\ell$. Here is the class of moves we will model geometrically in section 9.4; strings of these moves will be used in section 9.5 to prove the Chain Lemma.

Definition 9.24. A move of type C_n on a sequence $a_1, ..., a_n$ in $k^\times$ is a transformation of the kind:

$$\text{Type } C_n: \qquad (a_1, ..., a_n) \mapsto (a_1, ..., a_{n-2}, a_n \Psi_{n-1}(\mathbf{x}), a_{n-1}).$$

Here Ψ_{n-1} is a function of $\ell^{n-1} - \ell^{n-2}$ new variables $\mathbf{x}_1 = \{\mathbf{x}_{1,1}, ..., \mathbf{x}_{1,\ell-1}\}$, with each $\mathbf{x}_{1,j}$ a family of ℓ^{n-2} variables, and the function field of this move is $k(\mathbf{x}_1)$.

A move of type B_n on a sequence $a_1, ..., a_n$ in $k^\times$ is the result of switching a_{n-1} and a_n, and then doing ℓ moves of type C_n, applied to the sequence $a_1, ..., a_{n-2}, a_n, a_{n-1}$:

$$\text{Type } B_n: \qquad (a_1, ..., a_n) \mapsto (a_1, ..., a_{n-2}, \gamma_{n-1}, \gamma'_{n-1}).$$

The ith move of type C_n uses a fresh set of $\ell^{n-1} - \ell^{n-2}$ variables $\mathbf{x}_i = \{\mathbf{x}_{i,j}\}$, so a move of type B_n uses $\ell^n - \ell^{n-1}$ new variables.

By Lemma 9.23, $\{a_1, ..., a_n\} = -\{a_1, ..., a_{n-2}, a_n \Psi_{n-1}(\mathbf{x}), a_{n-1}\}$ in the larger group $K_n^M(k(\mathbf{x}_1))/\ell$. That is, each move of type C_n changes the symbol by a sign. Therefore each move of type B_n leaves the symbol unchanged:

$$\{a_1, ..., a_n\} = \{a_1, ..., a_{n-2}, \gamma_{n-1}, \gamma'_{n-1}\} \qquad (9.25)$$

in $K_n^M(F'_n)/\ell$, where $F'_n = k(\mathbf{x}_1, ..., \mathbf{x}_\ell)$ is a function field in $\ell^n - \ell^{n-1}$ variables. The functions γ_{n-1} and γ'_{n-1} in 9.25 are the ones appearing in the Chain Lemma 9.1. In section 9.5, we will define a variety S_{n-1} with function field F'_n. This is the initial step in the construction of the tower (9.19).

The next step uses a set $\mathbf{x}_2$ of $\ell^{n-1} - \ell^{n-2}$ more new variables to do a move of type B_{n-1} on $(a_1, ..., a_{n-2}, \gamma_{n-1})$ to get the sequence $(a_1, ..., a_{n-3}, \gamma_{n-2}, \gamma'_{n-2}, \gamma'_{n-1})$. The field of this move is $F'_{n-1} = F'_n(\mathbf{x}_2)$, a function field in $\ell^n - \ell^{n-2}$ variables over k. In section 9.5, we will define a variety S_{n-2} with this function field, together with a morphism $S_{n-2} \to S_{n-1}$.

Next, apply a move of type B_{n-2} over the field $F'_{n-2} = F'_{n-1}(\mathbf{x}_2)$, then a move of type B_{n-3}, and so on, ending with a move of type B_2 over the field F'_2 in $\ell^n - \ell^{n-2}$ variables over k. We have the sequence $(\gamma_1, \gamma'_1, \gamma'_2, ..., \gamma'_{n-1})$ in $\ell^n - \ell$ variables $\mathbf{x}_1, ..., \mathbf{x}_{\ell-1}$. Moreover, we see from Lemma 9.23 that

$$\{a_1, \ldots, a_n\} = \{\gamma_1, \gamma'_1, \gamma'_2, ..., \gamma'_{n-1}\} \quad \text{in } K_n^M(k)/\ell. \tag{9.26}$$

Let S be any variety containing $U = \mathbb{A}^{\ell^n - \ell}$ as an affine open, so that $k(S) = k(\mathbf{x}_1, ..., \mathbf{x}_{\ell-1})$, each $\mathbf{x}_i$ is ℓ^{n-1} variables $x_{i,j}$, and all invertible sheaves on U are trivial.

Parts (1), (2), and hence (3) of the Chain Lemma 9.1 are immediate from (9.25) and (9.26).

Now the only thing to do is to construct $S = S_1$, extend the invertible sheaves (and forms) from U to S, and prove parts (4) and (6) of 9.1.

9.4 THE TOWER OF VARIETIES P_r AND Q_r

In this section, we construct a tower of varieties P_r and Q_r over a fixed base scheme S', with ℓ-forms on invertible sheaves over them. Each P_r (resp., Q_r) will produce a model of the forms Ψ_r (resp., Φ_r) in (9.20) and (9.21). This tower, depicted in (9.26), is defined in 9.28 below.

$$P_{n-1} \to \cdots \to P_r \to Q_{r-1} \to P_{r-1} \to \cdots \to Q_1 \to P_1 \to Q_0 = S' \tag{9.26}$$

The passage from S' to the variety P_{n-1} is a model for the move of type C_n defined in 9.24.

Recall that if $\mathcal{E}_j$ is a sheaf over X_j and $\pi_j : \prod X_i \to X_j$ are the projections, the external product $\mathcal{E}_1 \boxtimes \cdots \boxtimes \mathcal{E}_m$ is defined to be $\pi_1^*(\mathcal{E}_1) \otimes \cdots \otimes \pi_m^*(\mathcal{E}_m)$.

Definition 9.27. Let X be a geometrically irreducible variety over some fixed base S'. Given invertible sheaves K, L on X, we can form the sheaf $V = \mathcal{O} \oplus L$, the $\mathbb{P}^1$-bundle $\mathbb{P}(V)$ over X, and $\mathcal{O}(-1)$. Taking products over S', set

$$P = \prod_1^{\ell-1} \mathbb{P}(\mathcal{O} \oplus L); \quad Q = X \times_{S'} P.$$

If X has relative dimension d over S' then P and Q have relative dimensions $(\ell-1)(d+1)$ and $\ell d + \ell - 1$, respectively.

On P and Q, we have the external products of the tautological sheaves:

$$\mathcal{O}_P(-1, \ldots, -1) = \mathcal{O}(-1) \boxtimes \mathcal{O}(-1) \boxtimes \cdots \boxtimes \mathcal{O}(-1) \text{ on } P,$$
$$K \boxtimes \mathcal{O}(-1, \ldots, -1) \text{ on } Q.$$

Given ℓ-forms φ and σ on K and L, respectively, the sheaf $\mathcal{O}(-1)$ has the ℓ-form ϵ, as in Example 9.7, and the sheaves $\mathcal{O}(-1,\ldots,-1)$ and $K \boxtimes \mathcal{O}(-1,\ldots,-1)$ are equipped with the product ℓ-forms $\Psi = \prod \epsilon$ and $\Phi = \varphi \otimes \Psi$.

Remark 9.27.1. Let $x = (x_1,\ldots,x_{\ell-1})$ denote the generic point of $X^{\ell-1}$. The function fields of P and Q are $k(P) = k(x)(y_1,\ldots,y_{\ell-1})$ and $k(Q) = k(P) \otimes k(x_0)$. We may represent the generic point of P in coordinate form as a $(\ell-1)$-tuple $\{(1 : y_i)\}$, where the y_i generate L over x_i. Then $y = \boxtimes_{i=1}^{\ell-1}(1, y_i)$ is a generator of $\mathcal{O}(-1,\ldots,-1)$ at the generic point, and $\Psi(y) = \prod(1 - \sigma(y_i))$ by 9.7. If v_0 is a generator of K at the generic point x_0 of X, then $\Phi(y) = \varphi(v_0)\Psi(y)$.

Example 9.27.2. An important special case arises when we begin with two invertible sheaves H on S', K on X, with ℓ-forms α and φ. In this case, we set $L = H \boxtimes K$ and equip it with the product form $\sigma(u \otimes v) = \alpha(u)\varphi(v)$. At the generic point q of Q we can pick a generator $u \in H|_q$ and set $y_i = u \otimes v_i$; the forms are the forms of (9.20) and (9.21), with Φ_r and a_{r+1} replaced by φ and α:

$$\Psi(y) = \prod\bigl(1 - \alpha(u)\varphi(v_i)\bigr), \quad \Phi(y) = \varphi(v_0)\,\Psi(y).$$

Remark 9.27.3. Suppose a group G acts on S', X, K, and L, and K_0, L_0 are nontrivial 1-dimensional representations so that at every fixed point x of X

$$k(x) = k \quad \text{and} \quad L_x \cong L_0.$$

Then G acts on P (resp., Q) with $2^{\ell-1}$ fixed points y over each fixed point of $X^{\ell-1}$ (resp., of X^{ℓ}), each with $k(y) = k$, and each fiber of $\mathcal{O}(-1,\ldots,-1)$ (resp., $K \boxtimes \mathcal{O}(-1,\ldots,-1)$) is the representation L_0^j (resp., $K_0 \otimes L_0^j$) for some j $(0 \le j < \ell)$. Indeed, G acts nontrivially on each term $\mathbb{P}^1$ of the fiber $\prod \mathbb{P}^1$, so that the fixed points in the fiber are the points $(y_1, ..., y_{\ell-1})$ such that each y_i is either $(0 : 1)$ or $(1 : 0)$.

We now set $Q_0 = S'$ and recursively define the tower (9.26) of P_r and Q_r over a fixed base S', an invertible sheaf K_r on Q_r, and an ℓ-form Φ_r on K_r. We start with invertible sheaves $H_1,\ldots,H_r$, and $K_0 = \mathcal{O}_{S'}$ on S'. Each H_i has a nonzero ℓ-form α_i, and K_0 has the ℓ-form $\Phi_0(t) = t^\ell$.

Definition 9.28. Given a variety Q_{r-1} over $S' = Q_0$ and an invertible sheaf K_{r-1} on Q_{r-1}, we form the varieties $P_r = P$ and $Q_r = Q$ using the construction in Definition 9.27, with $X = Q_{r-1}$, $K = K_{r-1}$, and $L = H_r \boxtimes K_{r-1}$ as in 9.27.2. To emphasize that P_r only depends upon S' and $H_1,\ldots,H_r$, we will sometimes write $P_r(S'; H_1,\ldots,H_r)$. As in 9.27, P_r has the invertible sheaf $\mathcal{O}(-1,\ldots,-1)$, and Q_r has the invertible sheaf $K_r = K_{r-1} \boxtimes \mathcal{O}(-1,\ldots,-1)$.

Constructed	Input
$Q_r = Q_{r-1} \times P_r$	$P_r \quad Q_{r-1}$
$K_r = K_{r-1} \boxtimes \mathcal{O}(-1, \ldots, -1)$	K_{r-1}
$\Phi_r = \Phi_{r-1} \Psi_r$	$\Psi_r \quad \Phi_{r-1}$

Inductively, the invertible sheaf K_{r-1} on Q_{r-1} is equipped with an ℓ-form Φ_{r-1}. As described in 9.27 and 9.27.2, the invertible sheaf $\mathcal{O}(-1, \ldots, -1)$ on P_r acquires an ℓ-form Ψ_r from the ℓ-form $\alpha_r \otimes \Phi_{r-1}$ on $L = H_r \otimes K_{r-1}$, and K_r acquires an ℓ-form $\Phi_r = \Phi_{r-1} \otimes \Psi_r$.

Example 9.28.1. $Q_1 = P_1$ is $\prod_1^{\ell-1} \mathbb{P}^1(\mathcal{O} \oplus H_1)$ over S', equipped with the invertible sheaf $K_1 = \mathcal{O}(-1, \ldots, -1)$. If H_1 is a trivial invertible sheaf with ℓ-form $\alpha_1(t) = a_1 t^\ell$ then Φ_1 is the ℓ-form Φ_1 of Example 9.22.
  P_2 is $\prod_1^{\ell-1} \mathbb{P}^1(\mathcal{O} \oplus H_2 \otimes K_1)$ over $Q_1^{\ell-1}$, and $K_2 = K_1 \boxtimes \mathcal{O}(-1, \ldots, -1)$.

Lemma 9.29. *If $r > 0$ then* $\dim(P_r/S') = (\ell^r - \ell^{r-1})$ *and* $\dim(Q_r/S') = \ell^r - 1$.

Proof. Set $d_r = \dim(Q_r/S')$. The lemma follows easily by induction from the formulas $\dim(P_{r+1}/S') = (\ell-1)(d_r + 1)$, $\dim(Q_{r+1}/S') = \ell d_r + \ell - 1$. □

Choosing generators u_i for H_i at the generic point s of S', we get units $a_i = \alpha_i(u_i)$ in $k(S')$, and $\{\alpha_1, \ldots, \alpha_r\}|_s = \{a_1, \ldots, a_r\}$. Recall that the inductive Definition 9.28 begins with the ℓ-form $\Phi_0(t) = t^\ell$ over $Q_0 = S'$.

Lemma 9.30. *For all $r > 0$, the ℓ-forms Ψ_r and Φ_r of 9.28 agree at the generic points of P_r and Q_r with the forms defined in (9.20) and (9.21).*

Proof. This follows by induction on r, using the analysis of Example 9.27.2. Given a point $q = (q_1, \ldots, q_\ell)$ of $Q_{r-1}^{\ell-1}$ and a point $\{(1 : y_i)\}$ on P_r over it, $y = \boxtimes_1^{\ell-1}(1, y_i)$ is a nonzero point on $\mathcal{O}(-1, \ldots, -1)$ and $y_i = 1 \otimes v_i$ for a section v_i of K_{r-1}. Since $\epsilon(1, y_i) = 1 - a_r \Phi_{r-1}(v_i)$ and $\Psi_r(y) = \prod \epsilon(1, y_i)$, the forms Ψ_r agree. Similarly, if v_0 is the generator of K_{r-1} over the generic point q_0 then $y' = v_0 \otimes y$ is a generator of K_r and

$$\Phi_r(y') = \Phi_{r-1}(v_0)\Psi_r(y),$$

which is also in agreement with the formula in (9.21). □

Recall that K_0 is the trivial invertible sheaf, and that Φ_0 is the standard ℓ-form $\Phi_0(v) = v^\ell$ on K_0. Every point of $P_r = \prod \mathbb{P}(\mathcal{O} \oplus L)$ has the form $w = (w_1, \ldots, w_{\ell-1})$, and the projection $P_r \to \prod Q_{r-1}$ sends $w \in P_r$ to a point $x = (x_1, \ldots, x_{\ell-1})$.

Proposition 9.31. *Let $s \in S'$ be a point such that $\alpha_1|_s, \ldots, \alpha_r|_s \neq 0$, and let w be a point of P_r over s.*

1. If $\Psi_r|_w = 0$, then $\{\alpha_1, \ldots, \alpha_r\}|_w$ vanishes in $K_r^M(k(w))/\ell$.
2. If $\Phi_r|_q = 0$ for some $q = (x_0, w) \in Q_r$, then $\{\alpha_1, \ldots, \alpha_r\}|_q$ vanishes in $K_r^M(k(q))/\ell$.

Proof. Since $\Phi_r = \Phi_{r-1} \otimes \Psi_r$, the assumption that $\Psi_r|_w = 0$ implies that $\Phi_r|_q = 0$ for any $x_0 \in Q_{r-1}$ over s. Conversely, if $\Phi_r|_q = 0$ then either $\Psi_r|_w = 0$ or $\Phi_{r-1}|_{x_0} = 0$. Since $\Phi_0 \neq 0$, we may proceed by induction on r and assume that $\Phi_{r-1}|_{x_j} \neq 0$ for each j, so that $\Phi_r|_q = 0$ is equivalent to $\Psi_r|_w = 0$.

By construction, the ℓ-form on $L = H_r \otimes K_{r-1}$ is $\sigma(u_r \otimes v) = a_r \Phi_{r-1}(v)$, where u_r generates the vector space $H_r|_s$, $a_r = \alpha_r(u_r)$, and v is a section of K_{r-1}. Since $\Psi_r|_w$ is the product of the forms $\epsilon|_{w_j}$, some $\epsilon|_{w_j} = 0$. Lemma 9.9 implies that $a_r \Phi_{r-1}(v)$ is a ℓth power in $k(x_j)$, and hence in $k(w)$, for any generator v of $K_{r-1}|_{x_j}$. By Lemma 9.23, $\{a_1, \ldots, a_{r-1}, \Phi_{r-1}\} = 0$ and hence

$$\{\alpha_1, \ldots, \alpha_r\}|_w = \{a_1, \ldots, a_r\} = \{a_1, \ldots, a_{r-1}, a_r \Phi_{r-1}\} = 0$$

in $K_r^M(k(w))/\ell$, as claimed. $\square$

We conclude this section with some identities in $CH^*(P_r)/\ell$, given in 9.34. To simplify the statements and proofs below, we write $\mathrm{ch}^*(X)$ for $CH^*(X)/\ell$ and $\mathrm{ch}_0(X)$ for $CH_0(X)/\ell$, and adopt the following notation.

Notation 9.32. Set $\eta = c_1(H_r) \in \mathrm{ch}^1(S')$, and $\gamma = c_1(\mathcal{O}(-1, \ldots, -1)) \in \mathrm{ch}^1(P_r)$. Writing $\mathbb{P}$ for the bundle $\mathbb{P}(\mathcal{O} \oplus H_r \otimes K_{r-1})$ over Q_{r-1}, with tautological sheaf $\mathcal{O}_{\mathbb{P}}(-1)$, let $c \in \mathrm{ch}(\mathbb{P})$ denote $c_1(\mathcal{O}_{\mathbb{P}}(-1))$ and let $\kappa \in \mathrm{ch}(Q_{r-1})$ denote $c_1(K_{r-1})$. We write c_j, κ_j for the images of c and κ in $\mathrm{ch}(P_r)$ under the jth coordinate pullbacks $\mathrm{ch}(Q_{r-1}) \to \mathrm{ch}(\mathbb{P}) \to \mathrm{ch}(P_r)$.

Lemma 9.33. *Suppose that $H_1, \ldots, H_{r-1}$ are trivial. Then*

  (a) $\gamma^{\ell^r} = \gamma^{\ell^{r-1}} \eta^d$ *in* $\mathrm{ch}(P_r)$, *where* $d = \ell^r - \ell^{r-1}$;
  (b) if in addition H_r is trivial, then $\gamma^d = -\prod c_j \kappa_j^e$, *where* $e = \ell^{r-1} - 1$;
  (c) if $S' = \mathrm{Spec}\, k$ then the zero-cycles $\kappa^e \in \mathrm{ch}_0(Q_{r-1})$ *and* $\gamma^d \in \mathrm{ch}_0(P_r)$ *have degrees*

$$\deg(\kappa^e) \equiv (-1)^{r-1} \quad and \quad \deg(\gamma^d) \equiv -1 \quad modulo\ \ell.$$

Proof. First note that because K_{r-1} is defined over the e-dimensional variety $Q_{r-1}(\mathrm{Spec}\, k; H_1, ..., H_{r-1})$, the element $\kappa = c_1(K_{r-1})$ satisfies $\kappa^{\ell^{r-1}} = 0$. Thus $(\eta + \kappa)^{\ell^{r-1}} = \eta^{\ell^{r-1}}$ and hence $(\eta + \kappa)^d = \eta^d$. Now the element $c = c_1(\mathcal{O}_{\mathbb{P}}(-1))$ satisfies the relation $c^2 = c(\eta + \kappa)$ in $\mathrm{ch}(\mathbb{P})$ and hence

$$c^{\ell^r} = c^{\ell^{r-1}}(\eta + \kappa)^d = c^{\ell^{r-1}} \eta^d$$

in $\mathrm{ch}^{\ell^r}(\mathbb{P})$. Now recall that $P_r = \prod \mathbb{P}$. Then $\gamma = \sum c_j$ and (a) holds:

$$\gamma^{\ell^r} = \sum c_j^{\ell^r} = \sum c_j^{\ell^{r-1}} \eta^d = \gamma^{\ell^{r-1}} \eta^d.$$

When H_r is trivial we have $\eta = 0$ and hence $c^2 = c\kappa$. Setting $b_j = c_j^{\ell^{r-1}} = c_j \kappa_j^e$, we have $\gamma^d = \gamma^{\ell^{r-1}(\ell-1)} = (\sum b_j)^{\ell-1}$. To evaluate this, we use the algebra trick that since $b_j^2 = 0$ for all j and $\ell = 0$ we have $(\sum b_j)^{\ell-1} = (\ell-1)! \prod b_j = -\prod b_j$. Thus (b) holds.

For (c), note that if $S' = \operatorname{Spec} k$ then $\eta = 0$ and γ^d is a zero-cycle on P_r. By the projection formula for $\pi : P_r \to \prod Q_{r-1}$, part (b) yields $\pi_* \gamma^d = (-1)^\ell \prod \kappa_j^e$. Since each Q_{r-1} is an iterated projective space bundle, the projective bundle formula [Har77, A11] yields $\operatorname{ch}(\prod Q_{r-1}) = \otimes_1^{\ell-1} \operatorname{ch}(Q_{r-1})$, and the degree of $\prod \kappa_j^e$ is the product of the degrees of the κ_j^e. By induction on r, these degrees are all the same, and nonzero, so $\deg(\prod \kappa_j^e) \equiv 1 \pmod{\ell}$. Hence $\deg(\gamma^d) \equiv -1 \pmod{\ell}$.

It remains to establish the formula for $\deg(\kappa^e)$ by induction on r, the case $r = 0$ being clear. Since the κ in $\operatorname{ch}(Q_r)$ is $c_1(K_r)$, and the Q_i are projective space bundles, it suffices to compute that $c_1(K_r)^{\ell^r - 1} = \kappa^e \gamma^d$ in $\operatorname{ch}(Q_r) = \operatorname{ch}(Q_{r-1}) \otimes \operatorname{ch}(P_r)$. Since $\kappa^{e+1} = 0$ and $c_1(K_r) = \kappa + \gamma$ we have

$$c_1(K_r)^{\ell^{r-1}} = \kappa^{e+1} + \gamma^{\ell^{r-1}} = \gamma^{\ell^{r-1}},$$

and hence $c_1(K_r)^d = \gamma^d$. Since $\gamma^{d+1} = 0$, this yields the desired calculation:

$$c_1(K_r)^{\ell^r - 1} = c_1(K_r)^e c_1(K_r)^d = (\kappa + \gamma)^e \gamma^d = \kappa^e \gamma^d. \qquad \square$$

Corollary 9.34. *If $H_1, \ldots, H_{r-1}$ are trivial, there is a ring homomorphism* $\mathbb{F}_\ell[\lambda, z]/(z^\ell - \lambda^{\ell-1} z) \to \operatorname{ch}(P_r)$, *sending λ to $\eta^{\ell^{r-1}}$ and z to $\gamma^{\ell^{r-1}}$.*

9.5 MODELS FOR MOVES OF TYPE C_N

In this section we construct maps $S_{n-1} \to S_n$ which model the ℓ moves of type C_n defined in 9.24. Each of the moves of type C_n introduces $\ell^{n-1} - \ell^{n-2}$ new variables, and will be modelled by a map $Y_r \to Y_{r-1}$ of relative dimension $\ell^{n-1} - \ell^{n-2}$, using the P_{n-1} construction in 9.28 over the base $S' = S_n$, starting with $Y_0 = S_n$. The result (Definition 9.35) will be a family of sheaves L_i lying over varieties Y_i fitting into a tower of the form:

$$
\begin{array}{ccccccccc}
J_{n-1} = L_\ell & & L_{\ell-1} & & L_2 & & L_1 & & L_0 = J_n \\
S_{n-1} = Y_\ell & \xrightarrow{f_\ell} & Y_{\ell-1} & \longrightarrow \cdots \longrightarrow & Y_2 & \xrightarrow{f_2} & Y_1 & \xrightarrow{f_1} & Y_0 = S_n.
\end{array}
$$

Fix $n \geq 2$, a variety S_n, and invertible sheaves $H_1, \ldots, H_{n-2}$, $L_0 = J_n$ and L_{-1} on S_n. The first step in building the tower is to form $Y_1 = P_{n-1}(S_n; H_1, \ldots, H_{n-2}, L_0)$ as in 9.28, with invertible sheaf $L_1 = L_{-1} \boxtimes \mathcal{O}_{Y_1}(-1, \ldots, -1)$. In forming the other Y_r, the base S' in the P_{n-1} construction 9.28 will become Y_{r-1} and the role of L_0 will be played by L_{r-1}. Here is the formal definition.

Definition 9.35. For $r > 1$, we define morphisms $f_r : Y_r \to Y_{r-1}$ and invertible sheaves L_r on Y_r as follows. Inductively, we are given a morphism $f_{r-1} : Y_{r-1} \to Y_{r-2}$ and invertible sheaves L_{r-1} on Y_{r-1}, L_{r-2} on Y_{r-2}. Set

$$Y_r = P_{n-1}(Y_{r-1}; H_1, \ldots, H_{n-2}, L_{r-1}) \xrightarrow{f_r} Y_{r-1},$$
$$L_r = f_r^* f_{r-1}^*(L_{r-2}) \otimes \mathcal{O}_{Y_r}(-1, \ldots, -1),$$

where $\mathcal{O}_{Y_r}(-1, \ldots, -1)$ is defined in 9.27.

We define S_{n-1} to be Y_ℓ; by Lemma 9.29, $\dim(Y_r/Y_{r-1}) = \ell^{n-1} - \ell^{n-2}$ and hence $\dim(S_{n-1}/S_n) = \ell^n - \ell^{n-1}$. Finally, we set

$$J_{n-1} = L_\ell, \qquad J'_{n-1} = f_\ell^*(L_{\ell-1}).$$

For example, when $n = 2$, this tower is exactly the tower of 9.10: we have $Y_r = P_1(Y_{r-1}; L_{r-1}) = \prod \mathbb{P}^1(\mathcal{O} \oplus L_{r-1})$.

Remark 9.35.1. The invertible sheaves J_{n-1} and J'_{n-1} will be the invertible sheaves of the Chain Lemma 9.1. The rest of the tower of S_i displayed in (9.19) will be obtained in Definition 9.37 by repeating this construction and setting $S = S_1$.

The ℓ-forms

We now define the ℓ-forms on the invertible sheaves J_{n-1} and J'_{n-1} of Definition 9.35. Suppose that the invertible sheaves L_{-1} and $L_0 = J_n$ on S_n are equipped with the ℓ-forms β_0 and β_{-1}. Initially, β_0 and β_{-1} are α_{n-1} and α_n. Then Ψ_{n-1} is an ℓ-form on $\mathcal{O}_{Y_i}(-1, \ldots, -1)$, depending upon β_0, and we endow the sheaf L_1 in Definition 9.35 with the ℓ-form $\beta_1 = f^*(\beta_{-1}) \otimes \Psi_{n-1}$; inductively, the form β_{r-1} determines a form Ψ_{n-1} and we endow the invertible sheaf L_r with the ℓ-form

$$\beta_r = f^*(\beta_{r-2}) \otimes \Psi_{n-1}.$$

Example. When $n = 2$, we saw that the tower 9.35 is exactly the tower of 9.10. In addition, the ℓ-form $\beta_r = \Psi_1$ (depending upon β_{r-1}) on $\mathcal{O}_{P_1}(-1, \ldots, -1)$ agrees with the ℓ-form φ_r of 9.16.

Lemma 9.36. *If $\beta_0 = \alpha_{n-1}$ and $\beta_{-1} = \alpha_n$, then (at the generic point s of Y_1) the ℓ-form β_1 agrees with the form $a_n \Psi_{n-1}$ in (9.24), while β_ℓ and $\beta_{\ell-1}$ agree with the ℓ-forms γ_{n-1} and γ'_{n-1} in (9.25). Moreover, $\{\alpha_1, \ldots, \alpha_{n-2}, \beta_0, \beta_{-1}\}$ agrees with $-\{\alpha_1, \ldots, \alpha_{n-2}, \beta_\ell, \beta_{\ell-1}\}$.*

Proof. By Lemma 9.30, the form $\Psi_{n-1}|_s$ agrees with the form Ψ_{n-1} of (9.20); this proves that $\beta_1|_s$ agrees with $a_n \Psi_{n-1}$. By Definition 9.24, the first move of type C_n replaces β_0, β_{-1} by β_1, β_0. Applying Lemma 9.30 ℓ times, we obtain (9.25). The final assertion now follows from 9.26. $\qquad\square$

9.6 PROOF OF THE CHAIN LEMMA

This section is devoted to proving the Chain Lemma 9.1.

Scheme of the proof *9.37.* The tower (9.19) of varieties S_i is obtained by downward induction, starting with $S_n = \operatorname{Spec}(k)$ and sheaves $H_1, \ldots, H_n$. Construction 9.35 with $L_0 = H_n$ and $L_{-1} = H_{n-1}$ yields S_{n-1} ($= Y_\ell$), sheaves J_{n-1} and J'_{n-1}, and ℓ-forms $\gamma_{n-1} = \beta_\ell$ and $\gamma'_{n-1} = \beta_{\ell-1}$. Inductively, we repeat construction 9.35 for i to produce the scheme S_i, the sheaves J_i and J'_i, and ℓ-forms on them, starting with the output S_{i+1} and sheaves $H_1, \ldots, H_{i-1}$, $L_0 = J_{i+1}$, and $L_{-1} = H_i$ of the previous step. Finally, we set $S = S_1$.

Since $\dim(S_i/S_{i+1}) = \ell^{i+1} - \ell^i$ we have $\dim(S_i/S_n) = \ell^n - \ell^i$. In particular, since $\dim(S_n) = 0$ and $S = S_1$ we have $\dim(S) = \ell^n - \ell$. This proves part (1) of the Chain Lemma.

By downward induction in the tower (9.19), each J_i and J'_i carries an ℓ-form, which we call γ_i and γ'_i, respectively. By 9.36, these forms agree with the forms γ_i and γ'_i of (9.25) and (9.26). Part (2) now follows from Lemma 9.36, (9.25), and (9.26); part (3) follows. Part (4) was proven in Proposition 9.31; part (5) follows. Thus we have established all but part (6) of the Chain Lemma.

The rest of this section is devoted to proving the final part (6), that the degree of the zero-cycle $c_1(J_1)^{\dim S}$ is relatively prime to ℓ. This will be achieved in Theorem 9.43. In preparation, we need to compare the degrees of the zero-cycles $c_1(J_{n-1})^{\dim S_{n-1}}$ on S_{n-1} and $c_1(J_i)^{\dim S_i}$ on S_i for $i < n-1$. In order to do so, we introduce the following algebra.

Definition 9.38. We define the graded $\mathbb{F}_\ell$-algebras A_r and $\bar{A}_r = A_r / \lambda_{-1} A_r$ by:

$$A_r = \mathbb{F}_\ell[\lambda_{-1}, \lambda_0, \ldots, \lambda_r, z_1, \ldots, z_r]/(\{z_i^\ell - \lambda_{i-1}^{\ell-1} z_i, \lambda_i - \lambda_{i-2} - z_i \mid i = 1, \ldots, r\}).$$

The variables λ_i and z_i all have degree 1.

Remark 9.38.1. Suppose that $H_1, \ldots, H_{n-2}$ are trivial. By Corollary 9.34 with $r = n-1$, there is an algebra homomorphism $A_\ell \xrightarrow{\rho} \operatorname{ch}(Y_\ell)$, sending each λ_i to $c_1(L_i)^{\ell^{n-2}}$ and each z_i to $c_1(\mathcal{O}_{Y_i}(-1, \ldots, -1))^{\ell^{n-2}}$. When L_{-1} is trivial, and hence $\lambda_{-1} = c_1(L_{-1}) = 0$, ρ factors through $\bar{A}_\ell$.

Lemma 9.39. *In $\bar{A}_r$, every element u of degree 1 satisfies $u^{\ell^2} = u^\ell \lambda_0^{\ell^2 - \ell}$.*

Proof. We will show that $\bar{A}_r$ embeds into a product of graded rings of the form $\Lambda_k = \mathbb{F}_\ell[\lambda_0][v_1, \ldots, v_k]/(v_1^\ell, \ldots, v_k^\ell)$. In each entry, $u = a\lambda_0 + v$ with $v^\ell = 0$ and $a \in \mathbb{F}_\ell$, so $u^\ell = a\lambda_0^\ell$ and $u^{\ell^2} = a\lambda_0^{\ell^2}$, whence the result.

Since $\bar{A}_{r+1} = \bar{A}_r[z]/(z^\ell - \lambda_r^{\ell-1} z)$ is flat over $\bar{A}_r$, it embeds by induction into a product of graded rings of the form $\Lambda' = \Lambda_k[z]/(z^\ell - \lambda^{\ell-1} z)$, $\lambda \in \Lambda_k$. If $\lambda \neq 0$, there is an embedding of Λ' into $\prod_{i=0}^{\ell-1} \Lambda_k$ whose i^{th} component sends z to $i\lambda$. If $\lambda = 0$, then $\Lambda' \cong \Lambda_{k+1}$.   □

Remark 9.39.1. It follows that if $k, m > 0$ and $(\ell^2 - \ell) \mid m$ then $u^{k\ell + m} = \lambda_0^m u^{k\ell}$.

Proposition 9.40. *In* $\bar{A}_\ell$, $\lambda_\ell^{\ell^N - \ell} = \lambda_0^{\ell^N - \ell^2} (\prod z_i^{\ell - 1} + T\lambda_0)$ *for all* $N \geq 2$, *where* T *is a homogeneous polynomial of degree* $\ell^2 - \ell - 1$.

Proof. By Definition 9.38, $\bar{A}_\ell$ is free over $\mathbb{F}_\ell[\lambda_0]$, with a basis consisting of the elements $\prod z_i^{m_i}$ $(0 \leq m_i < \ell)$. Thus any term of degree $\ell^N - \ell$ is a linear combination of $F = \lambda_0^{\ell^N - \ell^2} \prod z_i^{\ell - 1}$ and terms of the form $\lambda_0^{m_0} \prod z_i^{m_i}$, where $\sum m_i = \ell^N - \ell^2$ and $m_0 > \ell^N - \ell^2$. It suffices to determine the coefficient of F in $\lambda_\ell^{\ell^N - \ell}$. Since $\lambda_\ell^{\ell^N - \ell} = \lambda_0^{\ell^N - \ell^2} \lambda_\ell^{\ell^2 - \ell}$ by Remark 9.39.1, it suffices to consider $N = 2$, when $F = \prod z_i^{\ell - 1}$.

As in the proof of Proposition 9.15, if $\ell \geq r \geq 2$ we compute in the ring $\bar{A}_r$ that

$$\lambda_r^{r(\ell - 1)} = (z_r + \lambda_{r-2})^{\ell(r-1) + (\ell - r)} = (z_r^\ell + \lambda_{r-2}^\ell)^{r-1} \cdot (z_r + \lambda_{r-2})^{\ell - r}$$

$$= (z_r \lambda_{r-1}^{\ell - 1} + \lambda_{r-2}^\ell)^{r-1} (z_r + \lambda_{r-2})^{\ell - r} = z_r^{\ell - 1} \lambda_{r-1}^{(r-1)(\ell - 1)} + T,$$

where T is a homogeneous polynomial in $\bar{A}_{r-1}[z_r]$ of total degree $< \ell - 1$ in z_r. By induction on r, the coefficient of $(z_1 \cdots z_r)^{\ell - 1}$ in $\lambda_r^{r(\ell - 1)}$ is 1 for all r.   □

Lemma 9.41. *If* $S_n = \mathrm{Spec}(k)$ *and* $c = c_1(J_{n-1}) \in CH^1(S_{n-1})$, *then*

$$\deg(c^{\dim S_{n-1}}) \equiv -1 \pmod{\ell}.$$

Proof. Set $d = \dim(S_{n-1}) = \ell^n - \ell^{n-1}$; under the map $A_\ell \xrightarrow{\rho} \mathrm{ch}(S_{n-1})$ of 9.38.1, the degree $\ell^2 - \ell$ part of A_ℓ maps to $CH^d(S_{n-1})$. Since $S_n = \mathrm{Spec}(k)$, all the sheaves $H_1, \ldots, H_{n-2}, H_n$ and $L_0 = J_n$ are trivial. This implies that ρ factors through $\bar{A}_\ell$ and that $\rho(\lambda_0) = 0$. By Proposition 9.40, the zero-cycle $c^d = \rho(\lambda_\ell)^{\ell^2 - \ell}$ equals the product of the $\rho(z_i)^{\ell - 1} = c_1(\mathcal{O}_{Y_i}(-1, \ldots, -1))^{d/\ell}$. Because $S_{n-1} = Y_\ell$ is a product of iterated projective space bundles, $CH_0(Y_\ell)$ is the tensor product of their CH_0 groups, and the degree of c^d is the product of the degrees of the $c_1(\mathcal{O}_{Y_i}(-1, \ldots, -1))^{d/\ell}$, each of which is -1 by Lemma 9.33. It follows that $\deg(c^d) \equiv -1 \pmod{\ell}$.   □

Theorem 9.42. *Fix* $i \leq n - 1$. *If* S_i *has dimension* $\ell^s - \ell^r$ *and* $H_1, \ldots, H_{i-2}$ *and* H_i *are trivial, then the zero-cycles* $c_1(J_{i-1})^{\dim S_{i-1}} \in CH_0(S_{i-1})$ *and* $c_1(J_i)^{\dim S_i} \in CH_0(S_i)$ *have the same degree modulo* ℓ:

$$\deg(c_1(J_{i-1})^{\dim S_{i-1}}) \equiv \deg(c_1(J_i)^{\dim S_i}) \pmod{\ell}.$$

Proof. By Remark 9.38.1, there is a homomorphism $A_\ell \xrightarrow{\rho} \mathrm{ch}(S_{i-1})$, factoring through $\bar{A}_\ell$, sending λ_j to $c_1(L_j)^{\ell^{i-2}}$ and z_j to $c_1(\mathcal{O}_{Y_j}(-1,\ldots,-1))^{\ell^{i-2}}$.

Set $N = s - i + 2$ and $y = \lambda_0^{\ell^N - \ell^2}$, so $\rho(y) = c_1(J_i)^{\dim S_i} \in \mathrm{ch}_0(S_i)$. From Proposition 9.40 we have $\lambda_\ell^{\ell^N - \ell} \equiv y \prod z_j^{\ell-1}$, since

$$\rho(\lambda_0^{\ell^N - \ell^2 + 1}) = 0 \qquad \text{in } f^*\mathrm{ch}(S_i) \subset \mathrm{ch}(S_{i-1}).$$

By Lemma 9.14, the degree of this element equals the degree of y modulo ℓ. $\quad\square$

If we combine Lemma 9.41 and Theorem 9.42, we obtain the following result.

Theorem 9.43. *For each* $i < n$, $\deg(c_1(J_i)^{\dim S_i}) \equiv -1 \pmod{\ell}$.

Theorem 9.43 establishes the final part (6) of the Chain Lemma 9.1, that the degree of $c_1(J_1)^{\dim S_1}$ is relatively prime to ℓ.

9.7   NICE G-ACTIONS

In this section, we extend the Chain Lemma 9.1 to include an action by the group $\mu_\ell^n = (\mu_\ell)^n$ on S, J_i, and J_i' leaving γ_i and γ_i' invariant. We will assume that $1/\ell \in k$ and $\mu_\ell \subset k^\times$, so that we may regard μ_ℓ as a discrete group. We also show that the action is nice, in the following sense.

Recall that if a finite group G acts on X then the (reduced) fixed point subscheme $\mathrm{Fix}_G(X)$ is $\{x \in X : (\forall g \in G) gx = x \text{ and } g = 1 \text{ on } k(x)\}$.

Definition 9.44. (Rost, cf. [Ros98b, p. 2]) Let G be a finite group acting on a k-variety X. We say that the action is *nice* if $\mathrm{Fix}_G(X)$ is 0-dimensional, and consists of k-points.

When G also acts on an invertible sheaf L over X, G acts on the corresponding geometric bundle $\mathbb{L}$. We say that the action on L is *nice* if G acts nontrivially on $L|_x$ for every fixed point $x \in X$. In this case, $\mathrm{Fix}_G(\mathbb{L})$ is the zero-section of $\mathbb{L}$ over $\mathrm{Fix}_G(X)$.

Suppose that G acts nicely on each of several invertible sheaves L_i over X, giving a canonical representation $G \to \prod \mathrm{Aut}(L_i|_x) = \prod k^\times$. We say that G *acts nicely* on $\{L_1, \ldots, L_r\}$ if for each fixed point $x \in X$ the image of the canonical representation is $\prod G_i$, with each G_i nontrivial.

Remark 9.44.1. If $X_i \to S$ are equivariant maps and the X_i are nice ($i = 1, 2$), then G also acts nicely on $X_1 \times_S X_2$. However, even if G acts nicely on invertible sheaves L_i it may not act nicely on $L_1 \boxtimes L_2$, because the representation over (x_1, x_2) is the product representation $L_1|_{x_1} \otimes L_2|_{x_2}$, and a tensor product of nontrivial representations can be trivial.

Example 9.45. Suppose that G acts nicely on an invertible sheaf L over X. Then the induced G-action on $\mathbb{P} = \mathbb{P}(\mathcal{O} \oplus L)$ and its tautological sheaf $\mathcal{O}(-1)$ is nice. Indeed, if $x \in X$ is a fixed point then the fixed points of $\mathbb{P}|_x$ consist of the two k-points $\{[\mathcal{O}], [L]\}$, and if $L|_x$ is the representation ρ then G acts on $\mathcal{O}(-1)$ at these fixed points as ρ and ρ^{-1}, respectively.

By 9.44.1, G also acts nicely on the products $P = \prod \mathbb{P}(\mathcal{O} \oplus L)$ and $Q = X \times_{S'} P$ of Definition 9.27, but it may not act nicely on $\mathcal{O}_P(-1, \dots, -1)$.

Example 9.46. If an ℓ-group G acts nicely on L, it also acts nicely on the projective space $\mathbb{P}(\mathcal{A})$ of the Kummer algebra sheaf $\mathcal{A} = \mathcal{A}(L)$ of 9.8. Indeed, an elementary calculation shows that $\mathrm{Fix}_G \mathbb{P}(\mathcal{A})$ consists of the ℓ sections $[L^i]$, $0 \le i < \ell$ over $\mathrm{Fix}_G(X)$. In each fiber, the (vertical) tangent space at each fixed point is the representation $\rho \oplus \cdots \oplus \rho^{\ell-1}$. If $G = \mu_\ell$, this is the reduced regular representation (because $\mu_\ell \subset k^\times$).

Over any fixed point $x \in X$, $L|_x$ is trivial, and the cyclic group of order ℓ acts on the sheaf $\mathcal{A}|_x$ by $L^i \mapsto L^{i+1}$, rotating the fixed points. This induces G-isomorphisms between the tangent spaces at these points.

Example 9.46.1. In contrast, the action of G on $Y = \mathbb{P}(\mathcal{O} \oplus \mathcal{A})$ is not nice, because $\mathrm{Fix}_G(Y)$ is not 0-dimensional. In this case, an elementary calculation shows that $\mathrm{Fix}_G(Y)$ consists of the points $[L^i]$ of $\mathbb{P}(\mathcal{A})$, $0 < i < \ell$, together with the projective line $\mathbb{P}(\mathcal{O} \oplus \mathcal{O})$ over every fixed point x of X. For each x, the (vertical) tangent space at $[L^i]$ is $1 \oplus \rho \oplus \cdots \oplus \rho^{\ell-1}$; if $G = \mu_\ell$, this is the regular representation.

When $G = \mu_\ell^n$, the following lemma allows us to assume that the action on $L|_x$ is induced by the standard representation $\mu_\ell \subset k^\times$, via a projection $G \to \mu_\ell$.

Lemma 9.47. *Any nontrivial 1-dimensional representation ρ of $G = \mu_\ell^n$ factors as the composition of a surjection $G \xrightarrow{\pi} \mu_\ell$ with the standard representation of μ_ℓ.*

Proof. The representation ρ is an element of $(\mathbb{Z}/\ell)^n = G^* = \mathrm{Hom}(\mu_\ell^n, \mathbb{G}_m)$, and π is the Pontryagin dual of the induced map $\mathbb{Z}/\ell \to G^*$ sending 1 to ρ. $\square$

The construction of the P_r and Q_r in 9.28 is natural in the given invertible sheaves $H_1, \dots, H_r$ and K_0 over S'. Therefore the construction of the Y_r and S_{n-1} in 9.35 is natural in the given invertible sheaves $H_1, \dots, H_{n-2}, L_0$ and L_{-1}. The same is true for the construction of the S_i and $S = S_1$ in 9.37 above. Since $\prod_{i=1}^n \mathrm{Aut}(H_i)$ acts on the H_i, this group (and the subgroup $G = \mu_\ell^n$) will act on the variety S_{n-1}, and on the invertible sheaves J_{n-1} and J'_{n-1}. Hence G acts on the variety S of the Chain Lemma. We will show that it acts nicely on S.

Recall from Definition 9.28 that P_r and Q_r are defined by the construction 9.27 using the invertible sheaf $L_r = H_r \otimes K_{r-1}$ over Q_{r-1}.

Lemma 9.48. *If $S' = \mathrm{Spec}(k)$, then $G = \mu_\ell^r$ acts nicely on L_r, P_r, and Q_r.*

This implies that any subgroup of $\prod_{i=1}^r \mathrm{Aut}(H_i)$ containing μ_ℓ^r also acts nicely.

Proof. We proceed by induction on r, the case $r=1$ being 9.45, so we may assume that μ_ℓ^{r-1} acts nicely on Q_{r-1}. By 9.44.1, it suffices to show that $G=\mu_\ell^r$ acts nicely on $\mathbb{P}(\mathcal{O} \oplus L_r)$, where $L_r = H_r \otimes K_{r-1}$. Since the final component μ_ℓ of G acts trivially on K_{r-1} and Q_{r-1} and nontrivially on H_r, $G=\mu_\ell^{r-1} \times \mu_\ell$ acts nicely on L_r. By Example 9.45, G acts nicely on $\mathbb{P}(\mathcal{O} \oplus L_r)$. $\qquad\square$

The proof of Lemma 9.48 goes through in slightly greater generality.

Corollary 9.49. *Suppose that $G=\mu_\ell^n$ acts nicely on S' and on the invertible sheaves $\{H_1,\ldots,H_r\}$ over it. Then G acts nicely on L_r, P_r, and Q_r.*

Proof. Without loss of generality, we may replace S' by a fixed point $s \in S'$, in which case G acts nicely on $\{H_1,\ldots,H_r\}$ through the surjection $\mu_\ell^n \to \mu_\ell^r$. Now we are in the situation of Lemma 9.48. $\qquad\square$

Example 9.49.1. Suppose that G acts nicely on Y_{r-1} and on the invertible sheaves $H_1,\ldots,H_{n-2},L_{r-1}$. By Corollary 9.49, G acts nicely on $Y_r = P_{n-1}(Y_{r-1}; H_1,\ldots,H_{n-2},L_{r-1})$ and on the sheaf L_{n-1}. Since the last factor μ_ℓ of $G=\mu_\ell^n$ acts solely on L_{r-2}, it follows that the group $\mu_\ell^n = \mu_\ell^{n-1} \times \mu_\ell$ acts nicely on the family of invertible sheaves $H_1,\ldots,H_{n-2},L_{r-1}$, and $L_r = L_{r-2} \otimes \mathcal{O}(-1,\ldots,-1)$ over Y.

We can now process the tower of varieties Y_r defined in 9.35.

Proposition 9.50. *Suppose that $G=G_0 \times \mu_\ell^n$ acts nicely on S_n and (via $G \to \mu_\ell^n$) on $\{H_1,\ldots,H_{n-2},L_0,L_{-1}\}$. Then G acts nicely on each Y_r, and on its invertible sheaves $\{H_1,\ldots,H_{n-2},L_r,L_{r-1}\}$.*

Proof. The question being local, we may replace S' by a fixed point $s \in S'$, and G by μ_ℓ^n. We proceed by induction on r, the case $r=1$ being Example 9.49.1, since $L_1 = H_n \otimes \mathcal{O}(-1,\ldots,-1)$. Inductively, suppose that G acts nicely on Y_r and on $\{H_1,\ldots,H_{n-2},L_r,L_{r-1}\}$. Thus there is a factor of G isomorphic to μ_ℓ which acts nontrivially on L_r but acts trivially on $\{H_1,\ldots,H_{n-2},L_r\}$. Hence this factor acts trivially on $Y_{r+1} = P_{n-1}(Y_r; H_1,\ldots,H_{n-2},L_r)$ and its invertible sheaf $\mathcal{O}(-1,\ldots,-1)$, and nontrivially on $L_{r+1} = L_{r-1} \otimes \mathcal{O}(-1,\ldots,-1)$. The assertion follows. $\qquad\square$

Corollary 9.51. *$G=\mu_\ell^n$ acts nicely on S and J.*

Proof. By Definition 9.35, $S_{n-1}=Y_\ell$, $J_{n-1}=L_\ell$, and $J'_{n-1}=L_{\ell-1}$. By 9.50 with $r=\ell$, G acts nicely on S_{n-1} and on $\{H_1,\ldots,H_{n-2},J_{n-1},J'_{n-1}\}$. By downward induction, $G=\mu_\ell^{n-i} \times \mu_\ell^i$ acts nicely on S_i and $\{H_1,\ldots,H_{i-1},J_i,J'_i\}$ for all $i \le n$. The case $i=1$ is the conclusion, since $(S,J)=(S_1,J_1)$. $\qquad\square$

9.8   CHAIN LEMMA, REVISITED

In this section we prove a variation of the Chain Lemma, Theorem 9.52, which needs the extra assumption that $\mathrm{BL}(n-1)$ holds, and will be used to prove the multiplication principle 11.5. To state Theorem 9.52, we need some notation.

Recall from the Chain Lemma 9.1 that the variety S is equipped with an invertible sheaf J and an ℓ-form γ on J. As in Definition 9.8, there is a sheaf of Kummer algebras $\mathcal{A} = \oplus_{i=0}^{\ell-1} J^i$ on S associated to γ; we write $S(\sqrt[\ell]{\gamma})$ for the scheme $\mathrm{Spec}(\mathcal{A})$, noting that it is finite and flat of degree ℓ over S.

The fiber of $S(\sqrt[\ell]{\gamma})$ over a point $s \in S$ is $\mathrm{Spec}(\mathcal{A}|_s)$ where $\mathcal{A}_s$ is described in Definition 9.8: if u is a nonzero element of J_s and $t = \gamma_s(u) \in k(s)$ then $\mathcal{A}|_s = k(s)[u]/(u^\ell - t)$; if $t \notin k(s)^\ell$, we write $k(s)(\sqrt[\ell]{\gamma(s)})$ for the field $k(s)(\sqrt[\ell]{t})$.

If $\mu_\ell \subset k^\times$, every field extension E/k of degree ℓ is a cyclic Galois extension of the form $k(\sqrt[\ell]{t})$ for some $t \in k$. Theorem 9.52 says that if E splits the symbol $\{a_1, ..., a_n\}$ then E comes from the above construction for some k-point s of S.

Theorem 9.52. *Let k be an ℓ-special field k of characteristic 0, and suppose that $BL(n-1)$ holds. Given a nonzero symbol $\underline{a} \in K_n^M(k)/\ell$ over k, let S be the variety constructed in the Chain Lemma 9.1 for $\underline{a}$.*

For each cyclic extension E of k of degree ℓ that splits $\underline{a}$, there exists a k-rational point $s \in S(k)$ such that $E = k(\sqrt[\ell]{\gamma(s)})$.

Before proving Theorem 9.52, we state a corollary in the form that will be used to prove the multiplication principle (Theorem 11.5).

Corollary 9.53. *Let k be an ℓ-special field of characteristic 0, and suppose that $BL(n-1)$ holds. Given a nonzero symbol $\underline{a} \in K_n^M(k)/\ell$ over k, and a sequence $E_1, \ldots, E_n$ of cyclic extensions of k of degree ℓ splitting $\underline{a}$, there exist $a'_1, \ldots, a'_n \in k^\times$ such that $\underline{a} = \{a'_1, \ldots, a'_n\}$ and each E_i splits the symbol $\{a'_1, \ldots, a'_i\}$.*

Proof of Corollary 9.53. We will prove by induction on j, $1 \le j \le n$, that we can find $a'_1, \ldots, a'_n$ so that the splitting condition holds for $1 \le i \le j$. By Theorem 9.52, there is a k-point $s \in S$ so that $E_1 = k(\sqrt[\ell]{\gamma(s)})$. Since $\underline{a} = \{\gamma(s), \ldots\}$ (by the Chain Lemma 9.1), the case $j = 1$ follows.

Inductively, suppose that the splitting condition holds for $j - 1$. After relabelling, we may assume that E_i splits $\{a_1, \ldots, a_{i-1}\}$ for $2 \le i \le j$. Again using Theorem 9.52, choose a k-rational point $s \in S$ such that $E_1 = k(\sqrt[\ell]{\gamma(s)})$; E_1 splits $\{\gamma(s)\}$. By part (2) of the Chain Lemma 9.1, $\underline{a} = \{\gamma(s), \gamma'_1(s), ..., \gamma'_{n-1}(s)\}$ and $\{a_1, \ldots, a_{i-1}, \gamma_i(s)\} = \{\gamma(s), \gamma'_1(s), \ldots, \gamma'_{i-1}(s)\}$. By induction, the latter symbol is split by E_i for $2 \le i \le j$. Setting $a'_1 = \gamma(s)$ and $a'_i = \gamma'_{i-1}(s)$ for $i = 2, \ldots, j$, we see that the splitting condition holds for j. $\qquad\square$

The proof of Theorem 9.52 will use the invariant $\eta(X/S)$, which is defined in 8.22, and Theorems 7.5 and 8.25, which require characteristic 0.

Lemma 9.54. *If $\mu_\ell \subset k$ and γ is nonzero at the generic point s_0 of S, then either $\gamma_{s_0}(L|_{s_0}) \subseteq k(s_0)^{\times \ell}$ or else $S(\sqrt[\ell]{\gamma})$ is a pseudo-Galois cover of S with group μ_ℓ. In this case, $\eta(S(\sqrt[\ell]{\gamma})/S)$ is defined in $\mathbb{Z}/\ell$ and equals the degree of the zero-cycle $c_1(L)^{\dim S}$, which is nonzero by the Chain Lemma 9.1(6).*

Proof. Let u be a nonzero element of $L|_{s_0}$ and consider the element $t = \gamma_{s_0}(u)$ of $k(s_0)$. The fiber of $S(\sqrt[\ell]{\gamma})$ over s_0 is $E_{s_0} = k(s)[u]/(u^\ell - t)$; if $t \notin k(s_0)^\ell$ then E_{s_0} is a field extension of $k(s_0)$ with Galois group μ_ℓ. In this case, $S(\sqrt[\ell]{\gamma})$ is a variety with function field E_{s_0}. The natural action of μ_ℓ on any line bundle, and in particular L, extends naturally to an action on $\mathcal{A}$ and hence on $S(\sqrt[\ell]{\gamma})$. Thus $S(\sqrt[\ell]{\gamma}) \to S$ is a pseudo-Galois cover in the sense of Definition 8.19.

If s is a closed point of S contained in $V(\gamma)$ then $k(s)$ splits $\underline{a}$ by 9.1(4) and hence ℓ divides $[k(s):k]$. Hence $\eta = \eta(S(\sqrt[\ell]{\gamma})/S)$ is defined (see 8.22). Since the invertible sheaf corresponding to this pseudo-Galois covering of S is the dual L^* of L, $\eta = \deg(c_1(L^*)^{\dim S})$. Since $c_1(L^*) = -c_1(L)$ and $\dim(S) = \ell^n - \ell$ is even, we have $\eta = (-1)^{\dim S} \deg c_1(L)^{\dim S} = \deg c_1(L)^{\dim S}$, and the result follows. $\square$

Lemma 9.55. *For every finite field extension F' of $k(S(\sqrt[\ell]{\gamma}))$ of degree prime to ℓ, there is a diagram of smooth projective varieties and dominant maps, where the degree of h is prime to ℓ and F' is the function field of $\tilde{S}(\sqrt[\ell]{\gamma})$:*

$$
\begin{array}{ccc}
\tilde{S}(\sqrt[\ell]{\gamma}) & \longrightarrow & \tilde{S} \\
\tilde{h} \downarrow & & h \downarrow \\
S(\sqrt[\ell]{\gamma}) & \longrightarrow & S.
\end{array}
$$

If $\operatorname{char}(k) = 0$ and $\mu_\ell \subset k$ then $\eta(\tilde{S}(\sqrt[\ell]{\gamma})/S) = \deg(h) \cdot \deg(c_1(J)^{\dim S})$.

Proof. Choose a Sylow ℓ-subgroup P of the Galois group of $F'/k(S)$, and let F be the fixed subfield of F' fixed by P. Then $[F':F] = \ell$ and $[F:k(S)]$ is prime to ℓ. Let $\tilde{S}_F$ be the normal closure of S over F, with canonical map $\tilde{S}_F \to S$; locally, the restriction of $\tilde{S}_F$ over an affine open $\operatorname{Spec}(A) \subset S$ is $\operatorname{Spec}(\tilde{A})$, where $\tilde{A}$ is the normalization of A in F. Let $\tilde{S}$ be a resolution of singularities of $\tilde{S}_F$ with projection $h: \tilde{S} \to S$. Finally, the ℓ-form γ on J lifts to an ℓ-form $h^*\gamma$ on $h^*(J)$, and the finite flat map $\tilde{S}(\sqrt[\ell]{\gamma}) \to \tilde{S}$ is compatible with $S(\sqrt[\ell]{\gamma}) \to S$.

The final assertion comes from Theorem 8.25 and Lemma 9.54. $\square$

Remark 9.55.1. By the Chain Lemma 9.1(2), the function field of $S(\sqrt[\ell]{\gamma})$ splits $\underline{a}$. If X is a norm variety for $\underline{a}$, there is a finite field extension F' of $k(S(\sqrt[\ell]{\gamma}))$ of degree prime to ℓ and an F'-point $\operatorname{Spec}(F') \to X$. Forming $\tilde{S}$ as in Lemma 9.55, this F'-point extends to a rational map $\phi: \tilde{S}(\sqrt[\ell]{\gamma}) \dashrightarrow X$.

Recall that the cyclic group $C_\ell = \langle \sigma \rangle$ acts on X^ℓ by $\sigma(x_1, ..., x_\ell) = (x_2, ..., x_\ell, x_1)$, and that $C^\ell X$ denotes the geometric quotient variety X^ℓ/C_ℓ. Let σ be a generator of C_ℓ, and let $\phi: \tilde{S}(\sqrt[\ell]{\gamma}) \dashrightarrow X$ be the rational map mentioned above. Choosing an isomorphism $C_\ell \cong \mu_\ell$, the rational maps $\phi\sigma^i$ assemble to form a C_ℓ-equivariant rational map $g = (\phi, \phi\sigma, ..., \phi\sigma^{\ell-1})$ from $\tilde{S}(\sqrt[\ell]{\gamma})$ to X^ℓ. Taking geometric quotients yields a rational map $\bar{g}$ from $\tilde{S}$ to $C^\ell X = X^\ell/C_\ell$. Thus we have the commutative diagram:

$$
\begin{array}{ccc}
\tilde{S}(\sqrt[\ell]{\gamma}) & \xrightarrow{\;\;g = (\phi,\, \phi\sigma,\, \ldots,\, \phi\sigma^{\ell-1})\;\;} & X^\ell \\
\downarrow & & \downarrow \\
\tilde{S} & \xrightarrow{\qquad\qquad \bar{g} \qquad\qquad} & C^\ell(X).
\end{array}
\tag{9.55.2}
$$

Proposition 9.56. *If* $\mathrm{char}(k) = 0$ *and* $\mu_\ell \subset k$, *the degree of* g *is prime to* ℓ.

Proof. By the Degree Formula 8.25,

$$
\deg(h) \cdot \eta(S(\sqrt[\ell]{\gamma})/S) \equiv \eta(\tilde{S}(\sqrt[\ell]{\gamma})/\tilde{S}) \equiv \deg(g) \cdot \eta(X^\ell/C^\ell X).
$$

The left side is nonzero by Lemmas 9.54 and 9.55, which implies that $\deg(g)$ is also nonzero modulo ℓ. $\square$

Proof of Theorem 9.52. Let X be an ℓ-generic splitting variety for $\underline{a}$; X exists by Theorem 10.17, because of our inductive hypothesis that $\mathrm{BL}(n-1)$ holds. Since E splits $\underline{a}$ and E is ℓ-special, there is an E-point $\psi : \mathrm{Spec}(E) \to X$. Let σ be a generator of the cyclic group $G = \mathrm{Gal}(E/k)$; each $\psi\sigma^i$ is also an E-point of X. The sequence of these points yields a G-equivariant map $p : \mathrm{Spec}(E) \to X^\ell$; taking geometric quotients yields a diagram

$$
\begin{array}{ccc}
\mathrm{Spec}(E) & \xrightarrow{\;\;p = (\psi,\, \psi\sigma,\, \ldots,\, \psi\sigma^{\ell-1})\;\;} & X^\ell \\
\downarrow & & \downarrow \\
\mathrm{Spec}(k) & \xrightarrow{\qquad\qquad z \qquad\qquad} & C^\ell(X).
\end{array}
$$

Since the diagonal copy of X in X^ℓ has no k-points, and is the singular locus of $C^\ell(X)$, the image of z is a smooth point of $C^\ell(X)$, lying in the unramified locus of $X^\ell \to C^\ell(X)$. It follows that the diagram is cartesian, i.e., that the fiber over z consists of a single E-point, namely the image of p.

Since the map g in (9.55.2) is dominant of degree prime to ℓ (by 9.56), so is the induced rational map $\bar{g} : \tilde{S} \to C^\ell(X)$. By Theorem 7.5, we can lift the smooth k-point z of $C^\ell(X)$ to a k-point $\tilde{s}$ of $\tilde{S}$ and hence a k-point s of S such that $g(\tilde{s}) = z$. Since $\gamma \neq 0$ in a neighborhood of every k-point of S, by the Chain Lemma 9.1(5), the map $\tilde{S}(\sqrt[\ell]{t}) \to \tilde{S}$ is unramified in a neighborhood of s. Because the diagram (9.55.2) is cartesian, the fiber of $\tilde{S}(\sqrt[\ell]{t})$ over $\tilde{s}$ is a single point with residue field E. Thus $k(\sqrt[\ell]{t}) = k(\tilde{s})(\sqrt[\ell]{t}) \cong E$. $\square$

9.9 HISTORICAL NOTES

The Chain Lemma for $n = 2$ and $\ell = 2$ is a reformulation of the *common slot lemma* for quaternionic division algebras: if the algebra (a_1, a_2) is split over $k(\sqrt{c})$ then $(a_1, a_2) \cong (a_1, \gamma) \cong (c, \gamma)$ for some $\gamma \in k$. The extension to $n = 2$ and $\ell = 3$, formulated in terms of cyclic division algebras of degree 3 (in characteristic $\neq 3$), was given by Rost in [Ros99]. The Chain Lemma in its present form, and its proof, was given by Rost in the 1998 preprint [Ros98a].

Our formulation of the Chain Lemma (Theorem 9.1) is taken from the Suslin–Joukhovitski paper [SJ06, 5.1]. The proof of the Chain Lemma presented here is taken from [HW09] and is based upon Rost's preprint [Ros98b], his website [Ros98a], and Rost's lectures at the Institute for Advanced Study in 1999-2000 and 2005.

Chapter Ten

Existence of Norm Varieties

THE MAIN GOAL of this chapter, achieved in Theorem 10.17, is to construct norm varieties for symbols $\underline{a} = \{a_1, ..., a_n\}$ over a field k of characteristic 0, and start the proof (completed in chapter 11 using the Norm Principle) that norm varieties are Rost varieties. In turn, Rost varieties are used in chapter 5 to produce Rost motives, which are used in chapter 4 to establish the H90(n) condition used to prove our main theorem, Theorem 1.11.

Section 10.1 recalls the definition of a norm variety for a symbol $\underline{a}$ in $K_n^M(k)/\ell$; if $n \geq 2$ and k is ℓ-special, norm varieties are geometrically irreducible. In section 10.2, we use the Chain Lemma 9.1 to produce a specific ν_{n-1}-variety $\mathbb{P}(\mathcal{A})$, and a pencil Q of splitting varieties over $\mathbb{A}^1 - \{0\}$ whose fibers Q_w are fixed point equivalent to $\mathbb{P}(\mathcal{A})$. Using the bordism result 8.17, we see that any equivariant resolution $Q(\underline{a})$ of Q_w is a ν_{n-1}-variety. In section 10.3 we use Rost's Degree Formula 8.9 to show that any norm variety for $\underline{a}$ is ν_{n-1} because $Q(\underline{a})$ is. A norm variety for $\underline{a}$ is constructed in the final section 10.4 by induction on n, making use of the global inductive assumption that BL(n − 1) holds, and the Norm Principle 10.18, whose proof we postpone to chapter 11.

The name "norm variety" reflects the fact that these varieties are birational to a hypersurface in a family of Kummer algebras defined by the equation $\text{Norm}(u) = a_n$. The current meaning in Definition 10.1 has evolved over the years; see the historical notes 10.5 for further details.

10.1   PROPERTIES OF NORM VARIETIES

We first recall what a norm variety over a field k is. A field F over k is a *splitting field* for a symbol $\underline{a} \in K_n^M(k)/\ell$ if $\underline{a} = 0$ in $K_n^M(F)/\ell$. The following definition was given in chapter 1; see Definition 1.13.

Definition 10.1. An (irreducible) variety X over k is called a *splitting variety* for a symbol $\underline{a} \in K_n^M(k)/\ell$ if its function field splits $\underline{a}$, i.e., if $\underline{a}$ vanishes in $K_n^M(k(X))/\ell$. A splitting variety X is called an *ℓ-generic* splitting variety if any splitting field F has a finite extension E of degree prime to ℓ with $X(E) \neq \emptyset$.

A *norm variety* for a nonzero symbol $\underline{a} \in K_n^M(k)/\ell$ is a smooth projective ℓ-generic splitting variety for $\underline{a}$ of dimension $\ell^{n-1} - 1$.

Example 10.1.1. If $n=1$ then Spec $k(\sqrt[\ell]{a})$ is both a splitting variety and a norm variety for $a \in k^{\times}$. This is because the element a of $K_1^M(k)/\ell = k^{\times}/k^{\times \ell}$ is split by a field F exactly when F contains $\sqrt[\ell]{a}$. Similarly, if E is any finite field extension of $k(\sqrt[\ell]{a})$ of degree prime to ℓ, then $\mathrm{Spec}(E)$ is a norm variety for $a \in k^{\times}$.

If $n=2$ and N is the norm form defined by the extension $k(\sqrt[\ell]{a_1})/k$, then the $(\ell-1)$-dimensional affine variety Y defined by $N(X_0, \ldots, X_{\ell-1}) = a_2$ is a geometrically irreducible splitting variety for $\underline{a} = \{a_1, a_2\}$, as we saw in the proof of Proposition 1.25.

If $n=2$ and k contains the ℓth roots of unity, then the Severi–Brauer variety for $\underline{a} = \{a_1, a_2\}$, defined in 1.14, is a norm variety for the symbol $\underline{a}$. Since $K_2^M(F)/\ell \cong {}_\ell\mathrm{Br}(F)$, if F is a splitting field for $\underline{a}$ then the central simple k-algebra A defining the Severi–Brauer variety X satisfies $A \otimes_k F \cong M_\ell(F)$ and hence $X(F) \neq \emptyset$.

Example 10.1.2. If k' is a finite separable field extension of k, of degree prime to ℓ, then any norm variety for $\underline{a}$ over k' is also a norm variety for $\underline{a}$ over k. The assumption that $[k':k]$ is prime to ℓ ensures that $\underline{a}$ is a nonzero symbol in $K_n^M(k')/\ell$, and that any field F over k has a prime-to-ℓ extension F' over k'.

To avoid the frequent passage to extension fields associated with the ℓ-generic splitting hypothesis, it will be useful to assume that k has no extension fields of degree prime to ℓ, i.e., that k is an ℓ-*special* field (Definition 1.9): $1/\ell \in k$ and ℓ divides the order of every finite field extension of k.

The assumption that k'/k is separable is necessary. If k is not perfect, there may be an inseparable extension k' of k and a norm variety X over k' which is not smooth over k—and therefore cannot be a norm variety over k.

Remark 10.1.3. Let X be a splitting variety for $\underline{a}$. Then $k(x)$ is a splitting field for $\underline{a}$ for every point $x \in X$, by specialization (see [Wei13, III.7.3]). It follows that if a field F has a finite extension E of degree prime to ℓ with $X(E) \neq \emptyset$ then F is a splitting field of $\underline{a}$. In particular, this implies that the degree of every closed point $x \in X$ is divisible by ℓ.

Recall that the *field of constants* of an irreducible k-variety X is the algebraic closure k_c of k in $k(X)$. It is well known that X is geometrically irreducible over k_c. Example 10.1.2 shows that the field of constants of a norm variety need not be k; we now show that every norm variety is either geometrically irreducible or arises in this way, provided that $n \geq 2$.

Proposition 10.2. *Let k be a field containing $1/\ell$. If X is an ℓ-generic splitting variety for $\underline{a} = \{a_1, \ldots, a_n\}$, and $n \geq 2$, then $[k_c : k]$ is prime to ℓ and X is a geometrically irreducible ℓ-generic splitting variety for $\underline{a}$ over k_c.*

If k is ℓ-special and $n \geq 2$, every norm variety over k is geometrically irreducible.

Proof. Let Y be a geometrically irreducible splitting variety for $\underline{a}$, i.e., such that $k(Y)$ splits the symbol $\underline{a}$. Such a variety always exists; for example, Y could be the affine variety $N = a_2$ of Example 10.1.1, or the Severi–Brauer variety Y of $\{a_1, a_2\}$ over $k(\sqrt[\ell]{a_1})$. (See Example 10.1.2.)

If X is an ℓ-generic splitting variety, there is a field extension F of $k(Y)$ with $[F : k(Y)]$ prime to ℓ, and a point in $X(F)$. Let k_c be the field of constants of X. Then $K = k(Y) \otimes_k k_c$ is a subfield of F, since Y is geometrically irreducible.

Now the degree $[K : k(Y)] = [k_c : k]$ divides $[F : k(Y)]$, which is prime to ℓ. This proves that $[k_c : k]$ is prime to ℓ. Since X is always geometrically irreducible over k_c, it suffices to observe that X is an ℓ-generic splitting variety over k_c, which is elementary. □

Remark 10.2.1. Proposition 10.2 fails for $n = 1$: $\mathrm{Spec}(k(\sqrt[\ell]{a_1}))$ is a norm variety for a_1, is not geometrically irreducible over k, and $[k_c : k] = \ell$.

Lemma 10.3. *Let k' be a finite field extension of k, and X a variety over k.*

i) *If $X_{k'} = X \times_k k'$ is a norm variety for $\underline{a}$ over k', and $[k' : k]$ is prime to ℓ, then X and $X_{k'}$ are also norm varieties for $\underline{a}$ over k.*

ii) *If X is a geometrically irreducible norm variety for $\underline{a}$ over k, then $X_{k'}$ is a norm variety over k'.*

Proof. For (i), suppose that $X_{k'}$ is a norm variety over k'. By Example 10.1.2, $X_{k'}$ is a norm variety over k. Since $[k'(X_{k'}):k(X)] = [k':k]$ is prime to ℓ, $K_*^M(k(X))/\ell$ is a summand of $K_*^M(k(X_{k'}))/\ell$, so $\underline{a} = 0$ in $K_*^M(k(X))/\ell$. The fact that X is an ℓ-generic splitting variety follows from the fact that if $X_{k'}(E) \neq \emptyset$ then $X(E) \neq \emptyset$ via the map $X_{k'}(E) \to X(E)$. Since $X_{k'}$ is smooth over k', X is smooth over k (see [Har77, III.10.2]). Thus X is a norm variety over k.

For (ii), we may suppose that $n \geq 2$, as the case $n = 1$ is vacuous. Since X is smooth projective over k, $X_{k'}$ is smooth projective over k'. As X is geometrically irreducible over k, $X_{k'}$ is irreducible; the function field $k'(X_{k'})$ splits $\underline{a}$ since $k(X)$ does. Suppose that F is a splitting field of $\underline{a}$ over k'. Since X is ℓ-generic, there is a prime-to-ℓ extension E/F and a map $\mathrm{Spec}(E) \to X$ over k. Since $X_{k'} = X \times_k k'$ there is a map $\mathrm{Spec}(E) \to X_{k'}$. This map implies that $X_{k'}$ is ℓ-generic over k' and hence a norm variety over k'. □

As pointed out after Definition 1.9, any maximal prime-to-ℓ algebraic extension k'' of k is an ℓ-special field. This prime-to-ℓ passage from k to an ℓ-special field k'' preserves norm varieties, by the following argument.

If $\underline{a}$ is a nonzero symbol over k then it is also nonzero over k''. Any norm variety for $\underline{a}$ over k'' is the basechange of a smooth geometrically irreducible variety X' defined over a finite extension k' of k with $[k':k]$ prime to ℓ. The argument of Lemma 10.3(i) applies to show that X' is a norm variety for $\underline{a}$ over k' and hence is a norm variety over k. Conversely, if X is a geometrically irreducible norm variety over k then $X_{k''}$ is a norm variety over k'', by Lemma 10.3(ii).

Proposition 10.4. *If k is a field of characteristic 0, the property of being a norm variety for $\underline{a}$ is a birational invariant.*

Proof. Suppose that X and X' are smooth projective, with X a norm variety for $\underline{a}$, and that $X \dashrightarrow X'$ is a birational morphism. Choose a tower $X_n \to \cdots \to X_1 \to X$ of blow-ups along smooth centers and a morphism $f : X_n \to X'$ resolving the indeterminacies of the birational map; such a tower exists by the Weak Factorization Theorem (and resolution of singularities). For every field F, the map $X_n(F) \to X(F)$ is onto by construction and there is a map $X_n(F) \to X'(F)$; if F is an ℓ-special splitting field for $\underline{a}$ then $X(F)$ and hence $X_n(F)$ and $X'(F)$ are not empty. This shows that X' is ℓ-generic, as required. $\qquad\square$

Corollary 10.5. *If k is any field of characteristic 0, or if $\mu_\ell \subset k^\times$, every nonzero symbol $\{a_1, a_2\}$ has a geometrically irreducible norm variety over k.*

Proof. If $\mu_\ell \subset k^\times$, the Severi–Brauer variety is a geometrically irreducible norm variety, by Example 10.1.1. If $\mathrm{char}(k) = 0$, let Y be the (geometrically irreducible) affine splitting variety of Example 10.1.1 defined by $N = a_2$. Then Y is smooth; let $\bar{Y}$ be a smooth projective completion of Y. If k' is obtained by adjoining an ℓth root of unity to k, then $\bar{Y}_{k'}$ is a norm variety for k' by Proposition 10.4, since $Y_{k'}$ is birational to the Severi–Brauer variety (as observed in the proof of 1.25). By Lemma 10.3(i), $\bar{Y}$ is a norm variety. $\qquad\square$

10.2   TWO ν_{N-1}-VARIETIES

Part of the definition of a Rost variety (Definition 1.24) is that it is a ν_{n-1}-variety, and that some ν_i-variety maps to it for each smaller i. In section 10.3, we shall prove that norm varieties have these properties when k is a field of characteristic 0. For this, we need a pair of reference ν_{n-1}-varieties, $\mathbb{P}(\mathcal{A})$ and $Q(\underline{a})$. In this section, we use the Chain Lemma 9.1 to produce these varieties. We require $n \geq 2$, so that S and $\mathcal{A}$ are defined.

Recall from Definition 1.17 that a ν_i-*variety* is a smooth projective variety X of dimension $d = \ell^i - 1$ with $s_d(X) \not\equiv 0 \pmod{\ell^2}$. Here $i \geq 1$ and $s_d(X)$ is the degree of the characteristic class $s_d(T_X) \in CH^d(X)$, where T_X is the tangent bundle of X.

For example, when $n = 2$ we saw in Example 1.18(1) and Proposition 1.25 that the projective space $\mathbb{P}^{\ell-1}$ is a ν_1-variety (but not a norm variety), while any Severi–Brauer variety of dimension $\ell - 1$ is not only a norm variety, and a ν_1-variety, but is also a Rost variety for its underlying symbol.

Given a nontrivial symbol $\{a_1, ..., a_{n-1}\}$ in $K_{n-1}^M(k)/\ell$, where k contains $1/\ell$ and $n \geq 2$, the Chain Lemma 9.1 states that there is a smooth projective S of dimension $\ell^{n-1} - \ell$, an invertible sheaf J on S, and an ℓ-form $\gamma : J^{\otimes \ell} \to \mathcal{O}_S$ satisfying the properties listed in loc. cit. Let $\mathcal{A} = \mathcal{A}_\gamma(J)$ denote the associated Kummer algebra (defined in 9.8); it is the locally free $\mathcal{O}_S$-module $\bigoplus_{i=0}^{\ell-1} J^{\otimes i}$ equipped with a product defined by γ.

Theorem 10.6. *Let S be the variety of the Chain Lemma 9.1 for some symbol in $K_{n-1}^M(k)/\ell$, $n \geq 2$, and $\mathcal{A} = \mathcal{A}_\gamma(J)$ the associated sheaf of Kummer algebras over S. Then the projective bundle $\mathbb{P}(\mathcal{A})$ is a ν_{n-1}-variety.*

Proof. Let $\pi : \mathbb{P}(\mathcal{A}) \to S$ be the projection. Since $\dim(S) = \ell^{n-1} - \ell$, the dimension of $\mathbb{P}(\mathcal{A})$ is $\ell^{n-1} - 1$. In the Grothendieck group $K_0(\mathbb{P}(\mathcal{A}))$, we have that

$$[T_{\mathbb{P}(\mathcal{A})}] = \pi^*([T_S]) + [T_{\mathbb{P}(\mathcal{A})/S}]$$

where $T_{\mathbb{P}(\mathcal{A})/S}$ is the relative tangent bundle. The characteristic class s_d is additive, and $s_d(\pi^*[T_S]) = 0$ because the dimension of S is less than d, so $s_d(\mathbb{P}(\mathcal{A})) = s_d(T_{\mathbb{P}(\mathcal{A})/S})$. Now $[T_{\mathbb{P}(\mathcal{A})/S}] = [\pi^*(\mathcal{A}) \otimes \mathcal{O}(1)_{\mathbb{P}(\mathcal{A})/S}] - 1$. Applying additivity again, together with the definition of the characteristic class s_d (given before 1.17) and the decomposition of $\mathcal{A}$ and hence $\pi^*(\mathcal{A})$ into invertible sheaves, we obtain

$$s_d(\mathbb{P}(\mathcal{A})) = \deg \sum_{i=0}^{\ell-1} c_1\left(\pi^* J^{\otimes i} \otimes \mathcal{O}_{\mathbb{P}(\mathcal{A})/S}(1)\right)^d.$$

The projective bundle formula presents the Chow ring $CH^*(\mathbb{P}(\mathcal{A}))$ as:

$$CH^*(\mathbb{P}(\mathcal{A})) = CH^*(S)[y] / \left(\prod_{i=0}^{\ell-1}(y - ix) \right)$$

where $x = -c_1(J) \in CH^1(S)$ and $y = c_1(\mathcal{O}(1)) \in CH^1(\mathbb{P}(\mathcal{A}))$. Then $s_d(\mathbb{P}(\mathcal{A}))$ is the degree of the following element of the ring $CH^*(\mathbb{P}(\mathcal{A}))$:

$$s_d'(\mathbb{P}(\mathcal{A})) = \sum_{i=0}^{\ell-1}(y - ix)^d = \sum_{i=0}^{\ell-1} a_i y^i x^{d-i}$$

for some integer coefficients a_i. Since $x \in CH^1(S)$, we have $x^r = 0$ for any $r > \dim(S) = \ell^{n-1} - \ell$. It follows that $s_d'(\mathbb{P}(\mathcal{A})) = a_{\ell-1} y^{\ell-1} x^{\dim(S)}$. By part (6) of the Chain Lemma 9.1, the degree of $x^{\dim(S)} = (-1)^{\dim(S)} c_1(J)^{\dim(S)}$ is prime to ℓ. In addition, $\pi_*(y^{\ell-1}) = \pi_*(c_1(\mathcal{O}(1))^{\ell-1}) = [S]$ in $CH^0(S)$. By the projection formula $s_d(\mathbb{P}(\mathcal{A})) = a_{\ell-1} \deg x^{\dim(S)}$. Thus to prove the theorem, it suffices to show that $a_{\ell-1} \equiv \ell \pmod{\ell^2}$; this algebraic calculation is achieved in Lemma 10.7 below. $\square$

Lemma 10.7. *In the ring $R = \mathbb{Z}/\ell^2[x, y]/(\prod_{i=0}^{\ell-1}(y - ix))$, the coefficient of $y^{\ell-1}$ in $u_m = \sum_{i=0}^{\ell-1}(y - ix)^{\ell^m - 1}$ is ℓx^b, with $b = \ell^m - \ell$.*

Proof. Since u_m is homogeneous of degree $\ell^m - 1$, it suffices to determine the coefficient of $y^{\ell-1}$ in u_m in the ring

$$R/(x - 1) = \mathbb{Z}/\ell^2[y] / \left(\prod_{i=0}^{\ell-1}(y - i) \right) \cong \prod_{i=0}^{\ell-1} \mathbb{Z}/\ell^2.$$

If $m = 1$, then $u_1 = \sum_{i=0}^{\ell-1}(y-i)^{\ell-1}$ is a polynomial of degree $\ell - 1$ with leading term $\ell y^{\ell-1}$. Inductively, we use the fact that for all $a \in \mathbb{Z}/\ell^2$, we have

$$a^{\ell^2-\ell} = \begin{cases} 0, & \text{if } \ell \mid a; \\ 1, & \text{else.} \end{cases}$$

Thus for $m \geq 2$, if we set $k = (\ell^{m-1} - 1)/(\ell - 1)$, then $a^{\ell^m - 1} = a^{(\ell-1)+k(\ell^2-\ell)} = a^{\ell-1} \in \mathbb{Z}/\ell^2$, and therefore

$$u_m = \sum_{i=0}^{\ell-1}(y-i)^{\ell^m-1} = \sum_{i=0}^{\ell-1}(y-i)^{\ell-1} = u_1$$

holds in $R/(x-1)$; the result follows.   □

The second class of ν_{n-1}-varieties is defined using the norm $N : \mathcal{A} \to \mathcal{O}_S$. Recall from 9.8 that N is homogeneous of degree ℓ, so it induces a morphism $N : \mathbb{P}_S(\mathcal{A}) \to \mathcal{O}_S/\mathcal{O}_S^{\times \ell}$. Given a point s of S, the norm $N : A \to k(s)$ on the $k(s)$-algebra $A = \mathcal{A}|_s$ induces a norm $N : A_F \to F$ for every field F over $k(s)$, where A_F denotes $A \otimes_k F$. Since $N(\lambda a) = \lambda^\ell N(a)$ for $\lambda \in F$ and $a \in A_F$, it induces a quotient function $N : \mathbb{P}(A_F) \to F/F^{\times \ell}$, sending $[a]$ to $N(a)$. It also induces a function

$$\mathbb{P}_S(A_F \oplus F) \setminus \{[a,t] : t = 0\} \to F, \qquad [a,t] \mapsto N(a)/t^\ell.$$

Definition 10.8. We define the variety Q over $S \times \mathbb{A}^1$, and its fiber Q_w over $w \in k$, by the equation $N(a) = t^\ell w$:

$$Q = \left\{ ([a:t], w) \in \mathbb{P}_S(\mathcal{A} \oplus \mathcal{O}) \times \mathbb{A}^1 \mid N(a) = t^\ell w \right\},$$

$$Q_w = \left\{ [a:t] \in \mathbb{P}_S(\mathcal{A} \oplus \mathcal{O}) \mid N(a) = t^\ell w \right\}, \qquad \text{for } w \in k.$$

Since $\dim(S) = \ell^{n-1} - \ell$ we have $\dim(Q_w) = \ell^{n-1} - 1$.

Remark 10.8.1. When $w \neq 0$, there is a canonical map $\pi : Q_w \longrightarrow \mathbb{P}_S(\mathcal{A})$, sending $[a:t]$ to $[a]$. This is a cover of degree ℓ over its image, since $\pi([a:t]) = \pi([a:\zeta t])$ for all $\zeta \in \mu_\ell$. To see that π is well defined, note that over each point of S, the point $[0:1]$ of $\mathbb{P}_S(\mathcal{A} \oplus \mathcal{O})$ is disjoint from Q_w, i.e., Q_w is disjoint from the section $\sigma : S \cong \mathbb{P}_S(\mathcal{O}) \to \mathbb{P}_S(\mathcal{A} \oplus \mathcal{O})$. Hence the projection $\mathbb{P}_S(\mathcal{A} \oplus \mathcal{O}) - \sigma(S) \to \mathbb{P}_S(\mathcal{A})$ from these points induces the morphism $Q_w \to \mathbb{P}_S(\mathcal{A})$.

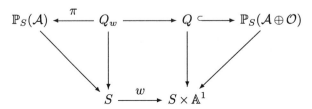

Lemma 10.9. *If $w \neq 0$, the projective variety Q_w is geometrically irreducible and the open subscheme Q_w^o where $t \neq 0$ is smooth.*

Proof. That Q_w^o is smooth over S (and hence smooth) is an easy consequence of the Jacobian criterion, or Remark 10.8.1.

Now extend the base to make k algebraically closed. If Q_w^o is reducible, then there exists an $s \in S$ such that the form $N_s - w$ is reducible over the UFD $\mathcal{O}_{S,s}$, or equivalently, the polynomial $N_s - wt^\ell$ is reducible in $R[t]$, where $R = \mathcal{O}_{S,s}[x_1, ..., x_\ell]$ and $N_s \in R$. If $F = \text{frac}(R)$, $N_s - wt^\ell$ must factor in $F[t]$ (as $\mu_\ell \subset F^\times$) and hence (by Gauss's Lemma) in $R[t]$ as $\prod(M - r_i t)$ for some linear form $M \in R$. Setting $t = 0$ yields $N_s = M^\ell$, i.e., N_s is the ℓth power of the linear form M. It follows that $\{a \in \mathcal{A}_s : N_s(a) = 0\}$, the degeneracy locus of N_s in $\mathcal{A}_s$, is a proper, nonzero $\mathcal{O}_{S,s}$-submodule of $\mathcal{A}_s$. However, this is impossible because $\mathcal{A}_s = \mathcal{O}_{S,s}[u]/(u^\ell - c)$, so either $u^\ell - c$ is irreducible (in which case $\mathcal{A}_s$ is a domain and $N_s(a) = 0$ for $a \neq 0$) or else $u^\ell - c$ factors (in which case $\mathcal{A}_s$ is a product of domains and the degeneracy locus is not a submodule). $\square$

Proposition 10.10. *For any nonzero $w \in k$, the variety Q_w is a splitting variety for the symbol $\{a_1, \ldots, a_{n-1}, w\}$.*

Proof. Let $x \in Q_w$ be a point, lying over the point $s \in S$. If $s \in V(\gamma)$, then by Theorem 9.1(4) the residue field $k(x)$ splits $\{a_1, \ldots, a_{n-1}\}$ and therefore splits $\{a_1, \ldots, a_{n-1}, w\}$. If s is not in $V(\gamma)$, then by the Chain Lemma 9.1(2) we can assume that $a_{n-1} = \gamma(u)$ in $k(s)$, and consider the Kummer algebra $A = F[u]/(u^\ell - a_{n-1})$ over $F = k(x)$. Then x determines an F-point $[a : 1]$ of $Q_w(F) \subset \mathbb{P}(A \oplus F)$, i.e., an element a of A such that $N_{A/F}(a) = w$ in F. Then we have $\{a_1, ..., a_{n-1}\}|_A = \{a_1, ..., u^\ell\} = 0$ over A, and (as in the proof of Lemma 9.18) we conclude:

$$\{a_1, ..., a_{n-1}, w\} = \{a_1, ..., a_{n-1}, Na\} = N\{a_1, ..., a_{n-1}, a\} = 0. \qquad \square$$

In order to show that Q_w is a ν_{n-1}-variety, we need to consider the action of the group $G = \mu_\ell^{n-1}$ on Q_w and $\mathbb{P}(\mathcal{A})$. For this, we recall some terminology from Definitions 8.15 and 9.44 about the action of G on k-varieties.

Two G-varieties X and Y are said to be *G-fixed point equivalent* over k if $\text{Fix}_G X$ and $\text{Fix}_G Y$ are 0-dimensional, lie in the smooth locus of X and Y, and there is a separable extension K of k and a bijection $\text{Fix}_G X_K \to \text{Fix}_G Y_K$ under which the families of tangent spaces at the fixed points are isomorphic as G-representations over K.

We say that G acts *nicely* on a k-variety X if the fixed locus $\text{Fix}_G X$ is a finite set of k-points; there is also a notion of G acting nicely on a line bundle.

In the case at hand, the group $G = \mu_\ell^{n-1}$ acts nicely on S, $\mathcal{O}_S$, J, and $\mathcal{A}$ by 9.51, and acts nicely on $\mathbb{P}(\mathcal{A})$ by 9.46. It also acts on $\mathbb{P}(\mathcal{A} \oplus \mathcal{O}_S)$ by 9.46.1. By inspection, G acts on Q; as the map $Q \to S \times \mathbb{A}^1$ is equivariant, each Q_w is a G-variety and the projection $\pi : Q_w \to \mathbb{P}(\mathcal{A})$ of Remark 10.8.1 is G-equivariant.

Theorem 10.11. *Assume that $1/\ell \in k$, $\mu_\ell \subset k^\times$, and $w \neq 0$. Then $G = \mu_\ell^{n-1}$ acts nicely on Q_w and $(\mathrm{Fix}_G Q_w) \cap (Q_w)_{sing} = \emptyset$. Moreover, Q_w and $\mathbb{P}(\mathcal{A})$ are G-fixed point equivalent over the field $k(\sqrt[\ell]{w})$.*

Proof. Since the maps $Q_w \xrightarrow{\pi} \mathbb{P}(\mathcal{A}) \to S$ are equivariant, π maps $\mathrm{Fix}_G Q_w$ to $\mathrm{Fix}_G \mathbb{P}(\mathcal{A})$, and both lie over the finite set $\mathrm{Fix}_G S$ of k-rational points. Since the tangent space $T_y \mathbb{P}(\mathcal{A})$ is the product of $T_s S$ and the tangent space of the fiber $\mathbb{P}(\mathcal{A}|_s)$, and similarly for Q_w, it suffices to consider a G-fixed point $s \in S$.

By Proposition 9.50 and Lemma 9.47, G acts nontrivially on $L = J|_s$ via a projection $G \to \mu_\ell$. By Example 9.46, G acts nicely on $\mathbb{P}(\mathcal{A})$. Thus there is no harm in assuming that $G = \mu_\ell$ and that L is the standard 1-dimensional representation.

Let $y \in \mathbb{P}(\mathcal{A})$ be a G-fixed point over s. By 9.45, the tangent space of $\mathbb{P}(\mathcal{A}|_s)$ at y is the reduced regular representation, and y is one of $[1], [L], \ldots, [L^{\ell-1}]$.

We saw in Example 9.46.1 that a fixed point $[a_0 : a_1 : \cdots : a_{\ell-1} : t]$ of G in $\mathbb{P}((\mathcal{A} \oplus \mathcal{O})|_s)$ is either one of the points $e_i = [\cdots 0 : a_i : 0 \cdots : 0]$, which do not lie on Q_w, or a point on the projective line $\{[a_0 : 0 : t]\}$, which lies on Q_w only when $a_0^\ell = w$. These points are defined over the field $K = k(\sqrt[\ell]{w})$, and $Q_w \otimes_k K$ meets the projective line in the K-points $[\zeta \sqrt[\ell]{w} : 0 : \cdots : 0 : 1]$, $\zeta \in \mu_\ell$. Each of these ℓ points is smooth on Q_w, and the tangent space (over s) is the reduced regular representation of G. $\square$

Remark 10.11.1. Since $\pi([\zeta \sqrt[\ell]{w} : 0 : \cdots : 0 : 1]) = [1 : 0 : \cdots : 0]$ for all $\zeta \in \mu_\ell$, $\mathrm{Fix}_G Q_w \xrightarrow{\pi} \mathrm{Fix}_G \mathbb{P}(\mathcal{A})$ is *not* a scheme isomorphism over $k(\sqrt[\ell]{w})$.

Now we need to invoke an equivariant cobordism result (Theorem 8.16, or rather Corollary 8.17). For this, we need to assume that k has characteristic 0.

Theorem 10.12. *Given a nonzero symbol $\underline{a}$ over a field k of characteristic 0, let $Q(\underline{a})$ denote a G-equivariant resolution of singularities of Q_{a_n}. Then $Q(\underline{a})$ is a geometrically irreducible ν_{n-1}-variety.*

Proof. Because k admits resolution of singularities and Q_{a_n} is geometrically irreducible by 10.9, $Q(\underline{a})$ exists and is geometrically irreducible. It is also a smooth projective completion of the (smooth) variety $Q_{a_n}^o$. By Theorem 10.11, $Q(\underline{a})$ is G-fixed point equivalent to $\mathbb{P}(\mathcal{A})$, which is a ν_{n-1}-variety by Theorem 10.6, i.e., $s_d \mathbb{P}(\mathcal{A}) \not\equiv 0 \pmod{\ell^2}$. Now Corollary 8.17 implies that $s_d Q(\underline{a}) \not\equiv 0 \pmod{\ell^2}$, i.e., that $Q(\underline{a})$ is a ν_{n-1}-variety. $\square$

10.3   NORM VARIETIES ARE ν_{N-1}-VARIETIES

In this section, we show that a norm variety is a ν_{n-1}-variety, and that a ν_i-variety maps to it for smaller i, at least when k is a field of characteristic 0. The proof will use two results which require characteristic 0: Rost's Degree Formula 8.9 and Theorem 10.12.

Construction 10.13. Suppose that a projective k-variety Y splits a symbol $\underline{a}$, and X is a projective ℓ-generic splitting variety for $\underline{a}$. Since $\underline{a}$ splits over the function field $k(Y)$, there is a finite extension $F/k(Y)$ of degree prime to ℓ and an F-point of X. We may choose a smooth projective model W for F having a morphism $f : W \to Y$ of degree prime to ℓ, extending $\mathrm{Spec}(F) \to \mathrm{Spec}(k(Y))$, and a morphism $g : W \to X$ extending $\mathrm{Spec}(F) \to X$.

Proposition 10.14. *When k is a field of characteristic 0 and $n \geq 2$, any norm variety X for a symbol $\underline{a} = \{a_1, \ldots, a_n\}$ is a ν_{n-1}-variety.*

Proof. Since $Q(\underline{a})$ splits the symbol, Construction 10.13 yields a smooth projective variety W and morphisms $f : W \to Q(\underline{a})$ and $g : W \to X$ such that f is generically finite of degree prime to ℓ.

$$X \xleftarrow{\;g\;} W \xrightarrow{\;f\;} Q(\underline{a})$$

By Proposition 10.12, $s_d(Q(\underline{a})) \not\equiv 0 \pmod{\ell^2}$. Because $Q(\underline{a})$ and X are splitting varieties for $\underline{a}$, ℓ divides the degree of every closed point of $Q(\underline{a})$ and X, by Remark 10.1.3. Hence we may apply Rost's Degree Formula (Theorem 8.9) to the morphisms f and g to conclude that

$$\deg(g)s_d(X) \equiv s_d(W) \equiv \deg(f)s_d(Q(\underline{a})) \pmod{\ell^2}.$$

Since the right side of this equation is nonzero modulo ℓ^2, both factors on the left-hand side are nonzero modulo ℓ^2 as well. Thus X is a ν_{n-1}-variety. $\qquad\square$

Finally, we prove that the ℓ-generic splitting variety X satisfies condition 1.24(b) to be a Rost variety. (Condition 1.24(b) is sometimes stated as saying that X is a "$\nu_{\leq n-1}$-variety," a notion we do not need.)

Proposition 10.15. *Let k be a field of characteristic 0, and suppose that X is a norm variety for the symbol $\underline{a} = \{a_1, \ldots, a_n\}$. Then for each i with $1 < i < n$, there exists a ν_{i-1}-variety W over k and a k-morphism $W \to X$.*

Proof. Fix $1 < i < n$ and let Y be a smooth projective geometrically irreducible ℓ-generic splitting variety of dimension $\ell^{i-1} - 1$ over k for the symbol $\{a_1, \ldots, a_i\}$. (At this point of our proof of the Bloch–Kato conjecture, we know that such a variety exists by induction on n.) Clearly, $\underline{a}$ splits over the function field $k(Y)$; since X is an ℓ-generic splitting variety for $\underline{a}$, Construction 10.13 yields a smooth projective variety W of dimension $d' = \ell^{i-1} - 1$ with a morphism $f : W \to Y$ of degree prime to ℓ and a morphism $g : W \to X$.

Because Y is a splitting variety for a non-split symbol, ℓ divides the degree of every 0-cycle on Y. Hence Rost's Degree Formula 8.9 applies to the morphism f, yielding $s_{d'}(W) \equiv \deg(f)s_{d'}(Y)$. Since $s_{d'}(Y) \not\equiv 0 \pmod{\ell^2}$, we conclude that $s_{d'}(W) \not\equiv 0 \pmod{\ell^2}$, i.e., that W is a ν_{i-1}-variety. $\qquad\square$

10.4 EXISTENCE OF NORM VARIETIES

In this section, we construct a norm variety for a given symbol $\underline{a} = \{a_1, \ldots, a_n\}$ over a field of characteristic 0. By Example 10.1.1 and Corollary 10.5, we may assume that $n \geq 3$; part of our global inductive hypothesis is that $BL(n-1)$ holds and that the Norm Principle (Theorem 10.18) holds for symbols of length $n-1$. We write $S^m(T)$ for the mth symmetric power of a variety T.

By induction, there is a geometrically irreducible norm variety Y (of dimension $\ell^{n-2} - 1$) for the symbol $\{a_1, \ldots, a_{n-1}\}$. Let $p : Y \times S^{\ell-1}(Y) \to S^\ell(Y)$ be the natural morphism; it is finite and surjective of degree ℓ. Let U denote the smooth locus of $S^\ell(Y)$, and let $\mathcal{A} = p_* \mathcal{O}_{Y \times S^{\ell-1}(Y)}|_U$ be the induced sheaf of finite-dimensional algebras on U, with associated geometric vector bundle $\mathbb{V}(\mathcal{A}) \to U$; $\mathbb{V}(\mathcal{A})$ has relative dimension ℓ over U.

Recall that the norm $N : \mathcal{A} \to \mathcal{O}_U$ induces a morphism of the associated vector bundles, from $\mathbb{V}(\mathcal{A})$ to $\mathbb{V}(\mathcal{O}_U) \cong U \times \mathbb{A}^1$.

Definition 10.16. Let X^o denote the hypersurface $N = a_n$ in $\mathbb{V}(\mathcal{A})$. Since $\dim(U) = \ell^{n-1} - \ell$, we have $\dim(X^o) = \ell^{n-1} - 1$.

$$
\begin{array}{ccccc}
X^o & \hookrightarrow & \mathbb{V}(\mathcal{A}) & \longrightarrow & \mathbb{V}(p_* \mathcal{O}_{Y \times S^{\ell-1}(Y)}) \\
\downarrow & & {\scriptstyle N}\downarrow & & {\scriptstyle N}\downarrow \\
U & \xrightarrow{\ a_n\ } & U \times \mathbb{A}^1 & \hookrightarrow & S^\ell Y \times \mathbb{A}^1
\end{array}
$$

Assuming that $\mathrm{char}(k) = 0$, resolution of singularities implies that the smooth hypersurface X^o has a smooth projective completion, which we write as $X(\underline{a})$. Note that $X(\underline{a})$ is only well defined up to birational equivalence.

For the next result, we recall that we are inductively assuming that $BL(n-1)$ holds, and that Y is a geometrically irreducible norm variety for $\{a_1, \ldots, a_{n-1}\}$, so that $X(\underline{a})$ exists.

Theorem 10.17. *If $\mathrm{char}(k) = 0$ and $n \geq 3$, the variety $X(\underline{a})$ is a geometrically irreducible norm variety for the symbol $\underline{a}$.*

This is the main result of this chapter; it fails for $n=2$ (see Example 10.20.1). Before proving it, we state and prove several lemmas which are needed in the proof. We will also need the following result, known as the *Norm Principle*, which will be proved in chapter 11. We state it here because we will use it in the proof of Theorem 10.17.

Theorem 10.18 (Norm Principle). *Let k be an ℓ-special field of characteristic 0, and $n \geq 3$. If Y is a norm variety for a nonzero symbol $\{a_1, ..., a_{n-1}\}$ in $K_{n-1}^M(k)/\ell$, then each unit in the image of $N : H_{-1,-1}(Y) \to k^\times$ is of the form $N_{k(y)/k}(\alpha)$, where $y \in Y$ is a closed point of degree ℓ and $\alpha \in k(y)^\times$.*

Lemma 10.19. X^o *is smooth over* U *and geometrically irreducible.*

Proof. That X^o is smooth over U (and hence smooth) is an easy consequence of the Jacobian criterion. Also, U is geometrically irreducible since $n \geq 2$. We may now argue as we did in the proof of Lemma 10.9.

Suppose that X^o is not geometrically irreducible. Replacing k by its algebraic closure, we may assume that X^o is reducible. That is, there is a point $u \in U$ such that X_u^o is reducible, i.e., the homogeneous polynomial $g(t) = N_u - a_n t^\ell$ is reducible in $R[t]$, where $R = \mathcal{O}_{U,u}[x_1, ..., x_\ell]$ and $N_u \in R$. If $F = \mathrm{frac}(R)$, then $F[t]/(g)$ is not a field. Since $\mu_\ell \subset F^\times$, $g(t)$ must factor into linear terms in $F[t]$. Since R is a UFD, $g(t)$ factors as $\prod(M - r_i t)$ over R with $M \in R$ of degree 1, by Gauss's Lemma. This implies that $N_u = M^\ell$, and hence that the degeneracy locus of N_u in $\mathcal{A}_u$ is a proper, nonzero $\mathcal{O}_{U,u}$-submodule. However, this is impossible because $\mathcal{A}_u = \mathcal{O}_{U,u}[x]/(x^\ell - c)$, so either $x^\ell - c$ is irreducible in $\mathcal{O}_{U,u}[x]$ (in which case $\mathcal{A}_u$ is a domain and $N_u(a) \neq 0$ for $a \neq 0$) or else $x^\ell - c$ factors (in which case $\mathcal{A}_u$ is a product of domains and the degeneracy locus is not a submodule). $\square$

The following key lemma is taken from Lemma 2.3 of [SJ06].

Lemma 10.20. *The function field* $F = k(X^o)$ *splits the symbol* $\underline{a}$.

Proof. The generic point x of X^o induces an F-point u of $U \subset S^\ell(Y)$. This corresponds to a cycle $y_1 + \cdots + y_k$ of Y_F such that $\sum [F(y_i):F] = \ell$. Since $\mathbb{V}(\mathcal{A})_x \cong \mathbb{A}_F^\ell$, x also determines a point $(\lambda_1, \ldots, \lambda_k) \in \prod F(y_i)$ such that $a_n = N(x) = \prod N_{F(y_i)/F}(\lambda_i)$.

By construction, the y_i are $F(y_i)$-points of Y. Since Y is a norm variety for $\underline{b} = \{a_1, \ldots, a_{n-1}\}$ we have $\underline{b}_{F(y_i)} = 0$ in $K_{n-1}^M(F(y_i))/\ell$ for each i. Thus we have the desired vanishing in $K_n^M(F)/\ell$, namely:

$$\underline{a} = \{\underline{b}, a_n\} = \sum N_{F(y_i)/F}\{\underline{b}, \lambda_i\} = 0. \qquad \square$$

Example 10.20.1. It is instructive to consider the case $n = 2$. In this case $Y = \mathrm{Spec}(E)$, $E = k(\sqrt[\ell]{a_1})$, $U = S^\ell Y$, and X^o is the hypersurface $N(u) = a_2$ in $\mathbb{P}_U^\ell$. However, U is the disjoint union of many points, so X^o is not even connected. Over each point $u \in U$, we saw in Lemma 1.15 and Proposition 1.25 that the hypersurface $N(u) = a_2$ in $\mathbb{P}_{k(u)}^\ell$ is an affine open subset of a norm variety for $\underline{a} = \{a_1, a_2\}$ over $k(u)$.

Lemma 10.21. *Let* Z *a smooth irreducible variety over an* ℓ*-special field* F. *If* $\dim(Z) > 0$ *then the set* $Z(F)$ *is either empty or dense and infinite.*

Proof. Let z be an F-point of Z, and let U be a dense open in Z. By Bertini's Theorem, there is a curve C on Z, meeting U, such that z is a smooth point of C; if the smooth projective closure of C has infinitely many F-points then so does $C \cap U$ and hence U. Thus it suffices to suppose that Z is a smooth projective curve.

Suppose that $z_1, \ldots, z_k$ are distinct points in $Z(F)$; we need to show there exists a rational point distinct from $z_1, \ldots, z_k$. Choose a rational function f on Z that has a pole at each of the z_i, and such that the degree of the divisor $(f)_\infty$ of poles is prime to ℓ. Then the degree of the divisor of zeros $(f)_0$ is also prime to ℓ and, since F is ℓ-special, the support of $(f)_0$ contains an F-rational point (which is obviously distinct from $z_1, \ldots, z_k$). $\qquad\square$

Remark 10.21.1. There can be singular varieties Z over F with $Z(F)$ finite. If Z is any projective variety with a smooth F-rational point, the proof of Lemma 10.21 goes through to show that $Z(F)$ is infinite.

Proof of Theorem 10.17. The variety $X(\underline{a})$ is smooth, projective, and geometrically irreducible by Lemma 10.19, and $\dim(X) = \dim(X^o) = \ell^{n-1} - 1$. It is a splitting variety by Lemma 10.20. To show that it is ℓ-generic, it suffices to show that $X = X(\underline{a})$ has an F-point for every ℓ-special field F splitting $\underline{a}$. Because $X^o \subset X$, it suffices to show that X^o has an F-point.

Recall that Y is a norm variety for $\{a_1, \ldots, a_{n-1}\}$. If F splits this symbol then $Y(F)$ is nonempty. Since F is ℓ-special, Lemma 10.21 shows that $Y(F)$ is infinite. Choose ℓ distinct F-points $y_1, \ldots, y_\ell$ of Y; their sum $y_1 + \cdots + y_\ell$ determines an F-point u of the nonsingular part U of $S^\ell(Y)$, and hence $(u, (1 + \cdots + 1, a_n))$ is an F-point in the hypersurface X^o over U.

If F does not split $\{a_1, \ldots, a_{n-1}\}$, then we can apply Proposition 7.7 to Y_F, which is a norm variety for $\{a_1, \ldots, a_{n-1}\}$ over F: since $\{a_1, \ldots, a_n\} = 0$ in $K_n^M(F)/\ell$, a_n is $N(\beta)$ for some $\beta \in H_{-1,-1}(Y_F)$. Note that Proposition 7.7 assumes $BL(n-1)$. By the Norm Principle 10.18, which applies to Y_F by induction, there is a closed point y of Y_F of degree ℓ over F and a unit $\alpha \in F(y)^\times$ such that $a_n = N(\beta) = N_{F(y)/F}(\alpha)$. The degree ℓ zero-cycle y on Y_F determines an F-point of $U \subset S^\ell Y$; see [SV96, 6.8]. Together with $\alpha \in \mathcal{A}_y$, y determines an F-point of X^o. Hence, $X^o(F)$ is nonempty, as asserted. $\qquad\square$

Remark 10.22. If $n \geq 2$ and $\mathrm{char}(k) = 0$, there is a norm variety for $\underline{a}$ with infinitely many points of degree ℓ, and in fact infinitely many E-points, where $E = k(\sqrt[\ell]{a_1})$. When $n = 2$, it suffices by Example 10.1.1 to consider the affine variety Y defined by $N_{E/k}(X_0, \ldots, X_{\ell-1}) = a_2$. Each $(t_1, a_2/t_1, 1, \ldots)$ is an E-point of $Y_E \cong \mathrm{Spec}\, E[t_1, \ldots, t_\ell]/(\prod t_i - a_2)$, so Y has infinitely many E-points. For $n \geq 3$, let Y be a norm variety for $\{a_1, \ldots, a_{n-1}\}$ with infinitely many E-points, and form X^o as in Definition 10.16. For each set $y_1, \ldots, y_\ell$ of distinct E-points of Y, $s = \sum y_i$ is an E-point of $S^\ell Y$ and $\mathcal{A}_s \cong \prod E$, so X_s^o is isomorphic to $\mathrm{Spec}\, E[t_1, \ldots, t_\ell]/(\prod t_i = a_2)$. This clearly has infinitely many E-points.

We conclude this section by constructing a geometrically irreducible norm variety for $\underline{a} = \{a_1, \ldots, a_n\}$ over any perfect field k of characteristic $p \neq \ell$, assuming the Norm Principle for the symbol $\{a_1, \ldots, a_{n-1}\}$.

If k is a field of finite characteristic, de Jong proved in [dJ96, 4.1, 4.2] that there is a finite field extension $k \subset k_1$ and a variety $\bar{X}_1$ which is smooth and geometrically irreducible over k_1, together with a finite surjective map from $\bar{X}_1$ to a projective closure $\bar{X}_0$ of X^o, and that if k is perfect then $\bar{X}_1$ is étale over a dense open subscheme of X^o. Gabber showed that we may take $[k(\bar{X}_1):k(X^o)]$ and hence $[k_1:k]$ to be prime to ℓ; see [Ill09]. If we assume that k is ℓ-special, then $k = k_1$, and we write $X(\underline{a})$ for this $\bar{X}_1$.

Lemma 10.23. *Suppose that k is a perfect field of characteristic $p > 0$, Y is a norm variety for $\underline{a} = \{a_1, ..., a_{n-1}\}$ over k, and the Norm Principle 10.18 holds for Y. If $BL(n-1)$ holds, then $X(\underline{a})$ is a geometrically irreducible norm variety for $\underline{a}$ over k.*

Proof. Since Lemmas 10.19, 10.20, and 10.21 do not use the characteristic zero assumption, and neither does Proposition 7.7, the proof of Theorem 10.17 goes through to show that X^o has an F-point for every ℓ-special field F over k.

Let $V \subset X^o$ be a dense open subscheme over which $\pi : X(\underline{a}) \to \bar{X}_0$ is étale. If $x : \mathrm{Spec}(F) \to X^o$ is in V, then $X(\underline{a})$ has an F-point over x because F is ℓ-special, by Lemma 7.3. If not, we produce a curve C in X^o, smooth at x and meeting the étale locus; by Lemma 10.21, $C(F)$ meets V, so there is an F-point in V and hence in $X(\underline{a})$, again by Lemma 7.3. $\qquad\square$

10.5 HISTORICAL NOTES

The study of hypersurfaces of constant norm in algebra bundles was a natural development in the classical study of quadratic forms. Merkurjev and Suslin used Severi–Brauer hypersurfaces in the 1982 paper [MeS82] to show that the norm residue $K_2(k)/\ell \to H^2_{\mathrm{ét}}(k, \mu_\ell^{\otimes 2})$ was an isomorphism. This led to an intensive study of the properties of these hypersurfaces in the late 1980s, including Rost's 1990 preprint [Ros90].

The notion of a *generic splitting variety* arose out of attempts to generalize the 1982 paper [MeS82] by Merkurjev and Suslin; the Severi–Brauer variety X of a division algebra D has the property that if F splits D then X has an F-point. An explicit definition is given in Voevodsky's 1996 preprint [Voe96], where it was proven that Pfister quadrics are generic splitting varieties for $\ell = 2$. (This definition does not appear in the published version [Voe03a] but does appear in [Ros98b] and [Voe99, 4.16].) Rost's analysis [Ros98c] of the motive of a hypersurface defined by a Pfister form led to Voevodsky's proof that $K_n^M(k)/2 \cong H^n_{\mathrm{ét}}(k, \mu_2^{\otimes n})$ in [Voe03a] (the Milnor Conjecture).

There has always been an ambiguity about what the definition of a norm variety should be. The prototype was the variety of points in $E = k[\sqrt[\ell]{a_1}]$ whose norm is a_2; it is a splitting variety for $\{a_1, a_2\}$ (see 1.25). The term "norm variety" was used in the 1980s to refer to hypersurfaces of constant norm, usually having special (unspecified) properties (see [Sus91], [Ros90]). In the 1998 preprint [Ros98b], Rost used the term *norm variety* to refer to the variety Q_b of Definition 10.8. In his fall 1999 lectures and in [Ros02], Rost used the term *norm variety* to mean a smooth projective splitting variety of dimension $\ell^{n-1} - 1$ with $s_d(X) \not\equiv 0$ (mod ℓ^2). Suslin and Joukhovitski [SJ06] deliberately left the term undefined, and the present definition originated in the 2009 paper [HW09].

Our choice (smooth projective ℓ-generic splitting variety for $\underline{a}$) is based upon Rost's observation (Theorem 10.17) that the smooth projective varieties associated to hypersurfaces of constant norm enjoy these properties.

Our presentation is based on the Suslin–Joukhovitski paper [SJ06], Rost's 1998 preprint [Ros98b], his website [Ros98a], and Rost's lectures at the Institute for Advanced Study in 1999-2000 and 2005. This includes our proof of the Norm Principle in chapter 11.

Chapter Eleven

Existence of Rost Varieties

IN THIS CHAPTER, we will prove that the norm varieties constructed in chapter 10 are indeed Rost varieties; in other words, we will prove that Rost varieties exist. Fix a sequence $\underline{a} = (a_1, ..., a_n)$ of units in k, and recall from Definition 1.24 that a *Rost variety* for $\underline{a}$ is a ν_{n-1}-variety such that:

(a) $\{a_1, ..., a_n\}$ vanishes in $K_n^M(k(X))/\ell$;
(b) for each $i < n$ there is a ν_i-variety mapping to X; and
(c) the following motivic homology sequence is exact:

$$H_{-1,-1}^{BM}(X \times X) \xrightarrow{\pi_{0*} - \pi_{1*}} H_{-1,-1}^{BM}(X) \xrightarrow{N} H_{-1,-1}^{BM}(k) \quad (\cong k^\times).$$

Recall from Definition 1.22 that the reduced group $\overline{H}_{-1,-1}(X)$ is the quotient of $H_{-1,-1}^{BM}(X)$ by the difference of the two projections from $H_{-1,-1}^{BM}(X \times X)$. When Y is smooth and proper, we have $H_{-1,-1}(Y) \cong H_{-1,-1}^{BM}(Y)$.

One goal of this chapter is to prove the following theorem, which was already stated in Theorem 10.18. Let us say that an element of $\overline{H}_{-1,-1}(X)$ is a *Kummer element* if it has a representative $[x, \alpha]$, where $x \in X$ is a closed point of degree ℓ and $\alpha \in k(x)^\times$.

Theorem 11.1 (Norm Principle). *Suppose that k is an ℓ-special field of characteristic 0, and that X is a norm variety for some nontrivial symbol $\underline{a}$. Then each element of $\overline{H}_{-1,-1}(X)$ is a Kummer element.*

We will see that the Norm Principle 11.1 implies our main result:

Theorem 11.2. *Let k be an ℓ-special field of characteristic 0, and let $\underline{a}$ be a nonzero symbol. Then any norm variety for $\underline{a}$ is also a Rost variety for $\underline{a}$.*

The idea of the proof of the Norm Principle is to show (in 11.5) that the Kummer elements of $\overline{H}_{-1,-1}(X)$ form a subgroup, so that it suffices to consider the generators. Now every generator of $\overline{H}_{-1,-1}(X)$ is the image of a Kummer element under a pushforward map $\overline{H}_{-1,-1}(X_E) \to \overline{H}_{-1,-1}(X)$ for some finite field extension E of k, so the result follows once we show (in 11.11) that these pushforward maps preserve Kummer elements.

11.1   THE MULTIPLICATION PRINCIPLE

In preparation for the proof of the Norm Principle, we develop some basic facts about elements of $\overline{H}_{-1,-1}(X)$ supported on points x with $[k(x):k]=\ell$.

Here is the case $n=2$ of the Norm Principle, which we will use to prove the multiplication principle 11.5. Recall that if k contains a primitive ℓ^{th} root of unity then the symbol $\{a_1,a_2\}$ determines a central division algebra.

Theorem 11.3 (Norm Principle for $n=2$). *Let Y be the Severi–Brauer variety associated to a central division k-algebra D, of degree ℓ over a field k containing $1/\ell$. Then $H^{BM}_{-1,-1}(Y)=\overline{H}_{-1,-1}(Y)$, and for every $\theta \in H^{BM}_{-1,-1}(Y)$ there exists a point $y \in Y$ with $[k(y):k]=\ell$, and $\lambda \in k(y)^{\times}$ so that $\theta=[y,\lambda]$.*

Proof. Merkurjev and Suslin showed in [MeS82, 8.7.2] that $N:H^{BM}_{-1,-1}(Y) \to k^{\times}$ is an injection with image $\mathrm{Nrd}(D) \subseteq k^{\times}$, so $H^{BM}_{-1,-1}(Y)=\overline{H}_{-1,-1}(Y)$. Given $\theta \in H^{BM}_{-1,-1}(Y)$, its image $N(\theta)$ in $k^{\times}$ can be written as the reduced norm of an element $\lambda \in D$. The subfield $E=k(\lambda)$ of D has degree ℓ because $Y(E) \neq \emptyset$ and $Y(k)=\emptyset$. Thus there is a point $y \in Y$ with $k(y)=E$. Since $N([y,\lambda])=\mathrm{Nrd}(\lambda)=N(\theta)$ in $k^{\times}$, we have $[y,\lambda]=\theta$ in $H_{-1,-1}(Y)$. $\square$

We now suppose that k is an ℓ-special field, and that X is a norm variety for $\underline{a}$. Note that the kernel and cokernel of $N:\overline{H}_{-1,-1}(X) \to k^{\times}$ are ℓ-groups, by Lemma 7.2(4). The Norm Principle is concerned with reducing the degrees of the field extensions $k(x)$ used to represent elements of $\overline{H}_{-1,-1}(X)$. For this, the following definition is useful.

Definition 11.4. We write $\widetilde{A}_0(k)$ for the subset of Kummer elements of $\overline{H}_{-1,-1}(X)$, i.e., those represented by $[x,\alpha]$ where $k(x)=k$ or $[k(x):k]=\ell$. If E/k is a field extension, $\widetilde{A}_0(E)$ denotes the corresponding subset of $\overline{H}_{-1,-1}(X_E)$.

Example 11.4.1. If X has a k-point x, then by 7.2(3) the norm map N is an isomorphism $\widetilde{A}_0(k) \cong \overline{H}_{-1,-1}(X) \xrightarrow{\cong} k^{\times}$, split by $\alpha \mapsto [x,\alpha]$.

The following result is known as the *multiplication principle*. The proof needs the hypothesis that $\mathrm{char}(k)=0$ in order to use the Corollary 9.53 to Theorem 9.52, which in turn invokes Rost's Degree Formula 8.25.

Theorem 11.5 (Multiplication Principle). Let X be a norm variety for a nonzero symbol $\underline{a}$ over an ℓ-special field k of characteristic 0. Then $\widetilde{A}_0(k)$ is a subgroup of $\overline{H}_{-1,-1}(X)$.

Proof. The set $\widetilde{A}_0(k)$ is nonempty because $E=k[\sqrt[\ell]{a_1}]$ splits the symbol and therefore $X(E) \neq \emptyset$. It is closed under additive inverses because $[x,\alpha]+[x,\alpha^{-1}]=[x,1]=0$. Hence it suffices to prove that $\widetilde{A}_0(k)$ is closed under addition, i.e., for each $[x_1,\alpha_1]$ and $[x_2,\alpha_2]$ in $\widetilde{A}_0(k)$ the sum $[x_1,\alpha_1]+[x_2,\alpha_2]$ in $\overline{H}_{-1,-1}(X)$ equals $[x_3,\alpha_3]$ for some $[x_3,\alpha_3] \in \widetilde{A}_0(k)$.

Using Corollary 9.53, we may assume that $\underline{a} = \{a_1, \ldots, a_n\}$ where $k(x_1)$ splits $\{a_1\}$ and $k(x_2)$ splits $\{a_1, a_2\}$. Let Y be the Severi–Brauer variety for the algebra $A(a_1, a_2)$, that is, a norm variety for $\{a_1, a_2\}$. Then Y has a $k(x_1)$-point y_1 and a $k(x_2)$-point y_2. Applying Theorem 11.3 to $\theta = [y_1, \alpha_1] + [y_2, \alpha_2]$, we see that the multiplication principle holds for Y: there is a closed point y_3 of degree ℓ and an $\alpha_3 \in k(y)^\times$ such that $[y_1, \alpha_1] + [y_2, \alpha_2] = [y_3, \alpha_3]$ in $\overline{H}_{-1,-1}(Y)$.

Moreover $k(Y)$ splits $\underline{a}$ so, using Construction 10.13, there is a smooth projective variety $\widetilde{Y}$, a morphism $f : \widetilde{Y} \to Y$ of degree prime to ℓ, and a morphism $g : \widetilde{Y} \to X$. We can lift the y_i to points (of the same name) of $\widetilde{Y}$ with the same residue fields, by 7.5(a). By Theorem 7.5(c), the morphism $f_* : \overline{H}_{-1,-1}(\widetilde{Y}) \to \overline{H}_{-1,-1}(Y)$ is an isomorphism. Thus $[y_1, \alpha_1] + [y_2, \alpha_2] = [y_3, \alpha_3]$ in $\overline{H}_{-1,-1}(\widetilde{Y})$. Now apply $g_* : \overline{H}_{-1,-1}(\widetilde{Y}) \to \overline{H}_{-1,-1}(X)$ to this relation; since X has no k-points, the points $x_i' = g(y_i)$ have degree ℓ over k, because $[k(x_i') : k]$ divides $[k(y) : k] = \ell$. Hence $k(x_i') \cong k(y_i)$, and the $[x_i', \alpha_i]$ are elements of $\widetilde{A}_0(k) \subseteq \overline{H}_{-1,-1}(X)$. Since $g_*[y_i, \alpha_i] = [x_i', \alpha_i]$, Lemma 7.2(1) implies that

$$[x_1, \alpha_1] + [x_2, \alpha_2] = [x_1', \alpha_1] + [x_2', \alpha_2] = [x_3', \alpha_3]. \qquad \square$$

Corollary 11.6. *If k is ℓ-special and X is a norm variety, the restriction of $\overline{H}_{-1,-1}(X) \xrightarrow{N} k^\times$ to $\widetilde{A}_0(k)$ is an injective group homomorphism.*

Proof. By the multiplication principle 11.5, $\widetilde{A}_0(k) \to k^\times$ is a group homomorphism. Suppose that $[x, \alpha]$ represents an element θ of $\widetilde{A}_0(k)$ such that $N(\theta) = N_{k(x)/k}(\alpha) = 1$. Then $\alpha = \sigma(\beta)/\beta$ for some β by the classical Hilbert Theorem 90 [Wei94, 6.4.8], so $\theta = [x, \sigma(\beta)] - [x, \beta]$. But $[x, \sigma(\beta)] - [x, \beta] = 0$ (apply 7.2(1) to σ), so $\theta = 0$. $\qquad \square$

Let S be the variety of the Chain Lemma 9.1 for the symbol $\underline{a}$, and $\mathcal{A} = \mathcal{A}(J)$ the associated sheaf of Kummer algebras over S (see 9.8). If s is a k-point of S, then $\gamma|_s$ is an ℓ-ary form on the 1-dimensional k-vector space $J|_s$; if w is a nonzero element of $J|_s$ then $\mathcal{A}|_s = k[u]/(u^\ell - \gamma|_s(w))$. By abuse of notation, we shall use $\gamma(s)$ to denote $\gamma|_s(w)$ for some fixed choice of $w \neq 0$; the class of $\gamma(s)$ is well defined in $k^\times/k^{\times\ell}$.

Lemma 11.7. *For every k-point s of S, $\gamma(s) \notin k^{\times\ell}$. Hence $k[\sqrt[\ell]{\gamma(s)}]$ is a field.*

Proof. By the Chain Lemma 9.1(5), $\gamma(s) \neq 0$. Similarly, $\gamma(s)$ cannot be in $k^{\times\ell}$, because that would imply that $\underline{a} = \{\gamma(s), \ldots\} = 0$ in $K_n^M(k)/\ell$, by the Chain Lemma 9.1 and specialization from S. Hence $\gamma(s) \notin k^{\times\ell}$, so $\mathcal{A}|_s$ is a field. $\qquad \square$

Since the fiber $\mathbb{A}(\mathcal{A})_s$ over s of the vector bundle $\mathbb{A}(\mathcal{A})$ is $\mathrm{Spec}(\mathrm{Sym}_k \mathcal{A}|_s)$, there is a natural isomorphism $\mathbb{A}(\mathcal{A})_s(k) \cong \mathcal{A}|_s$. Thus we may identify the k-points of $\mathbb{A}(\mathcal{A})_s$ with elements of $\mathcal{A}|_s$, and speak of the norm $N : \mathbb{A}(\mathcal{A})_s(k) \to k$.

Proposition 11.8 (Multiplication Principle, geometric version). *Let k be an ℓ-special field of characteristic 0, and X a norm variety for a symbol $\underline{a}$. Then the subgroup $\widetilde{A}_0(k)$ of $k^\times$ is the set of nonzero values of the map $N : \mathbb{A}(\mathcal{A})(k) \to k$.*

Proof. Let α be a k-point of $\mathbb{A}(\mathcal{A})$ over $s \in S(k)$ with $N(\alpha) \neq 0$. By Lemma 11.7, we may regard α as a nonzero element of the field $\mathcal{A}|_s = k[\sqrt[\ell]{\gamma(s)}]$. Since $\mathcal{A}|_s$ splits $\underline{a}$, and k is ℓ-special, there is a point $x \in X$ with $k(x) \cong \mathcal{A}|_s$ and $[k(x):k] = \ell$. If $N(\alpha)$ is nonzero in k then it is the image of $[x, \alpha] \in \overline{H}_{-1,-1}(X)$.

Conversely, Construction 11.9 shows that every element of $\widetilde{A}_0(k)$ arises as $[x, \alpha]$ for some α with $N(\alpha) \neq 0$, so the set of nonzero values of $N : \mathbb{A}(\mathcal{A})(k) \to k$ is exactly the image of $\widetilde{A}_0(k)$, which is a group by Theorem 11.5. □

Construction 11.9. Each element $[x, \alpha]$ of $\widetilde{A}_0(k)$ is represented by a k-point of $\mathbb{A}(\mathcal{A})$, and also by a k-point $([\alpha : 1], w)$ of the variety Q defined in 10.8. Indeed, since $k(x)$ splits the symbol, Theorem 9.52 states that there is a k-point $s \in S$ such that $k(x)$ is the Kummer k-algebra $\mathcal{A}|_s = k[\sqrt[\ell]{\gamma(s)}]$. As remarked above, we may then identify the element α of $\mathcal{A}|_s$ with a k-point of $\mathbb{A}(\mathcal{A})$. The map $\mathbb{A}(\mathcal{A}) \to Q$ sends α to $([\alpha : 1], w)$.

11.2   THE NORM PRINCIPLE

In this section, we will prove Theorems 11.2 and 11.1 (existence of Rost varieties and the Norm Principle) modulo the following result, whose proof we postpone to section 11.5.

Theorem 11.10. *Let k be an ℓ-special field of characteristic 0, and let X be a norm variety over k for $\underline{a}$. Suppose that $E = k[\epsilon]$ is a field extension of k with $\epsilon^\ell \in k$ such that $\underline{a}_E \neq 0$. For each $[z, \alpha] \in \widetilde{A}_0(E)$, there exist a finite set of points $x_i \in X$ of degree ℓ over k, $t_i \in k$ and $b_i \in k(x_i)$ such that*

$$N_{E(z)/E}(\alpha) = \prod N_{E(x_i)/E}(b_i + t_i \epsilon) \text{ in } k^\times.$$

Recall from Proposition 10.2 that when k is ℓ-special and $n \geq 2$, every norm variety X over k is geometrically irreducible, and hence (by Lemma 10.3) X_E is a norm variety over E. We also know that $\widetilde{A}_0(E)$ is a subgroup of $\overline{H}_{-1,-1}(X_E)$.

Theorem 11.11. *If k is ℓ-special of characteristic 0 and $[E:k] = \ell^\nu$ then the pushforward $\overline{H}_{-1,-1}(X_E) \longrightarrow \overline{H}_{-1,-1}(X)$ sends $\widetilde{A}_0(E)$ to $\widetilde{A}_0(k)$.*

Proof. By Galois theory and the structure of finite ℓ-groups, there is a chain of subfields $k = E_0 \subset E_1 \subset \cdots E_\nu = E$ with $[E_{i+1} : E_i] = \ell$. Thus we may assume that $[E:k] = \ell$. By Kummer theory, there is a $c \in k$ so that $E = k[\epsilon]/(\epsilon^\ell - c)$.

If $\underline{a}$ vanishes in E then the generic splitting variety X has an E-point x, and Theorem 11.11 is immediate from Example 11.4.1. Indeed, in this case X_E has an E-point x' over x, every element of $\tilde{A}_0(E) \cong E^\times$ has the form $[x', \alpha]$, and $N_{E/k}[x', \alpha] = [x, \alpha]$. Hence we may assume that $\underline{a}_E \neq 0$. This has the advantage that $E(x_i) = E \otimes_k k(x_i)$ is a field for every point $x_i \in X$ of degree ℓ; otherwise, there would be a homomorphism $E \to E(x_i) \to k(x_i)$, forcing $E \cong k(x_i)$—and splitting the symbol.

Choose $\theta = [z, \alpha] \in \tilde{A}_0(E)$ and let $x_i \in X$, t_i and b_i be the data given by Theorem 11.10. Each x_i lifts to an $E(x_i)$-point $x_i \otimes E$ of X_E so we may consider the element

$$\theta' = \theta - \sum [x_i \otimes E, b_i + t_i \epsilon] \in \overline{H}_{-1,-1}(X_E).$$

By the multiplication principle 11.5 over E, θ' belongs to the subgroup $\tilde{A}_0(E)$. By Theorem 11.10, its norm is

$$N(\theta') = N_{E(z)/E}(\alpha) / \prod N_{E(x_i)/E}(b_i + t_i \epsilon) = 1.$$

By Corollary 11.6, $\theta' = 0$. Hence

$$N_{E/k}(\theta) = \sum \left[x_i, N_{E(x_i)/k(x_i)}(b_i + t_i \epsilon) \right]$$

in $\overline{H}_{-1,-1}(X)$. Since $\tilde{A}_0(k)$ is a group by 11.5, this is an element of $\tilde{A}_0(k)$. □

Proof of the Norm Principle (Theorem 11.1). By the multiplication principle 11.5, it suffices to show that every generator $[z, \alpha]$ of $\overline{H}_{-1,-1}(X)$ is in $\tilde{A}_0(k)$. Since $[k(z) : k] = \ell^\nu$ for $\nu > 0$, Galois theory and the structure of finite ℓ-groups imply that there is a field E with $k \subseteq E \subset k(z)$ and $[k(z) : E] = \ell$, and that z lifts to a point z' of X_E with $k(z) \cong k(z')$. By construction, $[z', \alpha] \in \tilde{A}_0(E)$ and $\overline{H}_{-1,-1}(X_E) \to \overline{H}_{-1,-1}(X)$ sends $[z', \alpha]$ to $[z, \alpha]$. By Theorem 11.11, $[z, \alpha]$ is in $\tilde{A}_0(k)$, i.e., is represented by an element $[x, \beta]$ with $[k(x) : k] = \ell$. □

Proof of Theorem 11.2. Let X be a norm variety for $\underline{a} = \{a_1, \ldots, a_n\}$. Because k is a field of characteristic 0, X is a ν_{n-1}-variety by Proposition 10.14, and for $1 < i < n$ there is a ν_{i-1}-variety mapping to X by Proposition 10.15. Since k is ℓ-special, the Norm Principle implies that $\tilde{A}_0(k) = \overline{H}_{-1,-1}(X)$, so Corollary 11.6 implies that the motivic homology sequence is exact. □

11.3   WEIL RESTRICTION

Because there are not many good references for Weil restriction, we pause here to collect the basic facts that we shall need. The original references are [Wei56] and Weil's 1961 lecture notes, published as [Wei82, §1.3].

Definition 11.12. If E is a finite field extension of k, Weil restriction is the right adjoint of the basechange functor $V \mapsto V \times \operatorname{Spec} E$. That is, if X is any variety defined over E, the *Weil restriction* $\operatorname{Res}_{E/k} X$ is a variety over k of dimension $[E:k]\dim(X)$ which is characterized by

$$\operatorname{Hom}_k(V, \operatorname{Res}_{E/k} X) \cong \operatorname{Hom}_E(V \times \operatorname{Spec} E, X).$$

The Weil restriction exists for any quasi-projective variety by [Wei82, 1.3.1]. In particular, if F is a field containing k then the F-points are given by: $\operatorname{Res}_{E/k} X(F) = X(E \otimes_k F)$.

Example 11.13. For visual reasons, we shall write $\mathbb{A}^E$ for the Weil restriction $\operatorname{Res}_{E/k} \mathbb{A}^1$ of the affine line. It is isomorphic to $\mathbb{A}^n$, $n = [E:k]$, as a k-variety, and is characterized by $\mathbb{A}^E(F) = E \otimes_k F$. In particular, $\mathbb{A}^E(k) = E$.

Using the normal basis theorem, we can find a k-basis of E permuted by the Galois group Γ. Thus the coordinate ring of $\mathbb{A}^E$ is a polynomial ring $k[\omega_1, \dots, \omega_n]$ (the indeterminates correspond to the k-basis of E), and the function field of $\mathbb{A}^E$ is $k(\omega) = k(\omega_1, \dots, \omega_n)$. The Galois group Γ permutes the indeterminates ω_i, and $\mathbb{A}^E \cong \prod_{g \in \Gamma} \mathbb{A}^1$. This motivates the following result.

Lemma 11.14. *If X is a variety over E, and E is Galois over k with group Γ, then $\operatorname{Res}_{E/k}(X) \times \operatorname{Spec}(E) \cong \prod_{g \in \Gamma} X$.*

Proof. If A is an E-algebra, there is a natural isomorphism $A \otimes_k E \xrightarrow{\cong} \prod_{g \in \Gamma} A$. This induces natural isomorphisms

$$\left[\operatorname{Res}_{E/k}(X) \times \operatorname{Spec}(E) \right](A) \cong X(A \otimes_k E) \cong X\left(\prod_\Gamma A \right) \cong \prod_\Gamma X(A).$$

Since the functors of points are isomorphic, so are the schemes. $\square$

Remark 11.14.1. The original 1956 construction of Weil restriction in [Wei56] used Lemma 11.14 and descent to construct $\operatorname{Res}_{E/k}(X)$. Grothendieck gave another construction in [Gro61, 221-19], identifying $\operatorname{Res}_{E/k}(X)$ as an open subscheme of the Hilbert scheme $\operatorname{Hilb}_{X/k}$, assuming X is quasi-projective over k.

It is obvious that $\operatorname{Res}_{E/k}(X_1 \times X_2) = \operatorname{Res}_{E/k}(X_1) \times \operatorname{Res}_{E/k}(X_2)$. Less obvious (but not hard) is the fact that if Z is closed in X then $\operatorname{Res}_{E/k}(Z)$ is closed in $\operatorname{Res}_{E/k}(X)$. It follows, for example, that the Weil restriction of $\mathbb{A}^1 - \{0\}$ is the closed subvariety of $\mathbb{A}^E \times \mathbb{A}^E$ whose k-points are all pairs $(a, b) \in E \times E$ such that $ab = 1$.

11.4   ANOTHER SPLITTING VARIETY

In this section, we construct a G-variety $\overline{Y}$, parametrized by $\mathbb{A}^1$, which will be used in section 11.5 to establish Theorem 11.10. As we saw in section 11.2, this is the key step in proving Theorems 11.2 and 11.1. We will show that the general fibers of $\overline{Y} \to \mathbb{A}^1$ are G-fixed point equivalent to disjoint unions of copies of $\mathbb{P}(\mathcal{A})^\ell$, where $\mathcal{A} = \mathcal{A}(J)$ is the sheaf of Kummer algebras for $\underline{a} = \{a_1, \ldots, a_n\}$ over the variety S of the Chain Lemma (see Definition 9.8 and Theorem 9.1).

To construct $\overline{Y}$, we fix a Kummer extension $E = k[\epsilon]$ of k, where $\epsilon^\ell \in k$. Let $\mathcal{B}$ be the rank $\ell + 1$ $\mathcal{O}_S$-submodule $(\mathcal{A} \otimes k) \oplus (\mathcal{O}_S \otimes_k k \cdot \epsilon)$ of $\mathcal{A}_E = \mathcal{A} \otimes_k E$ and let $N_\mathcal{B} : \mathcal{B} \to \mathcal{O}_S \otimes_k E$ be the sheaf map over S induced by the norm on $\mathcal{A}_E$.

Definition 11.15. Let U be the variety $\mathbb{P}(\mathcal{A}) \times_k \mathbb{P}(\mathcal{B})^{\times(\ell-1)}$ over $S^{\times \ell}$. We write L for the invertible sheaf on U associated to the external product $\mathbb{L}(\mathcal{A}) \boxtimes \mathbb{L}(\mathcal{B})^{\boxtimes(\ell-1)}$ of the tautological bundles over $\mathbb{P}(\mathcal{A})$ and $\mathbb{P}(\mathcal{B})$. The product of the various norms $N_\mathcal{A}$ and $N_\mathcal{B}$ defines a sheaf map $N : L \to \mathcal{O}_S \otimes E$.

Note that $\dim(S) = \ell^n - \ell$; since U has relative dimension $\ell^2 - 1$ over $S^{\times \ell}$, $\dim(U) = \ell^{n+1} - 1$.

Recall that $\mathbb{A}^E$ denotes the Weil restriction $\mathrm{Res}_{E/k}\mathbb{A}^1$ of $\mathbb{A}^1$ (Example 11.13).

Definition 11.16. Let $\overline{Y}$ denote the subvariety of $\mathbb{P}(L \oplus \mathcal{O}) \times \mathbb{A}^E$ whose F-points are all $([\alpha : t], w)$ such that $N(\alpha) = t^\ell w$ in $E \otimes_k F$; since $\dim \mathbb{P}(L \oplus \mathcal{O}) = \ell^{n+1}$ we have $\dim(\overline{Y}) = \ell^{n+1}$. We write $\overline{Y}_w$ for the (scheme-theoretic) fiber over a point $w \in \mathbb{A}^E$ under the projection $\overline{Y} \to \mathbb{A}^E$ onto the second factor. Note that $\dim(\overline{Y}_w) = \ell \cdot \dim(Q_w)$, where Q_w is the variety of dimension $\ell^n - 1$ defined in 10.8.

Notation 11.17. Here is a description of the k-points of $\overline{Y}_w$. Let $([\alpha : t], w)$ be a k-rational point on $\overline{Y}$, so that $w \in \mathbb{A}^E(k) = E$. We may regard $[\alpha : t] \in \mathbb{P}(L \oplus \mathcal{O})(k)$ as being given by a point $u \in U(k)$, lying over a point $(s_0, \ldots, s_{\ell-1})$ of $S(k)^{\times \ell}$, and a nonzero pair $(\alpha, t) \in L|_u \times k$ (up to scalars). From the definition of L, we see that (up to scalars) α determines an ℓ-tuple $(b_0, b_1 + t_1\epsilon, \ldots, b_{\ell-1} + t_{\ell-1}\epsilon)$, where $b_i \in \mathcal{A}|_{s_i}$ and $t_i \in k$. When $\alpha \neq 0$ we have $b_0 \neq 0$ and, for all $i > 0$, $(b_i, t_i) \neq (0, 0)$. Finally, writing A_i for $\mathcal{A}|_{s_i}$, the norm condition says that in E:

$$N_{A_0/k}(b_0) \prod_{i=1}^{\ell-1} N_{A_i \otimes E/E}(b_i + t_i\epsilon) = t^\ell w.$$

If $k \subseteq F$ is a field extension, then an F-point of $\overline{Y}$ is described as above, replacing k by F and E by $E \otimes_k F$ everywhere.

Lemma 11.18. *Let $u \in U$ be a point over $(s_0, s_1, \ldots, s_{\ell-1}) \in S^{\times \ell}$, and write A_i for the $k(s_i)$-algebra $\mathcal{A}|_{s_i}$. Then the following hold.*

1. *If $\underline{a}$ doesn't split at any of the points $s_0, \ldots, s_{\ell-1}$, then the norm map $N : L|_u \to k(u) \otimes E$ is nonzero.*
2. *If $\underline{a}|_{s_0} \neq 0$ in $K_n^M(k(s_0))/\ell$, then A_0 is a field.*
3. *For $i \geq 1$, if $\underline{a}|_{E(s_i)} \neq 0$ in $K_n^M(E(s_i))/\ell$ then $A_i \otimes E$ is a field.*

Proof. The first assertion follows from part (4) of the Chain Lemma 9.1, which says that $s \notin V(\gamma_1)$, since by 9.8 the norm on L is induced from the ℓ-form γ_1

on J. Assertions (2–3) follow from part (2) of the Chain Lemma, since $\underline{a} \neq 0$ implies that γ_1 is nontrivial.   □

Lemma 11.19. *If $\overline{Y}$ has a k-point with $t = 0$ then $\underline{a}|_E = 0$ in $K_n^M(E)/\ell$.*

Proof. We use the description of a k-point of $\overline{Y}$ from 11.17. Since $(\alpha, t) \neq (0, 0)$, if $t = 0$, then $\alpha \neq 0$. Therefore $b_0 \neq 0 \in A_0$ and $b_i + t_i \epsilon \neq 0 \in A_i \otimes E$ for all i. By Lemma 11.18, if $\underline{a}|_E \neq 0$ in $K_n^M(E)/\ell$ then A_0 and all the algebras $A_i \otimes E$ are fields, so that $N(\alpha) = N_{A_0/k}(b_0) \prod_{i=1}^{\ell-1} N_{A_i \otimes E/E}(b_i + t_i \epsilon) \neq 0$, a contradiction to $t^\ell w = 0$.   □

Combining Lemmas 11.18 and 11.19, we obtain the following consequence (in the notation of 11.17):

Corollary 11.20. *If $\underline{a} \neq 0$ in $K_n^M(E)/\ell$ and $w \in E^\times$ is such that $\overline{Y}_w$ has a k-point, then*

1. *A_0 and the $A_i \otimes E$ are fields and*
2. *w is a product of norms of an element b_0/t of A_0 and elements $b_i + t_i \epsilon$ in the subsets $A_i + k \cdot \epsilon$ of $A_i \otimes_k E$ for all i.*

Remark 11.20.1. In Theorem 11.23 we will see that if w is a general element of E then such a k-point exists.

The group $G = \mu_\ell^n$ acts nicely on S, J, and $\mathcal{A}$ by Corollary 9.51, and on $\mathbb{P}(\mathcal{A})$ by 9.46. Letting G act trivially on $\mathbb{A}^E$, G acts on $\mathcal{B}$, U, and $\overline{Y}$ (but not nicely; see 9.44.1).

In the notation of 11.17, if $([\alpha : t], w)$ is a fixed point of the (nice) G-action on $\overline{Y}$ then the points $s_i \in S$ and $u = (u_0, u_1, ...)$ in $U = \mathbb{P}(\mathcal{A}) \times \prod \mathbb{P}(\mathcal{B})$ are fixed. By definition (see 9.44), the fixed points of any nice action are k-rational, so u_0 and the s_i are k-rational. If u is defined over F, each point $(b_i : t_i)$ is fixed in $\mathcal{B}|_{s_i}$. Since G acts nicely on J, Example 9.46.1 shows that if $t = 0$ then, for all i, either $t_i \neq 0$ (and $b_i \in F \subset A_i \otimes F$) or else $t_i = 0$ and $0 \neq b_i \in J|_{s_i}^{\otimes r_i} \otimes F \subseteq A_i \otimes F$ for some r_i, $0 \leq r_i < \ell$.

Lemma 11.21. *For all $w \in \mathbb{A}^E$, $\mathrm{Fix}_G \overline{Y}_w$ is disjoint from the locus where $t = 0$.*

Proof. Suppose $([\alpha : 0], w)$ is a fixed point defined over a field F containing k. As explained above, $b_0 \neq 0$ and (for each $i > 0$) $b_i + t_i \epsilon \neq 0$ and either $t_i \neq 0$ or there is an r_i so that $b_i \in J|_{s_i}^{\otimes r_i} \otimes F$. Let $I \subseteq \{1, \ldots, n-1\}$ be the set of indices such that $t_i \neq 0$.

By Example 9.46, $b_0 \in J|_{s_0}^{\otimes r_0}$ for some r_0, and hence $N_{A_0}(b_0)$ is a unit in k, because the ℓ-form γ is nontrivial on $J|_{s_0}$. Likewise, if $i \notin I$, then $N_{A_i \otimes F/F}(b_i)$ is a unit in F.

Now suppose $i \in I$, i.e., $t_i \neq 0$, and recall that in this case $b_i \in F \subset A_i \otimes F$. If we write EF for the algebra $E \otimes F$, noting that $EF \cong F[\epsilon]/(\epsilon^\ell - c)$ for some

$c \in F$, then the norm from $A_i \otimes EF$ to EF is simply the ℓth power on elements in EF, so $N_{A_i \otimes EF/EF}(b_i + t_i\epsilon) = (b_i + t_i\epsilon)^\ell$ as an element in the algebra EF. Taking the product, and keeping in mind that $t = 0$, we get the equation

$$\prod_{i \in I} N_{A_i \otimes EF/EF}(b_i + t_i\epsilon) = \prod_{i \in I}(b_i + t_i\epsilon)^\ell = 0.$$

Because EF is a separable F-algebra, it has no nilpotent elements. We conclude that

$$\prod_{i \in I}(b_i + t_i\epsilon) = 0.$$

The left-hand side of this equation is a polynomial of degree at most $\ell - 1$ in ϵ; since $\{1, \epsilon, \ldots, \epsilon^{\ell-1}\}$ is a basis of $F \otimes E$ over F, that polynomial must be zero. This implies that $b_i = t_i = 0$ for some i, a contradiction. $\qquad\square$

Proposition 11.22. *If $w \in \mathbb{A}^E$ is general then $\mathrm{Fix}_G \overline{Y}_w$ lies in the open subvariety of $\overline{Y}_w$ where $t \prod_{i=1}^{\ell} t_i \neq 0$.*

The expression that w is "general" means that w does not lie on a certain proper closed subvariety of $\mathbb{A}^E$.

Proof. By Lemma 11.21, $\mathrm{Fix}_G \overline{Y}_w$ is disjoint from the locus where $t = 0$, so we may assume that $t = 1$. Since w is general, we may also take $w \neq 0$. So let $([\alpha : 1], w)$ be a fixed point of $\overline{Y}_w$ defined over $F \supseteq k$ for which $t_j = 0$. As in the proof of the previous lemma, we collect those indices i such that $t_i \neq 0$ into a set I, and write EF for $E \otimes_k F$. Recall that for $i \in I$, we have $b_i \in F$. Since $j \notin I$, we have that $|I| \leq \ell - 2$. For $i \notin I$,

$$N_{A_i \otimes EF/EF}(b_i + t_i\epsilon) = N_{A_i \otimes F/F}(b_i) \in F^\times$$

(the norm cannot be 0 as $t^\ell w = w \neq 0$ by assumption). So we get that

$$\prod_{i \in I}(b_i + t_i\epsilon)^\ell = \xi w$$

for some $\xi \in F^\times$. If we view ξw as a point in $\mathbb{P}(E)(F) = (EF - \{0\})/F^\times$, then we get an equation of the form

$$\left[\prod_{i \in I}(b_i + t_i\epsilon)^\ell\right] = [w].$$

But the left-hand side lies in the image of the morphism $\prod_{i \in I} \mathbb{P}^1 \to \mathbb{P}(E)$ which sends $[b_i : t_i] \in \mathbb{P}^1(F)$ to $[\prod(b_i + t_i\epsilon)^\ell] \in \mathbb{P}(E)(F)$. Since $|I| \leq \ell - 2$, this image is a proper closed subvariety, proving the assertion for general w. $\qquad\square$

Remark 11.22.1. The open subvariety of $\overline{Y}_w$ in 11.22 is G-isomorphic (by setting t and all t_i to 1) to a closed subvariety of $\mathbb{A}(\mathcal{A})^\ell$, namely the fiber over $w \in \mathbb{A}^E$ of the map $N_{\mathcal{A} \otimes E/E} : \mathbb{A}(\mathcal{A})^\ell \to \mathbb{A}^E$ defined by the formula of 11.17:

$$N(b_0, \ldots, b_{\ell-1}) = N_{A_0/k}(b_0) \prod_{i=1}^{\ell-1} N_{A_i \otimes E/E}(b_i + \epsilon).$$

Indeed, $\mathbb{A}(\mathcal{A})^\ell$ is G-isomorphic to an open subvariety of $\overline{Y}$ and $N_{A_i \otimes E/E}$ is the restriction of $\alpha \mapsto N(\alpha)$.

Recall from Definition 8.15 that two G-varieties are fixed point equivalent if their fixed loci are 0-dimensional smooth points and that (after separable base-change) the tangent spaces are isomorphic as G-representations.

Theorem 11.23. *For a general closed point $w \in \mathbb{A}^E$, $\overline{Y}_w$ is G-fixed point equivalent to the disjoint union of $(\ell-1)!$ copies of $\mathbb{P}(\mathcal{A})^\ell$.*

Proof. Since both $\overline{Y}_w$ and $\mathbb{P}(\mathcal{A})^\ell$ lie over $S^{\times \ell}$, it suffices to consider a G-fixed point $s = (s_0, \ldots, s_{\ell-1})$ in $S(k)^\ell$ and prove the assertion for the fixed points over s. Because G acts nicely on S and J, $k(s) = k$ and (by Lemma 9.47) G acts on J_s via a projection $G \to \mu_\ell$ as the standard representation of μ_ℓ. Note that $J_s = J_{s_i}$ for all i.

By Example 9.46, there are precisely ℓ fixed points on $\mathbb{P}(\mathcal{A})$ lying over a given fixed point $s_i \in S(k)$, and at each of these points the (vertical) tangent space is the reduced regular representation of μ_ℓ. Thus each fixed point in $\mathbb{P}(\mathcal{A})^\ell$ is k-rational, the number of fixed points over s is ℓ^ℓ, and each of their tangent spaces is the sum of ℓ copies of the reduced regular representation.

Since w is general, we saw in 11.22 that all the fixed points of $\overline{Y}_w$ satisfy $t \neq 0$ and $t_i \neq 0$ for $1 \leq i \leq \ell-1$. By Remark 11.22.1, they lie in the affine open $\mathbb{A}(\mathcal{A})^\ell$ of $\mathbb{P}(L \oplus \mathcal{O})$. Because μ_ℓ acts nicely on J_s, an F-point $b = (b_0, \ldots, b_{\ell-1})$ of $\mathbb{A}(\mathcal{A})^\ell$ is fixed if and only if each $b_i \in F$. That is, $\text{Fix}_G(\mathbb{A}(\mathcal{A})^\ell) = \mathbb{A}^\ell$. Now the norm map restricted to the fixed point set is just the map $\mathbb{A}^\ell \to \mathbb{A}^E$ sending b to $b_0^\ell \prod_{i=1}^{\ell-1} (b_i + \epsilon)^\ell$. This map is finite of degree $\ell^\ell(\ell-1)!$, and étale for general w, so $\text{Fix}_G \overline{Y}_w$ has $\ell^\ell(\ell-1)!$ geometric points for general w. This is the same number as the fixed points in $(\ell-1)!$ copies of $\mathbb{P}(\mathcal{A})$ over s, so it suffices to check their tangent space representations.

At each fixed point b, the tangent space of $\mathbb{A}(\mathcal{A})^\ell$ (or $\overline{Y}$) is the sum of ℓ copies of the regular representation of μ_ℓ. Since this tangent space is also the sum of the tangent space of $\mathbb{A}^\ell$ (a trivial representation of G) and the normal bundle of $\mathbb{A}^\ell$ in $\overline{Y}$, the normal bundle must then be ℓ copies of the reduced regular representation of μ_ℓ. Since the tangent space of $\mathbb{A}^\ell$ maps isomorphically onto the tangent space of $\mathbb{A}^E$ at w, the tangent space of $\overline{Y}_w$ is the same as the normal bundle of $\mathbb{A}^\ell$ in $\overline{Y}$, as required. $\square$

Remark 11.23.1. The fixed points in $\overline{Y}_w$ are not necessarily rational points, and we only know that the isomorphism of the tangent spaces at the fixed points holds over a separable extension of k. This is parallel to the situation with the fixed points in Q_w described in Theorem 10.11.

Corollary 11.24. *For a general closed point $w \in \mathbb{A}^E$, the variety $\overline{Y}_w$ is a ν_n-variety over any field k of characteristic 0.*

Proof. By Theorem 11.23, $\overline{Y}_w$ is G-fixed point equivalent to $(\ell - 1)!$ copies of $\mathbb{P}(\mathcal{A})$, which is a ν_n-variety by Theorem 10.6. As in the proof of Proposition 10.12, Theorem 8.16 implies that $\overline{Y}_w$ is also a ν_n-variety.   □

11.5   EXPRESSING NORMS

The purpose of this section is to prove Theorem 11.10, that if $E = k(\epsilon)$ is a fixed Kummer extension of k and $w \in E$ is a norm $N_{E(z)/E}(\alpha)$ for a Kummer point $z \in X_E$, then w is a product of norms of the form specified in Theorem 11.10. In the language of section 11.1, this means that $[z, \alpha] = \sum[z_i, b_i + t_i \epsilon]$ in $\overline{H}_{-1,-1}(X)$. Since our proof will depend upon the bordism results in chapter 8, we will need to assume that k has characteristic 0. We will also assume that k is ℓ-special in order to invoke Theorem 9.52.

Let S be the variety of the Chain Lemma 9.1 for $\underline{a} = \{a_1, \ldots, a_n\}$, and $\mathcal{A} = \mathcal{A}(J)$ the associated sheaf of Kummer algebras over S (see 9.8). Recall from Definition 10.8 that Q is the subvariety of $\mathbb{P}(\mathcal{A} \oplus \mathcal{O}) \times \mathbb{A}_k^1$ consisting of all points $([a:t], w)$ such that $N(a) = t^\ell w$. Extending the base field to E yields the subvariety Q_E of $\mathbb{P}(\mathcal{A} \oplus \mathcal{O})_E \times \mathbb{A}_E^1$, and we write RQ for the Weil restriction $\mathrm{Res}_{E/k}(Q_E)$; see Definition 11.12. The main property of Weil restriction we will need is that the k-points of RQ are naturally isomorphic to the E-points of Q, and may thus be written as $([a:t], w)$ with $w \in E$. Similarly, we write $\mathbb{A}^E$ for the Weil restriction $\mathrm{Res}_{E/k}(\mathbb{A}_E^1)$, noting that $\mathbb{A}^E(k) \cong E$. Since Weil restriction is a functor, the projection $q : Q \to \mathbb{A}_k^1$ induces a morphism

$$Rq = \mathrm{Res}_{E/k}(q_E) : RQ \to \mathbb{A}^E. \tag{11.25}$$

Let U be the variety over $S^{\times \ell}$ defined in Definition 11.15; there is an associated invertible sheaf L on U, equipped with a norm $N : L \to \mathcal{O}_{S_E}$. Recall from Definition 11.16 that $\overline{Y}$ is the subvariety of $\mathbb{P}(L \oplus \mathcal{O}) \times \mathbb{A}^E$ whose k-points are tuples $([\alpha : t], w)$ with $t \in k$ and $w \in E$ such that $N(\alpha) = t^\ell w$ in E. By 11.17, a k-point $([\alpha : 1], w)$ of $\overline{Y}_w$ determines an ℓ-tuple $(s_0, \ldots, s_{\ell-1})$ of k-points of S and an ℓ-tuple $(b_0, b_1 + t_1 \epsilon, \ldots)$ with b_i in $A_i = \mathcal{A}|_{s_i}$, so that

$$w = N_{A_0/k}(b_0) \prod_{i=1}^{\ell-1} N_{A_i \otimes E/E}(b_i + t_i \epsilon).$$

To prove Theorem 11.10 it therefore suffices to show that $\overline{Y}_w(k)$ is nonempty whenever $w = N_{E(z)/E}(\alpha)$ is nonzero (we use $t_0 = 0$). To do this, we will produce a correspondence $Z \to \overline{Y} \times_{\mathbb{A}^E} RQ$ that is dominant and invoke the DN Theorem 8.18 to see that the degree of Z over RQ is prime to ℓ.

The correspondence Z is constructed using the geometric version 11.8 of the multiplication principle.

Lemma 11.26. *Let F be the function field of $\overline{Y}$. Then there exists a finite extension K/F, of degree prime to ℓ, and a point $\xi \in RQ(K)$ lying over the generic point of $\mathbb{A}^E$.*

Proof. Let F' be a maximal prime-to-ℓ extension of F; then the fields F' and $EF' = E \otimes_k F'$ are ℓ-special. We may regard the generic point of $\overline{Y}$ as an element of $\overline{Y}(F)$. Applying the inclusion $F \subset F'$ to this element, followed by the projection $\overline{Y} \to \mathbb{A}^E$, we obtain an element v of $\mathbb{A}^E(F') = EF'$. By 11.17, v is a product of norms from $\mathbb{A}(\mathcal{A})(EF')$. By the multiplication principle 11.8, there exists $\beta \in \mathbb{A}(\mathcal{A})(EF')$ such that $N(\beta) = v$. Now let ξ be the point $([\beta:1], v)$ of $RQ(F')$. By (11.25), $Rq(\xi) = v$, and ξ is defined over some finite intermediate extension $F \subseteq K \subseteq F'$, with $[K:F]$ prime to ℓ. $\square$

Let η_K denote the point of $\overline{Y}(K)$ defined by the inclusion of $F = k(\overline{Y})$ in K, and let $\xi: \operatorname{Spec} K \to RQ$ be the K-point of Lemma 11.26. We can now define Z to be a (smooth, projective) model of $(\eta_K, \xi) \in (\overline{Y} \times_{\mathbb{A}^E} RQ)(K)$, equipped with maps $\overline{Y} \xleftarrow{f} Z \xrightarrow{g} RQ$.

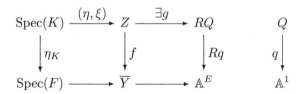

The following result uses the DN Theorem 8.18 of chapter 8 with $r = \ell$.

Theorem 11.27. *The morphism $g: Z \to RQ$ is proper and dominant (hence onto) and of degree prime to ℓ.*

Proof. Let ω be the generic point of $\mathbb{A}^E$. Using the normal basis theorem, we can write $E(\omega) = E(\omega_1, ..., \omega_\ell)$ for transcendentals ω_i permuted by the action of the cyclic Galois group $\Gamma = \operatorname{Gal}(E/k)$. We saw in Theorem 10.11 that the group $G = \mu_\ell^n$ acts on Q over $\mathbb{A}_k^1$, and acts nicely on Q_w for every $w \neq 0$ in k.

We will apply the DN Theorem 8.18 with base field $k' = E(\omega)$, group $G = \mu_\ell^n$, and symbols $u_i = \{a_1, \ldots, a_n, \omega_i\} \in K_{n+1}^M(k')/\ell$.

To define the X_i, choose a μ_ℓ^n-equivariant resolution of singularities $\widetilde{Q} \to Q$; since Q is smooth where $t \neq 0$ (by the Jacobian criterion), this is an isomorphism where $t \neq 0$. The map q extends to a map $\widetilde{Q} \to \mathbb{A}^1$, and we let X_i be the fiber $\widetilde{Q}_{\omega_i}$ over the point $\omega_i \in \mathbb{A}^1(E(\omega)) = E(\omega)$.

We claim that the hypotheses of the DN Theorem are satisfied. Since the Q_{ω_i} are projective, the $X_i = \widetilde{Q}_{\omega_i}$ are smooth projective varieties. By Proposition

10.10 and Lemma 10.19, X_i is a smooth, geometrically irreducible splitting variety for the symbol u_i of dimension $\ell^n - 1$. Thus, hypothesis (1) of the DN Theorem 8.18 is satisfied.

Let $R\widetilde{Q}$ denote the Weil restriction of $\widetilde{Q}_E$, and set $X = R\widetilde{Q}_{E(\omega)}$. Because our base field contains E, it follows from Lemma 11.14 that $RQ \times_{\mathbb{A}^E} \operatorname{Spec} E(\omega) \cong \prod_{\Gamma} Q_{E(\omega)}$ and

$$X = R\widetilde{Q}_{E(\omega)} = R\widetilde{Q} \times \operatorname{Spec} E(\omega) \cong \prod_{\Gamma} \widetilde{Q}_{E(\omega_i)}.$$

Finally, we let Y be some desingularization of $\overline{Y}_{E(\omega)}$, and let W be a model of $Z_{E(\omega)}$, which has a map to Y over $f_{E(\omega)}$ and a map to $X = R\widetilde{Q}_{E(\omega)}$ over $g_{E(\omega)}$.

By Theorem 10.11, Q_w and $\mathbb{P}(\mathcal{A})$ are G-fixed point equivalent over the field $k'(\sqrt[\ell]{w})$ for every $w \neq 0$ in k'. By Theorem 8.16, $t_{d,1}(X_i) = t_{d,1}(\mathbb{P}(\mathcal{A}))$; by Lemma 8.13, we conclude that $s_d(X_i) \equiv v s_d(\mathbb{P}(\mathcal{A})) \pmod{\ell^2}$ for some unit $v \in \mathbb{Z}/\ell$. Since $s_d(\mathbb{P}(\mathcal{A})) \not\equiv 0$ by Theorem 10.6, we conclude that $s_d(X_i) \not\equiv 0$, i.e., that hypothesis (3) of the DN Theorem 8.18 is satisfied.

Furthermore, $K = k'(X_1 \times \cdots \times X_{i-1})$ is contained in a rational function field over E; in fact, the field $E(\omega_j)(Q_{\omega_j})$ becomes a rational function field once we adjoin $\sqrt[\ell]{\gamma}$. Since E does not split $\underline{a}$, K does not split $\underline{a}$ either. It follows that K does not split $u_i = \underline{a} \cup \{\omega_i\}$, verifying hypothesis (2) of Theorem 8.18.

We have now checked the hypotheses (1–3) of Theorem 8.18. It remains to check that X and Y are G-fixed point equivalent up to a prime-to-ℓ factor. We proved in Theorem 10.11 that X_i is G-fixed point equivalent to $\mathbb{P}(\mathcal{A})$. We also proved in Theorem 11.23 that $\overline{Y}_{E(\omega)}$ is G-fixed point equivalent to $(\ell - 1)!$ copies of $\mathbb{P}(\mathcal{A})^\ell$, hence so is Y (since the fixed points lie in the smooth locus). Thus, Y is G-fixed point equivalent to $(\ell - 1)!$ copies of X. Therefore the DN Theorem applies to show that g is dominant and of degree prime to ℓ, as asserted.   □

Since k is ℓ-special, so is E. As stated in Corollary 11.6, the norm map $\widetilde{A}_0(E) \to E^\times$ is injective; we identify $\widetilde{A}_0(E)$ with its image. Thus $[z, \alpha] \in \widetilde{A}_0(E)$ is identified with $w = N_{E(z)/E}(\alpha) \in E^\times$.

Lemma 11.28. *Let k be an ℓ-special field of characteristic 0. Then the subgroup $\widetilde{A}_0(E)$ of $E^\times$ is equal to $q(Q(E)) - \{0\}$.*

Proof. Suppose we are given $[z, \alpha] \in \widetilde{A}_0(E)$ with $w = N_{E(z)/E}(\alpha)$ as above. By Construction 11.9, with k replaced by E, there is an E-point $s \in S_E$ such that $E(z) = E[\sqrt[\ell]{\gamma(s)}]$. Under the correspondence $E(z) \cong \mathbb{A}(\mathcal{A})_s(E)$, we identify α with an E-point of $\mathbb{A}(\mathcal{A})_E$ over $s \in S$. The map $\mathbb{A}(\mathcal{A})_E \to Q_E$ sends α to $([\alpha : 1], w)$, which we may regard as a k-point of RQ, and $w = Rq([\alpha : 1], w)$. This shows that $\widetilde{A}_0(E) \subseteq q(Q(E)) - \{0\}$.

Conversely, by Definition 10.8, an element u of $Q(E)$ over $s \in S$ with $q(u) \neq 0$ has the form $([\alpha : 1], w)$, where α is in $E' = E[\sqrt[\ell]{\gamma(s)}]$, and $q(u) = w = N_{E'/E}(\alpha)$. Such an element is in $\widetilde{A}_0(E)$.   □

Proof of Theorem 11.10. We have constructed a diagram $\overline{Y} \xleftarrow{f} Z \xrightarrow{g} RQ$ and proved that the degree of g is prime to ℓ (see 11.27). By blowing up Z if necessary we may assume that $g: Z \to RQ$ factors through $\tilde{g}: Z \to R\widetilde{Q}$, with the degree of $\tilde{g}$ prime to ℓ.

Let $[z, \alpha] \in \tilde{A}_0(E)$, and set $w = N_{E(z)/E}(\alpha)$. By Lemma 11.28, there exists a point $([\alpha : 1], w) \in RQ(k)$. Lift this to a point in $R\widetilde{Q}(k)$ (recall that $R\widetilde{Q} \to RQ$ is an isomorphism where $t \ne 0$). Since $Z \to R\widetilde{Q}$ is a morphism of smooth projective varieties of degree prime to ℓ and k is ℓ-special, we can lift $([\alpha : 1], w)$ to a k-point of Z, by Theorem 7.5(a). Applying $f: Z \to \overline{Y}$, we get a k-point in $\overline{Y}_w$. By the definition of $\overline{Y}$ and Corollary 11.20, this means that we can find Kummer extensions $k(x_i)/k$ (corresponding to points $s_i \in S$, and determining points $x_i \in X$ because X is an ℓ-generic splitting variety), elements $b_i \in k(x_i)$, and $t_i \in k$ such that $w = \prod_i N_{E(x_i)/E}(b_i + t_i \epsilon)$, as asserted. □

11.6   HISTORICAL NOTES

This chapter is based upon the preprints of Rost, except for the proof of the Norm Principle which Rost never wrote down. Our other sources are the papers [SJ06, HW09] and the lectures given by Rost at the Institute for Advanced Study in 1999-2000 and 2005.

The condition that $s_d(X) \not\equiv 0 \pmod{\ell^2}$ arose in [Voe03a, 3.2] as a sufficient condition for Margolis homology to vanish. The preprint [Ros96] introduced the notation $A_0(X, \mathcal{K}_1)$ for the group we now recognize as $H_{-1,-1}(X)$.

The term "Rost variety" is fairly recent, but the proof that such varieties exist was announced by Rost in 1998 and proven in [SJ06, Th. 0.1], modulo the proofs of the Chain Lemma 9.1 and the Norm Principle, Theorem 11.1. Proofs of these results appeared in [HW09].

Part III

Chapter Twelve

Model Structures for the $\mathbb{A}^1$-homotopy Category

IN ORDER TO work with objects in the Morel–Voevodsky $\mathbb{A}^1$-homotopy category **Ho**, it is useful to introduce a Quillen model structure on an appropriate category of spaces, one which is compatible with the passage in motivic cohomology from complexes of presheaves with transfers to the triangulated category $\mathbf{DM}_{\mathrm{nis}}^{\mathrm{eff}}$. This model structure will be used to set up the formalism underlying motivic Steenrod operations, symmetric powers, and motivic classifying spaces in the next few chapters.

The goal of this chapter is to provide such a framework, namely the $\mathbb{A}^1$-local projective model structure on the categories of simplicial presheaves and simplicial presheaves with transfers. These model categories, written as $\Delta^{\mathrm{op}}\mathrm{Pshv}(\mathbf{Sm})_{\mathbb{A}^1}$ and $\Delta^{\mathrm{op}}\mathbf{PST}(\mathbf{Sm})_{\mathbb{A}^1}$, are defined in 12.62 and 12.68. Their respective homotopy categories are $\mathbf{Ho}(\mathbf{Sm})$ and the full subcategory $\mathbf{DM}_{\mathrm{nis}}^{\mathrm{eff}\, \leq 0}$ of $\mathbf{DM}_{\mathrm{nis}}^{\mathrm{eff}}$; see 12.63 and 12.69.

Sections 12.2 and 12.4 introduce the notions of *radditive presheaves* and $\bar{\Delta}$-*closed classes*, and develop their basic properties. This material will be used in chapter 14 to deal with symmetric powers of motives, and in particular to show that symmetric power functors on termwise ind-representable simplicial presheaves preserve $\mathbb{A}^1$-weak equivalences. The theory of $\bar{\Delta}$-closed classes is needed because the extension of symmetric power functors to simplicial radditive presheaves is not a left adjoint.

We will use many of the basic ideas of Quillen model categories. Recall that a Quillen model category is a category equipped with three classes of morphisms (weak equivalences, fibrations, and cofibrations) satisfying 5 axioms, including a lifting axiom and a functorial factorization axiom; we refer the reader to the books by Hovey [Hov99] or Hirschhorn [Hir03] for more information about model categories.

Much of the material in this chapter is based upon the technique of Bousfield localization, which we recall in Definition 12.38. We will also need the fundamental notion of a Quillen adjunction, which we now recall.

12.0. If $\mathcal{A}, \mathcal{B}$ are model categories, an adjoint pair of functors (F, U) from $\mathcal{A}$ to $\mathcal{B}$ is called a *Quillen adjunction* if F preserves cofibrations and trivial cofibrations, or equivalently, if U preserves fibrations and trivial fibrations.

A Quillen adjunction induces a derived adjunction $(\mathbf{L}F, \mathbf{R}U)$ between the associated homotopy categories of $\mathcal{A}$ and $\mathcal{B}$; see [Hir03, 8.5.18]. For example,

$\mathbf{L}F(A)$ may be defined to be $F(QA)$, where $QA \xrightarrow{\simeq} A$ is a functorial cofibrant replacement; $\mathbf{R}U(B)$ may be defined using a functorial fibrant replacement of B.

We say that (F, U) is a *Quillen equivalence* if the derived adjunction defines an equivalence of homotopy categories. This is equivalent to the condition that (for all cofibrant A in $\mathcal{A}$ and all fibrant B in $\mathcal{B}$) a map $F(A) \to B$ is a weak equivalence in $\mathcal{B}$ if and only if its adjoint $A \to U(B)$ is a weak equivalence in $\mathcal{A}$. See [Hov99, 1.3.3] and [Hir03, 8.5.3] for more information.

Here is an example of a Quillen adjunction of interest to us. We will see in 12.28 that the forgetful functor from presheaves with transfers to presheaves on **Sm** has a left adjoint, $X \mapsto R_{\mathrm{tr}}(X^{\mathrm{rad}})$, and this defines a Quillen adjunction from $\Delta^{\mathrm{op}}\mathrm{Pshv}(\mathbf{Sm})_{\mathbb{A}^1}$ to $\Delta^{\mathrm{op}}\mathbf{PST}(\mathbf{Sm})_{\mathbb{A}^1}$. Similarly, the restriction functor from presheaves on the category **Norm** of normal varieties to the category **Sm** of smooth varieties defines a Quillen adjunction from $\Delta^{\mathrm{op}}\mathrm{Pshv}(\mathbf{Sm})_{\mathbb{A}^1}$ to $\Delta^{\mathrm{op}}\mathrm{Pshv}(\mathbf{Norm})_{\mathbb{A}^1}$. It will be studied in section 12.8.

12.1 THE PROJECTIVE MODEL STRUCTURE

We begin by considering the category $\mathrm{Pshv}(\mathcal{C})$ of presheaves on an arbitrary small category $\mathcal{C}$, i.e., contravariant functors from $\mathcal{C}$ to **Sets**, and the category $\Delta^{\mathrm{op}}\mathrm{Pshv}(\mathcal{C})$ of simplicial objects in $\mathrm{Pshv}(\mathcal{C})$. By abuse, we will apply this to skeletally small categories such as **Sm**.

Definition 12.1. A morphism $X \to Y$ of simplicial presheaves is called a *global weak equivalence* (resp., a *projective fibration*) if $X(C) \to Y(C)$ is a simplicial weak equivalence (resp., a Kan fibration) for every C in $\mathcal{C}$. It is a *trivial projective fibration* if it is both a projective fibration and a global weak equivalence.

Quillen showed that these determine the structure of a proper simplicial model category on $\Delta^{\mathrm{op}}\mathrm{Pshv}(\mathcal{C})$, called the *projective model structure*. As usual, the *projective cofibrations* are determined by the left lifting property. That is, a map $A \to B$ is a projective cofibration if for every trivial projective fibration $X \to Y$ and every solid diagram

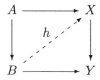

there is a map h making both triangles commute.

The pairing $X \boxtimes K$ of a simplicial presheaf X and a simplicial set K is the simplicial presheaf $n \mapsto \coprod_{K_n} X_n$, and the simplicial mapping space $\mathrm{Map}(X, Y)$ is the simplicial set $n \mapsto \operatorname{Hom}(X \boxtimes \Delta^n, Y)$. This pairing makes the model category $\Delta^{\mathrm{op}}\operatorname{Pshv}(\mathcal{C})$ into a *simplicial model category*. See Theorems 11.7.3 and 13.1.14 of [Hir03] for proofs.

Lemma 12.2. *Let C be an object of $\mathcal{C}$, and F a simplicial presheaf on $\mathcal{C}$. Then $F(C) \cong \operatorname{Map}(C, F)$ and for each n there is a natural isomorphism*

$$\operatorname{Map}_{\Delta^{\mathrm{op}}\operatorname{Pshv}(\mathcal{C})}(C \boxtimes \Delta^n, F) \cong \operatorname{Map}_{\Delta^{\mathrm{op}}\mathbf{Sets}}(\Delta^n, F(C)).$$

Proof. By direct computation, the summand $C \times [\iota]$ of $C \boxtimes \Delta^k$ corresponding to the canonical $\iota \in \Delta_k^k$ induces a natural isomorphism $F_k(C) \cong \operatorname{Hom}(C \boxtimes \Delta^k, F)$. As k varies, this shows that $F(C) \cong \operatorname{Map}(C, F)$. For the second asssertion, it suffices to show that $\operatorname{Hom}(C \boxtimes K, F) \cong \operatorname{Hom}(K, F(C))$ for $K = \Delta^n \times \Delta^k$ (as k varies). As with any simplicial set, we can write K as the colimit of its simplices $\sigma : \Delta^i \to K$. Since $C \boxtimes K$ is the colimit (over σ) of the $C \boxtimes \Delta^i$, we are reduced to the case $K = \Delta^i$, where $\operatorname{Hom}(\Delta^i, F(C)) \cong F_i(C) \cong \operatorname{Hom}(C \boxtimes \Delta^i, F)$. □

Given a functor $f : \mathcal{C} \to \mathcal{C}'$, there is a well-known adjunction (f^*, f_*) between presheaves on $\mathcal{C}$ and $\mathcal{C}'$. The direct image functor $f_* : \operatorname{Pshv}(\mathcal{C}') \to \operatorname{Pshv}(\mathcal{C})$ is $(f_* Y)(C) = Y(fC)$, and the inverse image functor $f^* : \operatorname{Pshv}(\mathcal{C}) \to \operatorname{Pshv}(\mathcal{C}')$ is given by the formula

$$(f^* X)(C') = \operatorname*{colim}_{C' \to fC} X(C).$$

It is well known (and an easy exercise) that if X is an object of $\mathcal{C}$, regarded as a presheaf, then $f^* X$ is represented by $f(X)$.

Clearly, f_* preserves projective fibrations and global weak equivalences. Therefore, (f^*, f_*) is a Quillen adjunction of projective model structures.

As with any Quillen adjunction, (f^*, f_*) induces a derived adjunction $(\mathbf{L}f^*, \mathbf{R}f_*)$ between the homotopy categories (see 12.0). Because f_* preserves global weak equivalences, it also defines a functor between homotopy categories; since $(f_* Y)(C) \xrightarrow{\simeq} (\mathbf{R}f_* Y)(C)$ is an equivalence for all Y and C, we have $f_* \cong \mathbf{R}f_*$. Thus we may regard f_* as the right adjoint of $\mathbf{L}f^*$.

Lemma 12.3. *Let $i : \mathcal{C} \subset \mathcal{C}'$ be a full embedding. Then:*

(1) the inverse image $i^* : \Delta^{\mathrm{op}}\operatorname{Pshv}(\mathcal{C}) \to \Delta^{\mathrm{op}}\operatorname{Pshv}(\mathcal{C}')$ is a full embedding as a full coreflective subcategory; that is, the unit of the adjunction $X \to i_* i^* X$ is an isomorphism for any X in $\Delta^{\mathrm{op}}\operatorname{Pshv}(\mathcal{C})$;

(2) the total inverse image $\mathbf{L}i^*$ embeds the homotopy category of $\Delta^{\mathrm{op}}\operatorname{Pshv}(\mathcal{C})$ into the homotopy category of $\Delta^{\mathrm{op}}\operatorname{Pshv}(\mathcal{C}')$ as a full (coreflective) subcategory.

Proof. For (1), it suffices to show that $X(C) \longrightarrow i_* i^* X(C) = (i^* X)(iC)$ is an isomorphism for every C in $\mathcal{C}$; this is clear from the definition of $i^* X(iC)$, as the identity map of C is initial among all maps $C \to C_1$ in $\mathcal{C}'$ with C_1 in $\mathcal{C}$.

For (2), we may assume that X is cofibrant; by (1), the unit map $X \to i_*(\mathbf{L}i^*X) \cong i_*i^*(X)$ is an isomorphism. This implies $X \to i_*\mathbf{L}i^*X$ is an isomorphism in the homotopy category, and hence that $\mathbf{L}i^*$ is a full embedding. $\square$

It is useful to have a broad family of examples of projectively cofibrant presheaves in $\Delta^{\mathrm{op}}\mathrm{Pshv}(\mathcal{C})$.

Example 12.4. For any C in $\mathcal{C}$, the representable presheaf $H = \mathrm{Hom}_{\mathcal{C}}(-, C)$ and the presheaves $H \boxtimes \Delta^n$ are projectively cofibrant in $\mathrm{Pshv}(\mathcal{C})$. This is because morphisms $H \boxtimes \Delta^n \to Y$ correspond to elements of $Y_n(C)$, and the liftings required by the left lifting property exist because any trivial Kan fibration $X(C) \to Y(C)$ is a surjection in each degree.

Any coproduct of representable presheaves is projectively cofibrant, since coproducts of cofibrant objects are cofibrant. It will be convenient to let $\mathcal{C}^{\amalg}$ denote the full subcategory of presheaves which consists of coproducts of representable presheaves; if $C = \amalg \mathrm{Hom}_{\mathcal{C}}(-, C_\alpha)$, we sometimes write $F(C)$ for $\mathrm{Hom}(C, F) = \prod F(C_\alpha)$.

It is worth noting that the Yoneda embedding does not preserve coproducts. Indeed, the presheaf coproduct $\amalg \mathrm{Hom}_{\mathcal{C}}(-, C_\alpha)$ in $\mathcal{C}^{\amalg}$ sends A to $\amalg \mathrm{Hom}(A, C_\alpha)$ while $\mathrm{Hom}_{\mathcal{C}}(-, \amalg C_\alpha)$ sends A to the typically larger set $\mathrm{Hom}(A, \amalg C_\alpha)$.

Recall from [AGV73, Vbis(5.1.1)] that a simplicial object Y in any category is called *split* if each Y_i has a subobject N_i such that the natural maps

$$\coprod \eta^* : \coprod_{\eta : [i] \twoheadrightarrow [j]} N_j \to Y_i$$

are isomorphisms. The coproduct here is over all surjections η in Δ with source i.

Theorem 12.5. *Every split simplicial presheaf in $\mathcal{C}^{\amalg}$ is projectively cofibrant.*

Since simplicial sets are split, presheaves of the form $C \boxtimes X_\bullet$ are split simplicial. Thus we have:

Corollary 12.5.1. *If C is an object of $\mathcal{C}$ and $X_\bullet$ is a simplicial set then $C \boxtimes X_\bullet$ is projectively cofibrant.*

Before giving the proof of 12.5, we need to introduce some terminology. Let $\Delta_{\leq n}$ denote the full subcategory of Δ with objects $[0], [1], \ldots, [n]$, and $i_n : \Delta_{\leq n} \subset \Delta$ the inclusion. The direct image functor i_{n*} is called n-*truncation*. In addition to the left adjoint i_n^*, the n-truncation functor i_{n*} also has a right adjoint $i_n^!$. The n-*coskeleton* of a simplicial F is defined to be $\mathrm{cosk}_n F = i_n^! i_{n*} F$, and the n-*skeleton* of F is defined to be $\mathrm{sk}_n F = i_n^* i_{n*} F$. There are natural 1–1 correspondences between maps $i_{n*}X \to i_{n*}Y$ of the n-truncated objects, maps $X \to \mathrm{cosk}_n Y$, and maps $\mathrm{sk}_n X \to Y$. Both $i_{n*}i_n^*$ and $i_{n*}i_n^!$ are the identity. There are natural maps $\mathrm{sk}_n X \to X \to \mathrm{cosk}_n X$.

Finally, the *boundary* $\partial\Delta^{n+1}$ of Δ^{n+1} is the simplicial subset generated by the faces of the unique nondegenerate $(n+1)$-simplex in Δ^{n+1}. That is, $\partial\Delta^{n+1} = i_n^* i_{n*} \Delta^{n+1}$.

Lemma 12.6. *If X is a representable presheaf and F is any simplicial presheaf, then presheaf maps $X \xrightarrow{a} (\mathrm{cosk}_n F)_{n+1}$ are in 1-1 correspondence with simplicial maps $\partial\Delta^{n+1} \longrightarrow F(X)$.*

Proof. Set $K = F(X)$. Then the left and right adjoints of i_{n*} yield:

$$\mathrm{Hom}(X, (\mathrm{cosk}_n F)_{n+1}) = (\mathrm{cosk}_n K)_{n+1} = \mathrm{Hom}(\Delta^{n+1}, \mathrm{cosk}_n K)$$
$$\cong \mathrm{Hom}(i_{n*}\Delta^{n+1}, i_{n*}K) \cong \mathrm{Hom}(\partial\Delta^{n+1}, F(X)). \qquad \square$$

Remark 12.6.1. Here is another proof. Giving a map from X to $(\mathrm{cosk}_n F)_{n+1}$ is equivalent to giving maps $a_0, \ldots, a_{n+1} : X \to F_n$ such that $\partial_i a_j = \partial_{j-1} a_i$ for $i < j$ (see [GJ99, VII.1.19]). Regarding $a_0, \ldots, a_{n+1}$ as a sequence of elements in $F_n(X)$, we have the data of a map $\partial\Delta^{n+1} \to F(X)$.

Proof of Theorem 12.5. Given a split simplicial C in $\mathcal{C}^{\mathrm{II}}$ and a trivial projective fibration $F \xrightarrow{\pi} B$, we need to show that any map $C \xrightarrow{b} B$ has a lift $C \xrightarrow{f} F$. By the Yoneda Lemma, we may regard the components $b_n : C_n \to B_n$ of b as elements of $B_n(C_n)$, and the components f_n of f as elements of $F_n(C_n)$.

We will construct a compatible family of maps $i_{n*}C \to i_{n*}F$ lifting the truncations of b; taking the inverse limit yields the desired lift $f : C \to F$ of b. The 0-truncation of b is the map b_0; since the set map $F_0(C_0) \to B_0(C_0)$ is onto, we may lift b_0 to an element f_0 of $F_0(C_0)$, representing a map $C_0 \to F_0$.

Inductively, suppose we are given a map $i_{n*}C \to i_{n*}F$ which lifts $i_{n*}b$, or equivalently by adjunction, a map $f' : C \to \mathrm{cosk}_n F$ lifting $C \to B \to \mathrm{cosk}_n B$. By Lemma 12.6, the component $f'_{n+1} : C_{n+1} \to (\mathrm{cosk}_n F)_{n+1}$ corresponds to a map $f' : \partial\Delta^{n+1} \to F(C_{n+1})$ fitting into the solid diagram

$$
\begin{array}{ccc}
\partial\Delta^{n+1} & \xrightarrow{\ f'\ } & F(C_{n+1}) \\
\downarrow & {\scriptstyle f_{n+1}} \nearrow & \ \ \downarrow {\scriptstyle \simeq}\, \pi(C_{n+1}) \\
\Delta^{n+1} & \xrightarrow[\ b_{n+1}\ ]{} & B(C_{n+1}).
\end{array}
$$

Since π is a trivial projective fibration, $\pi(C_{n+1})$ is a trivial fibration of simplicial sets, so there is a dotted arrow f_{n+1} making the diagram commute. By Lemma 12.6, f_{n+1} corresponds to a commutative diagram

$$
\begin{array}{ccc}
C_{n+1} & \longrightarrow & (\mathrm{cosk}_n C)_{n+1} \\
{\scriptstyle f_{n+1}} \downarrow & & \downarrow {\scriptstyle f'} \\
F_{n+1} & \longrightarrow & (\mathrm{cosk}_n F)_{n+1}.
\end{array}
$$

By assumption, C_{n+1} has a nondegenerate subobject N_{n+1}. By [AGV73, Vbis(5.1.3)], this diagram determines a map from the truncation $i_{n+1*}C$ of C to the truncation $i_{n+1*}F$ (equivalently, $\mathrm{cosk}_{n+1} C \to \mathrm{cosk}_{n+1} F$) which restricts to f' and lifts the $n+1$-truncation of b.   □

Construction 12.7. There is a canonical functor L_* from simplicial presheaves to globally cofibrant simplicial presheaves, as we assume that $\mathcal{C}$ has a set of objects. To construct it, we consider the discrete category $\mathcal{C}^\delta$ on the objects of $\mathcal{C}$. The inclusion $\iota \colon \mathcal{C}^\delta \subset \mathcal{C}$ induces an adjunction (ι^*, ι_*) of presheaf categories, with

$$\iota^* X(C') = \coprod_{C' \to C} X(C) = \coprod_C \mathrm{Hom}_\mathcal{C}(C', C) \times X(C) = \coprod_{C, X(C)} \mathrm{Hom}_\mathcal{C}(C', C),$$

for C' in $\mathcal{C}$ and X in $\mathrm{Pshv}(\mathcal{C}^\delta)$, and $L = \iota^* \iota_*$ is a cotriple on $\mathrm{Pshv}(\mathcal{C})$. Setting $L_n = L^{n+1}$, the map $L(F) \to F$ yields a simplicial map $L_* F \to F$ called the *cotriple resolution* of the presheaf F; see [Wei94, 8.6.4]. Recall that the nerve of $\mathcal{C}$ is the simplicial set $N_\bullet = N_\bullet(\mathcal{C})$ in which N_n is the set of all composable sequences $C_0 \to \cdots \to C_n$ in $\mathcal{C}$. Then we have the explicit formula

$$L_n F = \coprod_{N_n} \coprod_{F(C_n)} C_0. \tag{12.7.1}$$

By inspection, $L_* F$ is a split simplicial object of $\mathcal{C}^{\mathrm{II}}$, so by Theorem 12.5 it is projectively cofibrant. Moreover, F is the coequalizer of the diagram $L_1 F \rightrightarrows L_0 F$. If $F = F_\bullet$ is a simplicial presheaf we write $L_* F_\bullet$ for the diagonal of the bisimplicial presheaf $L_*(F_\bullet)$. Again by inspection, $L_* F_\bullet$ is a split simplicial object of $\mathcal{C}^{\mathrm{II}}$, so it is also projectively cofibrant.

Example 12.8.1. If $F = \mathrm{pt}$ is the constant one-point presheaf, then $(L_* \mathrm{pt})(C)$ is contractible for each C in $\mathcal{C}$. Indeed, $(L_* \mathrm{pt})(C)$ is the nerve of the comma category $C \backslash \mathcal{C}$, which has $1 \in \mathrm{Hom}(C, C)$ as initial object.

If C^δ is an object of $\mathcal{C}^\delta$, then $(\iota^* C^\delta)(C') = \mathrm{Hom}_\mathcal{C}(C', C)$.

Lemma 12.9. *The L_n commute with coproducts:* $L_n(\coprod_I F_i) = \coprod_I L_n(F_i)$.

Proof. This follows from (12.7.1), since $\coprod_{\mathrm{II} F_i(C_n)} C_0 = \coprod_I \coprod_{F_i(C_n)} C_0$.   □

Recall that two maps $f, g \colon A \to B$ between simplicial objects are *simplicially homotopic* if there are maps $h_i \colon A_n \to B_{n+1}$ ($i = 0, ..., n$) such that $\partial_0 h_0 = f$, $\partial_{n+1} h_n = g$, and the standard formulas for $\partial_i h_j$ and $\sigma_i h_j$ hold; see [Wei94, 8.3.11]. This gives rise to the notion of a simplicial homotopy equivalence in any category of simplicial objects, one which is preserved by functors.

Since the representable presheaf $F = C$ is $i^*(C^\delta)$, the general theory of cotriple resolutions (see [Wei94, 8.6.8–10]) implies that $L_* C \to C$ is a simplicial homotopy equivalence, and hence that $(L_* C)(C') \to C(C') = \mathrm{Hom}(C', C)$ is a homotopy equivalence for each C', i.e., $L_*C \to C$ is a global weak equivalence. Since the coproduct $\coprod L_*(C_\alpha) \to \coprod C_\alpha$ of simplicial homotopy equivalences is a simplicial homotopy equivalence, this argument establishes:

Corollary 12.9.1. *If F is a simplicial object in $\mathcal{C}^{\amalg}$ then $L_*F \to F$ is also a global weak equivalence.*

The following lemma will be used in 12.26 and (14.24.1).

Lemma 12.10. *Let $F_1 \to F_2$ be a monomorphism of pointed presheaves, with presheaf quotient $F_3(C) = F_2(C)/F_1(C)$. Then $L_*F_1 \to L_*F_2$ is a termwise split inclusion, and there is an isomorphism $L_*F_2/L_*F_1 \cong L_*F_3/L_*\mathrm{pt}$.*

Proof. For each object C of $\mathcal{C}$, choose a set-theoretic splitting of the surjection $F_2(C) \to F_3(C) = F_2(C)/F_1(C)$, i.e., a decomposition $F_2(C) = F_1(C) \vee F_3(C)$. For each C and C_0, this decomposition yields an isomorphism

$$\coprod_{F_2(C)} C_0 \cong \left(\coprod_{F_1(C)} C_0 \right) \amalg \left(\coprod_{F_3(C) \backslash *} C_0 \right).$$

As in the proof of Lemma 12.9, the explicit formula (12.7.1) shows that each $L_n F_2$ is the coproduct of $L_n F_1$ and the presheaf $Q_n = \coprod_{N_n} \coprod_{F_3(C) \backslash *} C_0$. Since $L_n F_3$ is the coproduct of $L_n \mathrm{pt}$ and Q_n, the result follows. $\square$

We conclude this section with a description of all cofibrant objects in Δ^{op} $\mathrm{Pshv}(\mathcal{C})$. It follows from the general theory of cellular model categories.

Definition 12.11. ([Hir03, 12.1]) A model category $\mathcal{M}$ is *cellular* if it has a set I of generating cofibrations whose sources and targets are compact objects, a set J of generating trivial cofibrations whose sources are small relative to I, and such that the cofibrations are effective monomorphisms. For example, $\Delta^{\mathrm{op}}\mathbf{Sets}$ is cellular: I is the set of all $\partial\Delta^n \to \Delta^n$ and J is the set of all horns $\Lambda^n_k \to \Delta^n$.

We refer the reader to [Hir03, 11.1] for the terminology in this definition. In particular, $\mathcal{M}$ is "cofibrantly generated" by I and J. Moreover, a map in $\mathcal{M}$ is a trivial fibration (resp., a fibration) iff it has the right lifting property with respect to I (resp., J).

Example 12.11.1. Hirschhorn observes that the projective model structure on $\Delta^{\mathrm{op}}\mathrm{Pshv}(\mathcal{C})$ is cellular in [Hir03, Prop. 12.1.5], assuming that $\mathcal{C}$ is small (or skeletally small); $I_{\mathcal{C}}$ is the set of all $C \boxtimes \partial\Delta^n \to C \boxtimes \Delta^n$ with C in $\mathcal{C}$, $J_{\mathcal{C}}$ is the set of all horns $C \boxtimes \Lambda^n_k \to C \boxtimes \Delta^n$ with C in $\mathcal{C}$, and the pairing $\boxtimes$ is defined at the end of 12.1.

A map which is a transfinite composition of morphisms which are pushouts of morphisms in a set I of maps in $\mathcal{C}$ is called a *relative I-cell complex*; see [Hir03, 10.5.8]. The following result is proven in [Hir03, 11.2.1] when I is $I_{\mathcal{C}}$.

Lemma 12.12. *Let $\mathcal{M}$ be a cellular model category. Then every cofibration is a retract of a relative $I_{\mathcal{C}}$-cell complex (which is also a cofibration).*

Let us say that a map $X \to Y$ of presheaves is *nice* if it is isomorphic to the projective cofibration $X \hookrightarrow X \amalg (\coprod C^i)$ for some family of representable presheaves C^i. From Example 12.4, the $X \to X \amalg (\coprod C^i \boxtimes \Delta^{n_i})$ are projective cofibrations as well as termwise nice. We now show that every projective cofibration is a retract of a termwise nice map.

Corollary 12.13. *If $f : X \to Y$ is a projective cofibration in $\Delta^{\mathrm{op}}\mathrm{Pshv}(\mathcal{C})$ then there is a termwise nice map $X \to \widetilde{Y}$ and a split inclusion $i : Y \to \widetilde{Y}$ whose splitting map p restricts to f on X.*

Proof. ([Voe10d, 3.27b]) The generating cofibrations of $\Delta^{\mathrm{op}}\mathrm{Pshv}(\mathcal{C})$ (maps in $I_{\mathcal{C}}$) are termwise nice by Example 12.11.1. Since pushouts of nice maps are nice, and transfinite compositions of nice morphisms are nice, relative $I_{\mathcal{C}}$-cell complexes are termwise nice cofibrations. By the small object argument, f factors as $X \xrightarrow{n} \widetilde{Y} \xrightarrow{p} Y$, where n is termwise nice and p has the right lifting property for $I_{\mathcal{C}}$. Since f is a retract of a relative $I_{\mathcal{C}}$-cell complex by Lemma 12.12, p has a section i with $n = if$. ☐

12.2   RADDITIVE PRESHEAVES

A convenient category of presheaves that we shall use for our motivic model structures is the category of *radditive* presheaves. One might think of radditive presheaves as intermediate between sheaves and presheaves since, on a Grothendieck site in which $\{U, V\}$ covers $U \amalg V$, every sheaf is radditive.

The use of radditive presheaves allows us to give a unified treatment of several Quillen adjunctions; see diagram (12.28.1) in section 12.3. In addition, the motivic Hurewicz functor R_{tr} is easily defined for radditive presheaves; see 12.28. Most of the material in this section was published in [Voe10d].

We assume throughout this section that $\mathcal{C}$ is a skeletally small category with finite coproducts (and hence an initial object $\emptyset$).

Definition 12.14. A presheaf of sets X on $\mathcal{C}$ is called *radditive* if $X(\emptyset)$ is a one-element set and the map $X(C \amalg C') \to X(C) \times X(C')$ is a bijection for all C, C' in $\mathcal{C}$. We write $\mathrm{rad}(\mathcal{C})$ for the full subcategory of $\mathrm{Pshv}(\mathcal{C})$ (presheaves on $\mathcal{C}$) consisting of all radditive presheaves.

Example 12.14.1. The representable presheaf $\mathrm{Hom}_{\mathcal{C}}(-, C)$ is radditive for each C in $\mathcal{C}$. Thus we may regard $\mathcal{C}$ as a full subcategory of $\mathrm{rad}(\mathcal{C})$, via the Yoneda embedding of $\mathcal{C}$ into $\mathrm{Pshv}(\mathcal{C})$. The following calculation shows that the embedding $\mathcal{C} \subset \mathrm{rad}(\mathcal{C})$ preserves coproducts: for radditive F,

$$\mathrm{Hom}_{\mathrm{rad}(\mathcal{C})}(\mathrm{Hom}_{\mathcal{C}}(-, C_1 \amalg C_2), F) = F(C_1 \amalg C_2) = F(C_1) \times F(C_2).$$

In contrast, the Yoneda embedding $\mathcal{C} \to \mathrm{Pshv}(\mathcal{C})$ does not preserve coproducts: if $C = C_1 \amalg C_2$ then $\mathrm{id}_C \in \mathrm{Hom}(C, C)$ is not in $\mathrm{Hom}_\mathcal{C}(C, C_1) \amalg \mathrm{Hom}_\mathcal{C}(C, C_2)$.

Example 12.14.2. Let **Sets** denote the category of finite sets and $*$ the one-element set. Any radditive presheaf F on **Sets** is completely determined by the set $F(*)$. Thus the Yoneda embedding gives an equivalence between **Sets** and $\mathrm{rad}(\mathbf{Sets})$. Similarly, the category $\mathbf{Sets}_+$ of pointed sets is equivalent to the category $\mathrm{rad}(\mathbf{Sets}_+)$ because F is completely determined by the data $* = F(*) \to F(S^0)$, i.e., by the pointed set $F(S^0)$.

Example 12.14.3. Suppose that $\mathcal{C}$ is either **Sm** or **Norm** (normal schemes over k), and let $\mathcal{C}_0$ denote the full subcategory of nonempty connected varieties in $\mathcal{C}$. Then a radditive presheaf on $\mathcal{C}$ is uniquely determined by its values on $\mathcal{C}_0$, because every object of $\mathcal{C}$ is uniquely a finite coproduct of objects in $\mathcal{C}_0$. In other words, the restriction ι_* of presheaves from $\mathcal{C}$ to $\mathcal{C}_0$ induces an equivalence between $\mathrm{rad}(\mathcal{C})$ and $\mathrm{Pshv}(\mathcal{C}_0)$. The inverse equivalence $\mathrm{Pshv}(\mathcal{C}_0) \to \mathrm{rad}(\mathcal{C})$ sends F to $F_0^{\amalg}(\amalg X_i) = \prod F(X_i)$, and $F \to F^{\mathrm{rad}} = (\iota_* F)^{\amalg}$ is the universal map to a radditive presheaf.

Example 12.14.4. If I is filtered and F_i are radditive, then the presheaf $\mathrm{colim}\, F_i$ is also radditive. Indeed, because filtered colimits commute with finite limits, $\mathrm{colim}_I \left(F_i(C) \times F_i(C') \right) = (\mathrm{colim}_I F_i(C)) \times (\mathrm{colim}_I F_i(C'))$.

Another important family consists of ind-objects of $\mathcal{C}$.

Definition 12.15. Let $\mathcal{C}^{\mathrm{ind}}$ denote the full subcategory of presheaves on $\mathcal{C}$ consisting of filtered colimits of representable presheaves; this is the same as the category of *ind-objects* of $\mathcal{C}$ (see [AGV73, I.8.2]). By Examples 12.14.1 and 12.14.4, ind-objects are radditive as presheaves. Thus $\mathcal{C}^{\mathrm{ind}}$ is a full subcategory of $\mathrm{rad}(\mathcal{C})$.

Example 12.15.1. By Example 12.14.1, the coproduct of $\mathrm{Hom}(-, C_1)$ and $\mathrm{Hom}(-, C_2)$ is $\mathrm{Hom}(-, C_1 \amalg C_2)$ in both $\mathrm{rad}(\mathcal{C})$ and in $\mathcal{C}^{\mathrm{ind}}$. Since the coproduct over an indexing set I is the filtered colimit of the coproducts over finite subcategories of I, it follows that arbitrary coproducts of objects of $\mathcal{C}$ exist in both $\mathrm{rad}(\mathcal{C})$ and $\mathcal{C}^{\mathrm{ind}}$, and are equal. We shall write $\amalg^{\mathrm{rad}} C_\alpha$ for the coproduct (in $\mathcal{C}^{\mathrm{ind}}$) of objects $\{C_\alpha : \alpha \in \mathcal{A}\}$ of $\mathcal{C}$, to distinguish it from the corresponding coproduct in $\mathrm{Pshv}(\mathcal{C})$. There is a canonical presheaf map $\eta : \coprod \mathrm{Hom}(-, C_\alpha) \to \amalg^{\mathrm{rad}} C_\alpha$ between coproducts.

We will prove in Proposition 12.17 that the inclusion $\mathrm{rad}(\mathcal{C}) \subset \mathrm{Pshv}(\mathcal{C})$ has a left adjoint $(-)^{\mathrm{rad}}$ for every $\mathcal{C}$. The key step is to show that reflexive coequalizers of radditive presheaves are radditive. Recall that a coequalizer diagram $A \rightrightarrows B$ in a category is called *reflexive* if both maps have a common right inverse $B \to A$.

Lemma 12.16. *If $E \rightrightarrows F$ is a reflexive coequalizer diagram of radditive presheaves, then the presheaf coequalizer is a radditive presheaf.*

Proof. If G is the presheaf coequalizer of $E \rightrightarrows F$, each set $G(C)$ is the set coequalizer of $E(C) \rightrightarrows F(C)$. In the category of sets, reflexive coequalizers commute with finite products (exercise!). Hence the set $G(C) \times G(C')$ is the coequalizer of $E(C \amalg C') \rightrightarrows F(C \amalg C')$. ∎

Proposition 12.17. *The inclusion* $\mathrm{rad}(\mathcal{C}) \subset \mathrm{Pshv}(\mathcal{C})$ *has a left adjoint* $(-)^{\mathrm{rad}}$, *and the universal map* $\eta_F : F \to F^{\mathrm{rad}}$ *is an isomorphism for every radditive* F.

Proof. (See [Voe10d, 3.6].) Recall from Construction 12.7 that any presheaf F is the coequalizer of the reflexive diagram $L_1 F \rightrightarrows L_0 F$. The formula (12.7.1) shows that each $L_n F$ is a coproduct of representable presheaves, so the coproduct $(L_n F)^{\mathrm{rad}}$ exists in $\mathrm{rad}(\mathcal{C})$ by Example 12.15.1. We define F^{rad} to be the presheaf coequalizer of the reflexive diagram $(L_1 F)^{\mathrm{rad}} \rightrightarrows (L_0 F)^{\mathrm{rad}}$; F^{rad} is radditive by Lemma 12.16. We also define $\eta_F : F \to F^{\mathrm{rad}}$ to be the coequalizer of the maps $\eta : L_i(F) \to (L_i F)^{\mathrm{rad}}$ in Example 12.15.1. It is now easy to check that F^{rad} is natural in F, and that η_F induces an isomorphism $\eta^* : \mathrm{Hom}(F^{\mathrm{rad}}, G) \cong \mathrm{Hom}(F, G)$ for any radditive G.

When F is radditive, this yields $\eta^* : \mathrm{Hom}_{\mathrm{rad}(\mathcal{C})}(F^{\mathrm{rad}}, -) \cong \mathrm{Hom}_{\mathrm{rad}(\mathcal{C})}(F, -)$ so η_F is an isomorphism by the Yoneda Lemma. ∎

Corollary 12.18. *The category* $\mathrm{rad}(\mathcal{C})$ *admits all small colimits; if* $\mathrm{colim}\, F_i$ *is the colimit of* $F : I \to \mathrm{rad}(\mathcal{C})$ *in* $\mathrm{Pshv}(\mathcal{C})$, *the colimit in* $\mathrm{rad}(\mathcal{C})$ *is* $(\mathrm{colim}\, F_i)^{\mathrm{rad}}$.

Construction 12.19. Recall from 12.7 that there is a canonical functor L_* from simplicial presheaves to cofibrant simplicial presheaves. If F is a simplicial radditive presheaf, we define **Lres** (F) to be $L_*(F)^{\mathrm{rad}}$. That is,

$$\mathbf{Lres}\,_n F = (L_n F)^{\mathrm{rad}} = \coprod_{C_0 \to \cdots \to C_n}^{\mathrm{rad}} \coprod_{F(C_n)}^{\mathrm{rad}} C_0$$

for all n. In particular, **Lres** (F) is an object of $\Delta^{\mathrm{op}}(\mathcal{C}^{\mathrm{ind}})$; see Example 12.15.1.

Alternatively, the adjunction (ι^*, ι_*) in 12.7 induces an adjunction $(\iota^{\mathrm{rad}}, \iota_*)$ between $\mathrm{Pshv}(\mathcal{C}^\delta)$ and $\mathrm{rad}(\mathcal{C})$. This gives rise to the cotriple $\iota^{\mathrm{rad}}\iota_* = L^{\mathrm{rad}}$ in $\mathrm{rad}(\mathcal{C})$, and **Lres** (F) is the resulting cotriple resolution of F in $\mathrm{rad}(\mathcal{C})$. As in Corollary 12.9.1, the general theory of cotriples implies that **Lres** $(C) \to C$ is a simplicial homotopy equivalence for every C.

Remark 12.19.1. ([Voe10d, 3.19(1)]) **Lres** commutes with filtered colimits: **Lres** $(\mathrm{colim}\, F_i) \cong \mathrm{colim}\, \mathbf{Lres}\, F_i$. Indeed, $\coprod_{\mathrm{colim}\, F_i(C_n)}^{\mathrm{rad}} C_0 = \mathrm{colim} \coprod_{F_i(C_n)}^{\mathrm{rad}} C_0$ for each chain $C_0 \to \cdots \to C_n$; see Lemma 12.9.

The following trick, modelled on the analysis for **Sets** and **Sets**$_+$ above, allows us to pass back and forth between the pointed and unpointed cases when $\mathcal{C}$ has a final object $*$. Let $\mathcal{C}_+ \subset \mathcal{C}$ denote the subcategory of pointed objects of $\mathcal{C}$ which are isomorphic to $C_+ = C \amalg *$ for some C, together with all pointed morphisms; $\mathcal{C}_+$ has initial object $*$ and has finite coproducts whenever $\mathcal{C}$ does.

Lemma 12.20. *If $\mathcal{C}$ has a final object, $\mathrm{rad}(\mathcal{C}_+)$ is equivalent to the category of pointed objects in $\mathrm{rad}(\mathcal{C})$.*

Proof. The canonical functor $\mathcal{C} \xrightarrow{b} \mathcal{C}_+$, defined by $b(C) = C_+$, preserves coproducts so it defines a functor $\mathrm{rad}(\mathcal{C}_+) \xrightarrow{b_*} \mathrm{rad}(\mathcal{C})$ sending G to $b_* G(C) = G(C_+)$; the presheaf $b_* G$ is pointed by the image of $* = G(*) \to G(C_+) = b_* G(C)$.

Conversely, given a pointed radditive presheaf F on $\mathcal{C}$, define $G : \mathcal{C}_+ \to \mathbf{Sets}$ on objects by letting $G(X)$ be the inverse image of the basepoint in the canonical map $F(X) \to F(*)$ associated to $* \to X$; if $X \cong C_+$ then $F(X) \cong F(C) \times F(*)$ and $G(X) \cong F(C)$. To a morphism $X \to X'$ in $\mathcal{C}_+$ we associate the composition

$$G(X') \hookrightarrow F(X') \to F(X) \cong F(C) \times F(*) \xrightarrow{\mathrm{pr}} F(C) = G(X).$$

It is routine to check that G is a presheaf; G is radditive because $G(*) = *$ and $G(C_+ \vee C'_+) = G((C \amalg C')_+) = F(C) \times F(C')$. Thus $F \mapsto G$ defines a functor from pointed radditive presheaves on $\mathcal{C}$ to $\mathrm{rad}(\mathcal{C}_+)$. By inspection, it is inverse to the given functor $b_* : \mathrm{rad}(\mathcal{C}_+) \to \mathrm{rad}(\mathcal{C})$. $\qquad\square$

Lemma 12.21. *If $\mathcal{C}$ is an additive category, $\mathrm{rad}(\mathcal{C})$ and $\mathrm{rad}(\mathcal{C}_+)$ are both equivalent to the category of contravariant additive functors from $\mathcal{C}$ to abelian groups.*

Proof. Clearly any additive contravariant functor is a radditive presheaf. Suppose conversely that F is a radditive presheaf on $\mathcal{C}$. For each C, the diagonal $\Delta : C \to C \oplus C$ gives an operation $F(C) \times F(C) \to F(C)$. The usual diagrams for a group object show that this operation is associative and commutative, that the basepoint 0 of $F(C)$ is an identity, and that an inverse operation on $F(C)$ is given by $F(-1_C)$; this makes $F(C)$ an abelian group. If F is the radditive presheaf underlying an additive functor to $\mathbf{Ab}$, this construction recovers the original additive structure on each $F(C)$. Given $B \to C$, the two maps $\Delta_C \circ f, (f \oplus f) \circ \Delta_B : B \to C \oplus C$ agree, showing that $\mathrm{Hom}_{\mathcal{C}}(B, C) \to \mathrm{Hom}_{\mathbf{Ab}}(F(C), F(B))$ is a homomorphism, i.e., that F is an additive functor. This establishes the equivalence between radditive presheaves and additive functors to $\mathbf{Ab}$.

Finally, since 0 is both initial and terminal in $\mathcal{C}$, any radditive functor F has $F(0) = 0$, and so is pointed; cf. Lemma 12.20. $\qquad\square$

Example 12.21.1. Let $\mathcal{S}$ be either the category $\mathbf{Sm}$ of smooth schemes, or the category $\mathbf{Norm}$ of normal schemes over k, and let $\mathbf{Cor}(\mathcal{S})$ denote the category of (finite) correspondences on $\mathcal{S}$, with coefficients in a ring R, as defined in [MVW, §1 and 1A]. We write $\mathbf{PST}(\mathcal{S})$ for the (abelian) category of contravariant additive functors $\mathbf{Cor}(\mathcal{S}) \to \mathbf{Ab}$; such functors are called *presheaves with transfers* on $\mathcal{S}$. By Lemma 12.21, $\mathbf{PST}(\mathcal{S})$ is the same as $\mathrm{rad}(\mathbf{Cor}(\mathcal{S}))$.

Given an arbitrary functor $f : \mathcal{C} \to \mathcal{C}'$, neither half of the adjunction (f^*, f_*) between presheaves will preserve radditive presheaves. We write f^{rad} for the functor $F \mapsto (f^*F)^{\mathrm{rad}}$ from $\mathrm{rad}(\mathcal{C})$ to $\mathrm{rad}(\mathcal{C}')$ (see 12.14.3). It agrees with f on representable presheaves; if X is an object of $\mathcal{C}$, regarded as a presheaf, then f^*X is $f(X)$, regarded as a (radditive) presheaf, and hence $f^{\mathrm{rad}}X = f^*X = f(X)$.

Lemma 12.22. *For any functor $f : \mathcal{C} \to \mathcal{C}'$, the functor $f^{\mathrm{rad}} : \mathrm{rad}(\mathcal{C}) \to \mathrm{rad}(\mathcal{C}')$ commutes with coproducts, filtered colimits, and reflexive coequalizers.*

Proof. ([Voe10d, 4.1]) The functor $(-)^{\mathrm{rad}} \circ f^*$ from $\mathrm{Pshv}(\mathcal{C})$ to $\mathrm{rad}(\mathcal{C}')$ commutes with all colimits because it is the composition of two left adjoints. Since the inclusion $\mathrm{rad}(\mathcal{C}) \subset \mathrm{Pshv}(\mathcal{C})$ preserves filtered colimits and reflexive coequalizers (the latter by Lemma 12.16), it follows that $f^{\mathrm{rad}}G$ preserves these special colimits as well.   □

12.3   THE RADDITIVE PROJECTIVE MODEL STRUCTURE

There is an analogue for radditive presheaves of the projective model structure for presheaves: we just restrict the notions of fibration and weak equivalence. As in the previous section, we assume throughout that $\mathcal{C}$ is a skeletally small category with finite coproducts.

Definition 12.23. A *global weak equivalence* (resp., *a projective fibration*) between simplicial radditive functors is just a global weak equivalence (resp., a projective fibration) between the underlying simplicial presheaves in the sense of Definition 12.1. We define *projective cofibrations* to be the maps having the left lifting property relative to trivial projective fibrations.

We will write $U \otimes K$ for $(U \boxtimes K)^{\mathrm{rad}}$ when U is a simplicial object in $\mathcal{C}^{\mathrm{ind}}$ and K is a simplicial set, so $(U \otimes K)_p = \mathrm{II}_{K_p}^{\mathrm{rad}} U_p$. Note that if each K_p is finite then $U \otimes K$ is in $\Delta^{\mathrm{op}} \mathcal{C}^{\mathrm{ind}}$. In particular, $C \otimes \Delta^n = (C \boxtimes \Delta^n)^{\mathrm{rad}}$ is in $\Delta^{\mathrm{op}} \mathcal{C}^{\mathrm{ind}}$.

Let $I_{\mathcal{C}}^{\mathrm{rad}}$ denote the set of all morphisms $C \otimes \partial \Delta^n \to C \otimes \Delta^n$ and $J_{\mathcal{C}}^{\mathrm{rad}}$ denote the morphisms $C \otimes \Lambda_k^n \to C \otimes \Delta^n$ with C in $\mathcal{C}$. These are the radditivizations of the generating cofibrations $I_{\mathcal{C}}$ and trivial cofibrations $J_{\mathcal{C}}$ for $\Delta^{\mathrm{op}} \mathrm{Pshv}(\mathcal{C})$ in Example 12.11.1.

Since radditivization is left adjoint to the inclusion $\mathrm{rad}(\mathcal{C}) \subset \mathrm{Pshv}(\mathcal{C})$, it follows formally from the cellular model structure on simplicial presheaves that $I_{\mathcal{C}}^{\mathrm{rad}}$ consists of cofibrations, and that a map in $\Delta^{\mathrm{op}} \mathrm{rad}(\mathcal{C})$ is a projective fibration (resp., trivial projective fibration) if and only if it has the right lifting property with respect to $J_{\mathcal{C}}^{\mathrm{rad}}$ (resp., $I_{\mathcal{C}}^{\mathrm{rad}}$).

We will see in Application 12.32 that the morphisms in $J_{\mathcal{C}}^{\mathrm{rad}}$ are global weak equivalences. Hence they are trivial cofibrations.

Lemma 12.24. *The global weak equivalences, projective fibrations, and projective cofibrations form a cellular model structure on $\Delta^{\mathrm{op}} \mathrm{rad}(\mathcal{C})$.*
We shall call it the projective model structure *on $\Delta^{\mathrm{op}} \mathrm{rad}(\mathcal{C})$.*

Proof. ([Voe10d, 3.25]) Given the criterion of Theorem 2.1.19 in [Hov99] for a cofibrantly generated model structure, and the remarks above, the proof is formal from the definitions, using $I_{\mathcal{C}}^{\mathrm{rad}}$ for the set of generating cofibrations, and $J_{\mathcal{C}}^{\mathrm{rad}}$ for the set of generating trivial cofibrations. □

Since the inclusion $\mathrm{rad}(\mathcal{C}) \subset \mathrm{Pshv}(\mathcal{C})$ preserves global weak equivalences and fibrations, the adjoint pair $(-^{\mathrm{rad}}, \mathrm{incl})$ of Proposition 12.17 defines a Quillen adjunction $\Delta^{\mathrm{op}}\mathrm{Pshv}(\mathcal{C}) \to \Delta^{\mathrm{op}}\mathrm{rad}(\mathcal{C})$. It follows that if $A \to B$ is a cofibration in $\mathrm{Pshv}(\mathcal{C})$ then $A^{\mathrm{rad}} \to B^{\mathrm{rad}}$ is a cofibration in $\mathrm{rad}(\mathcal{C})$. In particular, any projective cofibration in $\Delta^{\mathrm{op}}\mathrm{Pshv}(\mathcal{C})$ whose source and target are simplicial radditive is a projective cofibration in $\Delta^{\mathrm{op}}\mathrm{rad}(\mathcal{C})$, and any representable presheaf is projectively cofibrant in $\mathrm{rad}(\mathcal{C})$.

Example 12.24.1. Let $\mathbf{Cor}(\mathcal{S})$ be the category of (finite) correspondences on either **Sm** or **Norm**, and recall from Example 12.21.1 that $\mathrm{rad}(\mathbf{Cor}(\mathcal{S})) = \mathbf{PST}(\mathcal{S})$. By [MVW, 8.1], every projective object in $\mathbf{PST}(\mathcal{S})$ is a summand (i.e., retract) of a direct sum of representable presheaves. Since Kan fibrations of simplicial groups are just termwise surjections, projective cofibrant objects in $\Delta^{\mathrm{op}}\mathbf{PST}(\mathcal{S})$ are just termwise projective objects, and a projective cofibration $A \to A'$ is just an injection whose cokernel is termwise a projective object.

From this concrete description, it is easy to see that the model structure on $\Delta^{\mathrm{op}}\mathbf{PST}(\mathcal{S})$ is left proper, meaning that the pushout of a weak equivalence along a cofibration is a weak equivalence.

Remark 12.24.2. When $\mathcal{C}$ is either **Sm**, **Sch**, or **Norm**, the projective model structures on $\Delta^{\mathrm{op}}\mathrm{rad}(\mathcal{C})$ and $\Delta^{\mathrm{op}}\mathrm{rad}(\mathcal{C}_+)$ are proper. This is because radditive functors on $\mathcal{C}$ are the same as presheaves on $\mathcal{C}_0$, by Example 12.14.3. As illustrated in [Voe10d, Ex. 3.48], $\Delta^{\mathrm{op}}\mathrm{rad}(\mathcal{C})$ need not be proper for general $\mathcal{C}$.

The following lemma, taken from [Voe10d, 3.18], shows that **Lres** is a functorial cofibrant replacement functor for $\Delta^{\mathrm{op}}\mathrm{rad}(\mathcal{C})$.

Lemma 12.25. *For every radditive F, the map $\mathbf{Lres}\,(F) \to F$ is a global weak equivalence in $\Delta^{\mathrm{op}}\mathrm{rad}(\mathcal{C})$ and hence in $\Delta^{\mathrm{op}}\mathrm{Pshv}(\mathcal{C})$.*

Proof. We use the notation of Construction 12.7. For each C in $\mathcal{C}$, $\mathbf{Lres}\,_n F(C)$ is the coproduct of copies of $C_0(C)$, by (12.7.1). This is the same as evaluation at C of the presheaf $\iota_* \mathbf{Lres}\,_n (F)$ on $\mathcal{C}^{\delta}$. Since $\mathbf{Lres}\,(F)$ is the cotriple resolution associated to the cotriple $\iota^{\mathrm{rad}} \iota_*$ of 12.19, it follows from the general theory of cotriples (see [Wei94, 8.6.10]) that the augmented $\iota_* \mathbf{Lres}\,(F) \to \iota_* F$ is left contractible (in the sense of [Wei94, 8.4.6]) and hence is a simplicial homotopy equivalence. Evaluation yields a simplicial homotopy equivalence $\mathbf{Lres}\,(F)(C) \to F(C)$ for each C, showing that $\mathbf{Lres}\,(F) \to F$ is a global weak equivalence. □

Applying $(-)^{\mathrm{rad}}$ to Lemma 12.10, we see that if $F_1 \to F_2$ is a monomorphism of pointed radditive presheaves, with presheaf quotient $F_3(C) = F_2(C)/F_1(C)$, then $\mathbf{Lres}\,(F_1) \to \mathbf{Lres}\,(F_2)$ is a termwise split inclusion,

$$\mathbf{Lres}\,(F_2)/\mathbf{Lres}\,(F_1) \cong \mathbf{Lres}\,(F_3^{\mathrm{rad}})/\mathbf{Lres}\,(\mathrm{pt}),$$

and (by Example 12.8.1) $\mathbf{Lres}\,(\mathrm{pt}) \to \mathrm{pt}$ is a global weak equivalence.

Let us say that a map $f : A \to B$ in $\mathrm{rad}(\mathcal{C})$ is a *nice radditive map* if there is a nice presheaf map $\tilde{f} : A \to \widetilde{B} \cong A \amalg (\amalg C^i)$ (in the sense of 12.13) such that $B = \widetilde{B}^{\mathrm{rad}}$ and $f = \tilde{f}^{\mathrm{rad}}$. Here is the radditive analogue of Corollary 12.13.

Lemma 12.26. *(1) If $X \xrightarrow{f} Y$ is a projective cofibration in $\Delta^{\mathrm{op}}\mathrm{rad}(\mathcal{C})$ then there is a termwise nice radditive map $X \to \widetilde{Y}$ and a split inclusion $Y \xrightarrow{i} \widetilde{Y}$ whose splitting map p restricts to f on X.*

(2) Suppose that $\mathcal{C}$ is $\mathbf{Sm}$ or $\mathbf{Norm}$. If $E \to F$ is a projective fibration in $\Delta^{\mathrm{op}}\mathrm{Pshv}(\mathcal{C})$, then $E^{\mathrm{rad}} \to F^{\mathrm{rad}}$ is a projective fibration in $\Delta^{\mathrm{op}}\mathrm{rad}(\mathcal{C})$.

Proof. By Lemmas 12.12 and 12.24, cofibrations in $\Delta^{\mathrm{op}}\mathrm{rad}(\mathcal{C})$ are the same as retracts of relative $I_{\mathcal{C}}$-cell complexes. As in the proof of Corollary 12.13, pushouts and transfinite compositions preserve nice radditive maps, so each relative $I_{\mathcal{C}}$-cell complex in $\Delta^{\mathrm{op}}\mathrm{rad}(\mathcal{C})$ is a termwise nice radditive map. As in loc. cit., the map f factors as $X \xrightarrow{n} \widetilde{Y} \xrightarrow{p} Y$, where n is a termwise nice radditive map and p has the right lifting property for $I_{\mathcal{C}}$. Hence p has a section i with $n = if$.

For (2), note that any C is a finite coproduct of connected C_i, so the map from $E^{\mathrm{rad}}(C) = \prod E(C_i)$ to $F^{\mathrm{rad}}(C) = \prod F(C_i)$ is a finite product of the Kan fibrations $E(C_i) \to F(C_i)$. Hence $E^{\mathrm{rad}}(C) \to F^{\mathrm{rad}}(C)$ is a Kan fibration. $\square$

Lemma 12.27. *Suppose that $\mathcal{C}$ and $\mathcal{C}'$ have finite coproducts, and that $\mathcal{C} \xrightarrow{f} \mathcal{C}'$ commutes with finite coproducts. Then the adjunction (f^*, f_*) between presheaves restricts to an adjunction (f^*, f_*) between $\mathrm{rad}(\mathcal{C})$ and $\mathrm{rad}(\mathcal{C}')$.*

Similarly, we have a Quillen adjunction (f^*, f_*) between $\Delta^{\mathrm{op}}\mathrm{rad}(\mathcal{C})$ and $\Delta^{\mathrm{op}}\mathrm{rad}(\mathcal{C}')$, and an adjunction $(\mathbf{L}f^*, f_*)$ between their homotopy categories.

Proof. If Y is radditive, then so is $f_* Y$ because $f(C_1 \amalg C_2) = f(C_1) \amalg f(C_2)$. If X is radditive then so is $f^* X$, because filtered colimits of sets commute with finite products. By naturality, f^* and f_* also form an adjoint pair between the simplicial categories. Since f_* preserves global weak equivalences and projective fibrations, it is a Quillen adjunction. Finally, $f_* \cong \mathbf{R}f_*$. $\square$

Corollary 12.27.1. *If $i : \mathcal{C} \subset \mathcal{C}'$ preserves coproducts, the inverse image $i^* : \Delta^{\mathrm{op}}\mathrm{rad}(\mathcal{C}) \to \Delta^{\mathrm{op}}\mathrm{rad}(\mathcal{C}')$ is a full embedding as a coreflective subcategory.*

In addition, $\mathbf{L}i^*$ embeds the homotopy category of $\Delta^{\mathrm{op}}\mathrm{rad}(\mathcal{C})$ into the homotopy category of $\Delta^{\mathrm{op}}\mathrm{rad}(\mathcal{C}')$ as a full (coreflective) subcategory.

Proof. Immediate from Lemma 12.27 and Lemma 12.3. $\square$

Example 12.28. The standard embedding of $\mathbf{Sm} = \mathbf{Sm}/k$ into $\mathbf{Cor}(\mathbf{Sm}, R)$ preserves coproducts by definition [MVW, 1.1], so by Lemma 12.27 we have an adjunction (R_{tr}, u) between $\mathrm{rad}(\mathbf{Sm})$ and $\mathbf{PST}(\mathbf{Sm}, R) = \mathrm{rad}\,\mathbf{Cor}(\mathbf{Sm}, R)$ for each ring R. The notation reflects the fact that the right adjoint u is the functor from presheaves with transfers to the underlying (radditive) presheaves, and the composition of the left adjoint R_{tr} with the Yoneda embedding of $\mathbf{Cor}(\mathbf{Sm}, R)$ into $\mathbf{PST}(\mathbf{Sm}, R)$ sends X to the presheaf with transfers $R_{\mathrm{tr}}(X) = \mathbb{Z}_{\mathrm{tr}}(X) \otimes R$. Thus we have a pair of adjunctions:

$$\mathrm{Pshv}(\mathbf{Sm}) \xrightarrow{(-^{\mathrm{rad}}, \mathrm{incl})} \mathrm{rad}(\mathbf{Sm}) \xrightarrow{(R_{\mathrm{tr}}, u)} \mathbf{PST}(\mathbf{Sm}, R).$$

Hence there are Quillen adjunctions between the corresponding simplicial model categories. There is of course a similar diagram for pointed presheaves by Lemma 12.21, and a similar diagram with $\mathbf{Sm}/k$ replaced by $\mathbf{Norm}/k$.

We may compare these using the adjunctions (i^*, i_*) of 12.1 and Lemma 12.27 associated to the inclusion $i : \mathbf{Sm} \to \mathbf{Norm}$ and the induced inclusion $i : \mathbf{Cor}(\mathbf{Sm}) \to \mathbf{Cor}(\mathbf{Norm})$ of Example 12.21.1. By naturality and Example 12.28, these fit into a commutative diagram of projective model categories and Quillen adjunctions:

$$
\begin{array}{ccc}
\Delta^{\mathrm{op}}\mathrm{Pshv}(\mathbf{Sm}_+) & \xrightarrow{(i^*, i_*)} & \Delta^{\mathrm{op}}\mathrm{Pshv}(\mathbf{Norm}_+) \\
{\scriptstyle(-^{\mathrm{rad}}, \mathrm{incl})}\Big\downarrow & & \Big\downarrow{\scriptstyle(-^{\mathrm{rad}}, \mathrm{incl})} \\
\Delta^{\mathrm{op}}\mathrm{rad}(\mathbf{Sm}_+) & \xrightarrow{(i^*, i_*)} & \Delta^{\mathrm{op}}\mathrm{rad}(\mathbf{Norm}_+) \\
{\scriptstyle(R_{\mathrm{tr}}, u)}\Big\downarrow & & \Big\downarrow{\scriptstyle(R_{\mathrm{tr}}, u)} \\
\Delta^{\mathrm{op}}\mathbf{PST}(\mathbf{Sm}) & \xrightarrow{(i^*, i_*)} & \Delta^{\mathrm{op}}\mathbf{PST}(\mathbf{Norm}).
\end{array}
\qquad (12.28.1)
$$

By Corollary 12.27.1, the horizontal maps are fully faithful, even on the level of homotopy categories. Vertically, we have an adjunction $(\mathbf{L}R_{\mathrm{tr}}, u)$ of homotopy categories. By Example 12.19, we may choose the model $R_{\mathrm{tr}}(\mathbf{Lres}\,(F))$ for the total left derived functors $\mathbf{L}R_{\mathrm{tr}}$.

Example 12.28.2. If (X, x) is a pointed normal variety, regarded as a pointed presheaf on $\mathbf{Norm}$, we may form the simplicial wedge $X \vee \Delta^1$ with the 1-simplex by identifying x with one of the vertices of Δ^1. This is pointed (using the other vertex of Δ^1 as basepoint) and is a split simplicial object of $\Delta^{\mathrm{op}}\mathbf{Norm}_+$ because its degree n term is the coproduct of X and $n + 1$ copies of $\mathrm{Spec}(k)$. It is easy to see that $X \vee \Delta^1 \to (X, x)$ is a global weak equivalence.

The normalized chain complex corresponding to the simplicial presheaf with transfers $R_{\mathrm{tr}}(X \vee \Delta^1)$ is $R \xrightarrow{x} R_{\mathrm{tr}}(X)$ which is chain homotopic to the usual presheaf with transfers $R_{\mathrm{tr}}(X, x)$ in degree 0; see [MVW, p. 15].

In particular, for $\mathbb{G}_m = (\mathbb{A}^1 - \{0\}, 1)$ this gives a simplicial object V of $\mathbf{Norm}_+$ so that $R_{\mathrm{tr}}(V) = R_{\mathrm{tr}}(\mathbb{G}_m)$. Note that $R(1)[1] = C_* R_{\mathrm{tr}}(\mathbb{G}_m)$ by definition.

12.4 $\bar{\Delta}$-CLOSED CLASSES AND WEAK EQUIVALENCES

In this section we introduce Voevodsky's notion of a $\bar{\Delta}$-closed class [Voe10d], and use it to finish the proof that $\Delta^{\mathrm{op}}\mathrm{rad}(\mathcal{C})$ is a model category under the projective model structure of Definition 12.23, by showing that the morphisms in $J_{\mathcal{C}}^{\mathrm{rad}}$ are global weak equivalences (in Application 12.32); see Lemma 12.24.

We will also show that for any functor $\mathcal{C} \xrightarrow{f} \mathcal{C}'$, the inverse image f^* : $\Delta^{\mathrm{op}}\mathcal{C}^{\mathrm{ind}} \to \Delta^{\mathrm{op}}\mathcal{C}'^{\mathrm{ind}}$ will preserve global weak equivalences, as long as $\mathcal{C}$ and $\mathcal{C}'$ have finite coproducts. This has several useful consequences: in addition to the Application 12.32 just mentioned, it also implies that if X is a simplicial scheme then the canonical map from $\mathbf{L}R_{\mathrm{tr}}(X) = R_{\mathrm{tr}}(\mathbf{L}\mathrm{res}\,(X))$ to $R_{\mathrm{tr}}(X)$ is a global weak equivalence of simplicial presheaves with transfers; see 12.33.1. Third, it implies that the symmetric power functors (introduced in chapter 14) preserve global weak equivalences.

Definition 12.29 ($\bar{\Delta}$-closed classes). Let $\mathcal{E}$ be a category containing all small colimits, and E a class of morphisms in $\Delta^{\mathrm{op}}\mathcal{E}$. We say that E is a $\bar{\Delta}$-*closed class* if: (1) E contains all simplicial homotopy equivalences; (2) E has the 2-out-of-3 property (i.e., whenever $f \circ g$ exists and two out of f, g, and $f \circ g$ are in E then so is the third); (3) for each bisimplicial map $f : A_{\bullet\bullet} \to B_{\bullet\bullet}$ whose rows $f_j : A_{\bullet j} \to B_{\bullet j}$ are in E, the diagonal map $\mathrm{diag}(A) \to \mathrm{diag}(B)$ is in E; and (4) E is closed under finite coproducts and filtered colimits, i.e., if each $A_\alpha \to B_\alpha$ is in E for all α in a filtering diagram, then so is $\mathrm{colim}\, A_\alpha \to \mathrm{colim}\, B_\alpha$.

Let S be a class of morphisms in $\Delta^{\mathrm{op}}\mathrm{rad}(\mathcal{C})$ such that the Bousfield localization at S exists in the sense of Definition 12.38. Theorem 12.47 shows that the class of S-local equivalences is $\bar{\Delta}$-closed.

Remark 12.29.1. The intersection of $\bar{\Delta}$-closed classes in $\mathcal{E}$ is again a $\bar{\Delta}$-closed class. Thus each class of morphisms has a $\bar{\Delta}$-closure: the smallest $\bar{\Delta}$-closed class containing it. We will often use this observation, including in Theorems 12.31, 12.55, and 12.66.

Example 12.29.2. In the category of simplicial sets, the class W of weak homotopy equivalences is the smallest $\bar{\Delta}$-closed class in $\Delta^{\mathrm{op}}\mathbf{Sets}$. The fact that W is $\bar{\Delta}$-closed is well known. To see that every $\bar{\Delta}$-closed class E contains W, we use the fibrant replacement functor $K \mapsto K^{(\infty)} = \mathrm{colim}\, K^{(n)}$, obtained via the small object argument for the maps $\Lambda_k^n \to \Delta^n$.

Suppose that $K \to K'$ is a weak equivalence. Then $K^{(\infty)} \xrightarrow{\simeq} K'^{(\infty)}$ is a weak equivalence between Kan complexes; as such it is a simplicial homotopy equivalence, and hence in E. Now each $K^{(n)} \to K^{(n+1)}$ is in E because it is a pushout of a coproduct of the simplicial homotopy equivalences $\Lambda_k^n \to \Delta^n$ along maps $\Lambda_k^n \to K^{(n)}$, and pushouts preserve simplicial homotopy equivalences. By induction, each $K \to K^{(n)}$ is in E, and hence so is the colimit $K \to K^{(\infty)}$. The 2-out-of-3 property implies that $K \to K'$ is in E, as asserted.

Theorem 12.30. *Let* $\Phi : \mathcal{E} \to \mathcal{E}'$ *be a functor which commutes with filtered colimits, and* E' *a* $\bar{\Delta}$*-closed class of morphisms in* $\Delta^{\mathrm{op}}\mathcal{E}'$. *Let* E *denote the class of morphisms* η *in* $\Delta^{\mathrm{op}}\mathcal{E}$ *such that* $\Phi(\eta)$ *is in* E'. *Then* E *is* $\bar{\Delta}$*-closed.*

Proof. By assumption, E is closed under filtered colimits and has the 2-out-of-3 property. Since Φ preserves simplicial homotopies and diagonals of bisimplicial maps, the other conditions in Definition 12.29 are satisfied as well. $\square$

We now specialize to the case $\Delta^{\mathrm{op}}\mathcal{E} = \Delta^{\mathrm{op}}\mathcal{C}^{\mathrm{ind}}$, following [Voe10d, 3.16,21].

Theorem 12.31. *The global weak equivalences between objects of* $\Delta^{\mathrm{op}}(\mathcal{C}^{\mathrm{ind}})$ *form the smallest* $\bar{\Delta}$*-closed class of morphisms in* $\Delta^{\mathrm{op}}(\mathcal{C}^{\mathrm{ind}})$.

Proof. The class of global weak equivalences is $\bar{\Delta}$-closed in $\Delta^{\mathrm{op}}(\mathcal{C}^{\mathrm{ind}})$ because $X \to Y$ is a global weak equivalence exactly when each $X(U) \to Y(U)$ is a weak equivalence, and the class of weak equivalences is $\bar{\Delta}$-closed in $\Delta^{\mathrm{op}}\mathbf{Sets}$.

Conversely, we must show that if $f : X \to Y$ is a global weak equivalence of simplicial objects in $\mathcal{C}^{\mathrm{ind}}$, and E is any $\bar{\Delta}$-closed class of morphisms in $\Delta^{\mathrm{op}}\mathcal{C}^{\mathrm{ind}}$, then f is in E. If each X_n is representable, each $\mathbf{Lres}\,(X_n) \to X_n$ is a simplicial homotopy equivalence by Construction 12.19, and hence $\mathbf{Lres}\,(X) \to X$ is in E. In general, X is a filtered colimit of representable simplicial presheaves, by Construction 12.7, so (by Remark 12.19.1) $\mathbf{Lres}\,(X) \to X$ is in E. Form the diagram:

$$\begin{array}{ccc} \mathbf{Lres}\,(X) & \xrightarrow{\ \text{in } E\ } & X \\[4pt] {\scriptstyle \mathbf{Lres}\,(f)}\Big\downarrow {\scriptstyle \text{in } E} & & \Big\downarrow{\scriptstyle f} \\[4pt] \mathbf{Lres}\,(Y) & \xrightarrow{\ \text{in } E\ } & Y. \end{array} \qquad (12.31.1)$$

By the formula in 12.19, $\mathbf{Lres}\,(f) = \Delta(\mathbf{Lres}\,f_\bullet)$ is the diagonal of a bisimplicial map whose rows are coproducts of maps of the form $U_0 \otimes X(U_n) \to U_0 \otimes Y(U_n)$. Since each $X(U_n) \to Y(U_n)$ is a weak equivalence of simplicial sets, the rows are in E by Sublemma 12.31.2 and hence $\mathbf{Lres}\,(f)$ is in E. By the 2-out-of-3 property applied to diagram (12.31.1), f is in E. $\square$

Sublemma 12.31.2. *If* $K \to K'$ *is a weak equivalence of simplicial sets, then each* $U \otimes K \to U \otimes K'$ *is in* E, *the smallest* $\bar{\Delta}$*-closed class in* $\Delta^{\mathrm{op}}\mathcal{C}^{\mathrm{ind}}$.

Proof. We saw in Example 12.29.2 that $K^{(\infty)} \to K'^{(\infty)}$ is a simplicial homotopy equivalence. By functoriality, $U \otimes K^{(\infty)} \to U \otimes K'^{(\infty)}$ is a simplicial homotopy equivalence in $\Delta^{\mathrm{op}}\mathcal{C}^{\mathrm{ind}}$, and hence is in E.

$$\begin{array}{ccc} U \otimes K & \xrightarrow{\ \text{in } E\ } & U \otimes K^{(\infty)} \\[4pt] \Big\downarrow & & \Big\downarrow{\scriptstyle \text{s.h.e.}} \\[4pt] U \otimes K' & \xrightarrow{\ \text{in } E\ } & U \otimes K'^{(\infty)} \end{array}$$

Similarly, each $K^{(n)} \to K^{(n+1)}$ is the pushout of a coproduct of the simplicial homotopy equivalences $\Lambda_k^n \to \Delta^n$ (as we pointed out in loc. cit.). Therefore $U \otimes K^{(n)} \to U \otimes K^{(n+1)}$ is the pushout of a coproduct of the simplicial homotopy equivalences $U \otimes \Lambda_k^n \to U \otimes \Delta^n$; as such, these maps are in E. Taking the colimit over n, we see that $U \otimes K \to U \otimes K^{(\infty)}$ is in E, and similarly for $U \otimes K' \to U \otimes K'^{(\infty)}$. The 2-out-of-3 property implies that $U \otimes K \to U \otimes K'$ is in E. □

Application 12.32. The morphisms in $J_{\mathcal{C}}^{\mathrm{rad}}$ have the form $C \otimes \Lambda_k^n \to C \otimes \Delta^n$ with C in $\mathcal{C}$ (Example 12.11.1). By Sublemma 12.31.2 and Theorem 12.31, they are global weak equivalences. This fact was used in the proof of Lemma 12.24.

Now let $f : \mathcal{C} \to \mathcal{C}'$ be a functor. Since f^* is a left adjoint, it commutes with filtered colimits. Hence it sends $\mathcal{C}^{\mathrm{ind}}$ to $\mathcal{C}'^{\mathrm{ind}}$.

Corollary 12.33. *Let* $f : \mathcal{C} \to \mathcal{C}'$ *be a functor. Then the inverse image functor* $f^* : \Delta^{\mathrm{op}}(\mathcal{C}^{\mathrm{ind}}) \to \Delta^{\mathrm{op}}(\mathcal{C}'^{\mathrm{ind}})$ *takes global weak equivalences to global weak equivalences.*

Proof. ([Voe10d, Th.4.8]) Combine Theorems 12.31 and 12.30, with $\Phi = f^*$. If E is the class of maps η with $f^*\eta$ a global weak equivalence then E is $\bar{\Delta}$-closed, and hence E contains all global weak equivalences. □

Example 12.33.1. The functor $R_{\mathrm{tr}} : \Delta^{\mathrm{op}}\mathrm{Pshv}(\mathbf{Sm}_+) \to \Delta^{\mathrm{op}}\mathbf{PST}(\mathbf{Sm})$ is the inverse image functor associated to $\mathbf{Sm} \to \mathbf{Cor}(\mathbf{Sm}, R)$. Its derived functor $\mathbf{L}R_{\mathrm{tr}}$ sends f to $R_{\mathrm{tr}}(\mathbf{Lres}\,(f))$; see Example 12.7. Hence if $X_\bullet$ is a simplicial presheaf for which each X_n is an ind-object of $\mathbf{Sm}$, Corollary 12.33 applied to $\mathbf{Lres}\,(X_\bullet) \overset{\eta}{\to} X_\bullet$ implies that $\mathbf{L}R_{\mathrm{tr}}(X_\bullet) \to R_{\mathrm{tr}}(X_\bullet)$ is a global weak equivalence.

We conclude this section by describing the behavior of $\bar{\Delta}$-classes under various pushouts. This will be needed in section 12.6.

Definition 12.34. Suppose that $C \leftarrow A \to B$ is a diagram in $\Delta^{\mathrm{op}}\mathcal{E}$. By definition, the *homotopy pushout* is the (usual) pushout of the diagram

$$
\begin{array}{ccc}
A \amalg A & \xrightarrow{0,1} & A \otimes \Delta^1 \\
\downarrow & & \\
B \amalg C. & &
\end{array}
$$

By inspection, it is the diagonal of the bisimplicial object with rows $C \amalg A^{\amalg n} \amalg B$ obtained by formally adding degeneracies to $A \rightrightarrows B \amalg C$ (∂_0 and ∂_1 map A to B and C, respectively). (Cf. [BK72, XII.2].)

The mapping cylinder $cyl(f)$ of $f : A \to B$ is the special case of the homotopy pushout when $A = C$; it is also the pushout of $B \leftarrow A \to A \otimes \Delta^1$, and there are canonical maps $A \to B \amalg A \to cyl(f)$. It is a standard exercise, using the simplicial contraction of Δ^1, to show that $B \to cyl(f)$ is a simplicial homotopy equivalence.

Remark 12.34.1. It is a basic fact [BK72, XII.3.1(iv)] that the category of simplicial sets is left proper: the pushout $B \cup_A C$ of an injection $A \subset B$ of simplicial sets along any map $A \to C$ is weakly equivalent to the homotopy pushout.

The next lemma is based on [Voe10d, 2.10] and [Del09, p. 24].

Lemma 12.35. *Given a pushout square Q in $\Delta^{op}\mathcal{E}$*

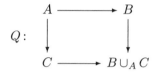

with each $A_n \to B_n$ a coprojection (i.e., $B_n = A_n \amalg B'_n$), the map from the homotopy pushout K_Q to $B \cup_A C$ is in every $\bar{\Delta}$-closed class in $\Delta^{op}\mathcal{E}$.

Proof. Let K_n denote the homotopy pushout of the square Q_n formed by the nth terms of Q. Since K_Q is the diagonal of the bisimplicial system with rows K_n, it suffices to show that each $K_n \to B_n \cup_{A_n} C_n$ is in E. Thus we may assume that $B = A \amalg B'$, so that $B \cup_A C = B' \amalg C$.

When $B = A \amalg B'$, the square Q is the coproduct of the two squares

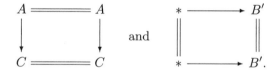

Thus the homotopy pushout K_Q is the coproduct of the mapping cylinders of $* \to B$ and $A \to C$. Thus $K \to B' \amalg C$ is in E. □

Proposition 12.36. *Let E be a $\bar{\Delta}$-closed set of morphisms in $\mathcal{E}$. Suppose that $Q \to Q'$ is a morphism of pushout squares in $\mathcal{E}$ of the form in Lemma 12.35, with each $A_n \to B_n$ and $A'_n \to B'_n$ a coprojection. If $A \to A'$, $B \to B'$, and $C \to C'$ are in E, then so is $B \cup_A C \to B' \cup_{A'} C'$.*

Proof. Let P and P' denote the homotopy pushouts of Q and Q'; by Lemma 12.35, $P \to B \cup_A C$ and $P' \to B' \cup_{A'} C'$ are in E. By the 2-out-of-3 property, it suffices to show that $P \to P'$ is in E. By Definition 12.34, P and P' are the diagonals of bisimplicial objects X and X' whose n^{th} rows are a coproduct of B, C and n copies of A, resp., of B', C' and n copies of A'. Since the rows of the bisimplicial map $X \to X'$ are in E, the diagonal map $P \to P'$ is also in E. □

Corollary 12.37. *Let E be a $\bar{\Delta}$-closed set of morphisms in $\mathcal{E}$. If $e : A \to B$ is a map in E, and each $A_n \to B_n$ is a coprojection, then E also contains the pushout $C \to B \cup_A C$ along any map $A \to C$ in $\mathcal{E}$.*

Proof. ([Voe10d, L. 2.13]) Apply Proposition 12.36 to the morphism of squares:

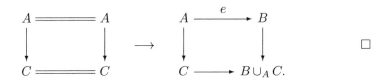

12.5   BOUSFIELD LOCALIZATION

In this section, we introduce the notion of Bousfield localization. The model structures on presheaves which are suitable for (Nisnevich) sheaf theory and $\mathbb{A}^1$-homotopy theory will be obtained by Bousfield localization of the projective model structure on presheaves.

Recall from [Hir03, 3.1] that if S is a class of maps in a model category $\mathcal{M}$ then a *left localization* of $\mathcal{M}$ with respect to S (if it exists) is a Quillen adjunction $(j, r) : \mathcal{M} \to \mathcal{M}_S$ such that j takes S to weak equivalences, and is universal with respect to this property.

Definition 12.38. ([Hir03, 3.1.4]) The *S-local objects* in $\mathcal{M}$ are the fibrant objects L such that $\mathrm{Map}(B, L) \to \mathrm{Map}(A, L)$ is a weak equivalence for every map $A \to B$ in S. The *S-local equivalences* are the maps $X \to Y$ in $\mathcal{M}$ such that $\mathrm{Map}(Y, L) \to \mathrm{Map}(X, L)$ is a weak equivalence for every S-local L.

The *left Bousfield localization* of $\mathcal{M}$ with respect to S (if it exists) is a left localization in which $\mathcal{M}$ and $\mathcal{M}_S$ have the same underlying category and the same class of cofibrations (the Quillen adjunction is the identity), the weak equivalences in $\mathcal{M}_S$ are the S-local equivalences in the sense of Definition 12.38, and the fibrations are determined by the right lifting property.

Since $r : \mathcal{M}_S \to \mathcal{M}$ preserves fibrations, every fibration in $\mathcal{M}_S$ is a fibration in $\mathcal{M}$, but not conversely. It turns out that the fibrant objects in $\mathcal{M}_S$ are exactly the S-local objects of $\mathcal{M}$ in the sense of Definition 12.38.

Remark 12.38.1. Bousfield localization preserves Quillen equivalences in the following sense; see [Hir03, 3.3.20]. Let (F, U) be a Quillen equivalence between model categories $\mathcal{M}$ and $\mathcal{N}$. If S is a class of maps in $\mathcal{M}$ and $T = F(S)$, then (F, U) is also a Quillen equivalence between $\mathcal{M}_S$ and $\mathcal{N}_T$.

Remark 12.38.2. By Ken Brown's Lemma ([Hov99, 1.1.12] or [Hir03, 7.7.2]), every S-local equivalence between S-local objects, i.e., between fibrant objects of $\mathcal{M}_S$, is also a weak equivalence in $\mathcal{M}$. Dually, every weak equivalence in $\mathcal{M}$ between cofibrant objects is an S-local equivalence in $\mathcal{M}_S$.

A proof of the following theorem is given in [Hir03, Thm. 4.1.1]. Recall that a model category is *left proper* if weak equivalences are preserved by pushouts along cofibrations. For example, $\Delta^{\mathrm{op}}\mathrm{Pshv}(\mathcal{C}_+)$ is left proper, because pushouts are constructed objectwise; see [Hir03, 13.1.14].

Theorem 12.39. *If $\mathcal{M}$ is a left proper cellular model category, then the left Bousfield localization exists for every set S, and is again a left proper cellular model category. If $\mathcal{M}$ is a simplicial model category, so is $\mathcal{M}_S$.*

Remark 12.39.1. Let $\mathcal{M}$ be as in Theorem 12.39. If $\mathcal{M}$ is skeletally small then the Bousfield localization of $\mathcal{M}$ exists with respect to any class of morphisms S. Indeed, there is a set S_0 contained in S such that every s in S is isomorphic to some s_0 in S_0, and $\mathcal{M}_S$ is the Bousfield localization of $\mathcal{M}$ with respect to S_0. This remark will be used without comment below, for example, when applied to the projective model category structure on $\Delta^{\mathrm{op}}\mathrm{Pshv}(\mathbf{Sm})$.

We also need some facts about the fibrant replacement functor in $\mathcal{M}_S$. To formulate them, we need to recall several definitions from Hirschhorn [Hir03].

Definition 12.40. Let λ be an ordinal, viewed as a category. By a λ-*sequence* in a category $\mathcal{M}$ we mean a functor X

$$X_0 \xrightarrow{i_0} X_1 \xrightarrow{i_1} \cdots \longrightarrow X_\alpha \xrightarrow{i_\alpha} \cdots, \qquad \alpha < \lambda,$$

from λ to $\mathcal{M}$ such that for each limit ordinal $\kappa < \lambda$ the induced map $\mathrm{colim}_{\alpha<\kappa} X_\alpha \to X_\kappa$ is an isomorphism. When $\lambda = \omega$, this is the usual notion of a sequence indexed by $\mathbb{N}$. The colimit of X (if it exists in $\mathcal{M}$) is sometimes called the *transfinite composition* of the maps i_α.

Given a set Λ of morphisms in a simplicial model category $\mathcal{M}$, a *relative Λ-cell complex* is a map that can be constructed as a transfinite composition of pushouts of maps in Λ.

Given a cofibrant object A and a cofibration $f : A \rightarrowtail B$, the nth *horn on* f is the cofibration Λ_f^n defined by

$$(A \otimes \Delta^n) \coprod_{A \otimes \partial \Delta^n} (B \otimes \partial \Delta^n) \xrightarrow{\Lambda_f^n} B \otimes \Delta^n.$$

If S is a set of cofibrations $f_s : A_s \rightarrowtail B_s$ in $\mathcal{M}$ with each A_s cofibrant, the set $\Lambda(S)$ of horns of the f_s (for all $n \geq 0$) is a *full set of horns on S* in the sense of [Hir03, 4.2.1].

The following theorem is proven by Hirschhorn, using the small object argument, in [Hir03, 4.2.5, 4.2.9, and 4.3.1]. The map $X \xrightarrow{\simeq} L_S X$ in 12.41(4) may be taken to be the fibrant replacement of X in the localized model structure $\mathcal{M}_S$.

Theorem 12.41. *Let $\mathcal{M}$ be a (simplicial) left proper cellular model category, with generating cofibrations I and generating trivial cofibrations J. We assume that every j in J is a relative I-cell complex with cofibrant domain.*

If S is a set of cofibrations with cofibrant domains, then:

1. *the set $\Lambda(S)$ of horns on S consists of S-local equivalences;*
2. *every relative $J \cup \Lambda(S)$-cell complex is both a cofibration and an S-local equivalence;*
3. *an object X of $\mathcal{M}$ is fibrant in $\mathcal{M}_S$ if and only if X is fibrant in $\mathcal{M}$ and has the right lifting property relative to $\Lambda(S)$;*
4. *for every X in $\mathcal{M}$ there is a natural map $X \to L_S X$ with $L_S X$ fibrant in $\mathcal{M}_S$. The map is a relative $J \cup \Lambda(S)$-cell complex.*

We will apply Theorem 12.41 when $\mathcal{M}$ is either $\Delta^{\mathrm{op}}\mathrm{Pshv}(\mathcal{C})$ or $\Delta^{\mathrm{op}}\mathrm{rad}(\mathcal{C})$.

Remark 12.41.1. The trivial cofibrations $J_{\mathcal{C}}$ and $J_{\mathcal{C}}^{\mathrm{rad}}$ satisfy the hypotheses of Theorem 12.41 by Example 12.11.1 and Lemma 12.24.

12.6   BOUSFIELD LOCALIZATION AND $\bar{\Delta}$-CLOSED CLASSES

Suppose that S is a class of morphisms in $\Delta^{\mathrm{op}}\mathrm{rad}(\mathcal{C})$ such that the Bousfield localization at S exists, e.g., if S is a set. The goal of this section is to prove that the S-local equivalences in $\Delta^{\mathrm{op}}\mathcal{C}^{\mathrm{ind}}$ form a $\bar{\Delta}$-closed class (Theorem 12.47).

Lemma 12.42. *Let E be a $\bar{\Delta}$-closed class in a model category $\mathcal{M} = \Delta^{\mathrm{op}}\mathcal{E}$. If $f : A \to B$ is a map in E, and each $A_n \to B_n$ is a coprojection (i.e., $B_n = A_n \amalg B'_n$), then the horns on f are also in E.*

Proof. Set $P = (A \otimes \Delta^n) \coprod_{A \otimes \partial \Delta^n} (B \otimes \partial \Delta^n)$. Now $A \otimes \partial \Delta^n \xrightarrow{\ e\ } B \otimes \partial \Delta^n$ is in E, since it is the diagonal of a bisimplicial map whose rows are finite coproducts of copies of f. Each term of e is a coprojection, since this is the case for the terms $A_n \to B_n$ of f. Corollary 12.37 implies that the pushout $A \otimes \Delta^n \to P$ of e is in E. Since the composition $A \otimes \Delta^n \to P \to B \otimes \Delta^n$ is in E, the nth horn $P \to B \otimes \Delta^n$ of f is in E by the 2-out-of-3 property. $\qquad\square$

If S is a set of maps $s : A_s \to B_s$ in $\mathcal{M}$, and $cyl(s)$ is the mapping cylinder of s defined in 12.34, let $cyl\,S$ denote the set of maps $A_s \to cyl(s)$ for s in S. From the universal property of left localization, the Bousfield localizations $\mathcal{M}_S$ and $\mathcal{M}_{cyl\,S}$ have the same cofibrations, weak equivalences, and fibrations. Their fibrant replacements $L_S X$ and $L_{cyl\,S} X$ are different, but weakly equivalent.

Proposition 12.43. *Let $\mathcal{M} = \Delta^{\mathrm{op}}\mathcal{E}$ be a (simplicial) left proper cellular model category, with generating cofibrations I and generating trivial cofibrations J which are relative I-cell complexes with cofibrant domains. We assume that for each map $T \to T'$ in J the maps $T_n \to T'_n$ are coprojections.*

If S is a set of cofibrations $A_s \xrightarrow{\ s\ } B_s$ in $\mathcal{M}$ with cofibrant domains, and each $A_s \amalg A_s \to A_s \otimes \Delta^1$ is a cofibration, then the fibrant replacement maps $X \to L_{cyl\,S} X$ for $\mathcal{M}_{cyl\,S}$ belong to every $\bar{\Delta}$-closed class E containing J and S.

Proof. Let E be the smallest $\bar{\Delta}$-closed set containing J and S. Since each $B_s \to cyl(s)$ is a simplicial homotopy equivalence, E also contains $cyl\,S$. By Lemma 12.42, the horns on $cyl\,S$ are also in E, as each $(A_s)_n \to cyl(s)_n$ is a coprojection. Each $A_s \to cyl(s)$ is a cofibration, as it is the composition of $A_s \rightarrowtail A_s \amalg B_s$ and the pushout of $A_s \amalg A_s \to A_s \otimes \Delta^1$. By Theorem 12.41(4), each $X \to L_{cyl\,S} X$ is a relative $J \cup \Lambda(cyl\,S)$-cell complex. Because E is closed under transfinite compositions, it suffices to show that E contains the pushout of elements of $J \cup \Lambda(cyl\,S)$ along any map. This follows from Corollary 12.37; we have already noted that the $A_s \to cyl(s)$ are termwise coprojections, and the elements of J are termwise coprojections by hypothesis. $\square$

Example 12.44. Let $\mathcal{M}$ be either $\Delta^{op}\mathrm{Pshv}(\mathcal{C})$ or $\Delta^{op}\mathrm{rad}(\mathcal{C})$. The trivial cofibrations $J_{\mathcal{C}}$ and $J_{\mathcal{C}}^{rad}$ of $\mathcal{M}$ satisfy the hypotheses of Proposition 12.43 by Example 12.11.1 and Lemma 12.24. If S is any set of cofibrations with cofibrant domains, then the fibrant replacement $X \to L_{cyl\,S} X$ is in the $\bar{\Delta}$-closure of $J \cup S$.

Lemma 12.45. *Let $\mathcal{M}$ be a model category. Suppose that f is a morphism of pushout squares with the X_i and X_i' cofibrant for $i = 1, 2, 3$:*

$$
\begin{array}{ccc}
X_1 & \longrightarrow & X_2 \\
{\scriptstyle g}\downarrow & & \downarrow \\
X_3 & \longrightarrow & X_4
\end{array}
\qquad \xrightarrow{\ f\ } \qquad
\begin{array}{ccc}
X_1' & \xrightarrow{\ g\ } & X_2' \\
{\scriptstyle g'}\downarrow & & \downarrow \\
X_3' & \longrightarrow & X_4'
\end{array}
$$

such that g and g' are cofibrations. If $f_i : X_i \to X_i'$ is a weak equivalence for $i = 1, 2, 3$ then f_4 is also a weak equivalence.

Proof. This is Proposition 15.10.10 in [Hir03]. $\square$

In order to prove that the class of S-local equivalences is $\bar{\Delta}$-closed, we need to introduce a few standard constructions.

Definition 12.46. If one forgets the degeneracies in a simplicial object, one obtains a *semi-simplicial object*, that is, a contravariant functor from the subcategory Δ_s of injections in Δ. In fact, forgetting the degeneracies is the direct image i_* associated with the inclusion $i : \Delta_s \hookrightarrow \Delta$; see [Wei94, 8.1.9].

As pointed out before Lemma 12.3 in Section 12.1, i_* has a left adjoint i^*, and $(i^* i_* K)_p$ is the coproduct of the K_j for $j \le p$, indexed by the surjections $[p] \to [j]$ in Δ. The functor $i^* i_*$ is sometimes called the *wrapping functor*, and the adjunction counit $i^* i_* K \to K$ is a weak equivalence; see [Voe10d, 3.22] or [Wei94, 8.4.4].

We note that $i^* i_* K$ is split simplicial by construction (with $N_n = K_n$). By Theorem 12.5, if each K_j is a coproduct of representable presheaves then $i^* i_* K$ is projectively cofibrant in $\mathcal{M} = \Delta^{\mathrm{op}} \mathcal{E}$.

Lemma 12.46.1. *The skeleta of $i^* i_* K$ fit into the pushout diagram:*

$$
\begin{array}{ccc}
K_n \otimes \partial \Delta^n & \longrightarrow & \mathrm{sk}_{n-1}(i^* i_* K) \\
\downarrow & & \downarrow {\scriptstyle i_n} \\
K_n \otimes \Delta^n & \longrightarrow & \mathrm{sk}_n(i^* i_* K).
\end{array}
$$

Proof. This is just Proposition VII.1.13 in [GJ99], applied to the split simplicial (or "degeneracy free") object $i^* i_* K$ with $A = \emptyset$ and $N_n = K_n$. $\qquad\square$

Theorem 12.47. *Let S be a class of morphisms in $\Delta^{\mathrm{op}} \mathrm{rad}(\mathcal{C})$ such that the Bousfield localization at S exists. Then the class of S-local equivalences in $\Delta^{\mathrm{op}} \mathrm{rad}(\mathcal{C})$ or $\Delta^{\mathrm{op}} \mathcal{C}^{\mathrm{ind}}$ is $\bar{\Delta}$-closed.*

Proof. We check first that S-local weak equivalences are closed under arbitrary filtered colimits. For each small filtered category I there is a model structure on I-diagrams for which the weak equivalences (resp., fibrant objects) are the maps $A \to B$ such that each $A_i \to B_i$ is an S-local equivalence (resp., is S-local). Given a weak equivalence $A \to B$ of I-diagrams, we need to show that $\mathrm{colim}_I A \to \mathrm{colim}_I B$ is an S-local equivalence.

Since global weak equivalences are closed under arbitrary filtered colimits, by Theorem 12.30 and Example 12.29.2, we may assume that the I-diagrams A and B are cofibrant in this model structure. For any S-local fibrant presheaf X, the I^{op}–diagrams $\mathrm{Map}(A, X)$ and $\mathrm{Map}(B, X)$ are fibrant diagrams of simplicial sets and hence the induced map

$$
\mathrm{Map}(\mathrm{colim}_I B, X) \cong \varprojlim_I \mathrm{Map}(B, X) \to \varprojlim_I \mathrm{Map}(A, X) \cong \mathrm{Map}(\mathrm{colim}_I A, X)
$$

is a weak equivalence of simplicial sets. By Definition 12.38, the map

$$
\mathrm{colim}_I A \to \mathrm{colim}_I B
$$

is an S-local equivalence of simplicial presheaves.

The only other nontrivial thing to check is that if $K \to K'$ is a bisimplicial map with each column $K_{p,\bullet} \to K'_{p,\bullet}$ an S-local equivalence, then $\mathrm{diag}(K) \to \mathrm{diag}(K')$ is an S-local equivalence. Applying the functorial cofibrant replacement **Lres** of Construction 12.19 to each column (see 12.25), we may assume that each column $K_{p,\bullet}$ and $K'_{p,\bullet}$ is cofibrant and that each $K_{p,q}$ and $K'_{p,q}$ is in $\mathcal{C}^{\mathrm{II}}$.

Let $L_{\bullet,\bullet}$ (resp., $L'_{\bullet,\bullet}$) be the bisimplicial objects obtained by applying the wrapping functor 12.46 to each *row* $K_{\bullet,q}$ (resp., $K'_{\bullet,q}$). Each row $L_{\bullet,q}$ and $L'_{\bullet,q}$ is cofibrant. In addition, each of the columns of L (and L') is cofibrant, because the pth column $L_{p,\bullet}$ is a coproduct of the cofibrant $K_{j,\bullet}$ for $j \leq p$. Since each $L_{p,\bullet} \to K_{p,\bullet}$ is a global weak equivalence, each $L_{p,\bullet} \to L'_{p,\bullet}$ is an S-local equivalence. As global weak equivalences are $\bar{\Delta}$-closed, the maps $\mathrm{diag}(L) \to \mathrm{diag}(K)$ and $\mathrm{diag}(L') \to \mathrm{diag}(K')$ are global weak equivalences. Therefore it suffices to show that $\mathrm{diag}(L) \to \mathrm{diag}(L')$ is an S-local equivalence.

Let $\mathrm{sk}_n L$ denote the bisimplicial object whose q^{th} row is the n-skeleton $\mathrm{sk}_n L_{\bullet,q}$. We will use induction on n to show that each $\mathrm{diag}(\mathrm{sk}_n L)$ and each $\mathrm{diag}(\mathrm{sk}_n L')$ is cofibrant, and that each $\mathrm{diag}(\mathrm{sk}_n L) \to \mathrm{diag}(\mathrm{sk}_n L')$ is an S-local equivalence. Since $\mathrm{diag}(L)$ is the colimit of the $\mathrm{diag}(\mathrm{sk}_n L)$ as $n \to \infty$, and similarly for $\mathrm{diag}(L')$, Proposition 17.9.1 of [Hir03] implies that $\mathrm{diag}(L) \to \mathrm{diag}(L')$ is an S-local equivalence between cofibrant objects, as required.

For $n = 0$, the qth row of $\mathrm{sk}_0 L$ is the constant presheaf $L_{0,q} = K_{0,q}$, so $\mathrm{diag}(\mathrm{sk}_0 L)$ is the first column $L_{0,\bullet}$ of L; this is cofibrant by construction, and $\mathrm{diag}(\mathrm{sk}_0 L) \to \mathrm{diag}(\mathrm{sk}_0 L')$ is an S-local equivalence.

For $n > 0$, Lemma 12.46.1 yields pushout squares of simplicial objects:

$$
\begin{array}{ccc}
K_{n,q} \otimes \partial\Delta^n & \longrightarrow & \mathrm{sk}_{n-1}(L_{\bullet,q}) \\
\downarrow & & \downarrow{\scriptstyle i_n} \\
K_{n,q} \otimes \Delta^n & \longrightarrow & \mathrm{sk}_n(L_{\bullet,q})
\end{array}
\qquad
\begin{array}{l}
\text{for all } q \\[1em]
\text{and hence}
\end{array}
\qquad
\begin{array}{ccc}
K_{n,\bullet} \otimes \partial\Delta^n & \longrightarrow & \mathrm{diag}(\mathrm{sk}_{n-1} L) \\
\downarrow & & \downarrow{\scriptstyle i_n} \\
K_{n,\bullet} \otimes \Delta^n & \longrightarrow & \mathrm{diag}(\mathrm{sk}_n L).
\end{array}
$$

Since the left verticals are cofibrations, so are the right verticals i_n. Since $\mathrm{diag}(\mathrm{sk}_{n-1} L)$ is cofibrant by induction, it follows that $\mathrm{diag}(\mathrm{sk}_n L)$ is cofibrant. We have a morphism between the pushout squares of Lemma 12.46.1:

$$
\begin{array}{ccc}
K_{n,\bullet} \otimes \partial\Delta^n & \longrightarrow & \mathrm{diag}(\mathrm{sk}_{n-1} L) \\
\downarrow & & \downarrow{\scriptstyle i_n} \\
K_{n,\bullet} \otimes \Delta^n & \longrightarrow & \mathrm{diag}(\mathrm{sk}_n L)
\end{array}
\quad \longrightarrow \quad
\begin{array}{ccc}
K'_{n,\bullet} \otimes \partial\Delta^n & \longrightarrow & \mathrm{diag}(\mathrm{sk}_{n-1} L') \\
\downarrow & & \downarrow{\scriptstyle i'_n} \\
K'_{n,\bullet} \otimes \Delta^n & \longrightarrow & \mathrm{diag}(\mathrm{sk}_n L').
\end{array}
$$

As the $K_{n,\bullet} \to K'_{n,\bullet}$ are S-local equivalences, so are the maps $K_{n,\bullet} \otimes \Delta^n \to K'_{n,\bullet} \otimes \Delta^n$ and $K_{n,\bullet} \otimes \partial\Delta^n \to K'_{n,\bullet} \otimes \partial\Delta^n$. By Lemma 12.45 and induction on n, the maps $\mathrm{diag}(\mathrm{sk}_n L) \to \mathrm{diag}(\mathrm{sk}_n L')$ are S-local equivalences. $\square$

12.7 NISNEVICH-LOCAL PROJECTIVE MODEL STRUCTURE

In this section, we introduce a Bousfield localization of the projective model structure on presheaves which is suitable for (Nisnevich) sheaf theory. Recall that we are restricting our attention to the case where $\mathcal{C}$ is **Sch**, **Sm**, or **Norm**, and $\mathcal{C}_+$ is the category of varieties with a disjoint basepoint.

When $\mathcal{C}$ is **Sch**, **Sm**, or **Norm**, the Nisnevich topology on $\mathcal{C}$ and $\mathcal{C}_+$ may be defined in terms of "upper distinguished" squares; see [MV99, 3.1.3] or [MVW, 12.5]. By definition (see loc. cit.), a commutative square Q of the form

$$Q: \quad \begin{array}{ccc} U' & \longrightarrow & V' \\ \downarrow & & \downarrow f \\ U & \xrightarrow{\ i\ } & V \end{array} \qquad\qquad (12.48)$$

is called *upper distinguished* if i is an open immersion, $U' = U \times_V V'$, f is étale, and $(V' - U')_{\mathrm{red}} \to (V - U)_{\mathrm{red}}$ is an isomorphism.

The Nisnevich topology on $\mathcal{C}$ is the smallest topology such that $\{U, V'\}$ is a covering of V for each upper distinguished square (12.48). A presheaf F on $\mathcal{C}$ is a Nisnevich sheaf if and only if F takes upper distinguished squares to pullback squares; see [MV99, 3.1.4] or [MVW, 12.7].

Definition 12.49. A morphism $X \to Y$ in $\Delta^{\mathrm{op}}\mathrm{Pshv}(\mathcal{C}_+)$ (resp., in $\Delta^{\mathrm{op}}\mathrm{rad}(\mathcal{C}_+)$ or $\Delta^{\mathrm{op}}\mathbf{PST}(\mathcal{C})$) is called a *Nisnevich-local equivalence* if it induces an isomorphism on the Nisnevich sheaves of homotopy groups on each scheme U in $\mathcal{C}$, for any choice of basepoint $x \in X(U)$.

The *Nisnevich-local* model structure on $\Delta^{\mathrm{op}}\mathrm{Pshv}(\mathcal{C}_+)$ is defined to be the Bousfield localization of the projective model structure of Definition 12.1 at the class S of Nisnevich-local equivalences: the cofibrations are the projective cofibrations, and the fibrations are determined by the right lifting property. That this model structure exists was proved by Blander in [Bla01]; the Nisnevich-local equivalences are the same as the S-local equivalences by Lemma 12.50.

When $\mathcal{C}$ is **Sch**, **Sm**, or **Norm**, the Bousfield localizations of the projective model structures on $\Delta^{\mathrm{op}}\mathrm{rad}(\mathcal{C}_+)$ and $\Delta^{\mathrm{op}}\mathbf{PST}(\mathcal{C})$ at the classes of Nisnevich-local equivalences are also called Nisnevich-local model structures.

We will write $\Delta^{\mathrm{op}}\mathrm{Pshv}(\mathcal{C}_+)_{\mathrm{nis}}$, etc., for these model structures. We will sometimes refer to them as Nisnevich-local *projective* model structures to distinguish them from their corresponding injective model structures, which are discussed in Lemma 12.58.

The idea of using Bousfield localization to define local model structures is due to Blander, who proved the following result in [Bla01, Thm. 1.5, Lemma 4.1]. Although Blander's results are stated for $\Delta^{\mathrm{op}}\mathrm{Pshv}(\mathcal{C}_+)$, his proofs also work for the model structures on radditive presheaves and presheaves with transfers on $\mathcal{C}$.

Lemma 12.50. *In the Nisnevich-local model structures on* $\Delta^{\mathrm{op}}\mathrm{Pshv}(\mathcal{C}_+)$, $\Delta^{\mathrm{op}}\mathrm{rad}(\mathcal{C}_+)$, *and* $\Delta^{\mathrm{op}}\mathbf{PST}(\mathcal{C})$, *the weak equivalences are exactly the Nisnevich-local equivalences.*

Moreover, a simplicial presheaf F is Nisnevich-local in the sense of Bousfield localization (Definition 12.38) if and only if (i) $F \to *$ is a projective fibration and (ii) F converts upper distinguished squares to homotopy pullback squares.

Remark 12.50.1. Blander actually considered the Bousfield localization for any essentially small Grothendieck site on $\mathcal{C}$, i.e., localization at the class S of maps inducing an isomorphism on the sheaves of homotopy groups, and proved in [Bla01, 1.5] that S is always the class of S-local equivalences.

Remark 12.50.2. If there were only a set of Nisnevich-local equivalences, we could also conclude that the model structures exist by Theorem 12.39, as the projective model structures are left proper (see Remark 12.24.2 and Example 12.24.1) and cellular (by 12.11.1). Thus if we restrict to presheaves X with the cardinality of all $X(U)$ uniformly bounded by an appropriate cardinal, we would also get Nisnevich-local model structures.

By naturality of Bousfield localization [Hir03, 3.3.20], there are Quillen adjunctions:

$$\Delta^{\mathrm{op}}\mathrm{Pshv}(\mathcal{C}_+)_{\mathrm{nis}} \xrightarrow{(-^{\mathrm{rad}},\, \mathrm{incl})} \Delta^{\mathrm{op}}\mathrm{rad}(\mathcal{C}_+)_{\mathrm{nis}} \xrightarrow{(R_{\mathrm{tr}},\, u)} \Delta^{\mathrm{op}}\mathbf{PST}(\mathcal{C})_{\mathrm{nis}};$$

$$\Delta^{\mathrm{op}}\mathrm{Pshv}(\mathbf{Sm}_+)_{\mathrm{nis}} \xrightarrow{(i^*,\, i_*)} \Delta^{\mathrm{op}}\mathrm{Pshv}(\mathbf{Norm}_+)_{\mathrm{nis}};$$

$$\Delta^{\mathrm{op}}\mathbf{PST}(\mathbf{Sm}_+)_{\mathrm{nis}} \xrightarrow{(i^*,\, i_*)} \Delta^{\mathrm{op}}\mathbf{PST}(\mathbf{Norm}_+)_{\mathrm{nis}}.$$

It is easy to see that the squares in (12.28.1) also form a commutative diagram of Nisnevich-local model categories and Quillen adjunctions.

In section 12.8, we will need the following comparison between the Nisnevich-local model structures on $\Delta^{\mathrm{op}}\mathrm{Pshv}(\mathcal{C}_+)$ and $\Delta^{\mathrm{op}}\mathrm{rad}(\mathcal{C}_+)$.

Recall from Definition 12.38 that a presheaf L is Nisnevich-local if $\mathrm{Hom}(f, L)$ is a projective weak equivalence for every Nisnevich-local map f.

Corollary 12.51. *A morphism $X \xrightarrow{f} Y$ in $\Delta^{\mathrm{op}}\mathrm{rad}(\mathcal{C}_+)$ is a Nisnevich-local fibration (resp., Nisnevich-local equivalence) if and only if it is a Nisnevich-local fibration (resp., Nisnevich-local equivalence) in $\Delta^{\mathrm{op}}\mathrm{Pshv}(\mathcal{C}_+)$.*

A presheaf X in $\Delta^{\mathrm{op}}\mathrm{rad}(\mathcal{C}_+)$ is Nisnevich-local if and only if it is Nisnevich-local in $\Delta^{\mathrm{op}}\mathrm{Pshv}(\mathcal{C}_+)$.

Proof. It suffices to establish the result for (Nisnevich-)local fibrations, since the other assertions follow formally from Lemma 12.50. One direction is easy: if f is a local fibration in $\Delta^{\mathrm{op}}\mathrm{rad}(\mathcal{C}_+)$, then f is a local fibration in $\Delta^{\mathrm{op}}\mathrm{Pshv}(\mathcal{C}_+)$ by the Quillen adjunction.

Conversely, suppose that $f : X \to Y$ is a Nisnevich-local fibration in $\Delta^{\mathrm{op}}\mathrm{Pshv}(\mathcal{C}_+)$, i.e., f has the right lifting property for trivial cofibrations in $\Delta^{\mathrm{op}}\mathrm{Pshv}(\mathcal{C}_+)$. Let $j : A \xrightarrow{\simeq} B$ be a trivial cofibration in $\Delta^{\mathrm{op}}\mathrm{rad}(\mathcal{C}_+)$. We can factor j in $\Delta^{\mathrm{op}}\mathrm{Pshv}(\mathcal{C}_+)$ as $A \xrightarrow{\simeq} B' \xrightarrow{\simeq} B$, where the first map is a trivial cofibration and the second map is a trivial fibration; note that B' is not radditive. By Lemma 12.26(1), we can also factor j as the composition of a termwise nice

map $A \rightarrowtail \widetilde{B}$, the map $\widetilde{B} \rightarrow \widetilde{B}^{\mathrm{rad}}$, and a retraction $\widetilde{B}^{\mathrm{rad}} \rightarrow B$ of $B \xrightarrow{\iota} \widetilde{B}^{\mathrm{rad}}$. Given
the solid square below on the left, the map $B' \rightarrow B \rightarrow Y$ has a lift $\lambda : B' \rightarrow X$
by the lifting property for f relative to $A \xrightarrow{\simeq} B'$. Thus we may form the
right-hand diagram below, in which the top composite is a. Because $B' \twoheadrightarrow B$ is
a trivial local fibration in $\Delta^{\mathrm{op}}\mathrm{Pshv}(\mathcal{C}_+)$, there is a map $\beta : \widetilde{B} \rightarrow B'$ factoring the
first square on the right; the evident map $\lambda\beta$ factors the outer square on the
right.

Factoring $\lambda\beta$ as $\widetilde{B} \rightarrow \widetilde{B}^{\mathrm{rad}} \xrightarrow{\gamma} X$, the composite $B \xrightarrow{\iota} \widetilde{B}^{\mathrm{rad}} \xrightarrow{\gamma} X$ factors
the large square on the left. Thus f has the right lifting property for trivial
cofibrations in $\Delta^{\mathrm{op}}\mathrm{rad}(\mathcal{C}_+)$, i.e., f is a local fibration in $\Delta^{\mathrm{op}}\mathrm{rad}(\mathcal{C}_+)$. $\square$

Our next goal is to show (in Proposition 12.54) that the Nisnevich-local
model structure on $\Delta^{\mathrm{op}}\mathrm{Pshv}(\mathcal{C})$ is actually the Bousfield localization at a class
S of maps, indexed by the upper distinguished squares. The S-localization is
designed to turn upper distinguished squares into homotopy cartesian squares.

Lemma 12.52. *For each diagram $B \leftarrow A \rightarrow C$ in $\mathcal{C}$, the homotopy pushout K
is cofibrant in $\Delta^{\mathrm{op}}\mathrm{Pshv}(\mathcal{C})$, and $K \amalg K \rightarrow K \boxtimes \Delta^1$ is a cofibration.*

Proof. Recall that K is the pushout of the diagram $B \amalg C \leftarrow A \amalg A \rightarrow A \boxtimes \Delta^1$.
As pointed out in Example 12.11.1, $A \amalg A \rightarrow A \boxtimes \Delta^1$ is one of the generating
cofibrations in the projective model structure on $\Delta^{\mathrm{op}}\mathrm{rad}(\mathcal{C})$, so the pushout
$B \amalg C \rightarrow K$ is a cofibration. Since $B \amalg C$ is cofibrant (by Examples 12.4 and
12.14.1), K is cofibrant.

Similarly, $(A \boxtimes \Delta^1) \boxtimes \partial\Delta^1 \rightarrow (A \boxtimes \Delta^1) \boxtimes \Delta^1$ is a cofibration, and $K \amalg K \rightarrow$
$K \boxtimes \Delta^1$ is the pushout of this along $A \boxtimes \Delta^1 \rightarrow K$, so it is a cofibration. $\square$

Lemma 12.53. *Given a diagram $B \leftarrow A \rightarrow C$ of representable presheaves with
homotopy pushout K, and a globally fibrant presheaf F, the simplicial set
$\mathrm{Map}(K, F)$ is the homotopy pullback of the diagram $F(B) \leftarrow F(A) \rightarrow F(C)$.*

Proof. For each n, the homotopy pushout of $B \boxtimes \Delta^n \leftarrow A \boxtimes \Delta^n \rightarrow C \boxtimes \Delta^n$ is
$K \boxtimes \Delta^n$. Since $\mathrm{Map}(K, F)_n = \mathrm{Hom}(K \boxtimes \Delta^n, F)$, this implies that $\mathrm{Map}(K, F)$ is
the presheaf pullback of the diagram

$$\mathrm{Map}(A \times \Delta^1, F) \rightarrow \mathrm{Map}(A \amalg A, F) \leftarrow \mathrm{Map}(B \amalg C, F).$$

Here $A \amalg A$ and $B \amalg C$ are coproducts of presheaves. Since $F(A) = \mathrm{Map}(A, F)$ by Lemma 12.2, and similarly for $F(B)$, $F(C)$,

$$\mathrm{Map}(B \amalg C, F) = \mathrm{Map}(B, F) \times \mathrm{Map}(C, F) \cong F(B) \times F(C)$$

and similarly for $\mathrm{Map}(A \amalg A, F)$; see [Hir03, 9.2.3]. Since $\mathrm{Map}(A \boxtimes \Delta^1, F) \cong \mathrm{Map}_{\Delta^{\mathrm{op}}\mathbf{Sets}}(\Delta^1, F(A))$ by Lemma 12.2, we have the pullback diagram

$$
\begin{array}{ccc}
\mathrm{Map}(K, F) & \longrightarrow & \mathrm{Map}(\Delta^1, F(A)) \\
\downarrow & & \downarrow{\scriptstyle p} \\
F(B) \times F(C) & \longrightarrow & F(A) \times F(A).
\end{array}
$$

Since $F(A)$, $F(B)$, and $F(C)$ are fibrant, and p is a fibration, $\mathrm{Map}(K, F)$ is the homotopy pullback of $F(B) \leftarrow F(A) \rightarrow F(C)$; see [Hir03, 13.3.2]. □

For each upper distinguished square Q as in (12.48), with homotopy pushout K_Q, there is a canonical map $s_Q : K_Q \rightarrow V$.

Proposition 12.54. *The Bousfield localization of* $\Delta^{\mathrm{op}}\mathrm{Pshv}(\mathcal{C})$ *or* $\Delta^{\mathrm{op}}\mathrm{rad}(\mathcal{C})$ *at* $S = \{K_Q \rightarrow V\}$ *is the Nisnevich-local model structure.*

Proof. We give the proof for $\mathrm{Pshv}(\mathcal{C})$; the proof for $\mathrm{rad}(\mathcal{C})$ is similar. As with any Bousfield localization, the S-local equivalences are determined by the S-local objects (that is, the fibrant objects); see 12.38. Thus it suffices to show that an object is S-local if and only if it is Nisnevich-local. Since the localization $\Delta^{\mathrm{op}}\mathrm{Pshv}(\mathcal{C})_S \rightarrow \Delta^{\mathrm{op}}\mathrm{Pshv}(\mathcal{C})_{\mathrm{nis}}$ is a left Quillen functor, its adjoint preserves fibrant objects: every Nisnevich-local object is an S-local object.

Conversely, suppose that F is S-local; we need to show that it is Nisnevich-local. By Lemma 12.50, it suffices to show that it converts upper distinguished squares into homotopy pullback squares. Given an upper distinguished square Q as in (12.48), with homotopy pushout K_Q, the square

$$
\begin{array}{ccc}
\mathrm{Map}(K_Q, F) & \longrightarrow & F(V') \\
\downarrow & & \downarrow \\
F(U) & \longrightarrow & F(U')
\end{array}
$$

is a homotopy pullback square by Lemma 12.53, as F is globally fibrant by Definition 12.38. Since F is S-local and $K_Q \rightarrow V$ is in S, the map from $F(V) \cong \mathrm{Map}(V, F)$ to $\mathrm{Map}(K_Q, F)$ is a homotopy equivalence of simplicial sets (see 12.38). It follows that $F(Q)$ is a homotopy pullback square, as required. □

Let S denote the class of maps $s_Q : K_Q \rightarrow V$, as Q runs over the upper distinguished squares in $\mathcal{C}$, and let T denote the class of maps $t_Q : K_Q \rightarrow cyl(s_Q)$

(T is the class $cyl\,S$ of Proposition 12.43). Since each map s_Q is a Nisnevich-local equivalence, and each $V \to cyl(s_Q)$ is a simplicial homotopy equivalence, each t_Q is also a Nisnevich-local equivalence, and it is clear that the Bousfield localizations of $\Delta^{\mathrm{op}}\mathrm{rad}(\mathcal{C})$ at S and T are the same.

In fact, each $t_Q : K_Q \to cyl(s_Q)$ in T is a cofibration with a cofibrant domain. To see this, fix Q and recall from Lemma 12.52 that K_Q is cofibrant and that $K_Q \amalg K_Q \to K_Q \boxtimes \Delta^1$ is a cofibration. The pushout of this cofibration along $K_Q \amalg K_Q \to K_Q \amalg V$ is also a cofibration, namely, $t_Q \amalg v : K_Q \amalg V \to cyl(s_Q)$. Since V is cofibrant, $t_Q : K_Q \to cyl(s_Q)$ is a cofibration.

If X is a simplicial object in $\mathcal{C}^{\mathrm{ind}}$ then so is $L_{cyl\,T}X$, because $L_{cyl\,T}X$ is a relative cell complex for $J_{\mathcal{C}}^{\mathrm{rad}} \cup \Lambda(cyl\,T)$; $L_{cyl\,T}X$ is a transfinite composition of pushouts along maps in $J_{\mathcal{C}}^{\mathrm{rad}}$, which are in $\mathcal{C}^{\mathrm{ind}}$ by 12.32. See Theorem 12.41(4).

Remark 12.54.1. Given Proposition 12.54, Theorem 12.39 implies that these Nisnevich-local model structures are left proper and cellular. This was also proven by Blander (Lemmas 1.1 and 1.3 in [Bla01]).

Theorem 12.55. *Let S denote the class of morphisms $s_Q : K_Q \to V$, as Q runs over all upper distinguished squares (12.48). Then the class of Nisnevich-local equivalences in $\Delta^{\mathrm{op}}\mathcal{C}^{\mathrm{ind}}$ is the smallest $\bar{\Delta}$-closed class of morphisms containing S.*

Proof. ([Voe10d, 3.51]) Let E denote the smallest $\bar{\Delta}$-closed class of morphisms in $\Delta^{\mathrm{op}}\mathcal{C}^{\mathrm{ind}}$ containing S. By Theorem 12.47, the class of S-local equivalences is $\bar{\Delta}$-closed. Thus we have

$$E \subseteq \{S\text{–local equivalences}\} = \{\text{Nisnevich–local equivalences}\}.$$

(The equality is by Proposition 12.54.) We need to show that every S-local equivalence $f : X \to Y$ is in E. For each f, consider the following diagram.

$$
\begin{array}{ccc}
X & \xrightarrow{\;\;f\;\;} & Y \\
{\scriptstyle \text{in } E}\big\downarrow & & \big\downarrow{\scriptstyle \text{in } E} \\
L_{cyl\,T}(X) & \xrightarrow{\;\text{in } E\;} & L_{cyl\,T}(Y)
\end{array}
$$

By Proposition 12.43 and Example 12.44, the fibrant replacement maps $X \to L_{cyl\,T}X$ are in E. By the remarks before this theorem, E also contains the class T of maps $K_Q \to cyl(s_Q)$. By the 2-out-of-3 property, $L_{cyl\,T}(f) : L_{cyl\,T}(X) \to L_{cyl\,T}(Y)$ is an S-local equivalence between S-local fibrant objects. By Ken Brown's Lemma 12.38.2, $L_{cyl\,T}(f)$ is a global weak equivalence; by Theorem 12.31, it belongs to E. Again by the 2-out-of-3 property, f is in E. $\square$

Application 12.55.1. The functor $i^* : \Delta^{\mathrm{op}}\mathbf{Sm}^{\mathrm{ind}} \subset \Delta^{\mathrm{op}}\mathbf{Norm}^{\mathrm{ind}}$ preserves global weak equivalences, by Theorem 12.33. It follows from Theorem 12.55 that i^* also takes Nisnevich-local equivalences to Nisnevich-local equivalences.

We will see another application of Theorem 12.55 in Section 14.3, that the symmetric powers functors S^G preserve Nisnevich-local equivalences.

12.8 MODEL CATEGORIES OF SHEAVES

In this section, we compare our Nisnevich-local model structure on presheaves on $\mathcal{C}$ to the Morel–Voevodsky model structure on sheaves introduced in [MV99], when $\mathcal{C}$ is **Sch**, **Sm**, or **Norm**. The intermediary is a projective model structure on Nisnevich sheaves, due to Blander, which we now recall.

By [Bla01, 2.1] there is a proper simplicial cellular projective model structure on the category $\Delta^{\mathrm{op}}\mathrm{Sheaves}(\mathcal{C})$ of simplicial sheaves on $\mathcal{C}$. The weak equivalences and fibrations in this category are the Nisnevich-local equivalences and fibrations of the underlying simplicial presheaves. The usual sheafification adjunction (a^*, a_*) is a Quillen adjunction for both $\Delta^{\mathrm{op}}\mathrm{Pshv}(\mathcal{C}) \to \Delta^{\mathrm{op}}\mathrm{Sheaves}(\mathcal{C})$ and $\Delta^{\mathrm{op}}\mathrm{Pshv}(\mathcal{C})_\bullet \to \Delta^{\mathrm{op}}\mathrm{Sheaves}(\mathcal{C})_\bullet$ because the forgetful functor a_* preserves fibrations and weak equivalences.

Theorem 12.56. *We have Quillen equivalences for both the pointed and unpointed Nisnevich-local projective model categories:*

$$\Delta^{\mathrm{op}}\mathrm{Pshv}(\mathcal{C})_{\mathrm{nis}} \xrightarrow{\sim} \Delta^{\mathrm{op}}\mathrm{rad}(\mathcal{C})_{\mathrm{nis}} \xrightarrow{\sim} \Delta^{\mathrm{op}}\mathit{Sheaves}(\mathcal{C}).$$

Proof. Blander observed in [Bla01, 2.2] that sheafification (a^*, a_*) is a Quillen equivalence $\Delta^{\mathrm{op}}\mathrm{Pshv}(\mathcal{C}) \to \Delta^{\mathrm{op}}\mathrm{Sheaves}(\mathcal{C})$ for the Nisnevich-local projective model structure, because for any simplicial presheaf F the map $F \to a_*a^*F$ is a Nisnevich-local equivalence. This observation also holds in the pointed setting.

Example 12.14.3 shows that for any simplicial presheaf A the map $A \to A^{\mathrm{rad}}$ induces an isomorphism of Nisnevich sheaves. Using Corollary 12.51, it follows that for any simplicial radditive presheaf B, a map $A \to B$ is a local equivalence in $\Delta^{\mathrm{op}}\mathrm{Pshv}(\mathcal{C})$ if and only if the map $A^{\mathrm{rad}} \to B$ is a local equivalence in $\Delta^{\mathrm{op}}\mathrm{rad}(\mathcal{C})$. By definition (see 12.0), the Nisnevich-local Quillen adjunction $(-^{\mathrm{rad}}, \mathrm{incl})$ is a Quillen equivalence. $\qquad\square$

The inclusion $i : \mathbf{Sm} \subset \mathbf{Norm}$ allows us to compare homotopy categories. The following lemma is taken from [Voe10c, 2.43].

Lemma 12.57. *The total direct image* $\mathbf{L}i^*$ *embeds the Nisnevich-local homotopy category of* $\Delta^{\mathrm{op}}\mathrm{Pshv}(\mathbf{Sm})_{\mathrm{nis}}$ *into the Nisnevich-local homotopy category of* $\Delta^{\mathrm{op}}\mathrm{Pshv}(\mathbf{Norm})_{\mathrm{nis}}$ *as a full (coreflective) subcategory.*

Proof. By Lemma 12.3(1), $i^* : \Delta^{\mathrm{op}}\mathrm{Pshv}(\mathbf{Sm}) \to \Delta^{\mathrm{op}}\mathrm{Pshv}(\mathbf{Norm})$ is a full embedding as a coreflective subcategory (i.e., $X \cong i_* i^* X$ for all X). Because i_* preserves Nisnevich-local equivalences, the proof of Lemma 12.3 goes through to show that (i^*, i_*) is a Quillen adjunction between the corresponding Nisnevich-local model structures, and that $i_* \cong \mathbf{R}i_*$. As in the proof of Lemma 12.3(2), there is an induced adjunction $(\mathbf{L}i^*, \mathbf{R}i_*)$ between the homotopy categories, and $\mathbf{L}i^*$ is a full embedding as a coreflective subcategory: each $X \to i_*\mathbf{L}i^* X \cong i_* i^* X$ is an isomorphism in the homotopy category of $\Delta^{\mathrm{op}}\mathrm{Pshv}(\mathbf{Sm})_{\mathrm{nis}}$. $\square$

Injective model structures.   There is a different model structure on the category of simplicial sheaves, called the local *injective* model structure. Although the weak equivalences are still the Nisnevich-local equivalences of Definition 12.49, the cofibrations are the monomorphisms, and the fibrations are determined by the right lifting property. The injective model structure is due to Joyal (see [Jar86]) and is used by Morel and Voevodsky in [MV99].

Lemma 12.58. *The identity map is a Quillen equivalence, from the Nisnevich-local projective model structure on simplicial sheaves to the Nisnevich-local injective model structure on simplicial sheaves.*

Proof. The identity is a Quillen functor because every projective cofibration is a monomorphism and hence a cofibration for the Nisnevich-local injective structure (see 12.0). Because the homotopy categories of the two model structures are both the localization of $\Delta^{\mathrm{op}}\mathrm{Sheaves}(\mathcal{C})$ at the class of Nisnevich-local equivalences, they are canonically isomorphic. $\square$

Remark 12.58.1. (Blander) The Nisnevich-local injective model structure on simplicial sheaves has more cofibrations than the Nisnevich-local projective model structure. For example, if $U \to X$ is an open inclusion then the map of representable sheaves is a cofibration in the injective model structure (by definition), but not always a cofibration in the projective model structure, because it does not have the left lifting property with respect to a Kan fibration $F \to *$ unless $F(X) \to F(U)$ is onto.

We now turn to the corresponding results for sheaves with transfers.

Definition 12.59. Let $\mathbf{NST}(\mathcal{C}, R)$ denote the category of Nisnevich sheaves of R-modules with transfers (sheaves which are also presheaves with transfers). By [Jar03, 2.2], the Nisnevich-local equivalences and projective cofibrations determine a proper closed simplicial model structure on the simplicial category $\Delta^{\mathrm{op}}\mathbf{NST}(\mathcal{C}, R)$.

Because the sheafification $a^*_{\mathrm{nis}}(F)$ of a presheaf with transfers F is a sheaf with transfers (see [MVW, 13.1]), and the forgetful functor a_* to $\Delta^{\mathrm{op}}\mathbf{PST}$ $(\mathcal{C}, R)_{\mathrm{nis}}$ preserves fibrations and weak equivalences, sheafification determines a Quillen adjunction (a^*, a_*) from $\Delta^{\mathrm{op}}\mathbf{PST}(\mathcal{C}, R)_{\mathrm{nis}}$ to $\Delta^{\mathrm{op}}\mathbf{NST}(\mathcal{C}, R)$.

Theorem 12.60. *The adjoint pair* $(a_{\mathrm{nis}}^*, a_*)$ *defines a Quillen equivalence* $\Delta^{\mathrm{op}}\mathbf{PST}(\mathcal{C}, R)_{\mathrm{nis}} \xrightarrow{\simeq} \Delta^{\mathrm{op}}\mathbf{NST}(\mathcal{C}, R)$.

Moreover, both (equivalent) homotopy categories are equivalent to the full subcategory $\mathbf{D}^{\leq 0}(\mathbf{NST})$ *of the derived category* $\mathbf{D}^-(\mathbf{NST}(\mathcal{C}, R))$ *of Nisnevich sheaves of R-modules with transfers on $\mathcal{C}$.*

Proof. Let A be in $\Delta^{\mathrm{op}}\mathbf{PST}(\mathcal{C}, R)$ and let F be in $\Delta^{\mathrm{op}}\mathbf{NST}(\mathcal{C}, R)$. By Definition 12.49, a map $A \to a_* F$ is a Nisnevich-local equivalence if $a_{\mathrm{nis}}^*(A) \to F$ is a weak equivalence of simplicial sheaves, i.e., a Nisnevich-local equivalence in $\Delta^{\mathrm{op}}\mathbf{NST}(\mathcal{C}, R)$. Thus $(a_{\mathrm{nis}}^*, a_*)$ is a Quillen equivalence.

The embedding into the derived category follows from the observation that via the Dold–Kan correspondence, a Nisnevich-local equivalence between simplicial sheaves is the same as a quasi-isomorphism between the associated chain complexes of sheaves. □

Corollary 12.61. *The derived functor* $\mathbf{L}i^*$ *embeds the homotopy category* $\mathbf{D}^{\leq 0}(\mathbf{NST}(\mathbf{Sm}, R))$ *into the homotopy category* $\mathbf{D}^{\leq 0}(\mathbf{NST}(\mathbf{Norm}, R))$ *as a coreflective subcategory.*

Proof. Immediate from Lemma 12.3 and Theorem 12.60. □

12.9 A^1-LOCAL MODEL STRUCTURE

We are now ready to provide a model structure on presheaves associated to the Morel–Voevodsky pointed A^1-homotopy category $\mathbf{Ho}_\bullet$. For $\mathcal{C}$ either $\mathbf{Sm}$ or $\mathbf{Norm}$, consider the class S of maps $X \times \mathbb{A}^1 \to X$ (regarded as maps between constant simplicial presheaves on $\mathcal{C}$) and the corresponding class S_+ of maps $X \times \mathbb{A}^1_+ \to X_+$ in the pointed category $\Delta^{\mathrm{op}}\mathrm{Pshv}(\mathcal{C}_+)$.

Definition 12.62. We write $\Delta^{\mathrm{op}}\mathrm{Pshv}(\mathcal{C})_{\mathbb{A}^1}$ (resp., $\Delta^{\mathrm{op}}\mathrm{Pshv}(\mathcal{C}_+)_{\mathbb{A}^1}$) for the Bousfield localization of $\Delta^{\mathrm{op}}\mathrm{Pshv}(\mathcal{C})_{\mathrm{nis}}$ (resp., of $\Delta^{\mathrm{op}}\mathrm{Pshv}(\mathcal{C}_+)_{\mathrm{nis}}$) with respect to the class S of projections $X \times \mathbb{A}^1 \to X$ (resp., to the class S_+). They are called the A^1-*local projective model structures* on these categories, and the weak equivalences in this model structure are called A^1-*local equivalences*.

We can also form the Bousfield localization $\Delta^{\mathrm{op}}\mathrm{Sheaves}(\mathcal{C})_{\mathbb{A}^1}$ of the projective model structure on $\Delta^{\mathrm{op}}\mathrm{Sheaves}(\mathcal{C})$ with respect to S. By Remark 12.38.1 and Theorem 12.56, $\Delta^{\mathrm{op}}\mathrm{Pshv}(\mathcal{C})_{\mathbb{A}^1} \xrightarrow{\simeq} \Delta^{\mathrm{op}}\mathrm{Sheaves}(\mathcal{C})_{\mathbb{A}^1}$ is a Quillen equivalence. A similar assertion is true for the pointed categories.

Remark 12.62.1. We could also take the Bousfield localization at the class of inclusions $X \to X \times \mathbb{A}^1$. Since both the projections and inclusions define the same class of weak equivalences, the Bousfield localizations are equivalent.

The category $\Delta^{\mathrm{op}}\mathrm{Sheaves}(\mathcal{C})$ also has an A^1-*local injective* model structure. This is defined as the Bousfield localization of the *injective* model structure

(Lemma 12.58) with respect to the class of maps $X \to X \times \mathbb{A}^1$. It is the model category used by Morel and Voevodsky in [MV99, 3.2.1].

Definition 12.63. The $\mathbb{A}^1$-homotopy category $\mathbf{Ho}(\mathcal{C})$ is the homotopy category of the $\mathbb{A}^1$-local injective model structure on $\Delta^{\mathrm{op}}\mathrm{Sheaves}(\mathcal{C})$.

The pointed $\mathbb{A}^1$-homotopy category $\mathbf{Ho}_\bullet(\mathcal{C})$ is formed in the same way from pointed simplicial sheaves; see [MV99, p. 109].

Lemma 12.64. *The $\mathbb{A}^1$-local projective model structure and the $\mathbb{A}^1$-local injective model structure on $\Delta^{\mathrm{op}}\mathrm{Sheaves}(\mathcal{C})$ are Quillen equivalent. In particular, they have canonically equivalent homotopy categories. Thus we have Quillen equivalences:*

$$\mathbf{Ho}(\Delta^{\mathrm{op}}\mathrm{Pshv}(\mathcal{C})_{\mathbb{A}^1}) \xrightarrow{\simeq} \mathbf{Ho}(\Delta^{\mathrm{op}}\mathit{Sheaves}(\mathcal{C})_{\mathbb{A}^1}) \xrightarrow{\simeq} \mathbf{Ho}(\mathcal{C})$$

$$\mathbf{Ho}(\Delta^{\mathrm{op}}\mathrm{Pshv}(\mathcal{C}_+)_{\mathbb{A}^1}) \xrightarrow{\simeq} \mathbf{Ho}(\Delta^{\mathrm{op}}\mathit{Sheaves}(\mathcal{C}_+)_{\mathbb{A}^1}) \xrightarrow{\simeq} \mathbf{Ho}_\bullet(\mathcal{C}).$$

Proof. The class of maps $X \to X \times \mathbb{A}^1$ is the same class used in Remark 12.62.1 to form the $\mathbb{A}^1$-local projective model structure in Definition 12.62. The first assertion now follows from Lemma 12.58 and Remark 12.38.1. The final sentences follow from 12.0 and Definition 12.63. ☐

The $\mathbb{A}^1$-local equivalences are difficult to describe directly. By Definition 12.38, a map $E \to F$ is an $\mathbb{A}^1$-local equivalence if for any $\mathbb{A}^1$-local L the map $\mathrm{Map}(F, L) \to \mathrm{Map}(E, L)$ is a Nisnevich-local equivalence. This reduces the problem to describing $\mathbb{A}^1$-local objects. By Definition 12.38, an object L of $\Delta^{\mathrm{op}}\mathrm{Pshv}$ $(\mathcal{C}_+)$ is $\mathbb{A}^1$-*local* if and only if:

(i) L is Nisnevich-local, as characterized in Lemma 12.50, and
(ii) $\mathrm{Map}(U, L) \to \mathrm{Map}(U \times \mathbb{A}^1, L)$ is a weak equivalence for all U. By Lemma 12.2, $\mathrm{Map}(U, L) \cong L(U)$. Thus (ii) is equivalent to the condition:
(ii′) $L(U \times \mathbb{A}^1) \to L(U)$ is a weak equivalence for all U in $\mathcal{C}$.

Given Definition 12.62, Theorem 12.47 immediately implies:

Lemma 12.65. *The $\mathbb{A}^1$-local equivalences in $\Delta^{\mathrm{op}}(\mathcal{C}^{\mathrm{ind}})$ form a $\bar{\Delta}$-closed class.*

We now give another description of $\mathbb{A}^1$-equivalences, taken from [Voe10d, 3.49]. Here $\mathcal{C}$ is either **Sch**, **Sm**, or **Norm**, and "sheaves" means for the Nisnevich topology on $\mathcal{C}$.

Theorem 12.66.[1] *In either $\Delta^{\mathrm{op}}(\mathcal{C}^{\mathrm{ind}})$ or $\Delta^{\mathrm{op}}\mathit{Sheaves}(\mathcal{C})$, the $\mathbb{A}^1$-local equivalences form the smallest $\bar{\Delta}$-closed class containing the Nisnevich-local equivalences and the projections $X \times \mathbb{A}^1 \to X$.*

---

1. The proof of Theorem 12.66 is based on [Voe10d, 3.51] and [Del09, p. 389].

Proof. By Proposition 12.54 and Definition 12.62, the $\mathbb{A}^1$-local (projective) model structure is the Bousfield localization of the projective model structure at $S_0 \cup S_{\mathrm{nis}}$, where S_{nis} denotes the class of morphisms $s_Q : K_Q \to V$ described in Theorem 12.55, and S_0 denotes the class of zero-sections $X \to (X \times \mathbb{A}^1)$.

Let E denote the smallest $\bar{\Delta}$-closed class of morphisms in $\Delta^{\mathrm{op}}\mathcal{C}^{\mathrm{ind}}$ containing both S_{nis} and S_0; by Theorem 12.55, E contains all Nisnevich-local equivalences in $\Delta^{\mathrm{op}}\mathcal{C}^{\mathrm{ind}}$. The class of $\mathbb{A}^1$-local equivalences in $\Delta^{\mathrm{op}}(\mathcal{C}^{\mathrm{ind}})$ is $\bar{\Delta}$-closed, by Lemma 12.65, so it contains E. It remains to show that every $\mathbb{A}^1$-local equivalence in $\Delta^{\mathrm{op}}(\mathcal{C}^{\mathrm{ind}})$ is in E.

As in the proof of Theorem 12.55, let T_{nis} denote the class of maps $K_Q \to cyl(s_Q)$, let S_1 be the class $cyl(T)$, and set $S = S_0 \cup S_1$. Then E contains S, and the fibrant replacement maps $X \to L_S X$ of Theorem 12.41(4) also belong to E by Proposition 12.43.

For each $\mathbb{A}^1$-local equivalence $f : X \to Y$, consider the following diagram.

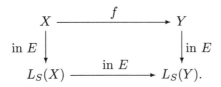

By the 2-out-of-3 property, the bottom map $L_S(f) : L_S(X) \to L_S(Y)$ is an $\mathbb{A}^1$-local equivalence between $\mathbb{A}^1$-local fibrant objects. By Ken Brown's Lemma 12.38.2, $L_S(f)$ is a global weak equivalence; by Theorem 12.31, $L_S(f)$ belongs to E. Again by the 2-out-of-3 property, f is in E. $\qquad\square$

Corollary 12.67. *The total direct image* $\mathbf{L}i^*$ *embeds the $\mathbb{A}^1$-local homotopy category* $\mathbf{Ho}_\bullet(\mathbf{Sm})$ *into the $\mathbb{A}^1$-local homotopy category* $\mathbf{Ho}_\bullet(\mathbf{Norm})$ *as a full (coreflective) subcategory.*

Proof. Because i_* commutes with products, it sends the maps $U \times \mathbb{A}^1_+ \to U_+$ to maps $(i_*U) \times \mathbb{A}^1_+ \to (i_*U)_+$, which are $\mathbb{A}^1$-local equivalences. Because i_* commutes with colimits, and preserves Nisnevich-local equivalences, Theorem 12.66 implies that i_* sends $\mathbb{A}^1$-local equivalences to $\mathbb{A}^1$-local equivalences, and $i_*Y \cong \mathbf{R}i_*Y$ for all Y. The proof of Lemma 12.3(2) now goes through, as it did for Lemma 12.57; that is, it suffices to assume X is $\mathbb{A}^1$-cofibrant, in which case the unit of the adjunction $X \to i_*\mathbf{L}X \cong i_*i^*X$ is an isomorphism. $\qquad\square$

Notation 12.67.1. When X and Y are smooth, morphisms $X \to Y$ in $\mathbf{Ho}(\mathbf{Sm}_+)$ are the same as morphisms in $\mathbf{Ho}(\mathbf{Norm}_+)$, by 12.67. This justifies writing $[X, Y]_{\mathbb{A}^1}$ for the morphisms from X to Y in the $\mathbb{A}^1$-homotopy category $\mathbf{Ho}(\mathbf{Norm}_+)$.

Presheaves with transfers. We now turn to the homological side of the story: the $\mathbb{A}^1$-model category structures on simplicial presheaves with transfers.

Definition 12.68. We write $\Delta^{\mathrm{op}}\mathbf{PST}(\mathcal{C})_{\mathbb{A}^1}$ for the Bousfield localization of $\Delta^{\mathrm{op}}\mathbf{PST}(\mathcal{C})_{\mathrm{nis}}$ with respect to the class of maps $R_{\mathrm{tr}}(X) \to R_{\mathrm{tr}}(X \times \mathbb{A}^1)$. We write $\mathbf{Ho}_{\mathrm{tr}}(\mathcal{C})$ for the associated homotopy category.

Let $\mathbf{DM}^{\leq 0}_{\mathrm{eff}}(\mathcal{C})$ denote the full subcategory of the triangulated category $\mathbf{DM}^{\mathrm{eff}}_{\mathrm{nis}}(\mathcal{C})$ of [MVW, 14.1] consisting of complexes concentrated in nonpositive cohomological degrees. Using Theorem 12.60, we immediately obtain the following result.

Corollary 12.69. *The homotopy category* $\mathbf{Ho}_{\mathrm{tr}}(\mathcal{C})$ *is equivalent to* $\mathbf{DM}^{\leq 0}_{\mathrm{eff}}(\mathcal{C})$.

Example 12.69.1. If $X_\bullet$ is a pointed simplicial object of $\mathcal{C}$ (or $\mathcal{C}^{\mathrm{ind}}$), then $\mathbf{L}R_{\mathrm{tr}}(X_\bullet) \xrightarrow{\simeq} R_{\mathrm{tr}}(X_\bullet)$ is an equivalence by 12.33.1. If $p \geq q$ then $R(q)[p]$ is a nonpositive cochain complex; regarding it as a simplicial object via the Dold–Kan correspondence, we have:

$$H^{p,q}(X_\bullet, R) = \mathrm{Hom}_{\mathbf{DM}}(R_{\mathrm{tr}}(X_\bullet), R(q)[p])$$
$$\cong \mathrm{Hom}_{\mathbf{Ho}_{\mathrm{tr}}}(\mathbf{L}R_{\mathrm{tr}}(X_\bullet), R(q)[p]) \cong [(X_\bullet)_+, uR(q)[p]].$$

This observation will be central to chapter 13.

Remark 12.69.2. Let L be a simplicial presheaf with transfers. Since each $X \to X \times \mathbb{A}^1$ is a map between cofibrant objects, it is well known that L is $\mathbb{A}^1$-local in $\Delta^{\mathrm{op}}\mathbf{PST}(\mathcal{C})$ if and only if uL is $\mathbb{A}^1$-local in $\Delta^{\mathrm{op}}\mathrm{Pshv}(\mathcal{C}_+)$. One reference for this fact is [Hir03, 3.1.12].

Remark 12.69.3. By naturality of Bousfield localization [Hir03, 3.3.20], the Quillen adjunctions (i^*, i_*) and (R_{tr}, u) of (12.28.1) are also Quillen adjunctions for the $\mathbb{A}^1$-local model structures, so that (12.28.1) induces a commutative diagram of $\mathbb{A}^1$-local model categories and Quillen adjunctions.

The corresponding diagram of $\mathbb{A}^1$-local homotopy categories uses the derived functors $\mathbf{L}R_{\mathrm{tr}}$ and $\mathbf{L}i^*$:

$$
\begin{array}{ccc}
\mathbf{Ho}_{\mathbb{A}^1}(\mathbf{Sm}_+) & \xrightarrow{(\mathbf{L}i^*,\, i_*)} & \mathbf{Ho}_{\mathbb{A}^1}(\mathbf{Norm}_+) \\
{\scriptstyle (\mathbf{L}R_{\mathrm{tr}},\, u)}\Big\downarrow & & \Big\downarrow{\scriptstyle (\mathbf{L}R_{\mathrm{tr}},\, u)} \\
\mathbf{Ho}_{\mathrm{tr}}(\mathbf{Sm}) & \xrightarrow{(\mathbf{L}i^*,\, i_*)} & \mathbf{Ho}_{\mathrm{tr}}(\mathbf{Norm}).
\end{array}
\qquad (12.70)
$$

Let M be an object in $\mathbf{DM}^{\leq 0}_{\mathrm{eff}}(\mathcal{C}) \simeq \mathbf{Ho}_{\mathrm{tr}}(\mathcal{C})$. Then uM is a simplicial group object, so its classifying object $B(uM)$ exists; $B(uM)$ is the simplicial object associated to $p \mapsto (uM)^p$. Since u preserves fibration sequences, such as $M \to \mathrm{cone}(M) \to M[1]$, we see that

$$u(M[1]) \simeq B(uM).$$

Similarly, since R_{tr} preserves cofibration sequences, such as the simplicial suspension sequence $V_+ \to \mathrm{cone}(V)_+ \to \Sigma V$, we have

$$R_{\mathrm{tr}}(\Sigma V) \simeq R_{\mathrm{tr}}(V)[1]. \tag{12.71}$$

If V is any simplicial radditive presheaf on $\mathbf{Norm}_+$, the natural transformation $R_{\mathrm{tr}}(i_* V) \to i_*(R_{\mathrm{tr}} V)$ is an $\mathbb{A}^1$-local equivalence of presheaves with transfers on $\mathbf{Sm}$, because it is a global weak equivalence by (12.28.1). Thus $R_{\mathrm{tr}}(i_* V)$ and $i_*(R_{\mathrm{tr}} V)$ are identified in $\mathbf{DM}_{\mathrm{nis}}^{\mathrm{eff}}$.

We can use (12.71) to obtain an interpretation of the motivic cohomology of any V in $\Delta^{\mathrm{op}}\mathbf{Norm}_+$. By (12.70) and Example 12.33.1, $\mathbf{Lres}\,(i_* V) \simeq i_*\mathbf{Lres}\,(V) \simeq i_* V$. Therefore

$$H^{p,q}(V, R) = \left[i_* V, uR(q)[p] \right]_{\mathbb{A}^1} \cong$$
$$\mathrm{Hom}_{\mathbf{DM}}(\mathbf{L}R_{\mathrm{tr}}(i_* V), R(q)[p]) = \mathrm{Hom}_{\mathbf{DM}}(i_* R_{\mathrm{tr}}(V), R(q)[p]).$$

Topological realization 12.72. There is a realization functor from $\mathbf{Sm}/\mathbb{C}$ to the category $\mathbf{Top}_{lc}$ of locally contractible topological spaces, sending X to $X(\mathbb{C})$ and $\mathbb{A}^1_\mathbb{C}$ to a contractible space. This implies that there is an inverse image functor, from $\Delta^{\mathrm{op}}\mathrm{Pshv}(\mathbf{Sm}/\mathbb{C})$ to simplicial sheaves on $\mathbf{Top}_{lc}$, sending $\mathbb{A}^1$-local equivalences to weak equivalences. By Morel–Voevodsky [MV99, 3.3.3], the inclusion of simplicial sets into simplicial sheaves on $\mathbf{Top}_{lc}$ induces an equivalence of homotopy categories (its inverse is evaluation of a simplicial sheaf on the topological simplices). Hence we have a functor $t^\mathbb{C}$ from the $\mathbb{A}^1$-homotopy category $\mathbf{Ho}(\mathbf{Sm}/\mathbb{C})$ to the usual homotopy category; see [MV99, 2.1.57, 2.3.17, and 3.3.4ff]. Up to homotopy, the topological realization functor sends $\mathbb{A}^n - 0$ to S^{2n-1} and $\mathbb{A}^n/(\mathbb{A}^n - 0)$ to S^{2n}.

The Quillen functor $u: \Delta^{\mathrm{op}}\mathbf{PST}(\mathbf{Sm}) \to \Delta^{\mathrm{op}}\mathrm{rad}(\mathbf{Sm}_+)$ of Remark 12.69.3 lands in the category of simplicial sheaves of R-modules. Composition with $t^\mathbb{C}$ gives a simplicial R-module, which we may identify with a chain complex using the Dold–Kan correspondence. As the composition sends $\mathbb{A}^1$-local equivalences to chain homotopy equivalences, it induces a functor on homotopy categories. By the previous paragraph, $t^\mathbb{C}$ sends $\mathbb{L}^a[b]$ to the reduced chain complex $\widetilde{C}_*(S^{2a+b}, R)$ and the reduced motive $\widetilde{S}^m(\mathbb{L}^a[b])$ of (14.1) to $\widetilde{C}_*(\widetilde{S}^m(S^{2a+b}), R)$.

12.10   HISTORICAL NOTES

The global projective model structure on simplicial presheaves was present in Quillen's foundational paper [Qui67] on homotopical algebra. Brown and Gersten introduced the local projective model structure in [BG73] as a descent tool for simplicial sheaves, and the injective model structure on simplicial sheaves is due to Joyal; see [Jar86]. These model structures were studied by Jardine, who pointed out in [Jar87] and [Jar00, App. B] that the model structures could profitably be lifted to presheaves.

The original model structure used by Morel and Voevodsky in [MV99, 2.1.4] to define the motivic homotopy category **Ho** was the global *injective* model structure on simplicial sheaves, and the $\mathbb{A}^1$-localization of this model structure. Blander pointed out in [Bla01] that this could also be accomplished using the projective model structure and its Bousfield localizations on either sheaves or presheaves, as we have done here; see also [RØ06]. Theorem 12.5 is new.

The radditive functors approach we have used is taken from [Voe10d]. Some of the characterizations are new, such as 12.26 and 12.51. Our definition 12.29 of a $\bar{\Delta}$-closed class is taken from [Voe10d, 2.18]. This differs slightly, both from the definition in [Del09]—which requires arbitrary coproducts—and the original definition in [Voe00a]. Theorem 12.47 is new.

The construction of the Hurewicz functor R_{tr} was done independently by many people; see [Spi01, 14.7], [Mor04, p. 241], [Wei04], and [Rio07]. It is also implicit in the 2000 preprint [Voe00a], and underlies the viewpoint presented in [Voe03c, 2.1]. The use of the adjunction (R_{tr}, u) to construct classifying spaces is taken primarily from [Voe10c], and partially from [Wei09].

Chapter Thirteen

Cohomology Operations

Fix a perfect field k. Although motivic cohomology was originally defined for smooth varieties over k, it is more useful to view it as a functor defined on the pointed $\mathbb{A}^1$-homotopy category $\mathbf{Ho_\bullet}$, originally constructed by Morel and Voevodsky in [MV99, 3.3.2] and discussed previously in section 12.9.

After defining cohomology operations and giving a few examples, we devote section 13.2 to an axiomatic treatment of the motivic Steenrod operations, following Voevodsky [Voe03c]. The motivic Milnor operations are presented in section 13.4. In section 13.6, we show that the sequence of Milnor operations Q_i is exact on the reduced cohomology of the suspension $\Sigma\mathfrak{X}$ attached to a Rost variety X, using the degree map $t_\mathcal{N}$ of section 13.5. We conclude with Voevodsky's motivic degree theorem in section 13.7, which is needed for the proof of Proposition 5.16.

13.1 MOTIVIC COHOMOLOGY OPERATIONS

For each abelian group A, the motivic cohomology groups $H^{p,q}(-,A)$ form a bigraded family of functors from smooth (simplicial) schemes to abelian groups. By definition, there is a chain complex $A(q)[p]$ of sheaves with transfers so that $H^{p,q}(X,A) = \operatorname{Hom}_{\mathbf{DM}}(R_{\mathrm{tr}}(X), A(q)[p])$; see [MVW, 3.1].

Suppose for simplicity that $p \geq q$. Then the cochain complex $A(q)[p]$ is zero in positive degrees, so it may be identified with a simplicial sheaf with transfers (by the Dold–Kan correspondence); we define $K(A(q),p)$ to be the underlying pointed simplicial sheaf $uA(q)[p]$. Via the adjunction (R_{tr}, u) in (12.70), we saw in Example 12.69.1 that for every smooth (simplicial) scheme X we have

$$H^{p,q}(X,A) = \operatorname{Hom}_{\mathbf{Ho_{tr}}}(R_{\mathrm{tr}}(X), A(q)[p]) \cong \operatorname{Hom}_{\mathbf{Ho_\bullet}}(X_+, K(A(q),p)).$$

(Here X_+ is X with a disjoint basepoint.) This formula allows us to extend motivic cohomology from smooth schemes to the pointed $\mathbb{A}^1$-homotopy category $\mathbf{Ho_\bullet}$.

Definition 13.1. Given a pointed motivic space M, we define the reduced motivic cohomology of M, $\widetilde{H}^{p,q}(M,A)$, to be $\operatorname{Hom}_{\mathbf{Ho_\bullet}}(M, K(A(q),p))$.

A *cohomology operation* ϕ is a natural transformation from $\widetilde{H}^{r,s}(-, A)$ to $\widetilde{H}^{p,q}(-, B)$ of contravariant functors $\mathbf{Ho_\bullet} \to \mathbf{Sets}$. It depends upon the integers r, s, p, q and the groups A and B; we say that ϕ has *bidegree* $(p-r, q-s)$.

By construction, $\widetilde{H}^{p,q}(-, A)$ is a representable cohomology theory on $\mathbf{Ho_\bullet}$. Using the natural isomorphism $H^{p,q}(X, A) \cong \widetilde{H}^{p,q}(X_+, A)$, a cohomology operation also determines a natural transformation $\phi_X : H^{r,s}(X, A) \to H^{p,q}(X, B)$ of functors defined on $\Delta^{\mathrm{op}}\mathbf{Sm}$.

For example, if R is a ring then $\widetilde{H}^{*,*}(M, R)$ is a ring, and any monomial $f(x) = cx^i$ with $c \in H^{p,q}(k, R)$ defines a cohomology operation on $\widetilde{H}^{r,s}(-, R)$, taking values in $\widetilde{H}^{p+ir,q+is}(-, R)$.

Lemma 13.1.1. *If $c \neq 0$, the operations $x \mapsto cx^i$ are nonzero on $H^{2n,n}(-, R)$.*

Proof. By the Projective Bundle Theorem ([MVW, 15.5]), there is a canonical line element $u \in H^{2,1}(\mathbb{P}^N, R)$ such that

$$H^{*,*}(\mathbb{P}^N, R) \cong H^{*,*}(k, R)[u]/(u^{N+1}).$$

When $N \geq ni$ and x is $u^n \in H^{2n,n}(\mathbb{P}^N, R)$, $cx^i = cu^{ni}$ is nonzero. $\qquad \square$

By the Yoneda Lemma, cohomology operations on $\widetilde{H}^{r,s}(-, A)$ are classified by the motivic cohomology of $K(A(s), r)$, with the identity operation $x \mapsto x$ corresponding to the canonical element $\alpha = \alpha_{r,s}^A \in \widetilde{H}^{r,s}(K(A(s), r), A)$. We record this:

Lemma 13.2. *The set of cohomology operations $\psi : \widetilde{H}^{r,s}(-, A) \to \widetilde{H}^{p,q}(-, B)$ is in 1–1 correspondence with elements of $\widetilde{H}^{p,q}(K(A(s), r), B)$, with ψ corresponding to $\psi(\alpha)$.*

Example 13.2.1 ($H^{2,1}$). The classifying space $K(\mathbb{Z}(1), 2)$ is represented by the pointed ind-scheme $(\mathbb{P}^\infty, *)$; see [MV99, 4.3.8] or [Voe03c, 2.1]. Since $H^{*,*}(\mathbb{P}^\infty, R) = \varprojlim \widetilde{H}^{*,*}(\mathbb{P}^N, R)$ is the power series ring $H^{*,*}(k, R)[[t]]$ for every ring R (by the Projective Bundle Theorem [MVW, 15.5]), motivic cohomology operations $\widetilde{H}^{2,1}(-, \mathbb{Z}) \to \widetilde{H}^{p,q}(-, R)$ are classified by sequences $(a_1, a_2, \ldots)$ in the finite sum $\oplus_{i=1}^q H^{p-2i,q-i}(k, R)$.

We can interpret the cohomology operation corresponding to $(a_1, \ldots)$ as the polynomial $x \mapsto f(x) = \sum a_i x^i$. Thus if t denotes a formal variable of bidegree $(2, 1)$ then cohomology operations $\widetilde{H}^{2,1}(-, \mathbb{Z}) \to \widetilde{H}^{p,q}(-, R)$ are just homogeneous polynomials $f(t)$ in $H^{*,*}(k, R)[t]$ of bidegree (p, q).

Example 13.2.2 ($H^{1,1}$). The pointed scheme $\mathbb{G}_m = (\mathbb{A}^1 - \{0\}, 1)$ represents the classifying space $K(\mathbb{Z}(1), 1)$. This follows from the homotopy fibration sequence $\mathbb{G}_m \to \mathbb{A}^\infty - \{0\} \to \mathbb{P}^\infty$ and Example 13.2.1; another proof will be given in 15.6. Since $\widetilde{H}^{p,q}(\mathbb{G}_m, R) \cong H^{p-1,q-1}(k, R)$, every motivic cohomology operation

$\tilde{H}^{1,1}(-,\mathbb{Z}) \xrightarrow{\lambda} \tilde{H}^{p,q}(-,R)$ has the form $\lambda(x) = ax$ for a unique $a \in H^{p-1,q-1}$ (k,R). The element a is determined by $\lambda(t) = at$, where $t \in \tilde{H}^{1,1}(\mathbb{G}_m, \mathbb{Z})$ is the canonical element, via the isomorphism $H^{p-1,q-1}(k,R) \cong \tilde{H}^{p,q}(\mathbb{G}_m, R)$ sending a to at.

Example 13.2.3 ($H^{0,0}$). Since $K(A,0)$ is the pointed set A, it is not hard to see that cohomology operations $\tilde{H}^{0,0}(-,A) \to \tilde{H}^{p,q}(-,B)$ correspond to pointed functions $f: A \to H^{p,q}(k,B)$ (sending 0 to 0). If X is a smooth connected scheme then the cohomology operation corresponding to f is the composition

$$H^{0,0}(X,A) = A \xrightarrow{f} H^{p,q}(k,B) \to H^{p,q}(X,B).$$

Operations in weight 0.   The case $s = q = 0$ may be classically understood as topological cohomology operations, classified in [Car54], because of the following well-known lemma. If X is a smooth scheme, let $\pi_0(X)$ denote the set of its components; if $X_\bullet$ is a smooth simplicial scheme, $\pi_0(X_\bullet)$ is a simplicial set.

Lemma 13.3. *For every smooth simplicial scheme $X_\bullet$, the motivic cohomology group $H^{p,0}(X_\bullet, A)$ is isomorphic to the topological cohomology $H^p_{\text{top}}(\pi_0 X_\bullet, A)$ of the simplicial set $\pi_0(X_\bullet)$.*

Proof. For smooth connected X we have $H^{p,0}(X,A) = H^p_{\text{zar}}(X,A) = 0$ for $p > 0$ and $H^{0,0}(X,A) = A$ almost by definition; see [MVW, 3.4]. Hence the spectral sequence $E_1^{p,q} = H^q(X_p, A) \Rightarrow H^{p+q,0}(X_\bullet, A)$ degenerates, as $E_1^{p,q} = 0$ for $q \neq 0$, and the $q = 0$ row $E_1^{*,0}$ is the chain complex $\text{Hom}(\pi_0(X_\bullet), A)$. As the cohomology of this complex is $H^*_{\text{top}}(\pi_0(X_\bullet), A)$, we are done.   □

Proposition 13.4. *Motivic cohomology operations $\tilde{H}^{r,0}(-,A) \to \tilde{H}^{p,0}(-,B)$ are in 1–1 correspondence with the classical topological cohomology operations $\tilde{H}^r_{\text{top}}(-,A) \to \tilde{H}^p_{\text{top}}(-,B)$.*

Proof. Let $K(A,r)$ denote the simplicial Eilenberg–Mac Lane space representing $H^r_{\text{top}}(-,A)$. As an object of $\mathbf{Ho}_\bullet$, $K(A(0),r)$ is represented by the pointed simplicial scheme which in degree i is the disjoint union of copies of $\text{Spec}(k)$ indexed by the elements of the set $K(A,r)_i$. In particular, $\pi_0(K(A(0),r)) = K(A,r)$. Lemma 13.3 implies that $\tilde{H}^{p,0}(K(A(0),r),B) \cong \tilde{H}^p_{\text{top}}(K(A,r),B)$.   □

Example 13.4.1. Consider the vector space V of all cohomology operations $H^{2a+1,0}(-,\mathbb{Z}/\ell) \to H^{2a\ell+2,0}(-,\mathbb{Z}/\ell)$ which vanish on suspensions. By Proposition 13.4, V is in 1–1 correspondence with the kernel of the map

$$H^{2a\ell+2}_{\text{top}}(K(\mathbb{Z}/\ell, 2a+1), \mathbb{Z}/\ell) \to H^{2a\ell+1}_{\text{top}}(K(\mathbb{Z}/\ell, 2a), \mathbb{Z}/\ell)$$

induced by $\Sigma K(\mathbb{Z}/\ell, 2a) \to K(\mathbb{Z}/\ell, 2a+1)$. By [Ser52] and [Car54], V is a 1-dimensional vector space spanned by βP^a_{top}. (For $\ell = 2$ we have $Sq^{2a+1}_{\text{top}} = \beta P^a_{\text{top}}$.) This observation is used in Corollary 6.32.

Remark 13.4.2. Recall that in topology the classical Eilenberg–Mac Lane space $BG = K(G, 1)$ of a group G represents cohomology in the sense that $H^1_{\text{top}}(X, G) = [X, BG]$. There is a canonical element $\alpha_1 \in H^1_{\text{top}}(BG, G)$, corresponding to the identity of BG, and the Yoneda Lemma yields a 1–1 correspondence between cohomology operations $\psi : H^1_{\text{top}}(-, G) \to H^p_{\text{top}}(-, \mathbb{Z}/\ell)$ and elements of the group $H^p_{\text{top}}(BG, \mathbb{Z}/\ell)$, given by $\psi \mapsto \psi(\alpha_1)$.

In motivic cohomology, we have the simplicial classifying space $B_\bullet(G)$, which is the simplicial set BG regarded as a simplicial scheme. In simplicial degree i, $B_\bullet(G)$ is the disjoint union over the indexing set G^{i+1} of copies of $\text{Spec}(k)$ in simplicial degree i, and the simplicial structure comes from the group structure of G. Now set $G = \mathbb{Z}/\ell$. By Lemma 13.3, $H^{1,0}(-, \mathbb{Z}/\ell)$ is represented by $B_\bullet(G)$ in the motivic homotopy category. Again by the Yoneda Lemma, there is a 1–1 correspondence between motivic cohomology operations $\psi : H^{1,0}(-, \mathbb{Z}/\ell) \to H^{p,q}(-, \mathbb{Z}/\ell)$ and elements of $H^{p,q}(B_\bullet G, \mathbb{Z}/\ell)$, given by $\psi \mapsto \psi(\alpha_1)$, where α_1 is the canonical element of $H^{1,0}(B_\bullet G, \mathbb{Z}/\ell)$. These facts were also mentioned in section 6.6, and used in Proposition 6.29.

Bi-stable operations. Recall from Definition 1.41 that a family of operations $\phi_{r,s} : H^{r,s}(-, A) \to H^{r+i,s+j}(-, B)$ of bidegree (i, j) is called *bi-stable* if it commutes with both the simplicial suspension and the Tate suspension isomorphisms. (It is *simplicially stable* if it commutes with the simplicial suspension.) For every $c \in H^{p,q}(k, \mathbb{Z})$, the left multiplication $\lambda_{r,s}(x) = c\,x$ is a bi-stable operation.

Since simplicially stable operations (and hence bi-stable operations) are additive by [Voe03c, 2.10], the non-additive operations in Examples 13.2.1 and 13.2.3 are not bi-stable. The additive operations $f(x) = x^\ell$ defined on $H^{2,1}$ do extend to bi-stable operations, namely the bi-stable operations P^1 defined in the next section. The operations f defined on $H^{0,0}(-, \mathbb{Z}/\ell)$ in Example 13.2.3 are only additive when f is a homomorphism, and in that case correspond to multiplication by $f(1) \in H^{p,q}(k, B)$.

The canonical example of a bi-stable operation is the family of *Bockstein* operations $\beta : H^{p,q}(X, \mathbb{Z}/\ell) \to H^{p+1,q}(X, \mathbb{Z}/\ell)$, which are the boundary maps in the long exact sequence associated to the exact coefficient sequence $0 \to \mathbb{Z}/\ell(q) \to \mathbb{Z}/\ell^2(q) \to \mathbb{Z}/\ell(q) \to 0$. It is the reduction modulo ℓ of the *integral Bockstein* $\tilde{\beta} : H^{p,q}(X, \mathbb{Z}/\ell) \to H^{p+1,q}(X, \mathbb{Z})$, the boundary map associated to the exact sequence $0 \to \mathbb{Z}(q) \xrightarrow{\ell} \mathbb{Z}(q) \to \mathbb{Z}/\ell(q) \to 0$. It is well known that $\beta^2 = 0$, because it is the composition

$$H^{p,q}(X, \mathbb{Z}/\ell) \xrightarrow{\tilde{\beta}} H^{p+1,q}(X, \mathbb{Z}) \to H^{p+1,q}(X, \mathbb{Z}/\ell) \xrightarrow{\beta} H^{p+2,q}(X, \mathbb{Z}/\ell).$$

The following result is stated without proof in [Voe03c, (8.1)].

Lemma 13.5. *The Bockstein is a derivation with respect to the cup product on* $H^{*,*}(X, \mathbb{Z}/\ell)$.

Proof. (Folklore) Recall that $H^{p,q}(X, \mathbb{Z}/\ell^\nu)$ is the pth hypercohomology of a chain complex $\mathbb{Z}(q)/\ell^\nu$ of Zariski sheaves. Choose flasque Godement-style resolutions $\mathbb{Z}(q)/\ell^2 \to \mathcal{I}(q)$ whose stalks are free (=injective) $\mathbb{Z}/\ell^2$-modules, and write $\bar{\mathcal{I}}(q)$ for $\mathcal{I}(q)/\ell$, so that $\mathbb{Z}(q)/\ell \to \bar{\mathcal{I}}(q)$ is also a flasque resolution. Given cycles $\bar{u}_i$ $(i = 1, 2)$ representing $u_i \in H^{p_i, q_i}(X, \mathbb{Z}/\ell)$, lift them to chains u_i' in $\mathcal{I}(q_i)$; then $\beta(u_i)$ is represented by u_i'', defined by $\delta(u_i') = i(u_i'')$.

The cup product $u_1 \cup u_2$ is represented by the image of $\bar{u}_1 \otimes \bar{u}_2$ under the map $m : \bar{\mathcal{I}}(q_1) \otimes \bar{\mathcal{I}}(q_2) \to \bar{\mathcal{I}}(q_1 + q_2)$ resolving $\mathbb{Z}/\ell(q_1) \otimes \mathbb{Z}/\ell(q_2) \to \mathbb{Z}/\ell(q_1 + q_2)$; see [MVW, 3.11]. Since the coboundary on $\mathcal{I}(q_1) \otimes \mathcal{I}(q_2)$ satisfies

$$\delta(u_1' \otimes u_2') = i(u_1'') \otimes u_2' + (-1)^{p_1} u_1' \otimes i(u_2'') = i(u_1'' \otimes \bar{u}_2) + (-1)^{p_1} i(\bar{u}_1 \otimes u_2'')$$

it follows that $\beta(u_1 \cup u_2)$ is represented by $m(u_1'' \otimes \bar{u}_2) + (-1)^{p_1} m(\bar{u}_1 \otimes u_2'')$, i.e., by $\beta(u_1) \cup u_2 + (-1)^{p_1} u_1 \cup \beta(u_2)$. $\qquad\qquad\qquad\qquad\qquad\qquad\qquad\square$

13.2   STEENROD OPERATIONS

One important family of bi-stable (and hence additive) cohomology operations are the *reduced power operations* P^i, which were constructed by Voevodsky in [Voe03c, p. 33], and mirror the classical Steenrod operations P^i_{top} in topology. In this section, we assume that $1/\ell \in k$ and give an axiomatic description of their salient properties; Voevodsky's construction of the P^i is sketched in the next section, following [Voe03c].

The operation P^i on $\widetilde{H}^{*,*}$ is bi-stable of bidegree $(2i(\ell-1), i(\ell-1))$. Thus

$$P^i : \widetilde{H}^{p,q}(X, \mathbb{Z}/\ell) \to \widetilde{H}^{p+2i(\ell-1), q+i(\ell-1)}(X, \mathbb{Z}/\ell)$$

for X in $\mathbf{Ho}_\bullet$. The following list of axioms are verified in [Voe03c, §9–§10].

Axioms for Steenrod operations 13.6. The operations P^i satisfy:

1. $P^0 x = x$ for all x, and $P^i x = x^\ell$ if x has bidegree $(2i, i)$.
2. $P^i x = 0$ if x has bidegree (p, q) with $q \le i$ and $p < q + i$.
3. If $\ell > 2$, the usual Cartan formula $P^n(xy) = \sum_{i=0}^{n} P^i(x) P^{n-i}(y)$ holds. The Cartan formula for $\beta P^n(xy)$ follows from this since β is a derivation.
4. If $\ell > 2$, the usual Adem relations hold (compare [Ste62, p. 77]). For $i < j\ell$:

$$P^i P^j = \sum_{t=0}^{[i/\ell]} (-1)^{i+t} \binom{(\ell-1)(j-t)-1}{i - t\ell} P^{i+j-t} P^t;$$

$$P^i \beta P^j = \sum_{t=0}^{[i/\ell]} (-1)^{i+t} \binom{(\ell-1)(j-t))}{i-t\ell} \beta P^{i+j-t} P^t$$

$$+ \sum_{t=0}^{[(i-1)/\ell]} (-1)^{i+t-1} \binom{(\ell-1)(j-t)-1}{i-t\ell-1} P^{i+j-t} \beta P^t.$$

The Adem relations for $\beta P^i P^j$ follow since β is a derivation.

If $\ell > 2$, it follows from the axioms that the P^i and the Bockstein β generate a bigraded ring, isomorphic to the topological Steenrod algebra $\mathcal{A}^*_{\text{top}}(\mathbb{Z}/\ell)$ described in [Ste62, VI]. In particular, every monomial in β and the P^i is a unique $\mathbb{Z}/\ell$-linear combination of the *admissible* monomials:

$$\beta^{\epsilon_0} P^{s_1} \beta^{\epsilon_1} \cdots P^{s_k} \beta^{\epsilon_k}, \quad \epsilon_i = 0, 1 \text{ and } s_i \geq \ell s_{i+1} + \epsilon_i.$$

The Adem relations show that the admissible monomials form a basis for the $\mathbb{Z}/\ell$-subalgebra of the ring of all bi-stable cohomology operations, isomorphic to $\mathcal{A}^*_{\text{top}}(\mathbb{Z}/\ell)$.

Axioms for Sq^a 13.7. When $\ell = 2$ we define Sq^{2i} to be P^i, and define Sq^{2i+1} to be βP^i. In particular, $Sq^0(x) = x$ and $Sq^1(x) = \beta(x)$ is the Bockstein. Thus Sq^{2i} has bidegree $(2i, i)$ and Sq^{2i+1} has bidegree $(2i+1, i)$. The operations Sq^i satisfy the following axioms, the first two of which are special cases of 13.6(1,2):

1. $Sq^0(x) = x$ for all x, and $Sq^{2i}(x) = x^2$ if x has bidegree $(2i, i)$.
2. $Sq^{2i}(x) = 0$ if x has bidegree (p, q) with $q \leq i$, $p < q + i$.
3. A modified Cartan formula holds:

$$Sq^{2n}(xy) = \sum_{i+j=n} Sq^{2i}(x) Sq^{2n-2i}(y) + \tau \sum_{i+j=n-1} Sq^{2i+1}(x) Sq^{2j+1}(y),$$

where τ is the nonzero element in $H^{0,1}(k, \mathbb{Z}/2) = \mu_2 \cong \mathbb{Z}/2$. The Cartan formula for $Sq^{2n+1}(xy)$ follows from this since β is a derivation.
4. Modified Adem relations hold; these are taken from [Voe03a, 10.2].

$$Sq^{2i} Sq^{2k} = \sum_{\substack{t=0 \\ \text{even}}}^{i} \binom{2k-t-1}{2i-2t} Sq^{2i+2k-t} Sq^t$$

$$+ \tau \sum_{\substack{t=0 \\ \text{odd}}}^{i} \binom{2k-t-1}{2i-2t} Sq^{2i+2k-t} Sq^t, \quad 0 < i < 2k$$

$$Sq^{2i} Sq^{2k+1} = \sum_{t=0}^{i} \binom{2k-t}{2i-2t} Sq^{2i+2k+1-t} Sq^t$$

$$+ \beta(\tau) \sum_{t=0 \text{ odd}}^{i} \binom{2k-t}{2i-2t} Sq^{2i+2k-t} Sq^t, \quad 0 < i < 2k+1.$$

The Adem relations for $Sq^{2i+1} Sq^b$ follow from this since β is a derivation.

Remark 13.7.1. When comparing with other formulations of the Adem relations for $\ell = 2$, it is useful to recall that if n is even and j is odd then $\binom{n}{j} \equiv 0 \pmod{2}$.

Formally setting $\tau = 1$ and $\beta(\tau) = 0$ transforms the Adem relations 13.7(4) for $\ell = 2$ into the topological Adem relations (see [Ste62, p. 2]): if $0 < a < 2b$ then

$$Sq^a Sq^b = \sum_{t=0}^{[a/2]} \binom{b-t-1}{a-2t} Sq^{a+b-t} Sq^t.$$

Remark 13.7.2. Let α_1 be the canonical element of $H^{1,0}(B_{\bullet}\mathbb{Z}/2, \mathbb{Z}/2)$. By Axiom 13.7(2), $Sq^2(\tau\alpha_1) = Sq^2(\tau) = 0$; by the Cartan formula 13.7(3), $\tau Sq^2(\alpha_1) = \tau\beta(\tau)\beta(\alpha_1)$. Now multiplication by τ is injective, as a consequence of the Milnor Conjecture (the main theorem of this book for $\ell = 2$). This proves that $Sq^2(\alpha_1) = \beta(\tau)\beta(\alpha_1)$, which is nonzero, while $Sq^3(\alpha_1) = \beta Sq^2(\alpha_1) = 0$.

Definition 13.8. Let $\mathcal{A}^{*,*} = \mathcal{A}^{*,*}(k, \mathbb{Z}/\ell)$ denote the subalgebra of the ring of all bi-stable cohomology operations generated by the P^i, β and left multiplication by elements of $H^{*,*}(k, \mathbb{Z}/\ell)$. The Adem relations show that it is free as a left $H^{*,*}(k, \mathbb{Z}/\ell)$-module, with the admissible monomials as a basis, even if $\ell = 2$; see [Voe03c, 11.5].

13.3 CONSTRUCTION OF STEENROD OPERATIONS

In this section, we briefly sketch Voevodsky's construction of the motivic cohomology operations. They are modelled on Steenrod's construction, which we recall.

Topological operations: The topological Steenrod operations P^i are constructed in [Ste62] as follows. Let us write $H^*(X)$ for $H^*_{\text{top}}(X, \mathbb{Z}/\ell)$ and let C_ℓ denote the cyclic group of order ℓ. Steenrod first constructs a *reduced power operation* in [Ste62, VII.2.3]:

$$P: H^n(X) \longrightarrow H^{n\ell}_{C_\ell}(X) \cong \bigoplus_{i+j=n\ell} H^i(X) \otimes H^j(BC_\ell).$$

If $\ell \neq 2$, $H^*(BC_\ell)$ is $\mathbb{Z}/\ell[[c, d]]/(c^2 = 0)$ with $\beta(c) = d$; if $\ell = 2$, $H^*(BC_\ell)$ is $\mathbb{Z}/\ell[[c]]$. In either case, $c \in H^1(BC_\ell)$ so there is a canonical element w_j in $H^j(BC_\ell)$ for all $j \geq 0$. Steenrod defines $P^i(x)$ for $x \in H^n(X)$ and $i \leq n/2$ to be the coefficient of $(-1)^i w_{(n-2i)(\ell-1)}$ (times a constant ν_n if $\ell \neq 2$); see [Ste62, VII.6.1].

Motivic operations: For every subgroup G of the symmetric group Σ_ℓ, and each $n > 0$, Voevodsky constructs a map $K(\mathbb{Z}/\ell(n), 2n) \wedge (B_{\mathrm{gm}}G)_+ \to K(\mathbb{Z}/\ell(n\ell), 2n\ell)$ in [Voe03c, 5.3 and p. 28], representing a *power map*

$$P : \widetilde{H}^{2n,n}(-) \longrightarrow \widetilde{H}^{2n\ell,n\ell}(- \wedge B_{\mathrm{gm}}G_+).$$

Here we abbreviate $\widetilde{H}^{*,*}(-, \mathbb{Z}/\ell)$ as $\widetilde{H}^{*,*}$. He then uses the computation, given in [Voe03c, 6.16] (and sketched in 15.16 and Corollary 15.17), that

$$H^{*,*}(B_{\mathrm{gm}}\Sigma_\ell) = \begin{cases} H^{*,*}(k)[[c,d]]/(c^2 = 0), & \ell \neq 2; \\ H^{*,*}(k)[[c,d]]/(c^2 + \tau d + \rho c), & \ell = 2. \end{cases}$$

Here $c \in H^{2\ell-3,\ell-1}(B_{\mathrm{gm}}\Sigma_\ell)$, $d \in H^{2\ell-2,\ell-1}(B_{\mathrm{gm}}\Sigma_\ell)$, and $\beta(c) = d$. Moreover, by the Künneth formula (see Proposition 15.30):

$$H^{*,*}(X \times B_{\mathrm{gm}}\Sigma_\ell) \cong \widetilde{H}^{*,*}(X_+ \wedge (B_{\mathrm{gm}}\Sigma_\ell)_+) \cong H^{*,*}(X) \otimes_{H^{*,*}(k)} H^{*,*}(B_{\mathrm{gm}}\Sigma_\ell).$$

For $x \in H^{2n,n}(X)$ and $a \leq n/2$, Voevodsky defines $P^i(x)$ to be the coefficient of d^{n-i} in $P(x)$; see [Voe03c, (9.1) and p. 33].

If $p \neq 2q$, Voevodsky defines P^i on $H^{p,q}(X)$ to make the following diagram commute for b large and $a = 2q - p + b$, so that $p + a + b = 2(q + b)$:

$$
\begin{array}{ccc}
H^{p,q}(X) & \xrightarrow{\ \cong\ } & H^{p+a+b,q+b}(S^a \wedge \mathbb{G}_m^b \wedge X) \\
\Big\downarrow{\scriptstyle P^i} & & \Big\downarrow{\scriptstyle P^i} \\
H^{p+2i(\ell-1),q+i(\ell-1)}(X) & \xrightarrow{\ \cong\ } & H^{p+a+b+2i(\ell-1),q+b+i(\ell-1)}(S^a \wedge \mathbb{G}_m^b \wedge X).
\end{array}
$$

This produces a bi-stable family of operations; see [Voe03c, Prop. 2.6]. Finally, the verification of the axioms 13.6–13.7 is given in sections 9–10 of [Voe03c].

13.4   THE MILNOR OPERATIONS Q_I

In this section, we introduce the sequence of *motivic Milnor operations* Q_i, $i \geq 0$. These are bi-stable motivic cohomology operations, starting with the Bockstein Q_0. For $\ell \neq 2$, they satisfy the same formulas as the Milnor operations Q_i^{top} in topology, but the formulas are slightly different when $\ell = 2$.

13.9. Topological Milnor operations: The topological Steenrod algebra $\mathcal{A}^* = \mathcal{A}_{\mathrm{top}}^*$ contains a distinguished family of cohomology operations Q_i^{top}, $i \geq 0$, called the *Milnor operations* after their discovery by Milnor in [Mil58]. By definition,

Q_0^{top} is the Bockstein; the rest are determined inductively by the commutator formula

$$Q_{i+1}^{\mathrm{top}} = [P_{\mathrm{top}}^{\ell^i}, Q_i^{\mathrm{top}}].$$

Thus Q_i^{top} has degree $2\ell^i - 1$ and $Q_1^{\mathrm{top}} = P_{\mathrm{top}}^1\beta - \beta P_{\mathrm{top}}^1$. Milnor proved in [Mil58, 4a] that the Q_i generate an exterior algebra under composition, i.e., $Q_i^2 = 0$ and $Q_iQ_j = -Q_jQ_i$, and in [Mil58, Lemma 9] that the Q_i are derivations.

Milnor's paper [Mil58] also defined a family of cohomology operations $P_{\mathrm{top}}^{\mathbf{r}}$, indexed by finite sequences $\mathbf{r} = (r_1, r_2, ...)$ of natural numbers, such that the usual operation P_{top}^r is $P_{\mathrm{top}}^{\mathbf{r}}$ when $\mathbf{r} = (r)$, and proved in [Mil58, 4a] that

$$P^{\mathbf{r}}Q_i - Q_iP^{\mathbf{r}} = Q_{i+1}P^{\mathbf{r}-(\ell^i,0,0)} + Q_{i+2}P^{\mathbf{r}-(0,\ell^i,0)} + \cdots . \qquad (13.9.1)$$

In fact, the finite products $Q_{i_1}^{\mathrm{top}} \cdots Q_{i_s}^{\mathrm{top}} P_{\mathrm{top}}^{\mathbf{r}}$ $(i_1 < \cdots < i_s)$ form a basis of the topological Steenrod algebra $\mathcal{A}^*$; see [Mil58, 4a].

By Theorem 1 of [Mil58], there is a coproduct $\Delta : \mathcal{A}^* \to \mathcal{A}^* \otimes \mathcal{A}^*$, making the graded dual $\mathcal{A}_*$ into a graded Hopf algebra. By Theorem 2 of [Mil58], $\mathcal{A}_*$ is the tensor product of an exterior algebra in variables $\{\tau_1, \tau_2, ...\}$ with a polynomial algebra in variables $\{\xi_1, \xi_2, ...\}$, with τ_i dual to Q_i; Milnor defines $P^{(r_1, r_2, ...)}$ to be dual to the product $\xi_{r_1}\xi_{r_2}\cdots$, so that ξ_j is dual to $P^{\mathbf{r}_j}$.

Definition 13.10. When $\ell \neq 2$, the Milnor operation Q_i on $H^{*,*}(X, \mathbb{Z}/\ell)$ is the cohomology operation of bidegree $(2\ell^i - 1, \ell^i - 1)$ defined inductively by setting $Q_0 = \beta$ (the Bockstein), $Q_1 = P^1\beta - \beta P^1$, and $Q_{i+1} = [P^{\ell^i}, Q_i]$.

When $\ell = 2$, we again set $Q_0 = Sq^1 = \beta$ (the Bockstein) and $Q_1 = P^1\beta - \beta P^1 = Sq^3 + Sq^2 Sq^1$. Voevodsky defines operations $P^{\mathbf{r}}$ using Milnor's formulas (see Remark 13.10.1) and shows in [Voe03c, 13.6] that the remaining Q_i may be defined inductively by the formula $Q_i = [\beta, P^{\mathbf{r}_i}]$, where $\mathbf{r}_i$ is the sequence $(0, ..., 0, 1)$ of length i.

Remark 13.10.1. It is convenient to write Q^E for $Q_{i_1} \cdots Q_{i_s}$, where E denotes the finite sequence $(i_1, ..., i_s)$. As in the topological situation 13.9, the finite products $Q^E P^{\mathbf{r}}$ form a basis of the motivic Steenrod algebra $\mathcal{A}^{*,*}(k)$ even when $\ell = 2$. In fact, Voevodsky constructs a coproduct Δ on $\mathcal{A}^{*,*}$ in [Voe03c, 11.8] and shows in [Voe03c, 12.6] that the dual $\mathcal{A}_{*,*}$ is a commutative algebra having a basis consisting of monomials $\tau^E \xi^{\mathbf{r}}$, where $\tau^E = \tau_{i_1} \cdots \tau_{i_s}$ and $\xi^{\mathbf{r}} = \prod \xi_i^{r_i}$; in [Voe03c, 13.2] he defines Q_i to be the dual of τ_i and defines $P^{\mathbf{r}}$ to be the dual of $\xi^{\mathbf{r}}$.

Lemma 13.11. *When $\ell \neq 2$, the Q_i generate an exterior algebra under composition, and each Q_i is a derivation:* $Q_i(xy) = Q_i(x)y + xQ_i(y)$. *Moreover,*

$$P^{\mathbf{r}}Q_i = Q_iP^{\mathbf{r}} + Q_{i+1}P^{\mathbf{r}-\ell^i}, \qquad Q_i = P^{\mathbf{r}_i}\beta - \beta P^{\mathbf{r}_i}.$$

Proof. As noted after 13.6, when $\ell \neq 2$ the subalgebra of bi-stable motivic operations generated by the Bockstein and the P^i is isomorphic to $\mathcal{A}^*_{\text{top}}$. Therefore the first sentence follows formally from 13.9. To see the final assertions, note that Milnor's formula (13.9.1) with $\mathbf{r} = (r, 0, \ldots)$ becomes $P^r Q_i - Q_i P^r = Q_{i+1} P^{r-\ell^i}$, and that Milnor's formula with $\mathbf{r} = \mathbf{r}_i = (0, \ldots, 0, 1)$ becomes $[P^{\mathbf{r}_i}, Q_0] = Q_i$. $\square$

The following consequence will be used in Lemma 5.14.

Corollary 13.12. *If $\ell \neq 2$ then*

$$\beta P^b = P^b \beta - P^{b-1} Q_1 + P^{b-1-\ell} Q_2 - P^{b-1-\ell-\ell^2} Q_3 + \cdots .$$

The analogue of Lemma 13.11 for $\ell = 2$ involves the class $\rho = \beta(\tau)$ of -1 in $k^\times / k^{\times 2} = H^{1,1}(k, \mathbb{Z}/2)$. This is illustrated by the formula $Q_2 = [P^2, Q_1] + \rho Q_0 Q_1 P^1$ of [Voe03c, 13.7], which is the special case $P^2 Q_1$ of our next formula.

Lemma 13.13. *When $\ell = 2$, the Q_i generate an exterior algebra under composition and we have*

$$P^r Q_k = Q_k P^r + Q_{k+1} P^{r-2^k} + \rho \, Q_{k-1} Q_k P^{r-2^{k-1}}$$
$$+ \rho^2 \, Q_{k-2} Q_{k-1} Q_k P^{r-3 \cdot 2^{k-2}} + \cdots + \rho^j \, Q_j \cdots Q_k P^{r-(2^j-1)2^{k-j}} + \cdots .$$

Proof. It suffices to write $P^r Q_k$ in terms of the standard basis $Q^E P^{\mathbf{r}}$ of $\mathcal{A}^{*,*}$ described in 13.10.1. This basis element cannot occur unless the coproduct of $\tau^E \xi^{\mathbf{r}}$ contains $\xi_1^r \otimes \tau_k$, by [Voe03c, (12.9)]. By inspection, the only term $\xi^{\mathbf{r}}$ which can occur is ξ_1^a, whose coproduct contains $\xi_1^a \otimes 1$. Following Milnor's method in [Mil58], we must look for terms τ^E whose coproduct contains $\xi^b \otimes \tau_k$. The two univariate terms $\tau^E = \tau_k, \tau_{k+1}$ contain $1 \otimes \tau_k$ and $\xi_1^{2^k} \otimes \tau_k$, leading to the terms $Q_k P^r + Q_{k+1} P^{r-2^k}$, and no other univariate terms are possible.

To check the remaining terms τ^E, we must check the τ_j whose coproduct contains $\xi_1^b \otimes \tau_i$ for $i < k$. The only possibilities are τ_j for $j \leq k$ since $\psi_*(\tau_0) = \tau_0 \otimes 1 + 1 \otimes \tau_0$, and if $j > 0$ then $\psi_*(\tau_j)$ contains $1 \otimes \tau_j + \xi_1^{2^{j-1}} \otimes \tau_{j-1}$. Since $\tau_j^2 = \rho \tau_{j+1} + \rho \tau_0 \xi_{j+1} + [-1] \xi_{j+1}$ by [Voe03c, 12.6], the term τ_j cannot occur for $j < k$ without τ_{j+1} also occurring. By induction on k with $E = (\epsilon_0, \ldots, \epsilon_k)$, $\psi_*(\tau^E)$ can only contain $\xi_1^b \otimes \tau_k$ when E is $E_j = (0, \ldots, 0, 1, 1, \ldots, 1)$ (j zeros), and in this case it contains $\rho^j \xi_1^b \otimes \tau_k$ for $b = 2^{k-j} + \cdots + 2^{k-1} = 2^{k-j}(2^j - 1)$. $\square$

The following consequence is needed to prove Lemma 5.14.

Corollary 13.14. *If $\ell = 2$ and $Q_j(x) = 0$ for $j = 0, \ldots, k-1$ then*

$$Q_{k+1} P^{r-2^k}(x) = P^r Q_k(x) + Q_k P^r(x), \quad \text{and hence} \quad Q_{k+1}(x) = [P^{2^k}, Q_k](x).$$

Proof. This follows from Lemma 13.13 by induction on k. If $j < k$ the hypothesis that $Q_j(x) = 0$ and the inductive formula for $Q_j P^a(x)$ shows that the term $Q_{j+1} \cdots Q_k(Q_j P^a)(x)$ contains $Q_{j+1}^2 = 0$. $\qquad\qquad\square$

Remark 13.14.1. If $\ell = 2$, the Q_i need not be derivations unless $\sqrt{-1} \in k$ (so $\rho = 0$). For example, $Q_1(xy) = Q_1(x)\, y + x\, Q_1(y) + \rho\,(\beta x)(\beta y)$. It is proven in [Voe03a, 13.4] that the general formula has the form

$$Q_i(xy) = Q_i(x)\, y + x\, Q_i(y) + \rho \sum c_{E,F} Q^E(x)\, Q^F(y),$$

where E, F are subsets of $\{0, ..., i-1\}$ and $c_{E,F} \in H^{*,*}(k, R)$.

Lemma 13.15. *The operations Q_i are $K_*^M(k)$-linear: if $y \in K_*^M(k)$ then*

$$Q_i(xy) = Q_i(x) \cdot y.$$

Proof. It suffices to consider $y \in k^\times$. Since $Q_j(y) = 0$ for all j, the result follows from Lemma 13.11 if $\ell \neq 2$, and from Remark 13.14.1 if $\ell = 2$. $\qquad\square$

13.5 Q_N OF THE DEGREE MAP

To compute the Margolis homology for Q_i in section 13.6, we need a motivic interpretation of the degree map $\deg : CH^d(Y) \to \mathbb{Z}$, where $CH^d(Y)$ is the group of zero-cycles on a smooth projective variety Y of dimension d.

At the motivic level, the degree map $\tau_{\mathrm{tr}} : \mathbb{L}^d \to \mathbb{Z}_{\mathrm{tr}}(Y)$ may be defined as the tensor product of $\mathbb{L}^d = \mathbb{Z}(d)[2d]$ with the dual $\mathbb{Z} \to \mathbb{Z}_{\mathrm{tr}}(Y)^*$ of the structure map, together with the duality $\mathbb{Z}_{\mathrm{tr}}(Y) \cong \mathbb{L}^d \otimes \mathbb{Z}_{\mathrm{tr}}(Y)^*$. Since $H^{2d,d}(\mathbb{L}^d, \mathbb{Z}) = \mathrm{Hom}(\mathbb{L}^d, \mathbb{L}^d) = \mathbb{Z}$ we have a map

$$\deg = \mathrm{Hom}(\tau_{\mathrm{tr}}, \mathbb{L}^d) : CH^d(Y) = H^{2d,d}(Y, \mathbb{Z}) \to \mathbb{Z}.$$

This is the usual degree map on zero-cycles, because for any closed point $S = \mathrm{Spec}(E)$ of Y the composition $\mathbb{Z} \to \mathbb{Z}_{\mathrm{tr}}(Y)^* \to \mathbb{Z}_{\mathrm{tr}}(S)^* \cong \mathbb{Z}_{\mathrm{tr}}(S)$ with the structure map is multiplication by $[E:k]$. We need to lift the construction of the degree map via τ_{tr} to the Morel–Voevodsky category of motivic spaces; see Definition 12.63.

Recall from [Voe03c, 4.3] that for any s-dimensional vector bundle E on Y the *Thom space* $Th_Y(E)$ is the pointed sheaf $E/(E-Y)$ and there is a *Thom class* t_E in $\widetilde{H}^{2s,s}(Th_Y(E), \mathbb{Z}) = \mathrm{Hom}(Th_Y(E), \mathbb{L}^s)$ and a *Thom isomorphism*

$$H^{*,*}(Y, \mathbb{Z}) \cong \widetilde{H}^{*,*}(Y_+, \mathbb{Z}) \xrightarrow{\ \simeq\ } \widetilde{H}^{*+2s,*+s}(Th_Y(E), \mathbb{Z}) \qquad (13.16)$$

defined by $a \mapsto a \cdot t_E$ (multiplication by t_E). For each $i > 0$, we set $T^i = \mathbb{A}^i/(\mathbb{A}^i - 0)$, so that $\mathbb{Z}_{\mathrm{tr}}(T^i) \cong \mathbb{L}^i$ and $\widetilde{H}^{2i,i}(T^i, \mathbb{Z}) \cong \mathbb{Z}$.

Let $\bar{t}_E$ denote the mod-ℓ reduction of the Thom class t_E. We pause to record the following result, proven in [Voe03c, 14.2(1)].

Lemma 13.17. *For any vector bundle E and any i we have $Q_i(\bar{t}_E) = 0$ in $H^{*,*}(Th_Y(E), \mathbb{Z}/\ell)$.*

Choose an s-dimensional vector bundle $\mathcal{N}$ on Y representing the stable normal bundle, and form its Thom space $Th_Y(\mathcal{N})$. The following theorem was proven in [Voe03a, Thm 2.11], and is the motivic analogue of Atiyah Duality [Ati61b, Thm. 3.3]; the topological analogue of the map τ in the theorem is the Pontryagin–Thom collapse map on an embedding of Y into a sphere.

Theorem 13.18. *There is a map $T^{d+s} \xrightarrow{\tau} Th_Y(\mathcal{N})$ such that the degree map $\deg : CH^d(Y) \to \mathbb{Z}$ coincides with the composition of the Thom isomorphism (13.16), τ^* and the cancellation isomorphism:*

$$H^{2d,d}(Y, \mathbb{Z}) \xrightarrow{t_\mathcal{N}} \widetilde{H}^{2(d+s),d+s}(Th_Y(\mathcal{N}), \mathbb{Z}) \xrightarrow{\tau^*} \widetilde{H}^{2(d+s),d+s}(T^{d+s}, \mathbb{Z}) \cong \mathbb{Z}.$$

Construction 13.19. The map τ of Theorem 13.18 determines a cofibration sequence:

$$T^{d+s} \xrightarrow{\tau} Th_Y(\mathcal{N}) \xrightarrow{p} Th_Y(\mathcal{N})/T^{d+s} \xrightarrow{\delta} \Sigma_s^1 T^{d+s}. \qquad (13.19.1)$$

Under the identification of $\mathbb{Z}_{\mathrm{tr}}(\Sigma_s^1 T^{d+s})$ with $\mathbb{L}^{d+s}[1]$, the map δ in (13.19.1) corresponds to a cohomology class

$$v = \mathbb{Z}_{\mathrm{tr}}(\delta) \in \widetilde{H}^{2(d+s)+1,d+s}(Th_Y(\mathcal{N})/T^{d+s}, \mathbb{Z}). \qquad (13.19.2)$$

For $d = \dim(Y) > 0$ we have $\widetilde{H}^{2s,s}(T^{d+s}, \mathbb{Z}) = \widetilde{H}^{2s-1,s}(T^{d+s}, \mathbb{Z}) = 0$ for weight reasons, so from the cohomology exact sequence associated to (13.19.1),

$$\widetilde{H}^{2s-1,s}(T^{d+s}) \to \widetilde{H}^{2s,s}(Th_Y(\mathcal{N})/T^{d+s})$$

$$\xrightarrow{p^*} \widetilde{H}^{2s,s}(Th_Y(\mathcal{N})) \xrightarrow{\tau^*} \widetilde{H}^{2s,s}(T^{d+s}),$$

we see that the Thom class $t_\mathcal{N} \in \widetilde{H}^{2s,s}(Th_Y(\mathcal{N}), \mathbb{Z})$ lifts to a unique class

$$\widetilde{t}_\mathcal{N} \in \widetilde{H}^{2s,s}(Th_Y(\mathcal{N})/T^{d+s}, \mathbb{Z}). \qquad (13.19.3)$$

If $\widetilde{t}$ and $\bar{t}_\mathcal{N}$ denote the reductions of $\widetilde{t}_\mathcal{N}$ and $t_\mathcal{N}$ modulo ℓ, then $p^*\widetilde{t}$ is $\bar{t}_\mathcal{N}$.

We now assume that $d = \ell^n - 1$, so that Q_n has bidegree $(2d+1, d)$, and $Q_n(\widetilde{t})$ has the same bidegree $(2d+2s+1, d+s)$ as v. Recall from section 1.3 that the characteristic number $s_d(Y)$ is defined as the degree of $s_d(T_Y)$, and that $s_d(Y) \equiv 0 \pmod{\ell}$.

Theorem 13.20. If $d = \ell^n - 1$ and $s_d(Y) = c \cdot \ell$ then $Q_n(\widetilde{t}) \equiv c \cdot v \pmod{\ell}$.

Proof. (Cf. [Voe11, 4.1]) Recall that $Q_n = P^{\mathbf{r}_n} \beta - \beta P^{\mathbf{r}_n}$ (for $\ell = 2$, from 13.10; for $\ell \neq 2$, from Lemma 13.11). Since $\widetilde{t}$ is the reduction of the integral class $\overline{t}_\mathcal{N}$ and β is the Bockstein, we have $\beta(\widetilde{t}) = 0$. Thus it is sufficient to show that

$$\beta \, P^{\mathbf{r}_n}(\widetilde{t}) \equiv c \cdot v \pmod{\ell}.$$

Since $p^* \widetilde{t} = \overline{t}_\mathcal{N}$, we have $p^* P^{\mathbf{r}_n}(\widetilde{t}) = P^{\mathbf{r}_n}(\overline{t}_\mathcal{N})$. The Thom isomorphism implies that $P_n^{\mathbf{r}}(\overline{t}_\mathcal{N})$ is $\overline{t}_\mathcal{N}$ times an element of $H^{2d,d}(Y, \mathbb{Z}/\ell) = CH^d(Y)/\ell$. By [Voe03c, Cor. 14.3], that element is the characteristic class $s_d(T_Y) \in CH^d(Y)$ of the tangent bundle T_Y of Y. Therefore $P_n^{\mathbf{r}}(\overline{t}_\mathcal{N})$ is the mod-ℓ reduction of the integral cohomology class $s_d(T_Y) \cdot t_\mathcal{N}$. This information is summarized by the following commutative square in the motivic category $\mathbf{DM}(k, \mathbb{Z})$:

$$
\begin{array}{ccc}
\mathbb{Z}_{\mathrm{tr}}(Th_Y(\mathcal{N})) & \xrightarrow{\quad p \quad} & \mathbb{Z}_{\mathrm{tr}}(Th_Y(\mathcal{N})/T^{d+s}) \\
{\scriptstyle s_d(T_Y) \cdot t_\mathcal{N}} \downarrow & & \downarrow {\scriptstyle P^{\mathbf{r}_n}(\widetilde{t})} \\
\mathbb{Z}(d+s)[2d+2s] & \xrightarrow{\text{mod-}\ell} & \mathbb{Z}/\ell(d+s)[2d+2s].
\end{array}
$$

Since $\mathbb{Z}_{\mathrm{tr}}(T^{d+s})$ is $\mathbb{L}^{d+s} = \mathbb{Z}(d+s)[2d+2s]$, this square extends to a morphism of distinguished triangles, where the top triangle is $\mathbb{Z}_{\mathrm{tr}}$ of (13.19.1):

$$
\begin{array}{ccccccc}
\mathbb{L}^{d+s} & \xrightarrow{\ \tau\ } & \mathbb{Z}_{\mathrm{tr}}(Th_Y(\mathcal{N})) & \xrightarrow{\ p\ } & \mathbb{Z}_{\mathrm{tr}}(Th_Y(\mathcal{N})/T^{d+s}) & \xrightarrow{\ \delta\ } & \mathbb{L}^{d+s}[1] \\
{\scriptstyle c'} \downarrow & & {\scriptstyle s_d(T_Y) \cdot t_\mathcal{N}} \downarrow & & {\scriptstyle P^{\mathbf{r}_n}(\widetilde{t})} \downarrow & & \downarrow {\scriptstyle c'} \\
\mathbb{L}^{d+s} & \xrightarrow{\ \ell\ } & \mathbb{L}^{d+s} & \longrightarrow & \mathbb{Z}/\ell(d+s)[2d+2s] & \xrightarrow{\ \widetilde{\beta}\ } & \mathbb{L}^{d+s}[1]
\end{array}
$$

for some morphism c'. Since $\mathrm{Hom}(\mathbb{L}^{d+s}, \mathbb{L}^{d+s}) = \mathbb{Z}$, c' is multiplication by some integer; the commutativity of the left square and Theorem 13.18 means that we have $c' \cdot \ell = \tau^*(s_d(T_Y) \cdot t_\mathcal{N}) = \deg(s_d(T_Y)) = s_d(Y)$, so $c' = c$. The commutativity of the right square means that we have the desired congruence:

$$c \cdot v = c \cdot [\delta] = \widetilde{\beta} \circ P^{\mathbf{r}_n}(\widetilde{t}),$$

which is equivalent to $\beta P^{\mathbf{r}_n}(\widetilde{t})$ modulo ℓ, as desired. $\square$

13.6   MARGOLIS HOMOLOGY

Since the cohomology operations Q_i satisfy $Q_i^2 = 0$, it is natural to consider $H^{*,*}(Y_\bullet, \mathbb{Z}/\ell)$ as a family of chain complexes with differential Q_i, for any simplicial variety $Y_\bullet$. The resulting homology is bigraded, and referred to as *Margolis* *homology* for Q_i because the analogous construction in topology is due to H. Margolis [Mar83]. We will show that the Margolis homology vanishes on some simplicial varieties, using the degree map $t_{\mathcal{N}}$ of Theorem 13.18.

Here is a simple example, involving the Bockstein $Q_0 = \beta$. Let X be any smooth variety, and $\mathfrak{X}$ the simplicial scheme $s \mapsto X^{s+1}$ of Definition 1.32. Recall from Definition 1.36 that the unreduced suspension $\Sigma\mathfrak{X}$ is defined to be the mapping cone of $\mathfrak{X}_+ \to \mathrm{Spec}(k)_+$.

Example 13.21. Suppose that X has a closed point x, such that $k(x)$ is a finite separable field extension of k. By Corollary 1.38, the groups $\widetilde{H}^{*,*}(\Sigma\mathfrak{X}, \mathbb{Z})$ have exponent $[k(x):k]$. If $[k(x):k]$ is not divisible by ℓ^2 then the $\widetilde{H}^{*,*}(\Sigma\mathfrak{X}, \mathbb{Z}_{(\ell)})$ have exponent ℓ. Thus the universal coefficient sequence for the integral Bockstein (localized at ℓ) is

$$0 \to \widetilde{H}^{p,q}(\Sigma\mathfrak{X}, \mathbb{Z}_{(\ell)}) \to \widetilde{H}^{p,q}(\Sigma\mathfrak{X}, \mathbb{Z}/\ell) \xrightarrow{\tilde{\beta}} \widetilde{H}^{p+1,q}(\Sigma\mathfrak{X}, \mathbb{Z}_{(\ell)}) \to 0,$$

and the sequence $\xrightarrow{\beta} \widetilde{H}^{*,*}(\Sigma\mathfrak{X}, \mathbb{Z}/\ell) \xrightarrow{\beta}$ is seen to be exact by splicing. That is, the Margolis homology of $\widetilde{H}^{*,*}(\Sigma\mathfrak{X}, \mathbb{Z}/\ell)$ for $\beta = Q_0$ vanishes.

Fix $i \geq 1$, $d = \ell^i - 1$ and a smooth d-dimensional variety Y such that $s_d(Y) \not\equiv 0 \pmod{\ell^2}$. For every smooth projective X admitting a map $Y \to X$, we construct a map $\Phi: \widetilde{H}^{p,q}(\Sigma\mathfrak{X}) \to \widetilde{H}^{p-2d-1,q-d}(\Sigma\mathfrak{X})$. Recall (from 12.1) that a map $F \to G$ of simplicial sheaves is called a global weak equivalence if $F(U) \to G(U)$ is a weak equivalence of simplicial sets for every U.

Construction 13.22. As in 13.18, choose a vector bundle $\mathcal{N}$ on Y representing the stable normal bundle. We observed just after Definition 1.32 in chapter 1 that, because $\mathcal{N}$ maps to Y and hence X, both $\mathfrak{X} \times \mathcal{N} \to \mathcal{N}$ and $\mathfrak{X} \times (\mathcal{N} - Y) \to (\mathcal{N} - Y)$ are global weak equivalences of sheaves.

Because $(\mathcal{N} - Y)_+ \to \mathcal{N}_+ \to Th_Y(\mathcal{N})$ is a cofibration sequence of pointed sheaves, and the smash product with $\mathfrak{X}_+$ preserves cofibration sequences, it follows that $\mathfrak{X}_+ \wedge Th_Y(\mathcal{N}) \simeq Th_Y(\mathcal{N})$. Since $\Sigma\mathfrak{X}$ is the cone of $\mathfrak{X}_+ \to \mathrm{Spec}(k)_+$, this implies that $\Sigma\mathfrak{X} \wedge Th_Y(\mathcal{N}) \simeq 0$. From the cofibration sequence (13.19.1) we deduce that

$$\Sigma\mathfrak{X} \wedge (Th_Y(\mathcal{N})/T^{d+s}) \xrightarrow{\simeq} \Sigma\mathfrak{X} \wedge \Sigma T^{d+s} \qquad (13.22.1)$$

is a (global) weak equivalence. We define Φ to be the composite map:

$$\widetilde{H}^{p,q}(\Sigma\mathfrak{X}) \xrightarrow{\tilde{t}_{\mathcal{N}}} \widetilde{H}^{p+2s,q+s}(\Sigma\mathfrak{X} \wedge (Th_Y(\mathcal{N})/T^{d+s}))$$

$$\xrightarrow{(13.22.1)} \widetilde{H}^{p+2s,q+s}(\Sigma\mathfrak{X} \wedge \Sigma T^{d+s}) \cong \widetilde{H}^{p-2d-1,q-d}(\Sigma\mathfrak{X}).$$

We now implicitly work with coefficients $\mathbb{Z}/\ell$, so that Q_i is defined, has bidegree $(2d+1, d)$, and both $Q_i \Phi$ and ΦQ_i have bidegree 0. This is illustrated by the following diagram, where we have written K for $Th_Y(\mathcal{N})/T^{d+s}$; the map

$\widetilde{t}$ is the reduction modulo ℓ of the lift $\widetilde{t}_N$ of the Thom class t_N, defined in (13.19.3). Thus the horizontal composites are the operation Φ. The lower right isomorphism sends $v \cdot x$ to x, where v is defined in (13.19.2).

$$
\begin{array}{ccccc}
\widetilde{H}^{p,q}(\Sigma \mathfrak{X}) & \xrightarrow{\ \widetilde{t}\ } & \widetilde{H}^{p',q'}(\Sigma \mathfrak{X} \wedge K) & \xrightarrow[\ (13.22.1)\ ]{\cong} & \widetilde{H}^{p-2d-1,q-d}(\Sigma \mathfrak{X}) \\
{\scriptstyle Q_i}\downarrow & & {\scriptstyle Q_i}\downarrow & & {\scriptstyle Q_i}\downarrow \\
\widetilde{H}^{p+2d+1,q+d}(\Sigma \mathfrak{X}) & \xrightarrow{\widetilde{t}} & \widetilde{H}^{p'+2d+1,q'+d}(\Sigma \mathfrak{X} \wedge K) & \xrightarrow[\ (13.22.1)\ ]{\cong} & \widetilde{H}^{p,q}(\Sigma \mathfrak{X})
\end{array}
$$

Here $p' = p + 2s$ and $q' = q + s$. We warn the reader that, although the right square commutes, the left square does *not* commute. This is quantified by Proposition 13.23, which is taken from [Voe03a, 3.3].

Proposition 13.23. *Suppose Y is smooth of dimension $d = \ell^i - 1$, and $s_d(Y) = c \cdot \ell$. Then for every smooth X and map $Y \to X$, $Q_i \Phi - \Phi Q_i$ is multiplication by c on $\widetilde{H}^{p,q}(\Sigma \mathfrak{X}, \mathbb{Z}/\ell)$.*

Proof. The isomorphism $\widetilde{H}^{p'+2d+1,q'+d}(\Sigma \mathfrak{X} \wedge K) \cong \widetilde{H}^{p,q}(\Sigma \mathfrak{X})$ of (13.22.1) sends $Q_i(\widetilde{t} \cdot x) - \widetilde{t} \cdot Q_i(x)$ to $Q_i \Phi(x) - \Phi Q_i(x)$, and $v \cdot x$ to x. The proposition now follows from the observation that, by Lemma 13.23.1 and Theorem 13.20,

$$
Q_i(\widetilde{t} \cdot x) - \widetilde{t} \cdot Q_i(x) = Q_i(\widetilde{t}) \cdot x = cv \cdot x. \qquad \square
$$

Lemma 13.23.1. *For all $x \in \widetilde{H}^{p,q}(\Sigma \mathfrak{X}, \mathbb{Z}/\ell)$, $Q_i(\widetilde{t} \cdot x) = Q_i(\widetilde{t}) \cdot x + \widetilde{t} \cdot Q_i(x)$.*

Proof. When $\ell \neq 2$, this is just the assertion that Q_i is a derivation (by Lemma 13.11). Thus we may suppose that $\ell = 2$. By Remark 13.14.1, we have the formula

$$
Q_i(\widetilde{t} \cdot x) = Q_i(\widetilde{t}) \, x + \widetilde{t} \, Q_i(x) + \rho \sum_{E,F} c_{E,F} \prod_{e \in E} Q_e(\widetilde{t}) \prod_{f \in F} Q_f(x),
$$

where E, F are subsets of $\{1, ..., i-1\}$. Thus it suffices to show that for all $e < i$ we have $Q_e(\widetilde{t}) = 0$ in $\widetilde{H}^{a,b}(Th_Y(\mathcal{N})/T^{d+s})$ (where $a = 2s + 2\ell^e - 1$ and $b = s + \ell^e - 1$). Consider the cohomology exact sequence of (13.19.1),

$$
\widetilde{H}^{a,b}(\Sigma_s^1 T^{d+s}) \xrightarrow{\ \delta^*\ } \widetilde{H}^{a,b}(Th_Y(\mathcal{N})/T^{d+s}) \xrightarrow{\ p^*\ } \widetilde{H}^{a,b}(Th_Y(\mathcal{N})).
$$

Now $\widetilde{H}^{a,b}(\Sigma_s^1 T^{d+s}) = 0$ for weight reasons when $e < i$, so it suffices to observe that $p^* Q_e(\widetilde{t}) = Q_e(\widetilde{t}_N)$ vanishes in $\widetilde{H}^{a,b}(Th_Y(\mathcal{N}))$, by Lemma 13.17. We now show that the Margolis homology vanishes for ν_i-varieties. $\qquad \square$

Theorem 13.24. *Suppose there is ν_i-variety X_i and a map $X_i \to X$. Then the Margolis sequence for Q_i is exact on $\widetilde{H}^{*,*}(\Sigma \mathfrak{X}, \mathbb{Z}/\ell)$.*

$$\xrightarrow{Q_i} \widetilde{H}^{*-2\ell^i+1,*-\ell^i+1} \xrightarrow{Q_i} \widetilde{H}^{*,*}(\Sigma \mathfrak{X}, \mathbb{Z}/\ell) \xrightarrow{Q_i} \widetilde{H}^{*+2\ell^i-1,*+\ell^i-1} \xrightarrow{Q_i}$$

Proof. Since the case $i = 0$ was handled in Example 13.21, we assume $i > 0$ and use the map Φ constructed in 13.22. By Proposition 13.23, $Q_i\,\Phi - \Phi\,Q_i$ must be multiplication by a nonzero constant $c \in \mathbb{Z}/\ell$. Thus the Margolis sequence is exact, because if $Q_i(x) = 0$ we have $x = Q_i(\Phi\,x)/c = Q_i(\Phi\,x/c)$. $\square$

Theorem 13.24 is one of the main reasons that ν_i-varieties are useful. This theorem was used in Proposition 3.15 to show that the cohomology operation $Q_{n-1} \cdots Q_0$ is an injection from $H^{n,n-1}(\mathfrak{X}, \mathbb{Z}/\ell)$ to $H^{2b\ell+2,b\ell+1}(\mathfrak{X}, \mathbb{Z})$, and may be viewed as the key topological step in the entire proof of Theorems A and B.

13.7   A MOTIVIC DEGREE THEOREM

Let X be a variety of dimension $d = \ell^n - 1$ such that $s_d(X) \not\equiv 0 \pmod{\ell^2}$, i.e., a ν_n-variety, set $R = \mathbb{Z}/\ell$ and let $\mathfrak{X}$ be the simplicial scheme introduced in 1.32. In Lemma 6.8, we identified $R_{\mathrm{tr}}(\mathfrak{X})$ with the unit motive $\mathbf{R}$, defined in 6.3. Also recall from section 13.5 that the degree map $CH^d(X) \to \mathbb{Z}$ may be interpreted motivically using a map $\tau_{\mathrm{tr}} : \mathbb{L}^d \to \mathbb{Z}_{\mathrm{tr}}(X)$, or (mod ℓ) as a map $\mathbf{R} \otimes \mathbb{L}^d \to R_{\mathrm{tr}}(X)$.

Proposition 5.16 asserts that (under certain hypotheses) a map λ (defined in Proposition 5.9) is such that $\lambda \circ \tau_{\mathrm{tr}} : R_{\mathrm{tr}}(\mathfrak{X}) \otimes \mathbb{L}^d \to S^{\ell-1}A$ is nonzero in $\mathbf{DM}_{\mathrm{nis}}^{\mathrm{eff}}$ $(\mathfrak{X}, R)$. Theorem 13.25 shows that it suffices to produce a nonzero element α with $Q_n(\alpha) = 0$, which is accomplished in the proof of 5.16.

Theorem 13.25.[1] *Let X be a ν_n-variety, and let α be a nonzero class in $H^{p,q}(\mathfrak{X}, R) = \mathrm{Hom}_{\mathbf{DM}}(\mathbf{R}, \mathbf{R}(q)[p])$, with $p > q$, such that $Q_n(\alpha) = 0$. Suppose that the structure map $R_{\mathrm{tr}}(X) \xrightarrow{\pi} \mathbf{R}$ factors as $R_{\mathrm{tr}}(X) \xrightarrow{\lambda} M \xrightarrow{y} \mathbf{R}$ with $\alpha \circ y = 0$ for some M.*
Then $\lambda \circ \tau_{\mathrm{tr}} : \mathbf{R} \otimes \mathbb{L}^d \to M$ is nonzero as a map in $\mathbf{DM}_{\mathrm{nis}}^{\mathrm{eff}}(\mathfrak{X}, R)$.

The situation in Theorem 13.25 and its proof is summarized by the following commutative diagram in $\mathbf{DM}$, with $\alpha \circ y = 0$.

---
1. Compare with [Voe11, 4.4].

Proof. We begin the proof with a reduction. Let M' be the motive defined by the distinguished triangle

$$\mathbf{R}(q)[p-1] \longrightarrow M' \longrightarrow \mathbf{R} \xrightarrow{\alpha} \mathbf{R}(q)[p].$$

Since we assume that $\alpha \circ y = 0$, y lifts to a morphism $M \to M'$ in the sense that y is the composite $M \to M' \to \mathbf{R}$. Therefore to prove the proposition it is sufficient to show that the composition

$$\mathbf{R} \otimes \mathbb{L}^d \to R_{\mathrm{tr}}(X) \to M \to M'$$

is nonzero. We may now forget about the original M and consider only M'.

By 4.5, $\mathrm{Hom}_{\mathfrak{X}}(\mathbf{R} \otimes \mathbb{L}^d, \mathbf{R}) = H^{-2d,-d}(\mathfrak{X}, R)$; this is zero as $-d < 0$. Thus the composition $\pi \circ \tau_{\mathrm{tr}} : \mathbf{R} \otimes \mathbb{L}^d \to \mathbf{R}$ is zero, and π extends to a morphism $\widetilde{\pi} :$ $\mathrm{cone}(\tau_{\mathrm{tr}}) \to \mathbf{R}$; $\widetilde{\pi}$ is uniquely determined because $\mathrm{Hom}(\mathbf{R} \otimes \mathbb{L}^d[1], \mathbf{R}) = 0$.

If the composition $\mathbf{R} \otimes \mathbb{L}^d \to M'$ were zero then λ would lift to $\widetilde{\lambda} : \mathrm{cone}(\tau_{\mathrm{tr}}) \to M'$ and $y \circ \widetilde{\lambda}$ would be a second lift of π. This would imply that

$$\alpha \circ \widetilde{\pi} = \alpha \circ y \circ \widetilde{\lambda} = 0.$$

Therefore, to finish the proof, it suffices to show that $\alpha \circ \widetilde{\pi}$ is nonzero.

Recall from Theorem 13.18 that the degree map τ_{tr} arises from the composition of the Thom isomorphism with a map $\tau : T^N \to Th_X(\mathcal{N})$, where $\mathcal{N}$ is an s-dimensional bundle representing the stable normal bundle on X, $N = d + s$, and $T^N = \mathbb{A}^N / (\mathbb{A}^N - 0)$. Smashing the sequence (13.19.1) with $\mathfrak{X}_+$, we see that τ determines a cofibration sequence

$$T^N \wedge \mathfrak{X}_+ \xrightarrow{\tau} Th_X(\mathcal{N}) \wedge \mathfrak{X}_+ \to (Th_X(\mathcal{N})/T^N) \wedge \mathfrak{X}_+ \xrightarrow{\delta} \Sigma_s^1 T^N \wedge \mathfrak{X}_+.$$

By (13.19.3), the Thom class $t_{\mathcal{N}}$ in $H^{2d,d}(Th_X(\mathcal{N}), \mathbb{Z})$ lifts to a class $\widetilde{t}_{\mathcal{N}}$ in $\widetilde{H}^{2d,d}(Th_X(\mathcal{N})/T^N, \mathbb{Z})$. By abuse of notation, we write $\widetilde{t}$ for the image of $\widetilde{t}_{\mathcal{N}}$ modulo ℓ in $\widetilde{H}^{2d,d}(Th_X(\mathcal{N})/T^N \wedge \mathfrak{X}_+, R)$. Since $Th_X(\mathcal{N})$ maps to X and hence $\mathfrak{X}$, we have $Th_X(\mathcal{N}) \wedge \mathfrak{X}_+ \xrightarrow{\sim} Th_X(\mathcal{N})$ by the discussion in section 4.2. Therefore we have a commutative diagram:

$$
\begin{array}{ccc}
\mathrm{Hom}(\mathrm{cone}(\tau_{\mathrm{tr}}), \mathbf{R}(q)[p]) & \xrightarrow{\quad p^* \quad} & H^{p,q}(X, R) \\
\downarrow {\scriptstyle y \mapsto y \cdot \widetilde{t}} & & \downarrow {\scriptstyle x \mapsto x \cdot t_{\mathcal{N}}} \\
\widetilde{H}^{p+2d,q+d}(Th_X(\mathcal{N})/T^N \wedge \mathfrak{X}_+, R) & \longrightarrow & \widetilde{H}^{p+2d,q+d}(Th_X(\mathcal{N}), R).
\end{array}
$$

In the left vertical map, we have written $y \cdot \widetilde{t}$ for the more pedantic product of $p^*(y) \in H^{p,q}(X, R)$ with $\widetilde{t}$.

Since π represents $1 \in H^{0,0}(X, R)$ and $\widetilde{\pi} \in \mathrm{Hom}(\mathrm{cone}(\tau_{\mathrm{tr}}), \mathbf{R})$ has $p^*(\widetilde{\pi}) = \pi$, we have $\widetilde{\pi} \cdot \widetilde{t} = \widetilde{t}$. Hence the left vertical map sends $\alpha \circ \widetilde{\pi}$ to $\alpha \cdot \widetilde{t}$. Hence it suffices to show that $\widetilde{t} \cdot \alpha \neq 0$.

We are going to show that $Q_n(\widetilde{t} \cdot \alpha) \neq 0$. First, we claim that

$$Q_n(\widetilde{t} \cdot \alpha) = Q_n(\widetilde{t}) \alpha. \tag{13.25.1}$$

For $\ell > 2$, this follows from the fact that Q_n is a derivation (Lemma 13.11) and $Q_n(\alpha) = 0$. For $\ell = 2$, Q_n is not a derivation, and there are additional terms on the right side of (13.25.1) which depend on $Q_i(\widetilde{t})$ for $i < n$, as described in Remark 13.14.1. As in the proof of Lemma 13.23.1, it follows from simple weight considerations that $Q_i(\widetilde{t}) = 0$ for $i < n$ and therefore (13.25.1) holds for $\ell = 2$ as well.

Theorem 13.20 shows that the right-hand side $Q_n(\widetilde{t})\alpha$ of (13.25.1) equals $c \cdot v \cdot \alpha$ where $c = s_d(X)/\ell$ and v is given in (13.19.2). Since X is a ν_n-variety, c is an invertible element of $\mathbb{Z}/\ell$. Hence it suffices to check that $v \cdot \alpha \neq 0$. Since $v = \delta^*(u)$, where u is the tautological generator of

$$H^{2N+1,N}(\Sigma_s^1 T^N, R) \cong H^{0,0}(\mathrm{Spec}(k), R) = R,$$

we have $v \cdot \alpha = \delta^*(u \cdot \alpha)$. Setting $p' = p + 2N$ and $q' = q + N$, the element $u \cdot \alpha$ lies in $H^{p'+1,q'}(\Sigma_s^1 T^N \wedge \mathfrak{X}_+, R) \cong H^{p,q}(\mathfrak{X}, R)$ and is nonzero because $\alpha \neq 0$. Because $Th_X(\mathcal{N}) \wedge \mathfrak{X}_+ \xrightarrow{\cong} Th_X(\mathcal{N})$, we have an exact sequence

$$H^{p',q'}(Th_X(\mathcal{N}), R) \xrightarrow{\tau^*} H^{p',q'}(T^N \wedge \mathfrak{X}_+, R) \xrightarrow{\delta^*} H^{p',q'}(Th_X(\mathcal{N})/T^N \wedge \mathfrak{X}_+, R),$$

and $H^{p',q'}(Th_X(\mathcal{N}), R) \cong H^{p+2d,q+d}(X, R) = 0$ by the Thom isomorphism and the cohomological dimension theorem. Hence δ^* is an injection; since $u \cdot \alpha \neq 0$, we have $v \cdot \alpha = \delta^*(u \cdot \alpha) \neq 0$, as required. $\qquad\square$

13.8 HISTORICAL NOTES

Steenrod's original 1947 paper constructing the Steenrod squares opened the doors for remarkable developments in algebraic topology; his Sq^1 was the Bockstein, first introduced by Meyer Bockstein in 1942. Henri Cartan discovered the Cartan formula almost immediately; Serre recognized the connection to Eilenberg–Mac Lane spaces and classified all mod–2 cohomomology operations in his 1952 note [Ser52]. Steenrod later constructed the reduced power operations P^i and Cartan used them to describe all mod–ℓ cohomomology operations in his 1954 note [Car54]. The 1962 book [Ste62] summarized the main developments of that era.

The idea that there should be cohomology operations in motivic cohomology, and that they should be defined in the $\mathbb{A}^1$-homotopy category, is due to

Voevodsky. His construction of the operations P^i is modelled on the work of Steenrod [Ste62]. The main structural theorems of [Voe03c] are summarized in section 13.2; some typos in loc. cit. have been corrected in 13.7.

Similarly, the results in section 13.4 are based on the 1958 paper [Mil58] by Milnor, which introduced the operations Q_i, and all the formulas for $\ell \neq 2$ are due to Milnor. Of course, the motivic definition of the Q_i and many of the other results are taken from [Voe03c]. The formulas 13.13 and 13.14 are new.

Sections 13.5 and 13.6 are taken from sections 2 and 3 of Voevodsky's paper [Voe03a] and from Riou's commentaries [Rio14] and [Rio12]. The final section 13.7 is taken from section 4 of [Voe11].

Chapter Fourteen

Symmetric Powers of Motives

IN THIS CHAPTER we develop the basic theory of symmetric powers of smooth varieties, which will play an important role in the construction of motivic classifying spaces in chapter 15.

The constructions in this chapter are based on an analogy with the corresponding symmetric power constructions in topology. Recall that if K is a set (or even a topological space) then the symmetric power $S^m K$ is defined to be the orbit space K^m/Σ_m, where Σ_m is the symmetric group. If K is pointed, there is an inclusion $S^m K \subset S^{m+1} K$ and $S^\infty K = \bigcup S^m K$ is the free abelian monoid on $K - \{*\}$.

When K is a connected topological space, the Dold–Thom theorem [DT58] says that $\tilde{H}_*(K, \mathbb{Z})$ agrees with the homotopy groups $\pi_*(S^\infty K)$. In particular, the spaces $S^\infty(S^n)$ have only one homotopy group ($n \geq 1$) and hence are the Eilenberg–Mac Lane spaces $K(\mathbb{Z}, n)$ which classify integral homology.

14.1 SYMMETRIC POWERS OF VARIETIES

If X is a quasi-projective variety over a perfect field, its mth symmetric power is the geometric quotient variety $S^m X = X^m/\Sigma_m$, where Σ_m is the symmetric group; locally, if $X = \mathrm{Spec}(A)$ and $B = A \otimes \cdots \otimes A$ then $S^m X = \mathrm{Spec}(B^{\Sigma_m})$. If G is a subgroup of Σ_m then we also set $S^G X = X^m/G$; locally $S^G X = \mathrm{Spec}(B^G)$. This is an abuse of notation, since $S^G X$ depends upon m.

The example $X = \mathbb{A}^2$ shows that symmetric powers of smooth varieties are not always smooth. However, if X is normal then X^m, $S^m X$, and $S^G X$ are also normal, because locally the coordinate rings of X^m and $S^G X$ are the normal domains $B = A^{\otimes m}$ and $B \cap k(X^m)^G$. For this reason, we shall work with the category **Norm** of normal quasi-projective varieties; S^m and S^G determine functors from **Norm** to itself.

Replacing X^m by the smash product $(X_+)^{\wedge m} = (X^m)_+$, the formula

$$\tilde{S}^G(X_+) = (X_+)^{\wedge m}/G = (S^G X)_+$$

defines "reduced" functors $\tilde{S}^G$ from the pointed category **Norm**$_+$ (normal varieties of the form $X \amalg *$) to itself. The unreduced functors S^m are defined on **Norm**$_+$ by $S^G(X_+) = \tilde{S}^G(X \amalg *)$.

By naturality, S^G and $\widetilde{S}^G$ also determine self-functors on $\Delta^{op}\mathbf{Norm}$ and $\Delta^{op}\mathbf{Norm}_+$. There is a natural decomposition:

$$S^m(X \amalg Y) = \coprod\nolimits_{i+j=m} S^i(X) \times S^j(Y). \tag{14.1}$$

Setting $Y = \mathrm{Spec}(k)$, formula (14.1) gives a decomposition of $S^m(X_+)$ as the coproduct of the $S^i(X)$ for $0 \le i \le m$. This decomposition is not natural for pointed maps, but it is related to $\widetilde{S}^m(X_+) = (S^m X)_+$ by a natural sequence of pointed objects, split for each X:

$$* \to S^{m-1}(X_+) \to S^m(X_+) \to \widetilde{S}^m(X_+) \to *. \tag{14.2}$$

We let $S^\infty(X_+)$ denote the filtered colimit of the pointed presheaves $S^m(X_+)$ along the maps in the split sequences (14.2). For example, for $S^0 = \mathrm{Spec}(k)_+$ we have $\widetilde{S}^m(S^0) = S^0$, $S^m(S^0) \cong \{0, 1, \dots, m\}$, and $S^\infty S^0 = \mathbb{N}$.

If X is a variety then the sets $(S^G X)(U)$ need not equal the symmetric powers $S^G(X(U))$ of the sets $X(U)$. For example, if $X = \mathbb{A}^1$ and $U = \mathrm{Spec}(k)$ then $S^2 X = \mathbb{A}^2$ and the natural map from $S^2(\mathbb{A}^1(k)) = \mathbb{A}^2(k)/\Sigma_2$ to $\mathbb{A}^2(k)$, sending $(a, b) = (b, a)$ to $(a + b, ab)$, is not onto when k has a quadratic field extension.

Definition 14.3. Let $G-\mathbf{Norm}$ denote the category of normal G-schemes and equivariant morphisms. The functor endowing a scheme with the trivial G-action has a left adjoint, the quotient functor $(/G) : T \mapsto T/G$ from $G-\mathbf{Norm}$ to $\mathbf{Norm}$. Its inverse image functor $(/G)^*$, from $\mathrm{Pshv}(G-\mathbf{Norm})$ to $\mathrm{Pshv}(\mathbf{Norm})$, is defined by $(/G)^* H(Y) = \mathrm{colim}_{Y \to T/G} H(T)$. Note that $(/G)^*$ commutes with colimits because it is left adjoint to $F \mapsto F(-/G)$.

It will be convenient to factor S^G as the composition of $(/G)^*$ and a functor P from presheaves on $\mathbf{Norm}$ to presheaves on $G-\mathbf{Norm}$, which we now define.

Definition 14.4. Let G be a subgroup of Σ_m. Given a G-scheme T, let $T^{\amalg m}$ denote the disjoint union of m copies of T, made into a G-scheme via the diagonal action, with quotient $T^{\amalg m}/G$. For any (simplicial) presheaf F on $\mathbf{Norm}$, we define the (simplicial) presheaf PF on the category of normal G-schemes by

$$(PF)(T) = F\left(T^{\amalg m}/G\right).$$

It is clear that P commutes with colimits: $P(\mathrm{colim}\, F_\alpha) \cong \mathrm{colim}(PF_\alpha)$. Since PF is natural in F, P is a functor from $\mathrm{Pshv}(\mathbf{Norm})$ to $\mathrm{Pshv}(G-\mathbf{Norm})$.

Similarly, we define $P : \mathrm{Pshv}(\mathbf{Norm}_+) \to \mathrm{Pshv}(G-\mathbf{Norm}_+)$ by sending a presheaf F on $\mathbf{Norm}_+$ to $(PF)(T) = F\left(T^{\vee m}/G\right)$, where $T^{\vee m} = T \vee \cdots \vee T$ with the diagonal action.

Lemma 14.5. *If X is a normal scheme, the presheaf PX is representable by the G-scheme X^m. That is, $(PX)(T) = \mathrm{Hom}_G(T, X^m)$.*[1]
Similarly, $P(X_+)$ is representable by the pointed G-scheme $(X^m)_+$.

Proof. Non-equivariantly, we have natural isomorphisms

$$\mathrm{Hom}(T, X^m) \cong \mathrm{Hom}(T, X)^m \cong \mathrm{Hom}(T^{\amalg m}, X).$$

A map $(f_1, ..., f_m) : T \to X^m$ is equivariant if $f_i(\gamma t) = f_{\gamma(i)}(t)$ for all $\gamma \in G$ and i. Since G acts trivially on X, this is also the condition for a map $T^{\amalg m} \to X$ to be equivariant. Thus we have $\mathrm{Hom}_G(T, X^m) \cong \mathrm{Hom}(T^{\amalg m}/G, X)$.
    The pointed version is left to the reader.   □

    The considerations of chapter 12 also apply to simplicial presheaves on $G-$**Norm**. For example, the notions of global weak equivalence, etc. make sense for simplicial presheaves on $G-$**Norm**. This includes the notion of a radditive presheaf (Definition 12.14).

Lemma 14.6. *P preserves global weak equivalences, and sends radditive presheaves to radditive presheaves.*

Proof. Let $F_1 \to F_2$ be a global weak equivalence of simplicial presheaves on **Norm**. Then for every normal G-scheme T we have a weak equivalence

$$(PF_1)(T) = F_1(T^{\amalg m}/G) \xrightarrow{\sim} F_2(T^{\amalg m}/G) = (PF_2)(T).$$

Hence $PF_1 \to PF_2$ is a global weak equivalence of simplicial presheaves on $G-$**Norm**. Similarly, if F is radditive on **Norm** then for every T_1 and T_2:

$$PF(T_1 \amalg T_2) = F\left((T_1^{\amalg m} \amalg T_2^{\amalg m})/G\right) = F((T_1^{\amalg m}/G) \amalg (T_2^{\amalg m}/G))$$
$$= F(T_1^{\amalg m}/G) \times F(T_2^{\amalg m}/G) = PF(T_1) \times PF(T_2).$$

Hence PF is also a radditive presheaf.   □

Proposition 14.7. *If F is a presheaf on **Norm**, resp., **Norm**$_+$, there is a presheaf isomorphism*

$$(S^G)^* F \cong (/G)^*(PF), \quad resp., \quad (\widetilde{S}^G)^* F \cong (/G)^*(PF).$$

Proof. Because all functors involved commute with colimits, and F is the coequalizer of $L_1 F \rightrightarrows L_0 F$ (see Construction 12.7), we may assume that F is represented by a normal scheme X. But by Lemma 14.5, we have

$$S^G(X) = (/G)^*(X^m) = (/G)^*(PX).$$ □

---

1. Taken from [Del09], Ex. 6.

14.2   SYMMETRIC POWERS OF CORRESPONDENCES

The symmetric power construction is compatible with elementary and finite correspondences, as defined in section 6.1. Indeed, if $f : X \to Y$ is a finite correspondence then so is $f^{\times m} : X^m \to Y^m$. If $f^{\times m}$ is an elementary correspondence then it induces an elementary correspondence $X^m/G \to Y^m/G$ for every $G \subseteq \Sigma_m$, almost by definition. In general, the finite correspondence $f^{\times m} : X^m \to Y^m/G$ descends to a finite correspondence from X^m/G to Y^m/G, which we call $S_{\mathrm{tr}}^G(f)$, by the case $Z = Y^m/G$ of the following formula of Suslin and Voevodsky; see [SV96, 5.16].

$$\mathbf{Cor}(X^m/G, Z) \xrightarrow{\cong} \mathbf{Cor}(X^m, Z)^G$$

Recall (say from [MVW]) that $\mathbf{Cor}(\mathbf{Norm})$ denotes the category of finite correspondences on $\mathbf{Norm}$, with coefficients in a fixed ring R. We will sometimes require that $|G|$ is invertible in R. For clarity, we shall write $R_{\mathrm{tr}}(X)$ for a normal scheme X, regarded as an object of $\mathbf{Cor}(\mathbf{Norm})$.

Definition 14.8. The endofunctor $S_{\mathrm{tr}}^G : \mathbf{Cor}(\mathbf{Norm}) \to \mathbf{Cor}(\mathbf{Norm})$ is defined for any subgroup G of Σ_m, on objects by $S_{\mathrm{tr}}^G(R_{\mathrm{tr}}(X)) = R_{\mathrm{tr}}(S^G X)$, and on morphisms by $f \mapsto S_{\mathrm{tr}}^G(f)$. When $G = \Sigma_m$ we write S_{tr}^m for $S_{\mathrm{tr}}^{\Sigma_m}$. By naturality, S_{tr}^G extends to an endofunctor on the simplicial category $\Delta^{\mathrm{op}}\mathbf{Cor}(\mathbf{Norm})$.

For X_+ in $\mathbf{Norm}_+$ we define $S_{\mathrm{tr}}^G R_{\mathrm{tr}}(X_+) = R_{\mathrm{tr}}(S^G(X_+))$, so that $S_{\mathrm{tr}}^G \circ R_{\mathrm{tr}} = R_{\mathrm{tr}} \circ S^G$ is a functor from $\mathbf{Norm}_+$ to $\mathbf{Cor}(\mathbf{Norm})$. In particular, $S_{\mathrm{tr}}^m R_{\mathrm{tr}}(X_+) = R_{\mathrm{tr}} S^m(X_+) = R_{\mathrm{tr}}((S^m X)_+)$. The next lemma is based on [Voe10c, 2.34].

Lemma 14.9. *For any (simplicial) normal varieties X, Y we have*

$$S_{\mathrm{tr}}^m(R_{\mathrm{tr}} X \oplus R_{\mathrm{tr}} Y) \cong \bigoplus_{i+j=m} S_{\mathrm{tr}}^i(R_{\mathrm{tr}} X) \otimes S_{\mathrm{tr}}^j(R_{\mathrm{tr}} Y).$$

Proof. Immediate from $R_{\mathrm{tr}}(X \coprod Y) = R_{\mathrm{tr}} X \oplus R_{\mathrm{tr}} Y$ and (14.1).   □

Lemma 14.10. *Let G be a finite group whose order is invertible in R. If G acts faithfully and algebraically on X then $\pi : X \to X/G$ induces an isomorphism $R_{\mathrm{tr}}(X)^G \xrightarrow{\cong} R_{\mathrm{tr}}(X/G)$.*

Proof. The transpose π^t is a finite correspondence from X/G to X, and is equivariant, so we may regard π^t as a map from $R_{\mathrm{tr}}(X/G)$ to $R_{\mathrm{tr}}(X)^G$. Since $\pi : X \to X/G$ is pseudo-Galois with group G (see Definition 8.19), the composition $\pi \circ \pi^t$ is multiplication by $|G|$ on $R_{\mathrm{tr}}(X/G)$—which is an isomorphism by assumption—and $\pi^t \circ \pi$ is $\sum_{g \in G} g$. (See [SV96, 5.17] or [SV00b, 3.6.7].) But $\sum g$ is an isomorphism on $R_{\mathrm{tr}}(X)^G$, since the idempotent $e = (1/|G|) \sum_{g \in G} g$ of $R[G]$ acts on $R_{\mathrm{tr}}(X)$ as projection onto the summand $R_{\mathrm{tr}}(X)^G$.   □

Corollary 14.11. *If $H \lhd G \subseteq \Sigma_n$ and $[G:H]$ is invertible in R, then G/H acts on $S^H X$ and $S^G_{\mathrm{tr}}(R_{\mathrm{tr}}(X)) \cong S^H_{\mathrm{tr}}(R_{\mathrm{tr}}(X))^{G/H}$.*

Proof. Apply Lemma 14.10 to the action of G/H on $S^H X$, using $S^G_{\mathrm{tr}} R_{\mathrm{tr}}(X) = R_{\mathrm{tr}}(S^G X)$ and $S^G X = (S^H X)/(G/H)$. $\qquad\qquad\qquad\qquad\qquad\qquad\qquad\qquad\qquad\qquad\qquad\square$

Examples 14.12. (a) Since $S^m(\operatorname{Spec} k) \cong \operatorname{Spec} k$, $S^m_{\mathrm{tr}}(R) \cong R$ for all m. If the s_i are the elementary symmetric polynomials in the x_j then $S^m(\mathbb{A}^1) = \operatorname{Spec}(k[s_1, ..., s_m]) \cong \mathbb{A}^m$.

(b) The functors S^m_{tr} extend to the idempotent completion of **Cor**, because if e is an idempotent finite correspondence then so is $S^G(e)$. For example, since $R_{\mathrm{tr}}(\mathbb{P}^1) \cong R \oplus \mathbb{L}^1$ and $S^m \mathbb{P}^1 \cong \mathbb{P}^m$, we see that $S^m_{\mathrm{tr}}(R \oplus \mathbb{L}^1) = R_{\mathrm{tr}}(\mathbb{P}^m) = R_{\mathrm{tr}}(\mathbb{P}^{m-1}) \oplus \mathbb{L}^m$. Lemma 14.9 yields $S^m_{\mathrm{tr}}(\mathbb{L}^1) \cong \mathbb{L}^m$.

(c) The map $\pi : S^m(\mathbb{A}^1) \to \mathbb{A}^1$, $(a_1, ..., a_m) \mapsto a_1 \cdots a_m$, and its restriction $S^m(\mathbb{A}^1 - \{0\}) \to \mathbb{A}^1 - \{0\}$, are vector bundles with fiber $\mathbb{A}^{m-1}$, split by $a \mapsto (a, 1, \ldots, 1)$. This is because π is the map $\operatorname{Spec}(k[s_1, ..., s_m]) \to \operatorname{Spec}(k[s_m])$. For later use, we note the consequence that $S^m(\mathbb{A}^1 - \{0\}) \cong (\mathbb{A}^1 - \{0\}) \times \mathbb{A}^{m-1}$, and hence the simplicial presheaf $S^m(\mathbb{A}^1 - \{0\})$ is $\mathbb{A}^1$-weakly equivalent to $(\mathbb{A}^1 - \{0\})$.

(d) Suppose that $m!$ is a unit of R, so that the symmetrizing idempotent $e = (\Sigma \sigma)/m!$ of $R[\Sigma_m]$ exists. Since $R[\Sigma_m]$ acts on $R_{\mathrm{tr}}(X^m)$, we can form the summand $e \cdot R_{\mathrm{tr}}(X^m)$. From Lemma 14.10 we see that the canonical map $R_{\mathrm{tr}}(X^m) \to R_{\mathrm{tr}}(S^m X) = S^m_{\mathrm{tr}}(R_{\mathrm{tr}} X)$ induces an isomorphism

$$S^m_{\mathrm{tr}}(R_{\mathrm{tr}} X) \cong R_{\mathrm{tr}}(X^m)^{\Sigma_m} = e \cdot R_{\mathrm{tr}}(X^m), \qquad \text{when } 1/m! \in R.$$

(e) Suppose that $1/m! \in R$. If T is such that the interchange τ on $T \otimes T$ is $\mathbb{A}^1$-homotopic to the identity (e.g., $T = \mathbb{L}^a[2b]$), then $S^m_{\mathrm{tr}}(T) \simeq T^{\otimes m}$, by (d). If $\tau \simeq -1$ (e.g., $T = \mathbb{L}^a[2b+1]$) and $m > 1$, then $S^m_{\mathrm{tr}}(T) \simeq 0$, again by (d). In particular,

$$S^m_{\mathrm{tr}}(\mathbb{L}^a[2b]) \simeq \mathbb{L}^{am}[2bm], \qquad S^m_{\mathrm{tr}}(\mathbb{L}^a[2b+1]) \simeq 0.$$

The following result is the analogue of a formula for augmented simplicial R-modules discovered by Steenrod in [Ste72]; it is taken from [Voe10c, 2.47].

Lemma 14.13. *(Steenrod's Formula) Applying R_{tr} to (14.2) (for X_+) yields a split exact sequence; the splitting is natural in **Norm$_+$**:*

$$0 \to R_{\mathrm{tr}} S^{m-1}(X_+) \to R_{\mathrm{tr}} S^m(X_+) \to R_{\mathrm{tr}} \widetilde{S}^m(X_+) \to 0.$$

Proof. By (14.2), the sequence is split exact for each X_+. Recall that, as in section 5.2, for each X and $i < m$ the transfer maps for $S^i(X) \times X^{m-i} \to S^m(X)$ and the structure map $\pi_X : X \to *$ induce maps $\tau_i : R_{\mathrm{tr}} S^m(X) \to R_{\mathrm{tr}} S^i(X)$ and hence from $S^m_{\mathrm{tr}} R_{\mathrm{tr}}(X_+) = R_{\mathrm{tr}}(S^m(X)_+)$ to $S^i_{\mathrm{tr}} R_{\mathrm{tr}}(X_+) = R_{\mathrm{tr}}(S^i(X)_+)$.

The alternating sum (over i) of the transfers τ_i defines a splitting map

$$\tau : S_{\mathrm{tr}}^m R_{\mathrm{tr}}(X_+) = R_{\mathrm{tr}}(S^m(X)_+) \to \oplus_{i=1}^m R_{\mathrm{tr}}(S^i(X)_+) = \oplus_{i=1}^m S_{\mathrm{tr}}^i R_{\mathrm{tr}}(X_+).$$

We claim that τ is natural for maps $f : Z_+ \to X_+$. To see this, note that Z_+ is the disjoint union of $f^{-1}(X)$ and $Y_+ = f^{-1}(*)$. Thus f factors as a composite

$$Z_+ \cong \left(f^{-1}(X) \amalg Y\right)_+ \xrightarrow{(f \amalg 1)_+} (X \amalg Y)_+ \xrightarrow{\pi_Y} X_+,$$

where π_Y sends Y to the basepoint. Since the transfers τ_i are natural for maps coming from **Norm**, such as $(f \amalg 1)_+$, it suffices to consider the projections $\pi_Y : (X \amalg Y)_+ \to X_+$.

We must show that for each $j > 0$ and $a < i$ the composition from the summand $S_{\mathrm{tr}}^i R_{\mathrm{tr}}(X) \otimes S_{\mathrm{tr}}^j R_{\mathrm{tr}}(Y)$ of $R_{\mathrm{tr}} S^m(X \amalg Y)$ to $S_{\mathrm{tr}}^a R_{\mathrm{tr}}(X)$ is zero.

This composition factors through each of the $\binom{j}{b}$ terms $S_{\mathrm{tr}}^a R_{\mathrm{tr}}(X) \otimes S_{\mathrm{tr}}^b R_{\mathrm{tr}}(Y)$ for $b = 0, \ldots, j$, and the result follows from $\sum_{b=0}^j (-1)^b \binom{j}{b} = 0$. $\square$

Corollary 14.14. *For any simplicial object $V_\bullet$ of* **Norm**$_+$ *there is an isomorphism*

$$R_{\mathrm{tr}}(S^\infty V_\bullet) = \lim_{m \to \infty} R_{\mathrm{tr}}(S^m V_\bullet) \cong \bigoplus_{i=1}^\infty S_{\mathrm{tr}}^i R_{\mathrm{tr}}(V_\bullet).$$

Proof. Since $R_{\mathrm{tr}} S^0(V_\bullet) = 0$, the formula $R_{\mathrm{tr}}(S^m V_\bullet) \cong \bigoplus_{i=1}^m S_{\mathrm{tr}}^i R_{\mathrm{tr}}(V_\bullet)$ follows by induction, using the splittings in Lemma 14.13. $\square$

Remark 14.14.1. We will now describe S_{tr}^m in terms of S_{tr}^ℓ when $R = \mathbb{Z}_{(\ell)}$. If G is any subgroup of Σ_m, and $n \geq 1$, the wreath product

$$G \wr \Sigma_n = G^n \rtimes \Sigma_n$$

acts on $\{1, \ldots, mn\}$ by decomposing it into n blocks of m elements, with G^n acting on the blocks and Σ_n permuting the blocks. Thus $G \wr \Sigma_n \subset \Sigma_{mn}$. It is easy to see that

$$S^{G \wr \Sigma_n}(X) = S^n(S^G(X)), \qquad S_{\mathrm{tr}}^{G \wr \Sigma_n}(R_{\mathrm{tr}} X) = S_{\mathrm{tr}}^n S_{\mathrm{tr}}^G(R_{\mathrm{tr}} X).$$

Similarly, if H is a subgroup of Σ_n and we embed $\Sigma_m \times \Sigma_n$ in Σ_{m+n} then $S^{G \times H}(X) = S^G(X) \times S^H(X)$ and $S_{\mathrm{tr}}^{G \times H}(R_{\mathrm{tr}}X) = S_{\mathrm{tr}}^G(R_{\mathrm{tr}}X) \otimes S_{\mathrm{tr}}^H(R_{\mathrm{tr}}X)$.

Proposition 14.15. *Consider the ℓ-adic expansion $m = m_0 + m_1\ell + \cdots + m_r\ell^r$, with $0 \le m_i < \ell$. The subgroup*

$$G = \Sigma_{m_0} \times (\Sigma_\ell \wr \Sigma_{m_1}) \times ((\Sigma_\ell \wr \Sigma_\ell) \wr \Sigma_{m_2}) \times \cdots \times ((\Sigma_\ell^{\ell^r}) \wr \Sigma_{m_r})$$
$$\subseteq \Sigma_{m_0} \times \Sigma_{m_1\ell} \times \Sigma_{m_2\ell^2} \times \cdots \times \Sigma_{m_r\ell^r}$$

of Σ_m contains a Sylow ℓ-subgroup of Σ_m. If $R = \mathbb{Z}_{(\ell)}$ or $\mathbb{Z}/\ell$, then for every simplicial V and $M = R_{\mathrm{tr}}(V)$, $S_{\mathrm{tr}}^m(M)$ is a direct summand of

$$S_{\mathrm{tr}}^G(M) = (S_{\mathrm{tr}}^{m_0} M) \otimes S_{\mathrm{tr}}^{m_1}(S_{\mathrm{tr}}^\ell M) \otimes S_{\mathrm{tr}}^{m_2}(S_{\mathrm{tr}}^\ell(S_{\mathrm{tr}}^\ell M)) \otimes \cdots \otimes S_{\mathrm{tr}}^{m_r}((S_{\mathrm{tr}}^\ell)^r M).$$

Proof. It is well known (and easy to check) that G contains a Sylow ℓ-subgroup of Σ_m, i.e., that $\ell \nmid d = [\Sigma_m : G]$. The displayed formula for $S_{\mathrm{tr}}^G(M)$ follows from the above remarks, and the map π from $S^G(V) = V^m/G$ to $S^m V = V^m/\Sigma_m$ is finite of degree d. The transpose π^t is a finite correspondence, and the composition $\pi \circ \pi^t$ is multiplication by d on $R_{\mathrm{tr}}(\widetilde{S}^m V) = S_{\mathrm{tr}}^m(M)$, and hence is an isomorphism. $\qquad\square$

14.3   WEAK EQUIVALENCES AND SYMMETRIC POWERS

The goal of this section is to show that the functors S^G, and in particular the symmetric powers S^m, preserve $\mathbb{A}^1$-local equivalences between filtered colimits of representable presheaves on **Norm**. As in Definition 12.15, we write **Norm**^{ind} for the category of *ind-normal varieties*, i.e., the category of filtered colimits of representable presheaves on **Norm** (see [AGV73, I.8.2]).

For example, consider the functorial cofibrant replacement **Lres**(F) of a simplicial radditive presheaf F, described in Example 12.19 and Remark 12.7. Each of the terms **Lres**$_n F$ is an infinite coproduct of representable presheaves, which are filtered colimits of finite coproducts. Since each finite coproduct is representable by Example 12.14.1, each **Lres**$_n F$ is a filtered colimit of representable presheaves. Hence **Lres**(F) is in Δ^{op}**Norm**^{ind} for every radditive presheaf F.

Lemma 14.16. *If X is a simplicial ind-normal variety, the cofibrant replacement **Lres**$(X) \xrightarrow{\sim} X$ of Example 12.19 induces global weak equivalences S^G**Lres**$(X) \xrightarrow{\sim} S^G X$ and $\widetilde{S}^G$**Lres**$(X_+) \xrightarrow{\sim} \widetilde{S}^G(X_+)$.*

Proof. The endofunctor S^G of $\mathcal{C} = $**Norm** has $(S^G)^* X = S^G X$. Since the map **Lres**$(X) \xrightarrow{\sim} X$ is a global weak equivalence in Δ^{op}**Norm**^{ind}, Corollary 12.33 states that S^G**Lres**$(X) \xrightarrow{\sim} S^G X$ is a global weak equivalence. The same argument, applied to **Norm**$_+$, yields the assertion for $\widetilde{S}^G(X_+)$. $\qquad\square$

For any radditive presheaf F, we set $(\mathbb{L}S^G)F = S^G \mathbf{Lres}\,(F)$. In this language, Lemma 14.16 states that $(\mathbb{L}S^G)X \xrightarrow{\sim} S^G X$ for all representable X.

Proposition 14.17. *The derived functor $\mathbb{L}S^G = S^G \mathbf{Lres}$ preserves global weak equivalences between simplicial radditive presheaves. That is, if $F_1 \to F_2$ is any global weak equivalence, then so is $\mathbb{L}S^G(F_1) \to \mathbb{L}S^G(F_2)$.*

The pointed functor $\mathbb{L}\widetilde{S}^G = \widetilde{S}^G \mathbf{Lres}$ also preserves global weak equivalences.

Proof. ([Voe10c, 2.11]) The map $\mathbf{Lres}\,(F_1) \xrightarrow{\simeq} \mathbf{Lres}\,(F_2)$ is a global equivalence because both of the $\mathbf{Lres}\,(F_i) \xrightarrow{\simeq} F_i$ are (by 12.25). Hence the induced map from $\mathbb{L}S^G(F_1) = S^G \mathbf{Lres}\,(F_1)$ to $\mathbb{L}S^G(F_2) = S^G \mathbf{Lres}\,(F_2)$ is a global weak equivalence, by Corollary 12.33. The argument for $\mathbb{L}\widetilde{S}^G$ is similar. $\qquad\square$

Next, we consider the behavior of symmetric powers under Nisnevich-local equivalences.

Proposition 14.18. *If $X \xrightarrow{f} Y$ is a Nisnevich-local equivalence between objects of $\Delta^{\mathrm{op}}\mathbf{Norm}^{\mathrm{ind}}$ then so is $S^G(f): S^G X \to S^G Y$.*

A similar assertion holds for $\widetilde{S}^G X \to \widetilde{S}^G Y$ when $X \to Y$ is in $\Delta^{\mathrm{op}}\mathbf{Norm}^{\mathrm{ind}}_+$.

Proof. Let E denote the class of morphisms f in $\Delta^{\mathrm{op}}\mathbf{Norm}^{\mathrm{ind}}_+$ such that $S^G f(W)$ is a weak equivalence for all hensel local W. It is a $\bar{\Delta}$-closed class, by Theorem 12.30. Thus it suffices to show that E contains the Nisnevich-local equivalences.

Fix a hensel local W and consider the functor $\Phi(X) = (S^G X)(W)$ from $\mathbf{Norm}^{\mathrm{ind}}$ to $\mathbf{Sets}$ (or from $\Delta^{\mathrm{op}}\mathbf{Norm}^{\mathrm{ind}}$ to $\Delta^{\mathrm{op}}\mathbf{Sets}$). Because S^G commutes with colimits, so does Φ.[2] It is also clear that if U is open in X then $S^G U$ is open in $S^G X$ and hence $\Phi(U) \to \Phi(X)$ is an injection.

Suppose that $f: V' \to V$ is an étale cover and $U \subset V$ is open, so that $(V' - U')_{\mathrm{red}} \to (V - U)_{\mathrm{red}}$ is an isomorphism, i.e., we have an upper distinguished square Q of the form (12.48). Let K_Q denote the associated homotopy pushout, defined in 12.34. Applying Φ, which commutes with colimits, we have two pushout squares of simplicial sets:

$$
\Phi(Q): \quad
\begin{array}{ccc}
\Phi(U') & \longrightarrow & \Phi(V') \\
\downarrow & & \downarrow f \\
\Phi(U) & \xrightarrow{\ i\ } & \Phi(V)
\end{array}
\qquad
\begin{array}{ccc}
\Phi(U') \amalg \Phi(U') & \hookrightarrow & \Phi(U' \otimes \Delta^1) \\
\downarrow & & \downarrow \\
\Phi(U) \amalg \Phi(V') & \longrightarrow & \Phi(K_Q).
\end{array}
$$

These squares show that $\Phi(V)$ and $\Phi(K_Q)$ are the respective pushout and homotopy pushout of $\Phi(U) \leftarrow \Phi(U') \to \Phi(V')$. Recall from Remark 12.34.1 that the pushout of an injection of simplicial sets is weakly equivalent to the homotopy pushout. Therefore $\Phi(V) \to \Phi(K_Q)$ is a weak equivalence.

---

2. Compare to [Del09, Thm.6].

Thus E contains the maps $K_Q \to V$ for all upper distinguished squares (12.48). By Theorem 12.55, every Nisnevich-local equivalence is in the $\bar{\Delta}$-closure of the class of maps $K_Q \to V$ for all upper distinguished squares (12.48). Therefore the Nisnevich-local equivalences are in E. □

Corollary 14.19. *The derived functor* $\mathbb{L}S^G = S^G\mathbf{Lres}$ *preserves Nisnevich-local equivalences.*

Proof. If $F_1 \to F_2$ is a Nisnevich-local equivalence of simplicial radditive presheaves, the cofibrant replacement $f: \mathbf{Lres}\,(F_1) \to \mathbf{Lres}\,(F_2)$ is a Nisnevich-local equivalence of objects in $\Delta^{\mathrm{op}}\mathbf{Norm}^{\mathrm{ind}}$. Therefore $S^G(f): \mathbb{L}S^G(F_1) \to \mathbb{L}S^G(F_2)$ is a Nisnevich-local equivalence by Proposition 14.18. □

We can now consider the behavior under $\mathbb{A}^1$-weak equivalences.

Lemma 14.20. *If* $\pi: V \to X$ *is a vector bundle, then* $S^G V \to S^G X$ *is an* $\mathbb{A}^1$-*local equivalence.*

Proof. If $X = \mathrm{Spec}(R)$ and $V = \mathrm{Spec}(A)$, $A = \mathrm{Sym}_R(P)$, then $V^m = \mathrm{Spec}(A^{\otimes m})$, and the graded algebra map

$$\phi_R : A^{\otimes m} \to A^{\otimes m}[u], \qquad s = a_1 \otimes \cdots \otimes a_m \mapsto s \cdot u^{e_1 + \cdots + e_m} \quad (a_i \in A_{e_i}),$$

is natural in R. The ϕ_R glue to define an equivariant map $V^m \times \mathbb{A}^1 \xrightarrow{\phi} V^m$, fitting into the equivariant diagram on the left below. Modding out by the G-action yields the commutative diagram on the right.

$$
\begin{array}{ccc}
V^m \xrightarrow[\sim]{u=0} V^m \times \mathbb{A}^1 & \quad & S^G(V) \xrightarrow[\sim]{u=0} S^G(V) \times \mathbb{A}^1 \xleftarrow[\sim]{u=1} S^G(V) \\
\pi^m \downarrow \qquad \phi \downarrow & \quad & S^G\pi \downarrow \qquad \phi/G \downarrow \qquad \| = \\
X^m \xrightarrow{t=0} V^m & \quad & S^G(X) \xrightarrow{t=0} S^G(V^m)
\end{array}
$$

From the right-hand diagram (and the 2-out-of-3 property), we see that ϕ/G is an $\mathbb{A}^1$-local equivalence. Since $S^G\pi$ is a retract of an $\mathbb{A}^1$-local equivalence, it follows that $S^G\pi$ is also an $\mathbb{A}^1$-local equivalence. □

Proposition 14.21. *If* $X \xrightarrow{f} Y$ *is an* $\mathbb{A}^1$-*local equivalence between objects of* $\Delta^{\mathrm{op}}\mathbf{Norm}^{\mathrm{ind}}$ *then so is* $S^G X \xrightarrow{S^G f} S^G Y$.

Proof. Let E denote the class of morphisms f in $\Delta^{\mathrm{op}}\mathbf{Norm}^{\mathrm{ind}}$ such that $S^G f$ is an $\mathbb{A}^1$-local equivalence. Since the class of $\mathbb{A}^1$-local equivalences is $\bar{\Delta}$-closed by Theorem 12.66 and S^G commutes with colimits, E is a $\bar{\Delta}$-closed class by Theorem 12.30.

Again by Theorem 12.66, the $\mathbb{A}^1$-local equivalences are the smallest $\bar{\Delta}$-closed class containing the Nisnevich-local equivalences and the maps $U \times \mathbb{A}^1 \to U$. Since S^G preserves Nisnevich-local equivalences by Proposition 14.18, and $S^G(X \times \mathbb{A}^1) \to S^G(X)$ is an $\mathbb{A}^1$-local equivalence by Lemma 14.20, E contains the $\mathbb{A}^1$-local equivalences. $\square$

Corollary 14.22. *The derived functor* $\mathbb{L}S^G = S^G\mathbf{Lres}$ *preserves* $\mathbb{A}^1$-*local equivalences.*

14.4 S^G OF QUOTIENTS X/U

In the next chapter, we will need to compute $\widetilde{S}^G(\mathbb{A}^n/U)$ and $\mathbb{L}\widetilde{S}^G(\mathbb{A}^n/U)$, where U is the complement of the origin and G is a subgroup of Σ_m. In this section we consider the more general problem of computing $\widetilde{S}^G(X/U)$, where $U \subset X$ is the open complement of a closed subvariety Z.

By definition, X/U is the pointed Nisnevich sheaf associated to the reflexive coequalizer $(X \cup_U *)$ of $(X \amalg U)_+ \rightrightarrows X_+$, which sends V to $X(V)/U(V)$. Since $(X \amalg U)_+$ is representable, $(X \cup_U *)$ is a radditive presheaf by Lemma 12.16.

Recall from section 12.1 that the functors $\widetilde{S}^G$ on $\mathbf{Norm}_+$ extend to inverse image functors $(\widetilde{S}^G)^*$ on $\mathrm{Pshv}(\mathbf{Norm}_+)$. Because $(\widetilde{S}^G)^*$ preserves colimits, $(\widetilde{S}^G)^*(X \cup_U *)$ is the presheaf coequalizer of $S^G(X \amalg U)_+ \rightrightarrows (S^GX)_+$; by Lemma 12.16, this presheaf is radditive. The following theorem is a combination of Example 5.8 in [Del09] and Proposition 2.12 in [Voe10c].

Theorem 14.23. *There is a Nisnevich-local equivalence*

$$(\widetilde{S}^G)^*(X/U) \simeq (\widetilde{S}^G)^*(X \cup_U *) \xrightarrow{f} S^GX/(S^GX - S^GZ).$$

Proof. Writing V for $S^GX - S^GZ$, S^GX/V is the pointed Nisnevich sheaf associated to the coequalizer $(S^GX) \cup_V *$ of $((S^GX) \amalg V)_+ \rightrightarrows (S^GX)_+$. We will define f by using a decomposition of $S^G(X \amalg U)$ to produce a morphism $h \colon S^G(X \amalg U)_+ \to ((S^GX) \amalg V)_+$, compatible with the coequalizer diagrams

$$
\begin{array}{ccccc}
S^G(X \amalg U)_+ & \rightrightarrows & (S^GX)_+ & \longrightarrow & (\widetilde{S}^G)^*(X \cup_U *) \\
\downarrow{\scriptstyle h} & & \| & & \downarrow{\scriptstyle f} \\
((S^GX) \amalg V)_+ & \rightrightarrows & (S^GX)_+ & \longrightarrow & (S^GX) \cup_V *
\end{array}
$$

and then show that f is a Nisnevich-local equivalence, by evaluating h at hensel local schemes.

Step 1: We first consider the case where $G = \Sigma_m$, using the decomposition $S^m(X \amalg U) = \amalg \; S^iX \times S^{m-i}U$ of (14.1). The identity map of S^mX and the concatenation maps

$$S^iX \times S^{m-i}U \to V, \qquad i < m,$$

assemble to define a natural map $h: S^m(X \amalg U) \to (S^m X) \amalg V$, compatible with the coequalizer diagrams. Thus h induces a morphism f as indicated above.

To show that f is a Nisnevich-local equivalence, it suffices (by the construction of coequalizers in the category of sets) to show that h is a surjection when evaluated at any hensel local scheme T.

Fix i and $j > 0$ such that $i + j = m$, and set $H = \Sigma_i \times \Sigma_j$, so that $S^i X \times S^j U = (X^i \times U^j)/H$ is an open subscheme of $S^i X \times S^j X = X^m/H$. Let Y_j denote the étale locus of $S^i X \times S^j U \to V$. We claim that the union of the étale maps $Y_j \to V$ $(j > 0)$ forms a Nisnevich cover of V. This implies that for any hensel local T, the subset $\amalg Y_j(T)$ of $\amalg (S^i X \times S^j X)(T)$ maps onto $V(T)$, which will finish the case $G = \Sigma_m$.

To demonstrate the claim, let B_j denote the locally closed subscheme $Z^i \times U^j$ of X^m. Because Z and U are disjoint, the stabilizer subgroups of points in B_j lie in H. By Lemma 14.23.1, applied to $G = \Sigma_m$ and $Y = X^m$, the étale locus of $S^i X \times S^j X \to S^m X$ contains $S^i Z \times S^j U = B_j/H$. As $S^i Z \times S^j U$ is also contained in $S^i X \times S^j U$, it is contained in Y_j.

As a scheme, V is the union of the images of the B_j $(0 < j \leq m)$. Therefore, it suffices to show that each morphism $S^i Z \times S^j U \to V$ is a locally closed immersion. Consider the Σ_m-closure of B_j in X^m, $A_j = \bigcup_{\sigma \in \Sigma_m} \sigma(B_j)$. Then A_j is the disjoint union of the $\sigma(B_j)$ as σ runs over a set of coset representatives of H in Σ_m. Hence A_j/Σ_m equals $B_j/H = S^i Z \times S^j U$. Since A_j is locally closed in X^m and the map $X^m \to X^m/\Sigma_m = S^m X$ is both open and proper, $S^i Z \times S^j U$ is locally closed in $S^m X$. This establishes the claim, and finishes the case $G = \Sigma_m$.

Step 2: When G is an arbitrary subgroup of Σ_m, we proceed by first analyzing $S^G(X \amalg U) = (X \amalg U)^m/G$. The G-scheme $(X \amalg U)^m$ breaks up as the disjoint union over $i + j = m$ of G-schemes $Y_{i,j}$, where $Y_{i,j}$ is the Σ_m-closure of $X^i \times U^j$; since the stabilizer of $X^i \times U^j$ is $\Sigma_i \times \Sigma_j$, $Y_{i,j}$ is the disjoint union over coset representatives:

$$Y_{i,j} = \coprod_{\sigma \in \Sigma_m/\Sigma_i \times \Sigma_j} \sigma(X^i \times U^j).$$

Note that $Y_{m,0} = X^m$, so $Y_{m,0}/G = S^G X$. For each pair (i, j) with $j > 0$ and for each σ, the images of the concatenations $\sigma(X^i \times U^j) \to X^m \to S^G X$ lie in V.

We define the morphism $h: S^G(X \amalg U) \to S^G X \amalg V$ as follows. We map the summand $Y_{m,0}/G = X^m/G$ to $S^G X$ by the identity map; the summands $Y_{i,j}/G$ with $j > 0$ are mapped to V by the mod G reduction of the concatenation maps defined above. It is clear that h is compatible with the coequalizer diagram.

As in the case $G = \Sigma_m$, f will be a Nisnevich-local equivalence if h is a surjection, i.e., if the étale locus of the morphisms $Y_{i,j}/G \to V$ $(j > 0)$ form a Nisnevich cover of V. To see this, we decompose each $Y_{i,j}$ even further using the double cosets $[\sigma]$ in $G\backslash\Sigma_m/(\Sigma_i \times \Sigma_j)$. Let $Y_{i,j}^{[\sigma]}$ denote the union of those components $\tau(X^i \times U^j)$ of $Y_{i,j}$ indexed by coset representatives τ in the G-orbit $G\sigma$ in $\Sigma_m/(\Sigma_i \times \Sigma_j)$. Then $Y_{i,j} = \coprod_{[\sigma]} Y_{i,j}^{[\sigma]}$, G acts on each factor $Y_{i,j}^{[\sigma]}$, and $S^G(X \amalg U) = \amalg Y_{i,j}^{[\sigma]}/G$.

The G-stabilizer of $\sigma \in \Sigma_m/(\Sigma_i \times \Sigma_j)$ is $H_{i,j}^{[\sigma]} = G \cap \sigma(\Sigma_i \times \Sigma_j)\sigma^{-1}$. Thus the orbit $G\sigma$ of $\Sigma_m/(\Sigma_i \times \Sigma_j)$ is isomorphic to $G/H_{i,j}^{[\sigma]}$, and $H_{i,j}^{[\sigma]}$ acts on $\sigma(X^i \times U^j)$. In summary, we have a natural isomorphism (induced by the inclusion)

$$\sigma(X^i \times U^j)/H_{i,j}^{[\sigma]} \xrightarrow{\cong} Y_{i,j}^{[\sigma]}/G,$$

and the map h restricted to the component $Y_{i,j}^{[\sigma]}/G$ of $S^G(X \amalg U)$ is

$$h_{i,j}^{[\sigma]} : \sigma(X^i \times U^j)/H_{i,j}^{[\sigma]} \to V.$$

Set $H = H_{i,j}^{[\sigma]}$ and $B = \sigma(Z^i \times U^j)$. Since B is closed in $\sigma(X^i \times U^j)$, B/H is closed in $\sigma(X^i \times U^j)/H$. Finally, let A denote the G-closure of B in X^m, $A = \bigcup_{\tau \in G} \tau(B)$. Now consider the following diagram, in which the top right map is an open immersion and both left horizontal maps are closed immersions.

$$
\begin{array}{ccccc}
B/H & \hookrightarrow & \sigma(X^i \times U^j)/H & \hookrightarrow & X^m/H \\
{\scriptstyle\cong}\downarrow & & {\scriptstyle\cong}\downarrow & & \downarrow \\
A/G & \hookrightarrow & Y_{i,j}^{[\sigma]}/G & \xrightarrow{h_{i,j}^{[\sigma]}} & V \subset X^m/G
\end{array}
$$

Because Z and U are disjoint, the stabilizer subgroups of points in B lie in H. By Lemma 14.23.1, applied to G and $Y = X^m$, the map $X^m/H \to X^m/G$ is étale in a neighborhood of B/H.

As a scheme, V is the union of the images of the B/H (over all (i,j) with $0 < j \leq m$ and all σ). That is, the images form a constructible stratification of V. Hence the union of the étale loci of the $Y_{i,j}^{[\sigma]}/G \to V$ form a Nisnevich cover of V. This implies that h is onto when evaluated at any hensel local scheme, and hence f is an isomorphism.   □

Lemma 14.23.1. *Let G be a finite group and H a subgroup. Suppose G acts faithfully on a quasi-projective variety Y, and let U be the (open) subscheme of points y whose (scheme-theoretic) stabilizer G_y is contained in H. Then $Y/H \to Y/G$ is étale on U/H.*

Proof. This is proven in V(2.2) of [SGA71], where the stabilizer G_y is called the *groupe d'inertie* of y.   □

We also need to determine $\mathbb{L}\widetilde{S}^G(X/U)$. By 14.19, it is Nisnevich-local equivalent to $\mathbb{L}\widetilde{S}^G(X \cup_U *) = S^G \mathbf{Lres}\,(X \cup_U *)$. We start with G-schemes.

Proposition 14.24. *Suppose that $V \subset T$ is an open inclusion of G-schemes, with pointed quotient presheaf $T \cup_V *$. Then we have a global weak equivalence $(/G)^* \mathbf{Lres}\,(T \cup_V *) \xrightarrow{\cong} (/G)^*(T \cup_V *)$.*

If T is $\mathbb{A}^1$-local equivalent to a point, $(/G)^*(T \cup_V *)$ is $\mathbb{A}^1$-local equivalent to the suspension of V/G.

Proof. Consider the map of squares, induced by **Lres** $(F) \to F$:

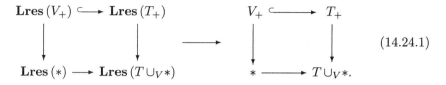

$$(14.24.1)$$

Since **Lres** is a functor, G acts on the terms in the left square. By Construction 12.19 (the radditive version of Lemma 12.10), the left top map is a termwise split monomorphism and the left square is cocartesian. By assumption, the right top map is a monomorphism and the right square is cocartesian. Since $(/G)^*$ is a left adjoint, it preserves pushouts. As $(/G)^*(V_+) = (V/G)_+$, $(/G)^*(T_+) = (T/G)_+$, and $(/G)^*(*) = *$, we have a map of cocartesian squares:

$$
\begin{array}{ccc}
(/G)^*\mathbf{Lres}\,(V_+) \hookrightarrow (/G)^*\mathbf{Lres}\,(T_+) & & (V/G)_+ \longrightarrow (T/G)_+ \\
\downarrow \qquad\qquad \downarrow & \longrightarrow & \downarrow \qquad\qquad \downarrow \\
(/G)^*\mathbf{Lres}\,(*) \longrightarrow (/G)^*\mathbf{Lres}\,(T \cup_V *) & & * \longrightarrow (/G)^*(T \cup_V *).
\end{array}
$$

$$(14.24.2)$$

Since each term **Lres** $(T_+)_n$ is termwise the coproduct of **Lres** $(V_+)_n$ and another coproduct of schemes, and $(/G)^*$ is computed termwise, the left top map is a monomorphism. Because $T \to T/G$ is an open map, $V/G \to T/G$ is an open inclusion. Thus $(V/G)(W) \to (T/G)(W)$ is a monomorphism for each W in **Norm$_+$**.

Recall from Remark 12.34.1 that if $K \to K'$ is a monomorphism of simplicial sets and $K \to L$ is any map then the pushout $K' \cup_K L$ is weakly equivalent to the homotopy pushout. It follows that both squares are homotopy pushouts when evaluated at any W.

Now the maps between the three corresponding top and left corners of (14.24.1) are global weak equivalences, by 12.19. By Theorem 12.33, applied to $(/G)^*$, the maps between the three corresponding top and left corners of (14.24.2) are also global weak equivalences. It follows that the homotopy pushout

$$(/G)^*\mathbf{Lres}\,(T \cup_V *) \to (/G)^*(T \cup_V *)$$

is also a global weak equivalence.

Now suppose that T is $\mathbb{A}^1$-local equivalent to a point. Since $\mathbb{A}^1$-localization preserves homotopy cocartesian squares, the pushout $(/G)^*(T \cup_V *)$ is $\mathbb{A}^1$-equivalent to the homotopy pushout of $(/G)^*(T_+) = (T/G)_+$ and $*$ along

$(/G)^*(V_+) = (V/G)_+$, i.e., to the cone of $(V/G)_+ \to S^0$. Hence $(/G)^*(T \cup_V *)$ is $\mathbb{A}^1$-local equivalent to the suspension of V/G (see Definition 1.36).   □

Remark. Proposition 14.24 is essentially Lemma 16 in [Del09], with $F = *$.

14.5   NISNEVICH G-LOCAL EQUIVALENCES

The goal of this section is to prove the following theorem, which is needed for Lemma 15.24.

Theorem 14.25. *Suppose that $U \subset X$ is an open inclusion of normal schemes. Then $\mathbb{L}\widetilde{S}^G(X \cup_U *) \xrightarrow{\sim} \widetilde{S}^G(X \cup_U *)$ is a Nisnevich-local equivalence.*

The proof will need the notion of a G-*local* hensel scheme, and a Nisnevich G-*local equivalence*. This notion first appeared in [Del09, p. 373].

Definition 14.26. Let G be a finite group scheme. A G-*local* hensel scheme T is a G-scheme whose underlying scheme is a finite disjoint union of hensel local schemes, and such that the scheme T/G is hensel local.

A map $F_1 \to F_2$ of simplicial presheaves on G–**Norm** is a *Nisnevich G-local equivalence* if $F_1(T) \to F_2(T)$ is a weak equivalence for every G-local hensel scheme T. A map $F_1 \to F_2$ of simplicial presheaves on G–**Norm**$_+$ is a *Nisnevich G-local equivalence* if $F_1(T_+) \to F_2(T_+)$ is a weak equivalence for every G-local hensel scheme T.

Remark 14.26.1. The G-local hensel schemes are the points of the G-Nisnevich topology on G–**Norm** of [Del09, §3.1]. In this topology, covers of T are collections of G-equivariant étale maps $Y_i \to T$ such that T admits a filtration by closed equivariant subschemes $\emptyset = T_0 \subset T_1 \subset \cdots \subset T_n = T$ such that for each j one of the maps $Y_i \to T$ admits a section over $T_j - T_{j-1}$.

Construction 14.27. The following construction produces G-local hensel schemes. Fix a hensel local scheme S and a G-scheme W. Given a G-local hensel scheme T with $T/G = S$, any G-map $T \to W$ induces a map $S \to W/G$. Conversely, given a map $f : S \to W/G$, the pullback $T = S \times_{W/G} W$ is a G-local hensel scheme, and the equivariant map $f_T : T \to W$ induces the original map f on the G-quotient schemes. Thus we have a bijection

$$\operatorname*{colim}_{T/G=S} \operatorname{Hom}_G(T, W) \xrightarrow{\cong} (W/G)(S) = \operatorname{Hom}(S, W/G).$$

The colimit is taken over the system of G-local schemes T over S with $T/G = S$. Note that a morphism $T_1 \to T_2$ in this system induces an isomorphism $\operatorname{Hom}_G(T_2, W) \xrightarrow{\cong} \operatorname{Hom}_G(T_1, W)$.

If F is a simplicial presheaf on $G-\mathbf{Norm}_+$, the system of G-local hensel T with $S = T/G$ is cofinal in the comma category of maps $S \to T'/G$. Hence

$$((/G)^*F)(S) = \operatorname*{colim}_{S \to T'/G} F(T') = \operatorname*{colim}_{S = T/G} F(T).$$

We will need two properties of Nisnevich G-local equivalences.

Lemma 14.28. *1) If $F_1 \to F_2$ is a Nisnevich G-local equivalence of radditive presheaves on $G-\mathbf{Norm}_+$, then $\mathbf{Lres}\, F_1 \to \mathbf{Lres}\, F_2$ is also a Nisnevich G-local equivalence.*

2) If $F_1 \to F_2$ is a Nisnevich G-local equivalence in $\Delta^{\mathrm{op}}(G-\mathbf{Norm}_+)$, then $(/G)^*F_1 \to (/G)^*F_2$ is a Nisnevich-local equivalence.

Proof. The first assertion follows from 12.25 for $\mathcal{C} = G-\mathbf{Norm}_+$, which asserts that $\mathbf{Lres}\, F_1 \to F_1$ and $\mathbf{Lres}\, F_2 \to F_2$ are global weak equivalences in $\Delta^{\mathrm{op}}\mathrm{rad}(\mathcal{C})$.

For the second assertion, we observe that for each hensel local S the map $(/G)^*F_1(S) \to (/G)^*F_2(S)$ is the colimit over the system of G-local hensel T with $S = T/G$ of the weak equivalences $F_1(T) \to F_2(T)$. $\square$

Lemma 14.29. *Let $U \subset X$ be an open inclusion of normal schemes, and set $Z = X - U$, $W = X^m - Z^m$. Then for every subgroup G of Σ_m there is a Nisnevich G-local equivalence*

$$P(X \cup_U *) \xrightarrow{f} (X^m \cup_W *).$$

Proof. By Lemma 14.5, $P(X_+) \cong (X^m)_+$. We claim there is a map h fitting into a coequalizer diagram in $\Delta^{\mathrm{op}}\mathrm{rad}(G-\mathbf{Norm})$ similar to that in Theorem 14.23:

$$
\begin{array}{ccccc}
P(X \amalg U)_+ & \rightrightarrows & P(X_+) & \longrightarrow & P(X \cup_U *) \\
\Big\downarrow h & & \cong \Big\downarrow & & \Big\downarrow f \\
(X^m \amalg W)_+ & \rightrightarrows & (X^m)_+ & \longrightarrow & X^m \cup_W *.
\end{array}
$$

The construction of h is parallel to the construction in the proof of Theorem 14.23. As a G-scheme, $P(X \amalg U)_+ \cong (X \amalg U)_+^m$ is the coproduct of G-schemes $Y_{i,j}$, $i + j = m$; each $Y_{i,j}$ is the coproduct of all terms $\sigma(X^i \times U^j)$ for σ in $\Sigma_m/(\Sigma_i \times \Sigma_j)$. As in the proof of Theorem 14.23, h maps $Y_{m,0} = X^m$ to X^m by the identity and maps $Y_{i,j}$ to W by concatenation when $j > 0$. As in loc. cit., h is compatible with the coequalizer diagram, and defines a map f. In fact, the map in Theorem 14.23 is obtained by applying $(/G)^*$ to the map f we have just constructed, since $V = (/G)^*W$.

Using the decomposition of the $Y_{i,j}$ into G-schemes $Y_{i,j}^{[\sigma]}$ in the proof of Theorem 14.23, we see that W has a G-invariant constructible stratification by locally closed subschemes $B_{i,j}^{[\sigma]}$ (the G-closure of the B in loc. cit.) such that the restriction of h to the étale loci of $Y_{i,j}^{[\sigma]} \to W$ splits over $B_{i,j}^{[\sigma]}$. (In the terminology

of Remark 14.26.1, the étale loci of $Y_{i,j} \to W$ form a G-Nisnevich cover of W.)
It follows that the maps

$$h : \coprod_{j>0} Y_{i,j}(T) \to W(T)$$

are surjective for every G-local hensel scheme T. By the construction of coequalizers in the category of sets, $P(X \cup_U *)(T) \to (X^m \cup_W *)(T)$ is a bijection. $\square$

Proof of Theorem 14.25.[3] Let us write C for the simplicial radditive presheaf $\mathbf{Lres}\,(X \cup_U *)$. By Lemmas 12.25 and 14.6, $PC \to P(X \cup_U *)$ is a global weak equivalence of simplicial radditive presheaves on $G\text{-}\mathbf{Norm}_+$. Applying the cofibrant resolution $\mathbf{Lres}\,H \xrightarrow{\simeq} H$ in $\Delta^{\mathrm{op}}\mathrm{rad}(G\text{-}\mathbf{Norm}_+)$ (which is a global weak equivalence of simplicial radditive presheaves by Lemma 12.25), we have a square of global weak equivalences in $\Delta^{\mathrm{op}}\mathrm{rad}(G\text{-}\mathbf{Norm}_+)$.

$$
\begin{array}{ccc}
\mathbf{Lres}\,(PC) & \xrightarrow{\ \simeq\ } & PC \\
{\scriptstyle\simeq}\big\downarrow & & \big\downarrow{\scriptstyle\simeq} \\
\mathbf{Lres}\,P(X \cup_U *) & \xrightarrow{\ \simeq\ } & P(X \cup_U *)
\end{array}
$$

With the exception of $P(X \cup_U *)$, the corners of this diagram are simplicial objects of $(G\text{-}\mathbf{Norm}_+)^{\mathrm{ind}}$, by the analysis before Lemma 14.16. Since $\widetilde{S}^G \cong (/G)^* P$ by Proposition 14.7, applying $(/G)^*$ to this diagram yields the following diagram of simplicial radditive presheaves on $\mathbf{Norm}_+$.

$$
\begin{array}{ccccc}
(/G)^*\mathbf{Lres}\,(PC) & \xrightarrow{\ \simeq\ } & (/G)^*PC & \xrightarrow{\ \cong\ } & \widetilde{S}^G(C) \\
{\scriptstyle\simeq}\big\downarrow & & \big\downarrow & & \big\downarrow \\
(/G)^*\mathbf{Lres}\,P(X \cup_U *) & \xrightarrow{\ \eta\ } & (/G)^*P(X \cup_U *) & \xrightarrow{\ \cong\ } & \widetilde{S}^G(X \cup_U *)
\end{array}
$$

The theorem asserts that the composition from the upper left to the lower right is a Nisnevich-local equivalence. By Theorem 12.33, the left vertical and top left horizontal maps are global weak equivalences. Thus to prove the assertion, it suffices to show that the lower left horizontal map

$$\eta : (/G)^*\mathbf{Lres}\,P(X \cup_U *) \to (/G)^*P(X \cup_U *)$$

is a Nisnevich-local equivalence.

We saw in Lemma 14.29 that $f : P(X \cup_U *) \to (X^m \cup_W *)$ is a Nisnevich G-local equivalence. Thus by Lemma 14.28, $\mathbf{Lres}\,P(X \cup_U *) \to \mathbf{Lres}\,(X^m \cup_W *)$

---

3. The proof of Theorem 14.25 is based on Proposition 42 in [Del09].

is a Nisnevich G-local equivalence. Applying $(/G)^*$, we obtain a commutative diagram

$$
\begin{array}{ccccc}
(/G)^*\mathbf{Lres}\,P(X\cup_U *) & \xrightarrow{\ \eta\ } & (/G)^*P(X\cup_U *) & \xrightarrow{\ \cong\ } & \widetilde{S}^G(X\cup_U *) \\
{\scriptstyle 14.28}\Big\downarrow\simeq & & {\scriptstyle (/G)^*f}\Big\downarrow & & {\scriptstyle 14.23}\Big\downarrow\simeq \\
(/G)^*\mathbf{Lres}\,(X^m\cup_W *) & \xrightarrow[14.24]{\ \simeq\ } & (/G)^*(X^m\cup_W *) & \xrightarrow{\ \cong\ } & \widetilde{S}^G(X_+)/V,
\end{array}
$$

where $V = S^G X - S^G(X-U)$. The bottom left horizontal map is a global weak equivalence by Proposition 14.24, the right vertical map is a Nisnevich-local equivalence by Theorem 14.23, and the left vertical map is a Nisnevich-local equivalence by Lemma 14.28. Thus the upper left horizontal map is a Nisnevich-local equivalence, as needed to finish the proof. $\square$

14.6 SYMMETRIC POWERS AND SHIFTS

In this section, we establish two distinguished triangles in **DM** (see Theorem 14.34); these will be used in Corollary 15.27 to describe $S_{\mathrm{tr}}^{\ell}(\mathbb{L}^a[b])$. We will assume that $(\ell-1)!$ is invertible in the coefficient ring R.

Definition 14.30. A motive X is called *even* if the transpose automorphism $\tau(X)$ of $X \otimes X$ is the identity; X is called *odd* if $\tau(X)$ is multiplication by -1.

It is easy to see that $\tau(X[n]) = (-1)^n \tau(X)$. This implies that if X is even then $X[1]$ is odd, and if X is odd then $X[1]$ is even. Since R is clearly even, this implies that $R[n]$ is even or odd, depending on whether n is even or odd.

Lemma 14.31. *The motives $\mathbb{L}^n$ are even.*

Proof. By [MVW, 15.8], $R(n)$ is even, and $\mathbb{L}^n = R(n)[2n]$. $\square$

Here is a useful decomposition result.

Theorem 14.32. *Let $f : X \to Y$ be a termwise split injection of simplicial ind-objects in* $\mathbf{Cor}(\mathbf{Norm}, R)$. *Then we have a collection of objects*

$$
S_{a,b}^n = S_{a,b}^n(f), \qquad 0 \le a \le b \le n,
$$

with $S_{0,n}^n = S_{\mathrm{tr}}^n(Y)$, such that for $0 \le a \le n$ there are isomorphisms

$$
S_{a,a}^n \cong S_{\mathrm{tr}}^a(X) \otimes S_{\mathrm{tr}}^{n-a}(Y/X).
$$

Moreover, for $0 \leq a \leq b < c \leq n$ we have distinguished triangles in **DM**:

$$S_{b+1,c}^n \to S_{a,c}^n \to S_{a,b}^n \to S_{b+1,c}^n[1].$$

These data are natural in $X \to Y$.

Proof. Set $Z = Y/X$ and choose termwise splittings $Y_p \cong X_p \oplus Z_p$, so that $S_{\mathrm{tr}}^n(Y_p) = \oplus_{i+j=n} S_{\mathrm{tr}}^i(X_p) \otimes S_{\mathrm{tr}}^j(Z_p)$ by Lemma 14.9. Let $S_{a,n}^n$ denote the sub-object of $S_{\mathrm{tr}}^n(Y)$ which termwise is the sum of the $S_{\mathrm{tr}}^i(X_p) \otimes S_{\mathrm{tr}}^j(Z_p)$ for $i \geq a$. Clearly $S_{0,n}^n = S_{\mathrm{tr}}^n(Y)$ and $S_{n,n}^n = S_{\mathrm{tr}}^n(X)$. Define $S_{a,b}^n$ to be the termwise cokernel of the termwise split injection $S_{b+1,n}^n \to S_{a,n}^n$. Because X is a simplicial subobject of Y, these are all simplicial ind-objects. The description of $S_{a,a}$ and the existence of the distinguished triangles are straightforward. $\square$

Corollary 14.33. *([Voe10c, 2.45]) Let $X \to Y$ be a termwise split injection of simplicial ind-objects in* **Cor(Norm**, R)*, with $Z = Y/X$. Then there are distinguished triangles*

$$X \otimes S_{\mathrm{tr}}^{n-1} Z \to S_{0,1}^n \to S_{\mathrm{tr}}^n Z \to X \otimes S_{\mathrm{tr}}^{n-1} Z[1],$$

$$S_{\mathrm{tr}}^{n-1}(X) \otimes Z \to S_{0,n-1}^n \to S_{0,n-2}^n \to S_{\mathrm{tr}}^{n-1}(X) \otimes Z[1].$$

Proof. These are the triangles of Theorem 14.32 for $(a,b,c) = (0,0,1)$ and $(a,b,c) = (0, n-2, n-1)$. $\square$

We now come to the main result in this section; it is taken from [Voe10c, 2.46].

Theorem 14.34. *Assume that $(\ell-1)!$ is invertible in R, and let X be a simplicial ind-object in* **Cor(Norm**, R)*.*

1. *If X is odd, there is a distinguished triangle*

$$X^{\otimes \ell}[\ell-1] \xrightarrow{\zeta} (S_{\mathrm{tr}}^{\ell} X)[1] \xrightarrow{\eta} S_{\mathrm{tr}}^{\ell}(X[1]) \longrightarrow X^{\otimes \ell}[\ell].$$

2. *If X is even, there is a distinguished triangle*

$$X^{\otimes \ell}[1] \xrightarrow{\zeta} (S_{\mathrm{tr}}^{\ell} X)[1] \xrightarrow{\eta} S_{\mathrm{tr}}^{\ell}(X[1]) \longrightarrow X^{\otimes \ell}[2].$$

Proof. Let CX denote the simplicial cone on X and consider the termwise split cofiber sequence $X \to CX \to SX$ in Δ^{op}**Cor(Norm**, R)^{ind}. Note that $CX \cong 0$ and $SX \cong X[1]$ in **DM**.

Assume first that X is odd. Then $Z = X[1]$ is even, so $S_{\mathrm{tr}}^{\ell-1}(X[1])$ is isomorphic to $(X[1])^{\otimes \ell-1} = X^{\otimes \ell-1}[\ell-1]$ by Example 14.12(e). Thus the first sequence of Corollary 14.33 with $n = \ell$, $Y = CX$, and $Z = X[1]$ becomes:

$$X^{\otimes \ell}[\ell-1] \to S_{0,1}^{\ell} \to S_{\mathrm{tr}}^{\ell}(X[1]) \to X^{\otimes \ell}[\ell].$$

Thus to prove assertion 14.34(1), it suffices to show that $S_{0,1}^\ell \cong S_{\mathrm{tr}}^\ell(X)[1]$. By Theorem 14.32 with $(a, b, c) = (0, 1, \ell)$, we have a distinguished triangle

$$S_{2,\ell}^\ell \to S_{0,\ell}^\ell \to S_{0,1}^\ell \xrightarrow{\delta} S_{2,\ell}^\ell[1].$$

By construction, $S_{0,\ell}^\ell$ is equal to $S_{\mathrm{tr}}^\ell(CX)$, which is zero. Hence δ is an isomorphism: $S_{0,1}^\ell \cong S_{2,\ell}^\ell[1]$. We are done if $\ell = 2$, since in that case $S_{2,2}^\ell \cong S_{\mathrm{tr}}^2(X)$.

By Theorem 14.32 with $(a, b, c) = (2, \ell - 1, \ell)$, we have a distinguished triangle

$$S_{\mathrm{tr}}^\ell X = S_{\ell,\ell}^\ell \to S_{2,\ell}^\ell \to S_{2,\ell-1}^\ell \to (S_{\mathrm{tr}}^\ell X)[1].$$

Therefore it remains to show that $S_{2,\ell-1}^\ell$ equals zero.

Now $S_{i,i}^\ell$ is isomorphic to $S_{\mathrm{tr}}^i(X) \otimes S_{\mathrm{tr}}^{\ell-i}(X[1])$. Since X is odd, $S_{\mathrm{tr}}^i(X) = 0$ for $1 < i < \ell$, by Example 14.12(e), so $S_{i,i}^\ell = 0$. We are now done if $\ell = 3$, since $S_{2,2}^\ell = 0$. From the distinguished triangles with $(a, b, c) = (2, i - 1, i)$ for $3 \le i \le \ell - 1$, $S_{i,i}^\ell = 0$ yields

$$S_{2,\ell-1}^\ell \xrightarrow{\cong} \cdots \xrightarrow{\cong} S_{2,3}^\ell \xrightarrow{\cong} S_{2,2}^\ell = 0.$$

This finishes the proof of the first part of Theorem 14.34.

Now assume that X is even. Then $S_{\mathrm{tr}}^{\ell-1}(X) \cong X^{\otimes \ell-1}$ by Example 14.12(e). Since $Z = X[1]$, $S_{\ell-1,\ell-1}^\ell \cong S_{\mathrm{tr}}^{\ell-1}(X) \otimes X[1] = X^{\otimes \ell}[1]$. Thus the second sequence of Corollary 14.33 with $n = \ell$, $Y = CX$, and $Z = X[1]$ becomes:

$$X^{\otimes \ell}[1] \to S_{0,\ell-1}^\ell \to S_{0,\ell-2}^\ell \to X^{\otimes \ell}[2].$$

The triangle for $(a, b, c) = (0, \ell - 1, \ell)$ yields $S_{0,\ell-1}^\ell \xrightarrow{\cong} (S_{\mathrm{tr}}^\ell X)[1]$. We are done if $\ell = 2$, as $S_{0,0}^2 = S_{\mathrm{tr}}^2(X[1])$. Since $X[1]$ is odd, $S_{\mathrm{tr}}^j(X[1]) = 0$ for $1 < j < \ell$ and hence $S_{i,i}^\ell = 0$ for $1 \le i \le \ell - 2$. Now the triangles for $(a, b, c) = (0, i - 1, i)$, $1 \le i \le \ell - 2$, yield

$$S_{0,\ell-2}^\ell \xrightarrow{\cong} \cdots \xrightarrow{\cong} S_{0,1}^\ell \xrightarrow{\cong} S_{0,0}^\ell = S_{\mathrm{tr}}^\ell(X[1]).$$

This finishes the proof.   □

Corollary 14.35. *Set $T = \mathbb{L}^a[b]$. Then there is a map $(S_{\mathrm{tr}}^\ell T)[1] \xrightarrow{\eta} S_{\mathrm{tr}}^\ell(T[1])$ whose cone is: $T^{\otimes \ell}[2]$ for b even, and $T^{\otimes \ell}[\ell]$ for b odd.*

Proof. By Lemma 14.31, $\mathbb{L}^a[b]$ is even or odd, according to the parity of b. Now apply Theorem 14.34.   □

By 12.72, we can refer to topology to calculate the maps in 14.34.

Example 14.36. When $T = R = \mathbb{Z}/\ell$, we have $S_{\mathrm{tr}}^\ell(R) = R_{\mathrm{tr}}(S^\ell \operatorname{Spec} k) = R$. We claim that $S_{\mathrm{tr}}^\ell(R[1]) = 0$, which by 14.34(2) implies that $S_{\mathrm{tr}}^\ell(R[2]) \cong R[2\ell]$.

To see the claim, we regard the discrete simplicial set S^1 as a union of copies of $\mathrm{Spec}\,k$ in each degree, so that $R \oplus R[1] \cong R_{\mathrm{tr}}(S^1)$. Since $R[1]$ is odd, $S^i_{\mathrm{tr}}(R[1]) = 0$ for $1 < i < \ell$ and we have $S^\ell_{\mathrm{tr}}(R \oplus R[1]) = R \oplus R[1] \oplus S^\ell_{\mathrm{tr}}(R[1])$. Nakaoka's simplicial calculation in [Nak57] shows that $S^\ell_{\mathrm{tr}}(R \oplus R[1]) = R \oplus R[1]$. It follows that $S^\ell_{\mathrm{tr}}(R[1]) = 0$, as claimed.

Example 14.37. For $T = R[b]$ and $b \geq 1$, we can compare with the cohomology of the Eilenberg–Mac Lane spaces $K(\mathbb{Z}, b)$. Recall that by Dold–Thom [DT58] we have a homotopy equivalence $K(\mathbb{Z}, b) \simeq S^\infty(S^b)$, where S^b is a (based) simplicial sphere, so we have $\widetilde{H}^*_{\mathrm{top}}(K(\mathbb{Z}, b), R) \cong \bigoplus \widetilde{H}^*_{\mathrm{top}}(\widetilde{S}^i(S^b), R)$ by Steenrod's formula in topology (cf. Corollary 14.14). If $\mathrm{char}(k) = 0$, the topological realization of Theorem 14.34 for $X = S^b \otimes \mathrm{Spec}(k)$ yields triangles in $D^-(R)$:

$$\widetilde{C}_*(\widetilde{S}^\ell(S^b), R)[1] \xrightarrow{t(\eta)} \widetilde{C}_*(\widetilde{S}^\ell(S^{b+1}), R) \longrightarrow \begin{cases} \widetilde{C}_*(S^{b\ell+2}), R) \\ \widetilde{C}_*(S^{b\ell+\ell}), R) \end{cases} \xrightarrow{t(\zeta)} .$$

Now take $R = \mathbb{Z}/\ell$. On cohomology, the map $\widetilde{H}^{*+1}_{\mathrm{top}}(\widetilde{S}^\ell(S^b)) \xrightarrow{t(\eta)} \widetilde{H}^*_{\mathrm{top}}(\widetilde{S}^\ell(S^{b+1}))$ is an injection, because it is a summand of the suspension map $\widetilde{H}^{*+1}_{\mathrm{top}}(K(\mathbb{Z}, b)) \to \widetilde{H}^*_{\mathrm{top}}(K(\mathbb{Z}, b+1))$, which Cartan showed was an injection in [Car54]. It follows that the map $t(\zeta)$ is zero on cohomology, and hence $t(\zeta) = 0$ in the semisimple triangulated category $D^-(\mathbb{Z}/\ell)$. Thus the triangles split and we have the formula in $D^-(\mathbb{Z}/\ell)$:

$$\widetilde{C}_*(\widetilde{S}^\ell(S^b), R) \cong \begin{cases} \bigoplus_{i=1}^{(b-1)/2} R[b + 2i(\ell - 1)] \otimes (R \oplus R[1]), & b \text{ odd}, \\ R[b\ell] \oplus \bigoplus_{i=1}^{b/2-1} R[b + 2i(\ell - 1)] \otimes (R \oplus R[1]), & b \text{ even}. \end{cases}$$

Note that the singular cohomology $\widetilde{H}^*_{\mathrm{top}}(\widetilde{S}^\ell(S^b), R)$ can be read off from this formula, and we get exactly the result obtained in 1957 by Nakaoka in [Nak57]; cf. [Milg, 4.2].

Corollary 14.38.[4] *When $\mathrm{char}(k) = 0$, $R = \mathbb{Z}/\ell$, and $b \geq 1$, we have*

$$\widetilde{S}^\ell_{\mathrm{tr}}(R[b]) \cong \begin{cases} \bigoplus_{i=1}^{(b-1)/2} R[b + 2i(\ell - 1)] \otimes (R \oplus R[1]), & b \text{ odd}, \\ R[b\ell] \oplus \bigoplus_{i=1}^{b/2-1} R[b + 2i(\ell - 1)] \otimes (R \oplus R[1]), & b \text{ even}. \end{cases}$$

Thus the canonical map $\widetilde{S}^\ell_{\mathrm{tr}}(R[b])[1] \to \widetilde{S}^\ell_{\mathrm{tr}}(R[b+1])$ is a split injection for all $b > 0$, and the zero map for $b = 0$.

Proof. We proceed by induction on b, the cases $b = 1, 2$ being covered in Example 14.36. Suppose that the formula holds for b and consider the map ζ in 14.34. If b is odd, $\zeta = 0$ because inductively $\widetilde{S}^\ell_{\mathrm{tr}}(R[b])$ is a sum of terms $R[c]$ with $c < b\ell$, so

---

4. Taken from [Wei09, 3.5.1].

that $\mathrm{Hom}(R[b\ell + \ell - 1], \widetilde{S}_{\mathrm{tr}}^{\ell}(R[b]))$ is zero. If b is even, the same considerations show that $\mathrm{Hom}(R[b\ell], \widetilde{S}^{\ell}(R[b])) \cong \mathrm{Hom}(R[b\ell], R[b\ell]) \cong \mathbb{Z}/\ell$. In this case, we see that the topological realization functor is an isomorphism on

$$\mathrm{Hom}(R[b\ell], \widetilde{S}_{\mathrm{tr}}^{\ell}(R[b])) \xrightarrow{\cong} \mathrm{Hom}_{D^-(\mathbb{Z}/\ell)}(\widetilde{C}_*(S^{b\ell}), \widetilde{C}_*(\widetilde{S}^{\ell}(S^b))) \cong \mathbb{Z}/\ell.$$

(The last isomorphism is from 14.37.) Since ζ maps to $\zeta_{\mathrm{top}} = 0$, we have $\zeta = 0$. Thus the triangle in 14.34 splits, and we obtain the inductive result for $\widetilde{S}^{\ell}(R[b+1])$. $\qquad\qquad\qquad\qquad\qquad\qquad\qquad\qquad\qquad\qquad\qquad\qquad\qquad\qquad\square$

14.7   HISTORICAL NOTES

Symmetric products were introduced by Dold and Thom in 1956, and published in the 1958 paper [DT58]. They played an important role in the development of algebraic topology and spectra, especially in the 1960s.

The use of symmetric products of varieties in motivic cohomology had its origins in the 1996 Suslin–Voevodsky paper [SV96], where effective correspondences from X to Y were shown to be in 1–1 correspondence with morphisms $X \to S^\infty Y_+$. Thus the connection between symmetric products and motivic cohomology was clear from the beginnings of the subject in the late 1990s.

Early approaches such as [MV99] and [Voe03c] used G-equivariant methods to model classifying spaces BG and to construct the Steenrod operations. The foundations of the symmetric product approach were presented by Voevodsky in a course in 2000–01 and published in [Del09] and [Voe10c].

The material in this chapter is based upon section 5 of [Del09] and section 2 of [Voe10c]. As we only need the notions of Nisnevich G-local equivalence and G-local hensel schemes (Definition 14.26), we have avoided a lengthy development of the G-Nisnevich topology, or the corresponding model structure on simplicial presheaves on $G-\mathbf{Norm}$ in [Del09].

Chapter Fifteen

Motivic Classifying Spaces

IN THIS CHAPTER, we first connect the motives $S_{\mathrm{tr}}^{\infty}(\mathbb{L}^n)$ to cohomology operations on $H^{2n,n}$ in Theorem 15.3, at least when $\mathrm{char}(k)=0$. This parallels the Dold–Thom theorem in topology, which identifies the reduced homology $\tilde{H}_*(X,\mathbb{Z})$ of a connected space X with the homotopy groups of the infinite symmetric product $S^{\infty}X$.

A similar analysis shows that $\mathbb{G}_m$ represents $H^{1,1}(-,\mathbb{Z})$, which allows us to describe operations on $H^{1,1}$ (section 15.2). In section 15.3 we introduce the notion of scalar weight operations on $H^{2n,n}$. In sections 15.4 and 15.5, we develop formulas for $S_{\mathrm{tr}}^{\ell}(\mathbb{L}^n)$. These formulas imply that $S_{\mathrm{tr}}^{\infty}(\mathbb{L}^n)$ is a proper Tate motive (15.28), so there is a Künneth formula for them (15.32). The chapter culminates in Theorem 15.38, where we show that βP^b is the unique cohomology operation of scalar weight 0 in its bidegree. This is used in part I, Theorem 6.33, to identify a cohomology operation ϕ with βP^n.

15.1 SYMMETRIC POWERS AND OPERATIONS

In this section, we relate cohomology operations to the symmetric powers $S_{\mathrm{tr}}^i(\mathbb{L}^n)$ (see Theorem 15.3). The key will be the following result of Suslin and Voevodsky from [SV96]. Recall that the algebraic *group completion* of an abelian monoid M is an abelian group A together with a map $M \to A$ which is universal for maps to abelian groups. Since $S^{\infty}(Y_+)$ is a presheaf of abelian monoids, its algebraic group completion is a presheaf of abelian groups.

Recall from Example 12.28 that $R_{\mathrm{tr}}(Y)$ is a presheaf with transfers (for any ring R and any scheme Y), and that $uR_{\mathrm{tr}}(Y)$ denotes the underlying presheaf.

Theorem 15.1. *If $\mathrm{char}(k)=0$ and Y is normal, there is a morphism $\eta_Y :$ $S^{\infty}(Y_+) \to u\mathbb{Z}_{\mathrm{tr}}(Y)$ of presheaves of abelian monoids, natural in Y, such that for each scheme X the following map is an algebraic group completion.*

$$\eta_Y(X): \mathrm{Hom}(X_+, S^{\infty}Y_+) \to u\mathbb{Z}_{\mathrm{tr}}(Y)(X)$$

Proof. Suslin and Voevodsky proved in [SV96, 6.8] that finite correspondences of degree $m \geq 0$ from X to Y correspond to morphisms from X to $S^m(Y)$, and effective finite correspondences correspond to elements of $\mathrm{Hom}(X_+, S^{\infty}Y_+)$.

Since elements of $\mathbb{Z}_{\mathrm{tr}}(Y)(X)$ correspond to all finite correspondences from X to Y, $\mathbb{Z}_{\mathrm{tr}}(Y)(X)$ is the group completion of $\mathrm{Hom}(X_+, S^\infty Y_+)$, and the group completion map $\eta_Y(X)$ is natural in X and Y. $\square$

Remark 15.1.1. If $\mathrm{char}(k) = p > 0$ and $R = \mathbb{Z}[1/p]$, the same proof (using [SV96, 6.8]) shows that $uR_{\mathrm{tr}}(Y)(X)$ is the group completion of the abelian monoid $\mathrm{Hom}(X_+, S^\infty Y_+)[1/p]$.

There is a topological notion of the group completion of an H-space, based on the observation that the realization of any simplicial abelian monoid is a commutative H-space. If B is a homotopy commutative H-space, $\pi_0(B)$ is an abelian monoid. A topological *group completion* of B is an H-map $f : B \to C$ such that $\pi_0(B) \to \pi_0(C)$ is the algebraic group completion, and f_* identifies $H_*(C)$ with the localization of the ring $H_*(B)$ at the multiplicative set $\pi_0(B)$. The topological group completion of B is unique up to homotopy equivalence.

If B is a simplicial abelian monoid, and A is its algebraic group completion, then $B \to A$ is also a topological group completion.

Corollary 15.2. *Let Y be a simplicial object of* **Norm**$_+$ *with* $\mathrm{char}(k) = 0$. *Assume that, for all normal X, either (i) the simplicial set* $\mathrm{Hom}(X_+, S^\infty Y)$ *is connected, or (ii) $\pi_0 \mathrm{Hom}(X_+, S^\infty Y)$ is an abelian group. Then the canonical simplicial morphism $S^\infty Y \to u\mathbb{Z}_{\mathrm{tr}}(Y)$ is a global weak equivalence.*

Proof. It suffices to show that each H-space $B = \mathrm{Hom}(X_+, S^\infty Y)$ is already topologically group complete. By assumption, $\pi_0(B)$ is an abelian group. The result follows because any simplicial H-space B has a homotopy inverse if and only if $\pi_0(B)$ is a group, by [Whi78, X.2.2]. $\square$

Remark 15.2.1. The hypothesis on π_0 is necessary in Corollary 15.2. For example, the morphism in 15.1 for $Y = S^0$ is the group completion $\mathbb{N} \to \mathbb{Z}$, which is not a weak equivalence.

Remark 15.2.2. If $\mathrm{char}(k) = p > 0$, we may define $S^\infty Y[1/p]$ to be the colimit of $S^\infty Y \xrightarrow{p} S^\infty Y \xrightarrow{p} \cdots$. Then for $R = \mathbb{Z}[1/p]$ we have a global weak equivalence $S^\infty Y[1/p] \xrightarrow{\simeq} uR_{\mathrm{tr}}(Y)$ for every simplicial object Y of **Norm**$_+$ satisfying (i) or (ii) from Corollary 15.2.

We now apply these general considerations to $\mathbb{L}^n = \mathbb{L}^n_R = R(n)[2n]$. Write K_n and $B_\bullet K_n$ for the underlying sheaves $u\mathbb{L}^n$ and $u\mathbb{L}^n[1]$ representing $H^{2n,n}(-, R)$ and $H^{2n+1,n}(-, R)$, respectively; see Lemma 13.2. The example $Y = S^0$ in Remark 15.2.1 shows that we need to restrict to $n > 0$.

Theorem 15.3. *Suppose that* $\mathrm{char}(k) = 0$. *For $n \geq 1$, there are $\mathbb{A}^1$-local equivalences in* $\Delta^{\mathrm{op}}\mathbf{PST}(\mathbf{Sm}, R)$:

$$R_{\mathrm{tr}}(K_n) \simeq S_{\mathrm{tr}}^{\infty}(\mathbb{L}_R^n) \simeq \bigoplus_{i=1}^{\infty} S_{\mathrm{tr}}^i(\mathbb{L}_R^n), \quad and$$

$$R_{\mathrm{tr}}(B_\bullet K_n) \simeq S_{\mathrm{tr}}^{\infty}(\mathbb{L}_R^n[1]) \simeq \bigoplus_{i=1}^{\infty} S_{\mathrm{tr}}^i(\mathbb{L}_R^n[1]).$$

Hence cohomology operations from $H^{2n,n}(-,\mathbb{Z})$ to $H^{p,q}(-,R)$, resp., from $H^{2n+1,n}(-,\mathbb{Z})$ to $H^{p,q}(-,R)$, are in 1-1 correspondence with elements of

$$\widetilde{H}^{p,q}(K_n, R) \cong \mathrm{Hom}_{\mathbf{DM}}(\oplus_{i=1}^{\infty} S_{\mathrm{tr}}^i(\mathbb{L}^n), R(q)[p]) = \prod_{i=1}^{\infty} \mathrm{Hom}(S_{\mathrm{tr}}^i(\mathbb{L}^n), R(q)[p]),$$

resp., with elements of

$$\widetilde{H}^{p,q}(B_\bullet K_n, R) \cong \mathrm{Hom}_{\mathbf{DM}}(\oplus_{i=1}^{\infty} S_{\mathrm{tr}}^i(\mathbb{L}^n[1]), R(q)[p])$$

$$= \prod_{i=1}^{\infty} \mathrm{Hom}(S_{\mathrm{tr}}^i(\mathbb{L}^n[1]), R(q)[p]).$$

Proof. First suppose that R is any commutative ring. Writing $\check{C}(\mathbb{A}^n)$ for the simplicial Čech scheme on $\mathbb{A}^n$, let V denote the cone of $(\mathbb{A}^n - 0)_+ \to \check{C}(\mathbb{A}^n)_+$, given by $(\mathbb{A}^n - 0) \subset \mathbb{A}^n$; V is an object of $\Delta^{\mathrm{op}}\mathbf{Sm}_+$. Note that V is $\mathbb{A}^1$-local equivalent to the cone of $(\mathbb{A}^n - 0) \to \mathbb{A}^n$, which in turn is Nisnevich-local equivalent to $\mathbb{A}^n/(\mathbb{A}^n - 0)$. By 12.69.3, R_{tr} preserves $\mathbb{A}^1$-local equivalences. Thus $\mathbb{L}_R^n = R_{\mathrm{tr}}(\mathbb{A}^n/(\mathbb{A}^n - 0))$ is $\mathbb{A}^1$-local equivalent to $R_{\mathrm{tr}}(V)$ in $\Delta^{\mathrm{op}}\mathbf{PST}(\mathbf{Sm}, R)$ for all R. Applying u, we see that $u\mathbb{L}_R^n$ is $\mathbb{A}^1$-local equivalent to $uR_{\mathrm{tr}}(V)$ in $\Delta^{\mathrm{op}}\mathrm{Pshv}(\mathbf{Sm}_+)$.

Now set $R = \mathbb{Z}$, and suppose that $\mathrm{char}(k) = 0$. Since $\check{C}(\mathbb{A}^n) \to \mathrm{Spec}(k)$ is a global weak equivalence, each $\mathrm{Hom}(X_+, S^m V)$ is connected. By Corollary 15.2, $S^{\infty}V \to u\mathbb{Z}_{\mathrm{tr}}(V)$ is a global weak equivalence. Hence we have $\mathbb{A}^1$-local equivalences $S^{\infty}V \simeq K_n$, and $S_{\mathrm{tr}}^{\infty}R_{\mathrm{tr}}(V) = R_{\mathrm{tr}}(S^{\infty}V) \simeq R_{\mathrm{tr}}(K_n)$. By Corollary 14.14, $R_{\mathrm{tr}}(K_n) \simeq \oplus S_{\mathrm{tr}}^m R_{\mathrm{tr}}(V)$. Finally, since $\mathbb{L}_R^n$ is $\mathbb{A}^1$-local equivalent to $R_{\mathrm{tr}}(V)$, Proposition 14.21 implies that each $S_{\mathrm{tr}}^m R_{\mathrm{tr}}(V)$ is $\mathbb{A}^1$-local equivalent to $S_{\mathrm{tr}}^m \mathbb{L}_R^n$.

Similarly, $B_\bullet K_n = u(\mathbb{L}^n[1]) \simeq u\mathbb{Z}_{\mathrm{tr}}(\Sigma V)$. In this case, the decomposition arises from $R_{\mathrm{tr}}(B_\bullet K_n) \simeq R_{\mathrm{tr}} S^{\infty}(\Sigma V)$, Proposition 14.21, and Corollary 14.14. □

Remark 15.3.1. If $\mathrm{char}(k) = p > 0$, the conclusions of Theorem 15.3 hold with $\mathbb{Z}$ replaced by $\mathbb{Z}[1/p]$. To see this, replace Corollary 15.2 with Remark 15.2.2 in the proof.

Example 15.3.2. For $n = 1$, Theorem 15.3 gives an alternative to the calculation in 13.2.1 that cohomology operations $\phi : H^{2,1}(-,\mathbb{Z}) \to H^{p,q}(-,R)$ are in 1-1 correspondence with homogeneous polynomials $f(t) = \sum_{i>0} a_i t^i$ of bidegree

(p, q) in $H^{*,*}(k, R)[t]$, where t has bidegree $(2, 1)$. The polynomial $\sum a_i t^i$ corresponds to the operation $\phi(x) = \sum a_i \bar{x}^i$.

Indeed, we saw in Example 14.12(b) that $S_{\mathrm{tr}}^i(\mathbb{L}^1) \cong \mathbb{L}^i$ for all $i > 0$. By Theorem 15.3, the classifying space $K_1 = u\mathbb{L}^1$ has cohomology:

$$H^{p,q}(K_1, R) = \prod_{i=1}^{\infty} \mathrm{Hom}_{\mathbf{DM}}(\mathbb{L}^i, R(q)[p]) = \prod_{i=1}^{\infty} H^{p-2i, q-i}(k, R).$$

Example 15.4. As pointed out before Lemma 13.2, the identity cohomology operation on $H^{2n,n}(-, R)$ corresponds to the canonical element $\alpha = \alpha_{2n,n}^R$ of $\widetilde{H}^{2n,n}(K_n, R)$. It follows from Theorem 15.3 and the Slice Lemma 6.15 that $\widetilde{H}^{2n,n}(K_n, R) \cong \mathrm{Hom}(\mathbb{L}^n, \mathbb{L}^n) \cong R$, and it is easy to see that this group is generated by α. This was first observed by Voevodsky in [Voe03c, 3.7].

Consider the ith power cohomology operation $x \mapsto x^i$ from $H^{2n,n}(X, R)$ to $H^{2ni,ni}(X, R)$. It is nonzero for every R, by Lemma 13.1.1, and has a natural interpretation in terms of the summand $S_{\mathrm{tr}}^i(\mathbb{L}^n)$ in the decomposition in Theorem 15.3. If $1/(\ell - 1)! \in R$ and $i < \ell$ then $S_{\mathrm{tr}}^i(\mathbb{L}^n) \cong \mathbb{L}^{ni}$ by Example 14.12(e), and the power map corresponds to the generator α^i of $\mathrm{Hom}(S_{\mathrm{tr}}^i(\mathbb{L}^n), \mathbb{L}^{ni}) = R$.

Consider the diagonal $V \to V^i \to S^i(V)$ for $V = \check{C}(\mathbb{A}^n)/(\mathbb{A}^n - 0)$. Applying R_{tr} yields the motivic power map $\mathbb{L}^n \to \mathbb{L}^{ni}$ (see [MVW, 3.11]), and the symmetrizing map $\mathbb{L}^{ni} \to S_{\mathrm{tr}}^i(\mathbb{L}^n)$, up to the $\mathbb{A}^1$-local equivalences $\mathbb{L}_R^{ni} \simeq R_{\mathrm{tr}}(V^i)$. Applying u to $\mathbb{L}^n \to \mathbb{L}^{ni}$ yields the map $K_n \to K_{ni}$ representing the power operation. By Theorem 15.3, the adjoint map $R_{\mathrm{tr}}(K_n) \to \mathbb{L}^{ni}$ may be identified with the map $S_{\mathrm{tr}}^{\infty}(\mathbb{L}^n) \to S_{\mathrm{tr}}^i(\mathbb{L}^n) \to \mathbb{L}^{ni}$.

15.2 OPERATIONS ON $H^{1,1}$

In this section we fulfill the promise of Example 13.2.2 to show that the classifying space $K(\mathbb{Z}(1), 1)$ is the pointed scheme $\mathbb{G}_m = (\mathbb{A}^1 - \{0\}, 1)$, at least when k has characteristic 0. As pointed out in loc. cit., this implies that every motivic cohomology operation $\phi : H^{1,1}(-, \mathbb{Z}) \to H^{p,q}(-, R)$ has the form $\phi(x) = ax$ for a unique element $a \in H^{p-1,q-1}(k, R)$. We begin with a useful general observation.

Lemma 15.5. *If (X, x) is a pointed normal variety, there are global weak equivalences $\mathbb{L}\widetilde{S}^m(X, x) \xrightarrow{\sim} (\widetilde{S}^m)^*(X, x) \xrightarrow{\sim} (S^m X, x^m)$.*
In particular, $\mathbb{L}\widetilde{S}^m(\mathbb{G}_m)$ is weakly equivalent to $(S^m(\mathbb{A}^1 - \{0\}), 1^m)$.

Proof. Since the presheaf (X, x) is the reflexive coequalizer of $S^0 \rightrightarrows X_+$, and $\widetilde{S}^m(S^0) = S^0$, the presheaf $(\widetilde{S}^m)^*(X, x)$ is the reflexive coequalizer of $S^0 \rightrightarrows \widetilde{S}^m(X_+) = (S^m X)_+$ by Lemma 12.16. Since the coequalizer identifies x^m with the basepoint, we have $(\widetilde{S}^m)^*(X, x) \cong (S^m X, x^m)$.

The global weak equivalence $X \vee \Delta^1 \to (X, x)$ of Example 12.28.2 induces a global weak equivalence $X \vee \Delta^1 \simeq \mathbf{Lres}\,(X \vee \Delta^1) \simeq \mathbf{Lres}\,(X, x)$ (see 12.25), so we have $S^m(X \vee \Delta^1) \simeq \mathbb{L}S^m(X \vee \Delta^1) \xrightarrow{\sim} \mathbb{L}S^m(X, x)$ by Lemma 14.16 and Proposition 14.17.

It follows from (14.1) that $S^m(X \vee \Delta^1)$ is the union of the sheaves $S^i X \times S^j \Delta^1$ along the subsheaves $S^i X \times x^j$ of $S^m X$ for $i + j = m$. The linear contractions $f_t(x_1, \ldots, x_j) = (tx_1, \ldots, tx_j)$ of the $(S^j \Delta^1)(U)$ provide a homotopy retraction of $S^m(X \vee \Delta^1)(U)$ onto $S^m X(U)$, sending the basepoint to x^m.   □

Theorem 15.6. *The pointed space* $\mathbb{G}_m = (\mathbb{A}^1 - \{0\}, 1)$ *represents* $H^{1,1}(-, \mathbb{Z})$, *at least when* $char(k) = 0$.

Proof. Recall from section 13.1 that $H^{1,1}(-, \mathbb{Z})$ is represented by $u\mathbb{Z}(1)[1]$. Because u preserves $\mathbb{A}^1$-local equivalences such as $\mathbb{Z}(1)[1] = C_* \mathbb{Z}_{\mathrm{tr}}(\mathbb{G}_m) \xrightarrow{\sim} \mathbb{Z}_{\mathrm{tr}}(\mathbb{G}_m)$ (see 12.70), it is also represented by $u\mathbb{Z}_{\mathrm{tr}}(\mathbb{G}_m)$. Also recall from Example 12.28.2 that $\mathbb{G}_m$ is weakly equivalent to the split simplicial variety $V = (\mathbb{A}^1 - \{0\}) \vee \Delta^1$, and hence $u\mathbb{Z}_{\mathrm{tr}}(V) \simeq u\mathbb{Z}_{\mathrm{tr}}(\mathbb{G}_m)$.

By Proposition 14.17, $S^\infty V \simeq \mathbb{L}S^\infty(V) \simeq \mathbb{L}S^\infty(\mathbb{G}_m)$. By Lemma 15.5 and Example 14.12(c), the maps $\mathbb{L}S^m(\mathbb{G}_m) \to (S^m(\mathbb{A}^1 - \{0\}), 1) \to \mathbb{G}_m$ and hence their colimit $\mathbb{L}S^\infty(\mathbb{G}_m) \to \mathbb{G}_m$ are $\mathbb{A}^1$-equivalences. This shows that there is an $\mathbb{A}^1$-local equivalence between $S^\infty V$ and $\mathbb{G}_m$.

We claim that the map $S^\infty V \to u\mathbb{Z}_{\mathrm{tr}}(V)$ is an $\mathbb{A}^1$-local equivalence. Assuming this, we have the asserted $\mathbb{A}^1$-local equivalence:

$$K(\mathbb{Z}(1), 1) = u\mathbb{Z}_{\mathrm{tr}}(\mathbb{G}_m) \simeq u\mathbb{Z}_{\mathrm{tr}}(V) \xleftarrow{\simeq} S^\infty V \xrightarrow{\simeq} \mathbb{G}_m.$$

To establish the claim, consider the diagram of presheaves of simplicial abelian monoids:

$$
\begin{array}{ccccccc}
S^\infty V & \xrightarrow[\sim]{\mathbb{A}^1} & S^\infty V(- \times \Delta^\bullet) & \longrightarrow & \mathbb{L}S^\infty \mathbb{G}_m(- \times \Delta^\bullet) & \xrightarrow[\sim]{\mathbb{A}^1} & \mathbb{G}_m \\
\downarrow{\scriptstyle \eta} & & \downarrow{\scriptstyle \eta} & & & & \\
u\mathbb{Z}_{\mathrm{tr}} V & \xrightarrow[\sim]{\mathbb{A}^1} & (u\mathbb{Z}_{\mathrm{tr}} V)(- \times \Delta^\bullet). & & & &
\end{array}
$$

The two left horizontal maps are $\mathbb{A}^1$-local equivalences by construction, the upper middle horizontal is a global weak equivalence by Proposition 14.17, and the upper right horizontal is an $\mathbb{A}^1$-local equivalence by Example 14.12(c). The vertical maps η are group completions, by Theorem 15.1. (This step requires $char(k) = 0$.) But the group completion of any simplicial abelian monoid H is a homotopy equivalence when $\pi_0(H)$ is a group; applying this to $H(U) = S^\infty V(U \times \Delta^\bullet)$, we see that the right vertical group completion map is a global weak equivalence. The claim follows.   □

15.3   SCALAR WEIGHT

In this section, we introduce the notion of *scalar weight* for motivic cohomology operations, and give some examples. After describing $S_{\mathrm{tr}}^{\ell}(\mathbb{L}^n)$ (Theorem 15.26), we will show that motivic cohomology operations $H^{2n,n}(-,\mathbb{Z}) \to H^{p,q}(-,\mathbb{Z}/\ell)$ are graded by scalar weight when $n > 0$ (section 15.7).

Definition 15.7. Let $\phi : H^{p,q}(-,R) \to H^{p',q'}(-,\mathbb{Z}/\ell)$ be a cohomology operation. We say that ϕ has *scalar weight 0* if $\phi(\ell \cdot x) = 0$ and $\phi(m \cdot x) = \phi(x)$ for every integer $m \not\equiv 0 \pmod{\ell}$, every X, and every $x \in H^{p,q}(X,R)$.

We say that ϕ has *scalar weight 1* if $\phi(m \cdot x) = m \cdot \phi(x)$ for every integer m, every X, and every $x \in H^{p,q}(X,R)$. More generally, if $1 \le s < \ell$, ϕ is said to have *scalar weight s* if $\phi(m \cdot x) = m^s \cdot \phi(x)$ for every integer m, every X, and every $x \in H^{p,q}(X,R)$. In particular, $\phi(m \cdot x) = \phi(m' \cdot x)$ if $m \equiv m'$ modulo ℓ. Hence the group $\mathbb{Z}/\ell^\times$ acts on the R-submodule of operations ϕ which have scalar weight s; if $m \in \mathbb{Z}/\ell^\times$ and $\alpha \in H^{p,q}(X,R)$ then $(\phi \cdot m)(\alpha) = m^s \phi(\alpha)$.

For example, the identity map and the Frobenius $\phi(x) = x^\ell$ have scalar weight 1 when $R = \mathbb{Z}/\ell$. More generally, any additive cohomology operation (such as P^b) has scalar weight 1.

Abstractly, cohomology operations need not have scalar weight, or even be a sum of operations having scalar weight.

Example 15.7.1. Scalar weight s cohomology operations ϕ on $H^{0,0}(-,\mathbb{Z})$ are completely determined by $\phi(1)$, where $1 \in H^{0,0}(k,\mathbb{Z}) \cong \mathbb{Z}$, since $\phi(m \cdot 1) = m^s \phi(1)$, and they are classified by the elements of $H^{*,*}(k,\mathbb{Z}/\ell)$. For example, there are exactly ℓ cohomology operations $H^{0,0}(-,\mathbb{Z}) \to H^{0,0}(-,\mathbb{Z}/\ell)$ of any scalar weight s.

In contrast, the description in Example 13.2.3 shows that there are uncountably many cohomology operations from $H^{0,0}(-,\mathbb{Z})$ to $H^{0,0}(-,\mathbb{Z}/\ell)$.

Example 15.7.2. We can create cohomology operations of all scalar weights by multiplication; if ϕ_i has scalar weight s_i then the monomial $x \mapsto \phi_1(x) \cdots \phi_m(x)$ has scalar weight s, where $s \equiv \sum s_i \pmod{\ell-1}$. In particular, if $a \in H^{*,*}(k,\mathbb{Z}/\ell)$ then $\phi(x) = ax^s$ has scalar weight s.

Example 15.7.3. Cohomology operations $H^{2,1}(-,\mathbb{Z}) \to H^{p,q}(-,\mathbb{Z}/\ell)$ are represented by homogeneous polynomials $f(t) = \sum_{i>0} a_i t^i$ of bidegree (p,q) in $H^{*,*}(k,\mathbb{Z}/\ell)[t]$ by either Example 13.2.1 or 15.3.2. That is, $\phi(x) = f(x)$. The operation $\phi(x) = ax^i$, corresponding to the monomial at^i, has scalar weight i modulo $(\ell - 1)$: $(\phi \cdot m)(x) = \phi(mx) = a(mx)^i = m^i \phi(x)$.

Thus every cohomology operation $H^{2,1}(-,\mathbb{Z}) \to H^{*,*}(-,\mathbb{Z}/\ell)$ is a unique sum $\phi_0 + \cdots + \phi_{\ell-2}$ of operations ϕ_s of scalar weight s. The operations of scalar weight 0 are exactly those represented by a polynomial in $t^{\ell-1}$, and every operation of scalar weight s $(0 < s < \ell - 1)$ corresponds to at^s or $t^s f(t^{\ell-1})$ for an operation $f(t^{\ell-1})$ of scalar weight 0.

Suppose that $n \geq 1$. By Theorem 15.3, $R_{\mathrm{tr}}(K_n) \simeq \bigoplus_1^\infty S_{\mathrm{tr}}^m(\mathbb{L}^n)$, so that every cohomology operation $H^{2n,n}(-, \mathbb{Z}) \to H^{*,*}(-, \mathbb{Z}/\ell)$ is a sum of operations corresponding to elements of $\mathrm{Hom}_{\mathbf{DM}}(S_{\mathrm{tr}}^m(\mathbb{L}^n), \mathbb{Z}/\ell(*)[*])$. Following Steenrod [Ste72], we will refer to these latter cohomology operations as having *rank m*.

Definition 15.8. A cohomology operation $H^{2n,n}(-, \mathbb{Z}) \to H^{p,q}(-, \mathbb{Z}/\ell)$ has rank m if it factors as $\mathbb{Z}_{\mathrm{tr}} K_n \to S_{\mathrm{tr}}^m(\mathbb{L}^n) \to \mathbb{Z}/\ell(q)[p]$.

In section 15.7, we will show that every motivic operation $H^{2n,n}(-, \mathbb{Z}) \to H^{*,*}(-, \mathbb{Z}/\ell)$ of rank m has scalar weight m. This is true for $1 \leq m < \ell$ by Examples 15.8.1 and 15.8.2. To handle the case $m = \ell$, we will need a description of $S_{\mathrm{tr}}^\ell(\mathbb{L}^n)$. That is the goal of the next two sections.

Example 15.8.1. Rank 1 cohomology operations $H^{2n,n}(-, \mathbb{Z}) \to H^{*,*}(-, \mathbb{Z}/\ell)$ are all of the form $\phi(x) = ax$, where

$$a \in H^{*-2n,*-n}(k, R) = \mathrm{Hom}(S_{\mathrm{tr}}^1(\mathbb{L}^n), R(*)[*]).$$

These operations are additive, and so have scalar weight 1.

Example 15.8.2. Similarly, if $1 < s < \ell$ then it follows from Example 15.4 that cohomology operations $H^{2n,n}(-, \mathbb{Z}) \to H^{p,q}(-, \mathbb{Z}/\ell)$ of rank s are all of the form $\phi(x) = ax^s$, and have scalar weight s. Here

$$a \in H^{p-2ns,q-ns}(k, R) = \mathrm{Hom}(S_{\mathrm{tr}}^s(\mathbb{L}^n), R(q)[p]).$$

15.4 THE MOTIVE OF $(V-0)/C$ WITH $V^C = 0$

The goal of this section is to describe the motive associated to the quotient $(V - \{0\})/C$, where C is the cyclic group of order ℓ, and V is a nontrivial representation over k with $V^C = \{0\}$. This description is preparation for the next section, whose goal is to derive a formula for $S_{\mathrm{tr}}^\ell(\mathbb{L}^n)$, where $\mathbb{L} = R(1)[2]$. This formula will then be used to establish the key uniqueness theorem 15.38.

By abuse of notation, we identify V with the affine space $\mathbb{A}(V) = \mathrm{Spec}$ $(\mathrm{Sym}(V^*))$, and let $(V-0)$ denote the algebraic variety $\mathbb{A}(V) - \{0\}$. We also write X for $(V-0)/C$. Since $V(k) \neq \{0\}$, both $(V-0)$ and X have a k-rational point. The choice of this point provides the motive $R_{\mathrm{tr}}(X)$ with a summand $R = R_{\mathrm{tr}}(\mathrm{Spec}(k))$, and provides $H^{*,*}(X, \mathbb{Z}/\ell)$ with a summand $H^{*,*}(k, \mathbb{Z}/\ell)$.

Our first step is to produce a non-constant map $u : R_{\mathrm{tr}}(X) \to R(1)[1]$, i.e., an element of $H^{1,1}(X, \mathbb{Z}/\ell)$ not in the summand $H^{1,1}(k, \mathbb{Z}/\ell) \cong k^\times / k^{\times\ell}$.

Lemma 15.9. *Let V be a nonzero representation of C with $V^C = 0$, and set $X = (V-0)/C$. If $\dim(V) \geq 2$ then $\mathcal{O}(X)^\times = k^\times$ and $\mathrm{Pic}(X) \cong \mu_\ell(k)$.*

Hence $H^{1,1}(X, \mathbb{Z}/\ell) \cong H^{1,1}(k, \mathbb{Z}/\ell) \oplus \mu_\ell(k)$.

Proof. Since $\dim(V) > 1$, $\mathcal{O}(X) = k[V-0] = k$. By [MVW, 4.9], $H^{1,1}(X, \mathbb{Z}/\ell) \cong H^1_{\text{ét}}(X, \mu_\ell)$, which is (non-canonically) the direct sum of $\mathcal{O}(X)^\times / \mathcal{O}(X)^{\times \ell} = k^\times / k^{\times \ell}$ and $\text{Hom}(\mathbb{Z}/\ell, \text{Pic}(X))$. Thus it suffices to show that $\text{Pic}(X) \cong \mu_\ell(k)$.

Since C is cyclic of prime order and $V^C = 0$, C acts freely on $(V-0)$. Hence $(V-0) \to X$ is a Galois cover with group C. There is a Hochschild–Serre spectral sequence with $E_2^{p,q} = H^p(C, H^q_{\text{ét}}(V-0, \mathbb{G}_m))$ converging to $H^*_{\text{ét}}(X, \mathbb{G}_m)$ [Mil80, III(2.20)]. Since $\text{Pic}(V-0) = 0$, the row $q = 1$ is zero, and we have

$$\text{Pic}(X) \cong E_2^{1,0} = H^1(C, k[V-0]^\times) = H^1(C, k^\times) \cong \mu_\ell(k). \qquad \square$$

Remark 15.9.1. If $\dim(V) = 1$, i.e., $\mathbb{A}(V) = \text{Spec}(k[t])$, then $X = (V-0)/C = \text{Spec}(k[t^\ell, t^{-\ell}])$ is isomorphic to $\mathbb{A}^1 - \{0\}$. In this case $\text{Pic}(X) = 0$ and $R_{\text{tr}}(X) \cong R \oplus R(1)[1]$.

Definition 15.10. Suppose that k contains a primitive ℓ^{th} root of unity ζ, and that V is a nonzero representation of C of dimension ≥ 2 with $V^C = 0$. By Lemma 15.9, the choice of a k-point of $X = (V-0)/C$ determines an isomorphism $H^{1,1}(X, \mathbb{Z}/\ell) \cong H^{1,1}(k, \mathbb{Z}/\ell) \oplus \mu_\ell$. Let u be the element of $H^{1,1}(X, \mathbb{Z}/\ell)$ corresponding to ζ, and let $v \in H^{2,1}(X, \mathbb{Z}/\ell)$ denote the Bockstein applied to u. Note that u determines a map $R_{\text{tr}}(X) \to R(1)[1]$ in $\mathbf{DM}(k, R)$, where $R = \mathbb{Z}/\ell$.

Each $v^i \in H^{2i,i}(X, \mathbb{Z}/\ell)$ determines a map $R_{\text{tr}}(X) \to \mathbb{L}^i$ in $\mathbf{DM}(k, R)$, and each $uv^i \in H^{2i+1,i+1}(X)$ determines a map $R_{\text{tr}}(X) \to \mathbb{L}^i(1)[1]$ in $\mathbf{DM}(k, R)$. The elements v^i, uv^i for $0 \leq i < \dim(V)$ assemble to form the map:

$$I_V = I_V(\zeta) : R_{\text{tr}}((V-0)/C) \longrightarrow \bigoplus_{i=0}^{\dim(V)-1} \left\{ \mathbb{L}^i \oplus \mathbb{L}^i(1)[1] \right\}.$$

The map $I_V(\zeta)$ in 15.10 depends on the choice of ζ. To see how, recall that the choice determines an isomorphism $(\mathbb{Z}/\ell)^\times \xrightarrow{\sim} \text{Aut}(\mu_\ell)$, sending s to $\zeta \mapsto \zeta^s$. For $s \in (\mathbb{Z}/\ell)^\times$, let $m(s)$ denote the isomorphism of $\bigoplus \mathbb{L}^i \oplus \mathbb{L}^i(1)[1]$ which is multiplication by s^i on $\mathbb{L}^i$ and by s^{i+1} on $\mathbb{L}^i(1)[1]$.

Lemma 15.11. *For each $s \in (\mathbb{Z}/\ell)^\times$, $I_V(\zeta^s)$ is the composition $m(s) \circ I_V(\zeta)$.*

Proof. Under the natural isomorphism $H^{1,1}(X, \mathbb{Z}/\ell) \cong H^1_{\text{ét}}(X, \mu_\ell)$ of Lemma 15.9, u corresponds to ζ and hence su corresponds to ζ^s. Since the components of the map $I_V(\zeta^s)$ are obtained from u by applying the Bockstein and taking products, the map $v : R_{\text{tr}}(X) \to \mathbb{L}(1)[1]$ is replaced by sv, v^i is replaced by $s^i v^i$, and uv^i is replaced by $s^{i+1} uv^i$. $\qquad \square$

Theorem 15.12. *Let V be a nonzero representation of C with $V^C = 0$. If $R = \mathbb{Z}/\ell$ and $\mu_\ell \subset k^\times$, the map I_V of Definition 15.10 is an isomorphism in $\mathbf{DM}(k, R)$:*

$$I_V : R_{\mathrm{tr}}\left((V-0)/C\right) \xrightarrow{\simeq} \oplus_{i=0}^{\dim(V)-1}\left\{\mathbb{L}^i \oplus \mathbb{L}^i(1)[1]\right\}.$$

To prove Theorem 15.12, note that the choice of ζ determines an isomorphism $C \cong \mu_\ell$, and allows us to consider representations of μ_ℓ as representations of C. We begin by considering the case in which V is an isotypical representation.

Lemma 15.13. *Let V_1 be the direct sum of $n \geq 2$ copies of the standard 1-dimensional representation of μ_ℓ. Then the map I_{V_1} of Definition 15.10 is an isomorphism in* $\mathbf{DM}(k, R)$:

$$I_{V_1} : R_{\mathrm{tr}}((V_1-0)/C) \xrightarrow{\simeq} \oplus_{i=0}^{n-1}\left\{\mathbb{L}^i \oplus \mathbb{L}^i(1)[1]\right\}.$$

Proof. ([Voe03c, 6.3]) Set $X = (V_1-0)/C$. Since the projection $(V_1-0) \to \mathbb{P}^{n-1}$ is equivariant with fiber $\mathbb{A}^1-\{0\}$, it factors through a map $X \to \mathbb{P}^{n-1}$ with fiber isomorphic to $\mathbb{A}^1-\{0\}$ (see Remark 15.9.1). As observed in [Voe03c, 6.3], X is isomorphic to the complement of the zero-section of the line bundle L on $\mathbb{P}^{n-1}$ associated to the sheaf $\mathcal{O}(-\ell)$. By [MVW, 15.15] we have a Gysin triangle

$$R_{\mathrm{tr}}(\mathbb{P}^{n-1})(1)[1] \longrightarrow R_{\mathrm{tr}}(X) \to R_{\mathrm{tr}}(L) \xrightarrow{\gamma} R_{\mathrm{tr}}(\mathbb{P}^{n-1})\otimes\mathbb{L} \longrightarrow R_{\mathrm{tr}}(X)[1],$$

and $R_{\mathrm{tr}}(L) \cong R_{\mathrm{tr}}(\mathbb{P}^{n-1}) \cong \oplus_{i=0}^{n-1}\mathbb{L}^i$. The components of the composition $R_{\mathrm{tr}}(X) \to R_{\mathrm{tr}}(L) \to \oplus_{i=0}^{n-1}\mathbb{L}^i$ are the maps v^i.

Now for any line bundle L over any Y, the Gysin map $\gamma : R_{\mathrm{tr}}(Y) = R_{\mathrm{tr}}(L) \to R_{\mathrm{tr}}(Y)\otimes\mathbb{L}$ is $1 \otimes [L]$, where $[L]$ is the class of L in $\mathrm{Pic}(Y)/\ell = \mathrm{Hom}(R, \mathbb{L})$. In the case at hand, $[\mathcal{O}(-\ell)] = 0$ in $\mathrm{Pic}(\mathbb{P}^{n-1})/\ell$. Therefore $\gamma = 0$, and the Gysin triangle splits. The result may be read off from this information. $\qquad\square$

This proves Theorem 15.12 when $\ell = 2$, since our assumption that $V^C = 0$ implies that $V \cong V_1$. If $\dim(V) = 1$, the isomorphism is given by Remark 15.9.1, so we may assume that $\dim(V) \geq 2$.

Next, we consider the case when V is isotypical for a different representation.

Corollary 15.14. *Let V_a be the direct sum of $n \geq 2$ copies of the twisted 1-dimensional representation $\zeta \cdot x = \zeta^a x$. Then the map $I_a = I_{V_a}$ of Definition 15.10 is an isomorphism in* $\mathbf{DM}(k, R)$, *and there is a commutative diagram:*

$$
\begin{array}{ccc}
R_{\mathrm{tr}}(V_1 - 0)/C & \xrightarrow[I_1]{\simeq} & \oplus_{i=0}^{n-1}\left\{\mathbb{L}^i \oplus \mathbb{L}^i(1)[1]\right\} \\
\cong\Big\downarrow & & \Big\downarrow\oplus(a^i, a^{i+1}) \\
R_{\mathrm{tr}}(V_a - 0)/C & \xrightarrow[I_a]{\simeq} & \oplus_{i=0}^{n-1}\left\{\mathbb{L}^i \oplus \mathbb{L}^i(1)[1]\right\}.
\end{array}
$$

Proof. The representation V_a is the pullback of the representation V_1 along the automorphism $\zeta \mapsto \zeta^a$ of $C = \mu_\ell$. Hence we may identify V_1 and $(V_1-0)/C$ with V_a and $(V_a-0)/C$, respectively. Since the automorphism acts as multiplication by a on $H^1(C, k^\times)$, which is $\mathrm{Pic}((V_1-0)/C)$ by Lemma 15.9, this identification multiplies both u and $v = \beta(u)$ by the unit a. $\qquad\square$

Remark 15.14.1. The power morphism $p_a : V_1 \to V_a$ sending $(x_1, \dots, x_n)$ to $(x_1^a, \dots, x_n^a)$ is C-equivariant, finite, and surjective of degree a^n. The same is true for the restriction $(V_1 - 0) \to (V_a - 0)$ and for the C-quotient $q_a : (V_1 - 0)/C \to (V_a - 0)/C$. Since the degree of q_a is prime to ℓ, it follows that $R_{\mathrm{tr}}(q_a)$ is a split surjection, and hence an isomorphism in $\mathbf{DM}(k, R)$ (by Corollary 15.14).

Proof of Theorem 15.12. Since $\mu_\ell \subset k^\times$, Maschke's theorem allows us to write V as a direct sum of 1-dimensional representations, and hence as a direct sum of the isotypical representations V_a as in Corollary 15.14, $1 \le a \le \ell - 1$ of dimension n_a, $\sum n_a = \dim(V)$. Let V' denote the sum of $\dim(V)$ copies of the standard representation, so that the direct sum of the equivariant morphisms p_a of Remark 15.14.1 is a finite C-equivariant map $p : V' \to V$ of degree $d = \prod a^{n_a}$. The same is true for the restriction $(V'-0) \to (V-0)$ of p, so the C-quotient $q : (V' - 0)/C \to (V - 0)/C$ is finite of degree d, prime to ℓ. It follows that $R_{\mathrm{tr}}(q)$ is a split surjection.

Since $q^* : \mathrm{Pic}((V-0)/C) \to \mathrm{Pic}((V'-0)/C)$ sends u_V to $d \cdot u_{V'}$ and hence v_V to $d \cdot v_{V'}$, it follows that $R_{\mathrm{tr}}(q) : R_{\mathrm{tr}}((V'-0)/C) \longrightarrow R_{\mathrm{tr}}((V-0)/C)$ is compatible with the maps $I_V, I_{V'}$ in Definition 15.10. Thus the diagram

$$
\begin{array}{ccc}
R_{\mathrm{tr}}(V' - 0)/C & \xrightarrow[I_{V'}]{\cong} & \oplus_{i=0}^{n-1}\left\{ \mathbb{L}^i \oplus \mathbb{L}^i(1)[1] \right\} \\
\downarrow{\scriptstyle R_{\mathrm{tr}}(q)} & & \cong \downarrow{\scriptstyle \oplus(d^i, d^{i+1})} \\
R_{\mathrm{tr}}(V - 0)/C & \xrightarrow[I_V]{} & \oplus_{i=0}^{n-1}\left\{ \mathbb{L}^i \oplus \mathbb{L}^i(1)[1] \right\}
\end{array}
$$

commutes. This implies that $R_{\mathrm{tr}}(q)$ is also a split injection, and hence an isomorphism. It follows that I_V is an isomorphism. $\qquad\square$

Proposition 15.15. *Let V be a representation of $G = C \rtimes \mathrm{Aut}(C)$ with $V^C = 0$ and $d = \dim(V) > 0$. If $R = \mathbb{Z}/\ell$ and $\mu_\ell \subset k^\times$, there is a natural isomorphism*

$$
R_{\mathrm{tr}}\left((V - 0)/G \right) \cong R_{\mathrm{tr}}\left((V - 0)/C \right)^{\mathrm{Aut}(C)}
$$

$$
\cong R \oplus \bigoplus_{\substack{0 < i < d \\ (\ell-1)|i}} \left(\mathbb{L}^i \oplus \mathbb{L}^{i-1}(1)[1] \right) \oplus
\begin{cases}
\mathbb{L}^{d-1}(1)[1], & (\ell - 1) \mid d; \\
0, & else.
\end{cases}
$$

Proof. Set $X = (V-0)/C$ and $A = \mathrm{Aut}(C)$. Since C acts trivially on X, the G-action on V induces an action of $A = G/C$ on X. Observe that $\mathrm{Aut}(C)$ acts faithfully on X. Lemma 14.10 yields the formula $R_{\mathrm{tr}}(X)^A \xrightarrow{\simeq} R_{\mathrm{tr}}(X/A) = R_{\mathrm{tr}}((V-0)/G)$. Given the direct sum decomposition of $R_{\mathrm{tr}}(X)$ in Theorem 15.12, the formula for $M = R_{\mathrm{tr}}(X)^A$ is immediate from Lemma 15.11. $\qquad\square$

We conclude this section by describing the motivic cohomology rings of $(V-0)/C$ and $(V-0)/G$, via the decompositions in Theorem 15.12 and Proposition 15.15. When $\mu_\ell \subset k^\times$, the reduced regular representation W_1 of C is the direct sum of the $\ell-1$ nontrivial 1-dimensional representations of $C \cong \mu_\ell$, and $W_1^C = 0$.

Let W_n denote the direct sum of n copies of W_1, so $d = \dim(W_n) = n(\ell-1)$, and set $X_n = (W_n - 0)/C$. Note that Σ_n acts on W_n by permuting its factors.

15.16. We briefly recall the computation of $H^{*,*}(B_{\mathrm{gm}}\mu_\ell)$ and $H^{*,*}(B_{\mathrm{gm}}\Sigma_\ell)$ in [Voe03c, §6]. Suppose that k has ℓ^{th} roots of unity, so that we may identify C with the algebraic group μ_ℓ. By [Voe03c, 6.1], the map $H^{p,q}(X_{n+1}, \mathbb{Z}/\ell) \to H^{p,q}(X_n, \mathbb{Z}/\ell)$ is an isomorphism if $n > q$. Taking the limit as $n \to \infty$ yields $H^{p,q}(B\mu_\ell, \mathbb{Z}/\ell) = \lim H^{p,q}(X_n, \mathbb{Z}/\ell)$ by [Voe03c, 6.2]. By [Voe03c, 6.10],

$$H^{*,*}(B\mu_\ell, \mathbb{Z}/\ell) = H^{*,*}(k, \mathbb{Z}/\ell)[[u, v]]/(u^2 = 0) \qquad (15.16.1)$$

for $\ell > 2$ and

$$H^{*,*}(B\mu_2, \mathbb{Z}/2) \cong H^{*,*}(k, \mathbb{Z}/\ell)[[u, v]]/(u^2 = \tau v + \rho u). \qquad (15.16.2)$$

Recall from 15.10 that $u \in H^{1,1}(X_n, \mathbb{Z}/\ell)$ and $v \in H^{2,1}(X_n, \mathbb{Z}/\ell)$.

Now consider the group $G = C \rtimes A$, $A = \mathrm{Aut}(C)$. For $\ell = 2$, when $A = 1$, we have $\mu_2 = G = \Sigma_2$ and hence $BG = B\mu_2$.

When $\ell > 2$, the group A acts by algebra maps, and $a \in A$ satisfies: $a \cdot u = au$, $a \cdot v = av$; see [Voe03c, 6.11]. Thus for $c = uv^{\ell-2}$ and $b = \beta(c) = v^{\ell-1}$ we have:

$$H^{*,*}(B_{\mathrm{gm}}G, R) \cong H^{*,*}(B_{\mathrm{gm}}\mu_\ell, R)^A \cong H^{*,*}(k, R)[[c, b]]/(c^2), \quad \ell \text{ odd}.$$

By [Voe03c, 6.13-6.14], c and b come from $H^{*,*}(B_{\mathrm{gm}}\Sigma_\ell, R)$. This implies that the canonical injection $H^{*,*}(B_{\mathrm{gm}}\Sigma_\ell, R) \hookrightarrow H^{*,*}(B_{\mathrm{gm}}G, R)$ is an isomorphism.

Corollary 15.17. *If $\mu_\ell \subset k^\times$, $H^{*,*}(X_n, \mathbb{Z}/\ell)$ is a free module over $H^{*,*} = H^{*,*}(k, \mathbb{Z}/\ell)$, and the elements v^i, uv^i for $0 \leq i < d = n(\ell-1)$ form a basis.*
(a) If $\ell \neq 2$ the ring structure is $H^{*,*}(X_n, \mathbb{Z}/\ell) \cong H^{*,*}[u, v]/(v^d = u^2 = 0)$.
(b) If $\ell = 2$ then the ring structure is

$$H^{*,*}(X_n, \mathbb{Z}/2) \cong H^{*,*}(k, \mathbb{Z}/2)[u, v]/(v^d, u^2 + \tau v + \rho u).$$

Here τ is the nonzero element of $H^{0,1}(k, \mathbb{Z}/2) = \mu_2(k) \cong \mathbb{Z}/2$ and ρ is the class of -1 in $H^{1,1}(k, \mathbb{Z}/2) \cong k^\times/k^{\times 2}$.

Proof. The first sentence is immediate from Theorem 15.12. Since $H^{*,*}(X_n, R)$ is a quotient of $H^{*,*}(B\mu_\ell, \mathbb{Z}/\ell)$, the algebra structure on $H^{*,*}(X_n, \mathbb{Z}/\ell)$ follows from (15.16.1) and (15.16.2). $\qquad\square$

The group $A = \mathrm{Aut}(C) \cong (\mathbb{Z}/\ell)^{\times}$ acts on $X_n = (W_n - 0)/C$ and hence on the ring $H^{*,*}(X_n, \mathbb{Z}/\ell)$, and $a \in A$ satisfies: $a \cdot u = au$, $a \cdot v = av$; see Lemma 15.11 or [Voe03c, 6.11]. Hence a acts on uv^{i-1} and v^i as multiplication by a^i; these elements are invariant if and only if $i \equiv 0 \pmod{(\ell-1)}$. In particular, the elements $c = uv^{\ell-2}$ and $b = v^{\ell-1}$ are invariant under this action. From Proposition 15.15 we immediately conclude:

Corollary 15.18. *If $\mu_\ell \subset k^{\times}$, $H^{*,*}(X_n/A, \mathbb{Z}/\ell) = H^{*,*}(X_n, \mathbb{Z}/\ell)^A$ is a free module over $H^{*,*} = H^{*,*}(k, \mathbb{Z}/\ell)$, and the elements b^j, cb^j for $0 \le j < n$ form a basis. If $\ell > 2$ the ring structure is*

$$H^{*,*}(X_n/A, \mathbb{Z}/\ell) \cong H^{*,*}[b, c]/(b^n = c^2 = 0).$$

The case $\ell = 2$ is given by Corollary 15.17(b), because in that case A is trivial.

15.5 THE MOTIVE $S_{\mathrm{tr}}^{\ell}(\mathbb{L}^n)$

In this section, we compute $S_{\mathrm{tr}}^{\ell}(\mathbb{L}^n)$. To do so, we identify the cyclic group $C = C_\ell$ with the Sylow ℓ-subgroup of Σ_ℓ on generator $(12 \cdots \ell)$, and identify $G = C \rtimes \mathrm{Aut}(C)$ with the normalizer $N_{\Sigma_\ell}(C)$, where $s \in (\mathbb{Z}/\ell)^{\times} \cong \mathrm{Aut}(C)$ corresponds to the permutation $\sigma(i) = si$ of $(12 \cdots \ell)$. Since $[\Sigma_\ell : C]$ is prime to ℓ, we may identify $S_{\mathrm{tr}}^{\ell}(\mathbb{L}^n)$ with a summand of both $S_{\mathrm{tr}}^C(\mathbb{L}^n)$ and $S_{\mathrm{tr}}^G(\mathbb{L}^n)$. As observed in Corollary 14.11, $S_{\mathrm{tr}}^G(\mathbb{L}^n) \cong S_{\mathrm{tr}}^C(\mathbb{L}^n)^{\mathrm{Aut}(C)}$, so the calculation is reduced to $S_{\mathrm{tr}}^C(\mathbb{L}^n)$.

We first extend the formulas for $R_{\mathrm{tr}}((V-0)/G)$ in section 15.4 to the case when $V^C \ne 0$ (15.22). For this, it is convenient to use an auxiliary space. Recall that the Thom space $Th(V)$ of an affine space V is defined to be the Nisnevich sheaf $V/(V-0)$. If a finite group G acts on V, $X = (V-0)/G$ is open in V/G.

Definition 15.19. If G is a finite group acting on V, we define $Th(V)/G$ to be the pointed Nisnevich sheaf quotient $(V/G)/X$, where $X = (V-0)/G$.

Since the radial $\mathbb{A}^1$-contraction $V \times \mathbb{A}^1 \to V$ is equivariant, it induces an $\mathbb{A}^1$-contraction $(V/G) \times \mathbb{A}^1 \to V/G$. By Proposition 14.24, $Th(V)/G$ is $\mathbb{A}^1$-equivalent to the suspension ΣX, i.e., the cone of $X_+ \to S^0$ (see 1.36). Thus there is a cofiber sequence

$$X_+ \to S^0 \to Th(V)/G \to \Sigma(X_+).$$

If $V \ne 0$ then $X(k) \ne \emptyset$, and $X_+ \to S^0$ splits, yielding $\Sigma X_+ \simeq \Sigma S^0 \vee Th(V)/G$; applying $R_{\mathrm{tr}}[-1]$ and using (12.71) yields an isomorphism

$$R_{\mathrm{tr}}((V - 0)/G_+) \cong R \oplus \mathbb{L} R_{\mathrm{tr}}(Th(V)/G)[-1], \quad V \ne 0. \qquad (15.20)$$

This display is (37) in [Voe10c]; cf. Lemma 16 in [Del09].

Lemma 15.21. *Let V be a nonzero representation of a finite group G with $V^G = 0$, and let $\mathbb{A}^n$ be a trivial representation of G.*

Then $Th(V \oplus \mathbb{A}^n)/G \cong Th(V)/G \wedge Th(\mathbb{A}^n)$, and

$$\mathbb{L}R_{\mathrm{tr}}(Th(V \oplus \mathbb{A}^n)/G) \cong \mathbb{L}R_{\mathrm{tr}}(Th(V)/G) \otimes \mathbb{L}^n \cong \{R_{\mathrm{tr}}((V-0)/G)/R\} \otimes \mathbb{L}^n[1].$$

The term in braces is the quotient by the basepoint inclusion of R.

Proof. There is a canonical isomorphism $Th(V) \wedge Th(\mathbb{A}^n) \xrightarrow{\simeq} Th(V \oplus \mathbb{A}^n)$, because both are the quotient of $V \oplus \mathbb{A}^n$ by $(V \oplus \mathbb{A}^n) - \{0\}$. (This was observed in [MV99, 3.2.17].) The same reasoning shows that there is a canonical isomorphism $Th(V)/G \wedge Th(\mathbb{A}^n) \xrightarrow{\simeq} Th(V \oplus \mathbb{A}^n)/G$, because both sheaves are the quotient of $V/G \times \mathbb{A}^n$ by $(V/G \times \mathbb{A}^n) - \{0\}$.

The final assertion is obtained by applying R_{tr}, using the formula (15.20) for $R_{\mathrm{tr}}(Th(V)/G) \oplus R[1]$ and observing that $R_{\mathrm{tr}}(Th(\mathbb{A}^n)) \cong \mathbb{L}^n$. □

Proposition 15.22. *Suppose that $\mu_\ell \subset k^\times$ and set $R = \mathbb{Z}/\ell$. If $V \neq 0$ is a representation of C with $V^C = 0$ and $d = \dim(V)$, the map of Definition 15.10 induces an isomorphism in $\mathbf{DM}(k, R)$ between $\mathbb{L}R_{\mathrm{tr}}(Th(V \oplus \mathbb{A}^n)/C)[-1] \oplus R$ and*

$$R_{\mathrm{tr}}((V \oplus \mathbb{A}^n - 0)/C) \xrightarrow{\simeq} R \oplus \mathbb{L}^n(1)[1] \oplus \bigoplus_{i=1}^{d-1} \{\mathbb{L}^{n+i} \oplus \mathbb{L}^{n+i}(1)[1]\}.$$

If V is a representation of $G = C \rtimes \mathrm{Aut}(C)$ with $V^C = 0$ and $d = \dim(V) > 0$,

$$R_{\mathrm{tr}}((V \oplus \mathbb{A}^n - 0)/G) \cong R_{\mathrm{tr}}((V \oplus \mathbb{A}^n - 0)/C)^{\mathrm{Aut}(C)}$$

$$\cong R \oplus \bigoplus_{\substack{0 < i < d \\ (\ell-1)|i}} (\mathbb{L}^{n+i} \oplus \mathbb{L}^{n+i-1}(1)[1]) \oplus \begin{cases} \mathbb{L}^{n+d-1}(1)[1], & (\ell-1) \mid d; \\ 0, & \text{else.} \end{cases}$$

Proof. ([Voe10c, 3.56]) Set $A = \mathrm{Aut}(C)$, $W = V \oplus \mathbb{A}^n$, $X = (W-0)/C$, and $X_0 = (V-0)/C$. Then (15.20) and Lemma 15.21 yield:

$$R_{\mathrm{tr}}(X)/R \cong \mathbb{L}R_{\mathrm{tr}}(Th(W)/C)[-1] \cong R_{\mathrm{tr}}(X_0)/R \otimes \mathbb{L}^n,$$

$$R_{\mathrm{tr}}(X/A)/R \cong \mathbb{L}R_{\mathrm{tr}}(Th(W)/G)[-1] \cong R_{\mathrm{tr}}(X_0/A)/R \otimes \mathbb{L}^n.$$

Now plug in the formulas for $R_{\mathrm{tr}}(X_0)/R$ and $R_{\mathrm{tr}}(X_0/A)/R$, which are given in Theorem 15.12 and Proposition 15.15. □

Since $\mathbb{L} = R(1)[2]$, combining Lemma 15.21 with Proposition 15.22 yields:

Corollary 15.23. *Suppose that V is a representation of $G = C \rtimes \mathrm{Aut}(C)$ with $V^C = 0$ and $d = \dim(V) > 0$ is divisible by $(\ell-1)$. Then $\mathbb{L}R_{\mathrm{tr}}(Th(V \oplus \mathbb{A}^n)/G)$ is* *isomorphic to*

$$R_{\mathrm{tr}}((V-0)/G)/R \otimes \mathbb{L}^n[1] \cong \bigoplus_{\substack{0<i<d \\ (\ell-1)|i}} \left(\mathbb{L}^{n+i} \oplus \mathbb{L}^{n+i}[1] \right) \oplus \mathbb{L}^{n+d}.$$

Here is our main application of Proposition 15.22. The symmetric group Σ_ℓ acts on the product $(\mathbb{A}^n)^\ell$ of ℓ copies of $\mathbb{A}^n$ by permuting the factors. When $n=1$, Σ_ℓ acts trivially on the diagonal $\mathbb{A}^1$ and we have $(\mathbb{A}^1)^\ell \cong \mathbb{A}^1 \oplus W_1$. The restriction (from Σ_ℓ to C_ℓ) of W_1 is the reduced regular representation of C_ℓ. Therefore $(\mathbb{A}^n)^\ell$ has the form $W_n \oplus \mathbb{A}^n$, where Σ_ℓ acts trivially on $\mathbb{A}^n$, and W_n is the direct sum of n copies of the representation W_1, as in Corollary 15.18.

Lemma 15.24. *For all $G \subseteq \Sigma_m$, $S_{\mathrm{tr}}^G \mathbb{L}^n \cong \mathbb{L} R_{\mathrm{tr}}(Th(\mathbb{A}^{n\ell})/G)$ in $\mathbf{DM}(k,R)$.*

Proof. By definition, $S_{\mathrm{tr}}^G(\mathbb{L}^n) = R_{\mathrm{tr}} \widetilde{S}^G \mathbf{Lres}\,(\mathbb{A}^n/(\mathbb{A}^n-0))$. By Theorems 14.25 and 14.23, there is a Nisnevich-local equivalence

$$\widetilde{S}^G \mathbf{Lres}\,(\mathbb{A}^n/(\mathbb{A}^n-0)) \xrightarrow{\simeq} (\widetilde{S}^G)^*(\mathbb{A}^n/(\mathbb{A}^n-0)) \xrightarrow{\simeq} S^G \mathbb{A}^n/(S^G \mathbb{A}^n - S^G 0).$$

Keeping in mind that $\widetilde{S}^G \mathbf{Lres}\,(\mathbb{A}^n/(\mathbb{A}^n-0))$ is in $\Delta^{\mathrm{op}} \mathbf{Norm}_+^{\mathrm{ind}}$, applying $\mathbb{L} R_{\mathrm{tr}}$ yields a Nisnevich-local equivalence

$$S_{\mathrm{tr}}^G(\mathbb{L}^n) = R_{\mathrm{tr}} \widetilde{S}^G \mathbf{Lres}\,(\mathbb{A}^n/(\mathbb{A}^n-0)) \xrightarrow{\simeq} \mathbb{L} R_{\mathrm{tr}}\left(S^G \mathbb{A}^n/(S^G \mathbb{A}^n - S^G 0)\right).$$

Finally, $S^G \mathbb{A}^n/(S^G \mathbb{A}^n - S^G 0)$ equals $Th(\mathbb{A}^{n\ell})/G$ because $S^G \mathbb{A}^n - S^G 0$ equals $\mathbb{A}^{n\ell}/G - 0 = (\mathbb{A}^{n\ell}-0)/G$. Applying $\mathbb{L} R_{\mathrm{tr}}$ gives the result. $\qquad \square$

Corollary 15.25. *If $\mu_\ell \subset k^\times$, $R = \mathbb{Z}/\ell$ and $G = C \rtimes \mathrm{Aut}(C)$ then*

$$S_{\mathrm{tr}}^C \mathbb{L}^n \cong \bigoplus_{i=n+1}^{n\ell-1} \left\{ \mathbb{L}^i \oplus \mathbb{L}^i[1] \right\} \oplus \mathbb{L}^{n\ell};$$

$$S_{\mathrm{tr}}^G \mathbb{L}^n \cong \bigoplus_{j=1}^{n-1} \left\{ \mathbb{L}^{n+j(\ell-1)} \oplus \mathbb{L}^{n+j(\ell-1)}[1] \right\} \oplus \mathbb{L}^{n\ell}.$$

Proof. Combine Lemma 15.24 and $\mathbb{A}^{n\ell} \cong W_n \oplus \mathbb{A}^n$ with Proposition 15.22 (for $S_{\mathrm{tr}}^C \mathbb{L}^n$) and Corollary 15.23 (for $S_{\mathrm{tr}}^G \mathbb{L}^n$), noting that $d = n(\ell-1)$. $\qquad \square$

Theorem 15.26. *When $R = \mathbb{Z}/\ell$, $S_{\mathrm{tr}}^\ell(\mathbb{L}^n)$ is $\mathbb{A}^1$-equivalent to $S_{\mathrm{tr}}^G(\mathbb{L}^n)$, i.e.,*

$$S_{\mathrm{tr}}^\ell(\mathbb{L}^n) \sim \bigoplus_{j=1}^{n-1} \left\{ \mathbb{L}^{n+j(\ell-1)} \oplus \mathbb{L}^{n+j(\ell-1)}[1] \right\} \oplus \mathbb{L}^{n\ell}.$$

Proof. ([Voe10c, 2.58]) The basis of $H^{*,*}(X_n/G)$, displayed in Corollary 15.18, corresponds to the factors of $S_{\mathrm{tr}}^G(\mathbb{L}^n)$, which are displayed in Corollary 15.25. Since $[\Sigma_\ell : G] = (\ell - 2)!$, $S_{\mathrm{tr}}^\ell(\mathbb{L}^n)$ is a summand of $S_{\mathrm{tr}}^G(\mathbb{L}^n)$, and $H^{*,*}(X_n/\Sigma_\ell)$ is a summand of $H^{*,*}(X_n/G)$. By 15.16, the canonical map $H^{*,*}(B_{\mathrm{gm}}\Sigma_\ell) \to H^{*,*}(B_{\mathrm{gm}}G)$ is an isomorphism. Hence the map on quotients, $H^{*,*}(X_n/\Sigma_\ell) \to H^{*,*}(X_n/G)$, is a surjection and hence an isomorphism. This implies that each summand of $S_{\mathrm{tr}}^G(\mathbb{L}^n)$ belongs to $S_{\mathrm{tr}}^\ell(\mathbb{L}^n)$. $\qquad\square$

For the next application, we need the topological realization functor $t^{\mathbb{C}}$ from $\mathbf{DM}^{\leq 0}$ to the subcategory $D^{\leq 0}(\mathbf{Ab})$ of the derived category of abelian groups, described in 12.72. This functor sends $R_{\mathrm{tr}}(\mathbb{A}^m/(\mathbb{A}^m - 0))$ to the reduced chain complex $\widetilde{C}_*(S^{2m}) \simeq R[-2m]$ of the based sphere S^{2m}, sends $\mathbb{L}^a[b]$ to $\widetilde{C}_*(S^{2a+b}, R)$ and $S_{\mathrm{tr}}^m(\mathbb{L}^a[b])$ to $\widetilde{C}_*(\widetilde{S}^m(S^{2a+b}))$. For the next result, we set

$$A = \bigoplus_{i=1}^a \left\{ \mathbb{L}^{a+i(\ell-1)}[1] \oplus \mathbb{L}^{a+i(\ell-1)}[2] \right\}.$$

Corollary 15.27. *When $R = \mathbb{Z}/\ell$, $a > 0$, and $\mathrm{char}(k) = 0$, there is a split injection $S_{\mathrm{tr}}^\ell(\mathbb{L}^a[b])[1] \xrightarrow{\eta} S_{\mathrm{tr}}^\ell(\mathbb{L}^a[b+1])$ for all b, and we have: $S_{\mathrm{tr}}^\ell(\mathbb{L}^a[1]) \cong A$;*

$$S_{\mathrm{tr}}^\ell(\mathbb{L}^a[b]) \cong A[b-1] \oplus \left\{ \mathbb{L}^{a\ell}[b] \oplus \mathbb{L}^{a\ell}[b+1] \right\} \otimes \bigoplus_{i=1}^k R[2i(\ell-1)], \quad b = 2k+1;$$

$$S_{\mathrm{tr}}^\ell(\mathbb{L}^a[b]) \cong S_{\mathrm{tr}}^\ell(\mathbb{L}^a[b-1])[1] \oplus \mathbb{L}^{a\ell}[b\ell], \quad b \geq 2 \text{ even.}$$

Proof. ([Wei09, 14.7.2]) Set $T = \mathbb{L}^a[b]$. We will assume the result is true for T and prove that it is true for $T[1]$. It holds for $b = 0$ by Theorem 15.26, so we assume it holds for b. For $b > 0$, consider the triangle of Theorem 14.34:

$$(S_{\mathrm{tr}}^\ell T)[1] \xrightarrow{\eta} S_{\mathrm{tr}}^\ell(T[1]) \to \mathrm{cone}(\eta) \xrightarrow{\delta} (S_{\mathrm{tr}}^\ell T)[2].$$

By Corollary 14.35, $\mathrm{cone}(\eta)$ is $T^{\otimes \ell}[2]$ for b even, and $T^{\otimes \ell}[\ell]$ for b odd. For b odd, δ is zero for weight reasons. For b even, the boundary map δ is an element of $\mathrm{Hom}(T^{\otimes \ell}, S_{\mathrm{tr}}^\ell T) \cong \mathbb{Z}/\ell$. Using the topological realization functor of 12.72, the topological calculations of Cartan [Car54] show that the boundary map δ is also zero for b even. In both cases, the triangle splits. The result now follows by induction on b, since:

$$S_{\mathrm{tr}}^\ell(\mathbb{L}^a[b+1]) \cong S_{\mathrm{tr}}^\ell(\mathbb{L}^a[b])[1] \oplus \mathrm{cone}(\eta). \qquad\square$$

Remark 15.27.1. The formulas in Corollary 15.27 also hold for $a = 0$, as $\mathbb{L}^0 = R$; we saw in Example 14.36 that $S_{\mathrm{tr}}^\ell(R[1]) = 0$, and $S_{\mathrm{tr}}^\ell(R[2]) \cong R[2\ell]$. The general formula for $\widetilde{S}_{\mathrm{tr}}^\ell(R[b])$ is given in Corollary 14.38.

15.6   A KÜNNETH FORMULA

A *proper Tate motive* is a direct sum of motives of the form $\mathbb{L}^a[b]$ with $b \geq 0$. The category of proper Tate motives over a field R is idempotent complete, and closed in **DM** under $\otimes$. Corollary 15.27 shows that the $S_{\mathrm{tr}}^\ell(\mathbb{L}^a[b])$ are proper Tate motives.

Theorem 15.28. *When $R = \mathbb{Z}/\ell$, $S_{\mathrm{tr}}^\infty(\mathbb{L}^n)$ is a proper Tate motive. For each q there are only finitely many terms of weight q.*

Proof. Combining 14.14, 14.15, 15.26, and 15.27 yields the theorem.   □

Lemma 15.29. *For all integers p, q, n, and i, and all simplicial schemes X:*

$$\mathrm{Hom}_{\mathbf{DM}}(R_{\mathrm{tr}}(X)(q)[p], R(i)[n]) = \begin{cases} H^{n-p, i-q}(X, R) & \text{if } q \leq i; \\ 0 & \text{if } q > i. \end{cases}$$

Proof. We may suppose that $p = 0$. Suppose first that $q \leq i$. By the Cancellation Theorem [MVW, 16.25] we have $\mathrm{Hom}(M(q), R(i)) = \mathrm{Hom}(M, R(i-q))$ for any M in **DM**. In particular, $\mathrm{Hom}(R_{\mathrm{tr}}(X)(q), R(i)[n]) = \mathrm{Hom}(R_{\mathrm{tr}}(X), R(i-q)[n]) = H^{n, i-q}(X, R)$. Similarly, the case when $q > i$ reduces to the case $i = 0$, $q > 0$. Here $R_{\mathrm{tr}}(X)(q)$ is a summand of $R_{\mathrm{tr}}(X \times \mathbb{P}^q)$ and $H^{p,0}(-, R) = H^p_{\mathrm{zar}}(-, R)$, so the result follows from $H^*_{\mathrm{zar}}(X, R) \cong H^*_{\mathrm{zar}}(X \times \mathbb{P}^q, R)$; see [Voe03c, 3.5].   □

Proposition 15.30. (Pure Künneth formula) *Let X and Y be pointed simplicial ind-schemes such that $R_{\mathrm{tr}}(Y)$ is a direct sum of motives $R(q_\alpha)[p_\alpha]$. Assume that for each q there are only finitely many α with $q_\alpha = q$. Then the Künneth homomorphism is an isomorphism:*

$$H^{*,*}(X, R) \otimes_{H^{*,*}(k, R)} H^{*,*}(Y, R) \to H^{*,*}(X \times Y, R).$$

Proof. By Lemma 15.29, $H^{n,i}(Y, R) = \mathrm{Hom}_{\mathbf{DM}}(R_{\mathrm{tr}}(Y), R(i)[n])$ is a free $H^{*,*}(k, R)$-module on finitely many generators γ_α in bidegrees (p_α, q_α). Similarly, $R_{\mathrm{tr}}(X \times Y)$ is the direct sum of the $R_{\mathrm{tr}}(X)(q_\alpha)[p_\alpha]$, so (by Lemma 15.29) $H^{n,i}(X \times Y, R) = \mathrm{Hom}_{\mathbf{DM}}(R_{\mathrm{tr}}(X \times Y), R(i)[n])$ is a free $H^{*,*}(X, R)$-module on the same set of generators γ_α in bidegrees (p_α, q_α). The result follows.   □

Recall that if $K \to H$ is a morphism of graded-commutative rings, $H \otimes_K H$ is an associative graded-commutative ring with product

$$(x \otimes y)(x' \otimes y') = (-1)^{|y|\,|x'|} xx' \otimes yy'.$$

The twist $\tau(x \otimes y) = (-1)^{|x|\,|y|} y \otimes x$ fixes K and defines an action of the symmetric group Σ_2 on $H \otimes_K H$, which extends to an action of Σ_m on the m-fold tensor product $H \otimes_K \cdots \otimes_K H$. We define $\mathrm{Sym}^m(H)$ to be the invariant subring

$(H \otimes_K \cdots \otimes_K H)^{\Sigma_m}$ of $H^{\otimes m}$. For example, if $|x| = 2$, $\mathrm{Sym}^m(K[x])$ is the polynomial ring $K[s_1, \ldots, s_m]$ on the symmetric polynomials s_i.

Corollary 15.31. *If $1/m! \in R$ and $R_{\mathrm{tr}}(Y)$ is a direct sum of $R(q_\alpha)[p_\alpha]$ with $q_\alpha = q$ for only finitely many α, then the $H^{*,*}(k, R)$-algebra $H^{*,*}(Y, R)$ satisfies*

$$\mathrm{Sym}^m H^{*,*}(Y, R) \cong H^{*,*}(S^m(Y), R).$$

Proof. By the Künneth formula 15.30, $H^{*,*}(Y, R)^{\otimes m} \to H^{*,*}(Y^m, R)$ is an isomorphism of free $H^{*,*}(k, R)$-modules. The symmetric group Σ_m acts on both sides and the isomorphism is equivariant, so the symmetric subrings are isomorphic. Finally, $H^{*,*}(S^m Y, R) = H^{*,*}(Y^m, R)^{\Sigma_m}$; see Example 14.12(d). $\square$

Recall from 15.8 that K_n is the sheaf underlying $\mathbb{L}^n$, representing $H^{2n,n}(-, R)$.

Corollary 15.32. *For all $n > 0$ the Künneth maps are isomorphisms:*

$$H^{*,*}(K_n, \mathbb{Z}/\ell) \otimes_{H^{*,*}} \cdots \otimes_{H^{*,*}} H^{*,*}(K_n, \mathbb{Z}/\ell) \xrightarrow{\simeq} H^{*,*}(K_n \times \cdots \times K_n, \mathbb{Z}/\ell),$$

or equivalently,

$$\tilde{H}^{*,*}(K_n, \mathbb{Z}/\ell) \otimes_{H^{*,*}} \cdots \otimes_{H^{*,*}} \tilde{H}^{*,*}(K_n, \mathbb{Z}/\ell) \xrightarrow{\simeq} \tilde{H}^{*,*}(K_n \wedge \cdots \wedge K_n, \mathbb{Z}/\ell).$$

Proof. We saw in the proof of Theorem 15.3 that there is a V in $\Delta^{\mathrm{op}}\mathbf{Sm}_+$ such that $R_{\mathrm{tr}}(V)$ is $\mathbb{A}^1$-local equivalent to $\mathbb{L}^n$ and $S^\infty V \simeq u\mathbb{L}^n = K_n$. By Theorem 15.28, $S^\infty \mathbb{L}^n$ is a proper Tate motive with only finitely many terms of any given weight. Thus

$$S_{\mathrm{tr}}^\infty R_{\mathrm{tr}}(V) = R_{\mathrm{tr}}(S^\infty V) \simeq R_{\mathrm{tr}}(K_n) \simeq S_{\mathrm{tr}}^\infty \mathbb{L}^n.$$

Proposition 15.30 implies that

$$H^{*,*}(X, \mathbb{Z}/\ell) \otimes_{H^{*,*}(k)} H^{*,*}(K_n, \mathbb{Z}/\ell) \cong H^{*,*}(X \times K_n, \mathbb{Z}/\ell).$$

The corollary follows by induction on the number of terms K_n. $\square$

15.7 OPERATIONS OF PURE SCALAR WEIGHT

With the description of $S_{\mathrm{tr}}^\ell(\mathbb{L}^n)$ in hand, we may now prove the following theorem, which shows that $H^{*,*}(K_n, \mathbb{Z}/\ell)$ is graded by the group $(\mathbb{Z}/\ell)^\times$. We assume that $\mathrm{char}(k) = 0$ and that $n \geq 1$. The rank of a cohomology operation $H^{2n,n}(-, \mathbb{Z}) \to H^{*,*}(-, \mathbb{Z}/\ell)$ is defined in 15.8.

Theorem 15.33. *The cohomology operations $H^{2n,n}(-, \mathbb{Z}) \to H^{*,*}(-, \mathbb{Z}/\ell)$ of rank m have scalar weight $m \mod (\ell - 1)$.*

The proof of Theorem 15.33 will occupy the rest of this section. Operations of rank $< \ell$ are covered by Example 15.8.1, so we may assume that $m \geq \ell$. The first lemma shows that operations coming from $S_{\mathrm{tr}}^m(\mathbb{L}^n)$ may be factored using the ℓ-adic expansion of m. By Example 15.7.2, this reduces the proof of Theorem 15.33 to $m = m_i \ell^i$.

Lemma 15.34. *Write $m = m_0 + m_1\ell + \cdots + m_r\ell^r$ with $0 \leq m_i < \ell$. Every cohomology operation ϕ of rank m is a sum of operations $x \mapsto \phi_0(x)\phi_1(x)\cdots\phi_r(x)$, where the ϕ_i have rank $m_i\ell^i$.*

Proof. Set $R = \mathbb{Z}/\ell$ and set $H_i = \Sigma_{m_i\ell^i}$. By Proposition 14.15, the subgroup $H = \prod H_i$ of Σ_m contains a Sylow ℓ-subgroup, and $S_{\mathrm{tr}}^m(\mathbb{L}^n)$ is a summand of $S_{\mathrm{tr}}^H(\mathbb{L}^n)$. Hence any rank m operation $\phi : K_n \to S_{\mathrm{tr}}^m(\mathbb{L}^n) \to R(*)[*]$ factors through a map $S_{\mathrm{tr}}^H(\mathbb{L}^n) \to R(*)[*]$. We saw before Proposition 14.15 that $S_{\mathrm{tr}}^H(\mathbb{L}^n) \cong \otimes S_{\mathrm{tr}}^{m_i\ell^i}(\mathbb{L}^n)$.

By the Künneth formula 15.30, we have

$$\mathrm{Hom}(S_{\mathrm{tr}}^H(\mathbb{L}^n), \mathbb{Z}/\ell(*)[*]) \cong \bigotimes_{i=0}^{r} \mathrm{Hom}(S_{\mathrm{tr}}^{m_i\ell^i}(\mathbb{L}^n), \mathbb{Z}/\ell(*)[*]),$$

so ϕ is induced by a sum of terms $\phi_0 \otimes \phi_1 \otimes \cdots \otimes \phi_r$. $\square$

The next two propositions handle the cases $m = \ell$ and $m = \ell^\nu$.

Proposition 15.35. *Every cohomology operation $H^{2n,n}(-, \mathbb{Z}) \xrightarrow{\phi} H^{p,q}(-, \mathbb{Z}/\ell)$ of rank ℓ has scalar weight 1.*

Proof. Set $T = \mathbb{Z}/\ell(q)[p]$, so ϕ comes from an element of $\mathrm{Hom}(S_{\mathrm{tr}}^\ell(\mathbb{L}^n), T)$. By Corollary 15.27, the map $S_{\mathrm{tr}}^\ell(\mathbb{L}^n)[1] \xrightarrow{\eta} S_{\mathrm{tr}}^\ell(\mathbb{L}^n[1])$ is a split injection, so $\mathrm{Hom}(S_{\mathrm{tr}}^\ell(\mathbb{L}^n[1]), T[1]) \xrightarrow{\eta^*} \mathrm{Hom}(S_{\mathrm{tr}}^\ell(\mathbb{L}^n), T)$ is onto. Hence ϕ lifts to a cohomology operation $\phi_1 : H^{2n+1,n}(-, \mathbb{Z}) \to H^{p+1,q}(-, \mathbb{Z}/\ell)$, in the sense that the suspension $\Sigma\phi(x)$ is $\phi_1(\Sigma x)$. If $x, y \in H^{2n,n}(X, \mathbb{Z})$ then by [Voe03c, 2.9], the cohomology operation ϕ_1 is additive on $H^{2n+1,n}(\Sigma X, \mathbb{Z})$, so:

$$\Sigma\phi(x+y) = \phi_1(\Sigma(x+y)) = \phi_1(\Sigma x + \Sigma y) = \phi_1(\Sigma x) + \phi_1(\Sigma y) = \Sigma\phi(x) + \Sigma\phi(y).$$

Since the suspension Σ is an isomorphism of groups, ϕ is additive. $\square$

We now establish the case $m = \ell^\nu$ by induction on ν, the case $\nu = 1$ being 15.35.

Proposition 15.36. *Every cohomology operation $H^{2n,n}(-, \mathbb{Z}) \xrightarrow{\phi} H^{p,q}(-, \mathbb{Z}/\ell)$ of rank ℓ^ν has scalar weight 1.*
If $0 < s < \ell$, then every operation of rank $s\ell^\nu$ has scalar weight s.

Proof. We proceed by induction on ν, the case $\nu = 0$ being in hand. By Proposition 14.15, every operation of rank $s\ell^\nu$ has the form $S^{s\ell^\nu}_{\text{tr}}(\mathbb{L}^n) \to S^s_{\text{tr}}(S^{\ell^\nu}_{\text{tr}}(\mathbb{L}^n))$ $\to T$. By 15.31, every element of $\text{Hom}(S^s_{\text{tr}}(S^{\ell^\nu}_{\text{tr}}(\mathbb{L}^n)), T)$ is a sum of monomials $\phi_1 \cdots \phi_s$ where the ϕ_i belong to $\text{Hom}(S^{\ell^\nu}_{\text{tr}}(\mathbb{L}^n), T)$. Once we show that the ϕ_i have scalar weight 1, it will follow from Example 15.7.2 that these monomials have scalar weight s. Thus we are reduced to the case $s = 1$.

Consider the subgroup $G = \Sigma_\ell \wr \cdots \wr \Sigma_\ell$ ($\nu - 1$ times) of $H = \Sigma_{\ell^{\nu-1}}$; as noted in Proposition 14.15, $G \wr \Sigma_\ell \subset H \wr \Sigma_\ell \subset \Sigma_{\ell^\nu}$, and $S^{\ell^\nu}_{\text{tr}}(\mathbb{L}^n)$ is a direct summand of both $S^{H\wr\Sigma_\ell}_{\text{tr}}(\mathbb{L}^n) = S^\ell_{\text{tr}}(S^H_{\text{tr}}(\mathbb{L}^n))$ and $S^{G\wr\Sigma_\ell}_{\text{tr}}(\mathbb{L}^n)$. Thus it suffices to treat cohomology operations of the form $S^{\ell^\nu}_{\text{tr}}(\mathbb{L}^n) \to S^\ell_{\text{tr}}(S^H_{\text{tr}}(\mathbb{L}^n)) \to T$. By Corollary 15.27, we may write $S^H_{\text{tr}}(\mathbb{L}^n) = S^{\ell^{\nu-1}}_{\text{tr}}(\mathbb{L}^n)$ as a sum of $\mathbb{L}^{a_i}[b_i]$. By Lemma 14.9, $S^\ell_{\text{tr}}S^{\ell^{\nu-1}}_{\text{tr}}(\mathbb{L}^n)$ is a sum of "linear" terms $S^\ell_{\text{tr}}(\mathbb{L}^{a_i}[b_i])$, which correspond to additive cohomology operations by Lemma 15.37 and "nonlinear" terms of the form

$$S^{r_1}_{\text{tr}}(\mathbb{L}^{a_{i_1}}[b_{i_1}]) \otimes \cdots \otimes S^{r_k}_{\text{tr}}(\mathbb{L}^{a_{i_k}}[b_{i_k}]), \quad \sum r_i = \ell.$$

By induction, each of the $\mathbb{L}^{a_i}[b_i]$ correspond to cohomology operations of scalar weight 1. Since $r_i < \ell$, we see from Example 14.12(d) that the $S^{r_i}(\mathbb{L}^{a_i}[b_i])$ correspond to operations $\phi_i(x)$ of scalar weight r_i. Hence the "nonlinear" terms $\phi_1(x) \cdots \phi_k(x)$ correspond to cohomology operations which have scalar weight $\sum r_i = \ell \equiv 1 \pmod{\ell - 1}$. $\qquad\square$

Lemma 15.37. *Let $\mathbb{L}^a[b]$ be a summand of $S^{\ell^{\nu-1}}_{\text{tr}}(\mathbb{L}^n)$. Then the motivic cohomology operations $H^{2n,n}(-, \mathbb{Z}) \to H^{*,*}(-, \mathbb{Z}/\ell)$ corresponding to elements of $\text{Hom}(S^\ell_{\text{tr}}(\mathbb{L}^a[b]), \mathbb{Z}/\ell(*)[*])$ are additive. Hence they have scalar weight 1.*

Proof. By Corollary 15.27, the map $S^\ell_{\text{tr}}(S^{\ell^{\nu-1}}_{\text{tr}}(\mathbb{L}^n))[1] \xrightarrow{\eta} S^\ell_{\text{tr}}(S^{\ell^{\nu-1}}_{\text{tr}}(\mathbb{L}^n)[1])$ is a split injection, and restricts to a split injection $S^\ell_{\text{tr}}(\mathbb{L}^a[b])[1] \xrightarrow{\eta} S^\ell_{\text{tr}}(\mathbb{L}^a[b+1])$. We may now argue as in the proof of Proposition 15.35. By Theorem 15.27, the map $\text{Hom}(S^\ell_{\text{tr}}(\mathbb{L}^a[b+1]), T[1]) \to \text{Hom}(S^\ell_{\text{tr}}(\mathbb{L}^a[b]), T)$ is onto. Hence ϕ lifts to a cohomology operation $\phi_1 : H^{2n+1,n}(-, \mathbb{Z}) \to H^{p+1,q}(-, \mathbb{Z}/\ell)$. But then ϕ is additive by [Voe03c, 2.9]. $\qquad\square$

This completes the proof of Theorem 15.33.

15.8   UNIQUENESS OF βP^N

The purpose of this section is to establish the uniqueness of certain cohomology operations from $H^{2n+1,n}(-, \mathbb{Z})$ to $H^{2n\ell+2,n\ell}(-, \mathbb{Z}/\ell)$. This is accomplished in Theorem 15.38.[1]

---

1. The n of this section is unrelated to the n in the norm residue homomorphism of the Bloch–Kato conjecture.

We saw in Definition 1.6 that the cohomology operations described in Theorem 15.38 correspond to elements of $H^{*,*}(B_\bullet K_n, \mathbb{Z}/\ell)$. We saw in section 14.2 (see 14.15) that a complete description of this cohomology is possible, but it is messy for $n > 1$.

Consider the cohomology operation $H^{2n+1,n}(-, \mathbb{Z}/\ell) \xrightarrow{\beta P^n} H^{2n\ell+2,n\ell}(-, \mathbb{Z}/\ell)$, where the Bockstein β and the Steenrod operation P^n are described in section 3.4. Since this operation is bi-stable, it is additive and commutes with simplicial suspension Σ by [Voe03c]. Since $P^n(y) = y^\ell$ for $y \in H^{2n,n}(X, \mathbb{Z}/\ell)$ by axiom 13.6(2), and the Bockstein β is a derivation, we have:

$$\beta P^n(\Sigma y) = \Sigma(\beta P^n y) = \Sigma \beta(y^\ell) = \ell \cdot \Sigma \beta(y) = 0.$$

Thus βP^n, or rather its composition with mod-ℓ reduction

$$H^{2n+1,n}(-, \mathbb{Z}) \to H^{2n+1,n}(-, \mathbb{Z}/\ell) \xrightarrow{\beta P^n} H^{2n\ell+2,n\ell}(-, \mathbb{Z}/\ell),$$

satisfies (1) and (2) of the following uniqueness theorem.

Theorem 15.38. *Let $\phi\colon H^{2n+1,n}(-, \mathbb{Z}) \to H^{2n\ell+2,n\ell}(-, \mathbb{Z}/\ell)$ be a cohomology operation such that for all X and all $x \in H^{2n+1,n}(X, \mathbb{Z})$:*

1. *$\phi(bx) = b\phi(x)$ for $b \in \mathbb{Z}$;*
2. *if $x = \Sigma y$ for $y \in H^{2n,n}(X, \mathbb{Z})$ then $\phi(x) = 0$.*

Then ϕ is a multiple of the composition of βP^n with mod-ℓ reduction.

Remark 15.38.1. In topology, βP^n is the image of a cohomology operation $H^{2n+1}(-, \mathbb{Z}) \to H^{2n\ell+2}(-, \mathbb{Z})$. We saw the relevance of this in chapter 5.

Lemma 15.39. *If $R(q)[p]$ is a summand of $S_{\mathrm{tr}}^m(\mathbb{L}^n)$, and $m \equiv s$ mod $(\ell - 1)$ for $m \geq 1$ and $0 \leq s < \ell - 1$, then:*

(a) *if $s \neq 0$ then $q \geq ns$, with equality iff $m < \ell$;*
(b) *if $s = 0$ then $q \geq n(\ell - 1)$, with equality iff $m = \ell - 1$;*
(c) *$p \geq 2q \geq 2n$.*

Proof. Recall from Proposition 15.30 that $R(q)[p] = \mathbb{L}^q[b]$ for $b \geq 0$, so $p = 2q + b$. Hence (a) and (b) imply (c). If $1 \leq m < \ell$ then $S_{\mathrm{tr}}^m(\mathbb{L}^n) \cong \mathbb{L}^{mn}$ and $q = mn$ by 14.12(e). This yields the equalities in (a) and (b). To prove the inequalities in (a) and (b), we suppose that $m \geq \ell$ and write $m = \sum m_i \ell^i$, noting that $\sum m_i > m_0$, $\sum m_i \equiv m$ mod $(\ell - 1)$. By Proposition 14.15 and Theorem 15.26, we also have

$$q \geq \sum m_i \left(n + i(\ell - 1)\right) > \left(\sum m_i\right)n.$$

Since $\sum m_i \geq s$, we have $q > ns$. If $s = 0$ then $\sum m_i \geq \ell - 1$ and we have $q > n(\ell - 1)$. $\square$

Remark 15.39.1. The equalities in Lemma 15.39 are the equations (2.6), (2.7), and (2.8) of [Voe11].

We now analyze the motivic cohomology $H^{2n\ell+2,n\ell}(B_\bullet K_n, \mathbb{Z}/\ell)$, where $B_\bullet K_n$ is the underlying sheaf $u\mathbb{L}^n[1]$. It is well known that the shift operator $[1]$ on chain complexes of abelian groups corresponds to the bar construction on simplicial abelian groups, under the Dold–Kan correspondence. Passing to sheaves of abelian groups, we see that since K_n is the simplicial sheaf corresponding to (a chain complex homotopy equivalent to) $\mathbb{L}^n$, the simplicial sheaf $B_\bullet K_n$ may be taken to be the simplicial classifying space $[r] \mapsto (K_n)^r$.

$$* \Longleftarrow K_n \rightleftarrows K_n \times K_n \cdots$$

There is a standard first-quadrant spectral sequence for the cohomology of any simplicial space V with coefficients in $\mathbb{Z}/\ell(n\ell)$; for $V = B_\bullet K_n$ it has

$$E_1^{r,s} = \widetilde{H}^{s,n\ell}(K_n^{\wedge r}, \mathbb{Z}/\ell) \Rightarrow \widetilde{H}^{r+s,n\ell}(B_\bullet K_n, \mathbb{Z}/\ell). \tag{15.40}$$

Now $\widetilde{H}^{s,w}(K_n, \mathbb{Z}/\ell) = 0$ for $w < n$ by Theorem 15.3. From the Künneth formula 15.32, it follows that $\widetilde{H}^{s,w}(K_n^{\wedge r}, \mathbb{Z}/\ell) = 0$ for $w < nr$. Thus if $n > 0$ we have $E_1^{r,s} = 0$ for $r > \ell$, and the spectral sequence (15.40) converges.

Proof of Theorem 15.38 when $\ell = 2$. If $\ell = 2$ then $\widetilde{H}^{2n,n}(K_n) \cong \mathbb{Z}/2$ on generator α; see Example 15.4. Hence $\alpha \wedge \alpha$ is the generator of $\widetilde{H}^{4n,2n}(K_n \wedge K_n) \cong \widetilde{H}^{2n,n}(K_n) \otimes \widetilde{H}^{2n,n}(K_n) \cong \mathbb{Z}/2$. Simple weight considerations, using the Künneth formula 15.32 and Theorem 15.26, show that $\widetilde{H}^{4n,2n}(K_n) = \mathbb{Z}/2$ on generator α^2, corresponding to P^n on $H^{2n,n}(-, \mathbb{Z})$, and $\widetilde{H}^{4n+1,2n}(K_n) = \widetilde{H}^{4n-1,2n}(K_n) = 0$. Since $E_1^{r,s} = 0$ for $r > 2$, the row $s = 4n$ of the spectral sequence (15.40) yields the exact sequence (with coefficients $\mathbb{Z}/2$):

$$0 \to \widetilde{H}^{4n+1,2n}(B_\bullet K_n) \to \widetilde{H}^{4n,2n}(K_n) \xrightarrow{d_1} \widetilde{H}^{4n+2,2n}(K_n^{\wedge 2}) \to \widetilde{H}^{4n+2,2n}(B_\bullet K_n) \to 0.$$

The first and last terms are nonzero because P^n and βP^n are nonzero cohomology operations on $H^{2n+1,n}(-, \mathbb{Z})$ (see [Voe03c] or section 13.2), so the differential d_1 is zero. Hence

$$H^{4n+1,2n}(B_\bullet K_n, \mathbb{Z}/2) \cong H^{4n+2,2n}(B_\bullet K_n, \mathbb{Z}/2) \cong \mathbb{Z}/2.$$

Thus every cohomology operation from $H^{2n+1,n}(-, \mathbb{Z})$ to $H^{4n+1,2n}(-, \mathbb{Z}/2)$ or $H^{4n+2,2n}(-, \mathbb{Z}/2)$ is a multiple of P^n or βP^n, respectively. $\qquad\square$

In order to prepare for the proof of Theorem 15.38 when $\ell > 2$, we consider the cohomology of $K_n \wedge \ldots \wedge K_n$ in scalar weight 1.

Lemma 15.41. *The scalar weight $s=1$ part of $H^{p,q}(K_n^{\wedge r}, \mathbb{Z}/\ell)$ vanishes if $q < n\ell$ and $r \geq 2$, and also if $q = n\ell$ and $p < 2n\ell$.*

Proof. ([Voe11, 2.7–2.8]) By 15.32 and 15.28, it suffices to consider $x_1 \otimes \cdots \otimes x_r$ where the x_i are in $\mathrm{Hom}(S_{\mathrm{tr}}^{a_i} \mathbb{L}^n, R(q_i)[p_i])$, $\Sigma p_i = p$, and $\Sigma q_i = q$. By 15.33, we may assume $\Sigma a_i \equiv 1 \mod (\ell - 1)$. If $q < n\ell$ then $a_i \neq 0$ by 15.39(b) and we must have $\Sigma a_i \geq \ell$, which is excluded by 15.39(a) as $q \geq n\Sigma a_i$. This establishes the case $q < n\ell$. When $q = n\ell$, the vanishing comes from 15.39(c).   □

Using Lemma 15.41, the relevant part of the spectral sequence looks like this:

$$
\begin{array}{cccccc}
 & 0 & & & & \\
s = 2n\ell + 1 & 0 & \widetilde{H}^{2n\ell+1,n\ell}(K_n) \longrightarrow & & & \\
s = 2n\ell & 0 & \widetilde{H}^{2n\ell,n\ell}(K_n) \xrightarrow{\ d_1\ } & \widetilde{H}^{2n\ell,n\ell}(K_n \wedge K_n) \to \widetilde{H}^{2n\ell,n\ell}(K_n^{\wedge 3}) \to & & \\
 & 0 & 0 & \text{(nothing in scalar weight 1 below here)} & & \\
 & r = 0 & r = 1 & r = 2 & r = 3. &
\end{array}
$$

Now $d_1(\alpha^\ell) = (\alpha \otimes 1 + 1 \otimes \alpha)^\ell - \alpha^\ell \otimes 1 - 1 \otimes \alpha^\ell = 0$, so α^ℓ is a permanent cycle in the spectral sequence.

Construction 15.42. Recall from [Voe03c, 3.7] that $H^{2n,n}(K_n, R) \cong R$ on the fundamental class $\alpha = \alpha_{2n,n}^R$. Since it is additive, α has scalar weight 1. Hence

$$
\gamma = \{(\alpha \otimes 1 + 1 \otimes \alpha)^\ell - \alpha^\ell \otimes 1 - 1 \otimes \alpha^\ell\}/\ell = \alpha^{\ell-1} \otimes \alpha + \cdots + \alpha \otimes \alpha^{\ell-1}
$$

is an element of $H^{2n\ell,n\ell}(K_n \wedge K_n, \mathbb{Z}/\ell)$ of scalar weight 1. A calculation shows that $d_1^{2,2n\ell}$ maps γ to zero in $H^{2n\ell,n\ell}(K_n \wedge K_n \wedge K_n, \mathbb{Z}/\ell)$. Formally this follows from $d_1(\alpha^\ell) = (\alpha \otimes 1)^\ell - (\alpha \otimes 1 + 1 \otimes \alpha)^\ell + (1 \otimes \alpha)^\ell = -\ell\gamma$ and $d_1 \circ d_1 = 0$.

Lemma 15.43. *Let D_r denote the subset of elements of scalar weight 1 in $H^{2n\ell,n\ell}(K_n^{\wedge r}, \mathbb{Z}/\ell)$. If $r > 1$ then D_r is the $\mathbb{Z}/\ell$-vector space generated by monomials of the form $\alpha^{i_1} \wedge \ldots \wedge \alpha^{i_r}$, where $i_1 + \cdots + i_r = \ell$ and each $i_j > 0$.*

When $\ell = 2$, we have already seen that $\widetilde{H}^{2n\ell,n\ell}(K_n^{\wedge r}, R) = 0$ for $r > 2$ and that $\widetilde{H}^{2n\ell,n\ell}(K_n \wedge K_n, R) \cong R$ on $\alpha \wedge \alpha$. Thus we may assume that $\ell > 2$.

Proof. ([Voe11, 2.9]) The monomials are linearly independent by the Künneth formula 15.32 and 13.1.1. Lemma 15.41 implies that D_r is generated by elements of the form $x_1 \otimes \cdots \otimes x_r$ where the $x_i \in \mathrm{Hom}(S^{a_i} \mathbb{L}^n, R(q_i)[p_i]) \subset H^{p_i,q_i}(K_n, R)$. By 15.39(b), if two of the x_i have scalar weight 0 then $n\ell = q = \sum q_i \geq 2n(\ell - 1)$, which cannot happen (as the case $\ell = 2$ has been ruled out).

Hence at most one x_i can have scalar weight 0. By 15.39(a), this can occur only if $r = 2$ and then only if (q_1, q_2) is $(n, n(\ell - 1))$ or $(n(\ell - 1), n)$. In the first case, x_1 is in $H^{2n,n}(K_n, R) \cong R$ (on α) and x_2 is in $\mathrm{Hom}(S^{\ell-1}\mathbb{L}^n, \mathbb{L}^{n(\ell-1)}) \cong R$

(on $\alpha^{\ell-1}$) by 15.39(b, c) and 14.12(e), so $x_1 \otimes x_2$ is a multiple of $\alpha \wedge \alpha^{\ell-1}$. The second case is similar.

Thus we are reduced to the case in which $r \geq 2$ and all $a_i \not\equiv 0 \mod (\ell-1)$. By 15.39(a, c) and $q = n\ell$ we must have $\Sigma a_i = \ell$, $q_i = na_i$, and $p_i = 2q_i$. Since $S_{\mathrm{tr}}^{a_i}(\mathbb{L}^n) = \mathbb{L}^{na_i}$ by 14.12(e) we must have $x_i = \alpha^{a_i}$ up to scalars. $\qquad\square$

Lemma 15.44. $E_2^{2,2n\ell}$ *is 1-dimensional on the class of* γ.

Proof. D_2 has basis $e_i = \alpha^i \wedge \alpha^{\ell-i}$, $1 \leq i \leq \ell - 1$. Let W be the subspace of D_3 on the monomials $f_i = \alpha^i \wedge \alpha \wedge \alpha^{\ell-i-1}$, $1 \leq i \leq \ell - 2$. The composition $D_2 \xrightarrow{d_1} D_3 \xrightarrow{\mathrm{proj}} W$ is onto because it sends e_1 to $-f_1$ and e_i to $-(i+1)f_i + i f_{i-1}$. This follows from the expansion

$$d_1(e_i) = 1 \wedge \alpha^i \wedge \alpha^{\ell-i} - \left(1 \wedge \alpha + \alpha \wedge 1\right) \wedge \alpha^{\ell-i}$$
$$+ \alpha^i \wedge \left(1 \wedge \alpha + \alpha \wedge 1\right)^{\ell-i} - \alpha^i \wedge \alpha^{\ell-i} \wedge 1.$$

Since $\dim(W) = \dim(D_2) - 1$, the kernel of d_1 is at most 1-dimensional; by 15.42, the kernel is nonzero as $d_1(\gamma) = 0$. $\qquad\square$

Proof of Theorem 15.38: We regard ϕ as an element of $\widetilde{H}^{2n\ell+2,n\ell}(B_\bullet K_n, \mathbb{Z}/\ell)$. Condition 15.38(1) says that ϕ has scalar weight 1. Condition 15.38(2) says that ϕ (like βP^n) is in the kernel of the map

$$\widetilde{H}^{2n\ell+2,n\ell}(B_\bullet K_n, \mathbb{Z}\ell) \to \widetilde{H}^{2n\ell+2,n\ell}(\Sigma K_n, \mathbb{Z}\ell) = \widetilde{H}^{2n\ell+1,n\ell}(K_n, \mathbb{Z}/\ell)$$

defined by the inclusion of ΣK_n into $B_\bullet K_n$ as the 1-skeleton in the B-direction. That is, ϕ and βP^n lie in the kernel of the edge map in the spectral sequence (15.40). Since they have scalar weight 1, Lemma 15.41 shows that they come from elements of $E_2^{2,2n\ell}$, which is 1-dimensional by Lemma 15.44. Since $\beta P^n \neq 0$, ϕ must be a multiple of βP^n. $\qquad\square$

Remark 15.45. Consider the simplicial scheme $B\mathbb{G}_a$ over $R = \mathbb{Z}/\ell$ and the cochain complex C^* associated to the cosimplicial abelian group $\mathcal{O}(B\mathbb{G}_a)$. Then $C^r = R[x_1, \ldots, x_r]$ is graded with the x_i in degree 1, and its cohomology $H^*(B\mathbb{G}_a, \mathbb{G}_a)$ was computed by Lazard in [Laz55, Thm. 1.21].

Define $C_{\deg=i}^p \to H^{2ni,ni}(K^p, \mathbb{Z}/\ell)$ by the rule $x_i \mapsto 1 \otimes \cdots \otimes \alpha \otimes \cdots 1$, where the α is in the ith place. It is observed in [Voe11] that this map induces a morphism of graded cochain complexes from $C_{\deg=i}^*$ to the row $s = 2ni$ of the E_1 page of the spectral sequence (15.40) with coefficients in $R(ni)$, and hence a homomorphism $H_{\deg=i}^r(B\mathbb{G}_a, \mathbb{G}_a) \to E_{2,\mathrm{sw}=i}^{r,2ni}$, taking the degree i subspace to subspace of cohomology having scalar weight i.

Lemma 15.43 implies that $C_{\deg=\ell}^r \cong D_r$ for $r \geq 2$ and hence that the induced map $H_{\deg=\ell}^r(B\mathbb{G}_a, \mathbb{G}_a) \to E_{2,\mathrm{sw}=\ell}^{r,2ni}$ is an isomorphism for $r \geq 3$ (and a surjection for $r = 2$). Therefore the rest of the row $E_{2,\mathrm{sw}=\ell}^{r,2n\ell}$ may also be read off from Lazard's calculations.

15.9   HISTORICAL NOTES

The present material is based upon Voevodsky's 2003 preprint [Voe03b], the 2007 preprint version of his paper [Voe10c] as well as [Wei09]. Section 15.4 is taken from Section 2.3 of [Voe10c]. Many of the arguments we use did not appear in the published versions of [Voe11] and [Voe10c].

Glossary

$\underline{a}$ $\{a_1, \ldots, a_n\}$ in $K_n^M(k)/\ell$, 7

$\mathbb{A}(\mathcal{A})$ vector bundle of Kummer algebra $\mathcal{A}$, 160

$\mathcal{A}_\phi(L)$ sheaf of Kummer algebras for ℓ-form ϕ, 121

$\widetilde{A}_0(k)$ the set of symbols in $\overline{H}_{-1,-1}(X)$, 159

$A^\dagger$ $\mathfrak{X}$-dual of A in $\mathbf{DM}^\varepsilon_{\mathrm{gm}}$, 58

$A_r, \bar{A}_r$ algebras used in Chain Lemma, 135

BL(n) Beilinson–Lichtenbaum hypothesis, 23

$\beta, \tilde{\beta}$ Bockstein, 20

$c_\#$ left adjoint to c^*, 79

C cyclic group C_ℓ of order ℓ, 264

$\mathcal{C}^{\amalg}$ coproducts of representable presheaves, 178

$\mathcal{C}^{\mathrm{ind}}$ filtered colimits of representable presheaves, 183

ch(X) Chow group mod ℓ, 132

$CH^*(X)$ Chow group of cycles modulo rational equivalence, 54

$\mathbf{Cor}(\mathcal{S})$ category of correspondences on $\mathcal{S}$, 187

$\check{C}(X)$ 0-coskeleton of X, 17

$cyl(f)$ mapping cylinder of $f : A \to B$, 192

$cyl\, S$ class of maps $A \to cyl(s)$, $s \in S$, 196

$\mathbf{D}(\mathbf{Sm}/k)$ derived category of sheaves, 32

δ element of $H^{n,n-1}(\mathfrak{X}, \mathbb{Z}/\ell)$, 50

Δ_s semi-simplicial category, 197

$\mathbf{DM}^\varepsilon_{\mathrm{gm}}$ subcategory of $\mathbf{DM}_{\mathrm{gm}}$ generated by $R_{\mathrm{tr}}(\mathfrak{X})$, 56

PF extension of presheaf F to normal G-schemes, 233

$\partial\Delta^m$ scheme boundary of m-simplex, 34

P^b reduced motivic power operations, 217

ϕ^V motivic cohomology operation $(=\beta P^b)$, 69

ϕ^V motivic cohomology operation, 90

ϕ_i motivic cohomology operations, 87

$\Phi_r(\mathbf{x})$ function used in symbol chain, 126

P_r tower of varieties in symbol chain, 130

$\Delta^{\mathrm{op}}\mathrm{Pshv}(\mathcal{C})_{\mathrm{nis}}$ Nisnevich-local model structure, 200

$\Delta^{\mathrm{op}}\mathrm{Pshv}(\mathcal{C})_{\mathbb{A}^1}$ $\mathbb{A}^1$-local model structure, 207

$\Delta^{\mathrm{op}}\mathrm{Pshv}(\mathcal{C})$ simplicial presheaves on $\mathcal{C}$, 176

$\psi, \psi_{\ell,n}$ homomorphism $\mathbb{L}_* \to \mathbb{F}_\ell[v]$ for ℓ-typical formal group laws, 110

Ψ_r ℓ-form for $n=2$, 125

$\Psi_r(\mathbf{x})$ function used in symbol chain, 126

$\mathbf{PST}(S)$ presheaves with transfers over S, 76

Q_i Milnor's cohomology operations, 221

Q_r tower of varieties in symbol chain, 130

Q_w splitting variety for $\{\underline{a}, w\}$, 149

$\mathrm{rad}(\mathcal{C})$ category of radditive presheaves, 182

$\mathrm{Res}_{E/k}X$ Weil restriction of X, 163

ρ class of -1 in $k^\times/k^{\times 2} = H^{1,1}(k, \mathbb{Z}/2)$, 263

R_{tr} motivic Hurewicz functor, 189

$S_{\mathrm{tr}}^m, S_{\mathrm{tr}}^G$ symmetric power for correspondences, 235

$S_{\mathrm{tr}}^\ell(\mathbb{L}^n)$ symmetric power of Lefschetz motive, 264

$s_n M, s_{<n} M, s_{\geq n} M$ slice functors on $\mathbf{DM}_{\mathrm{gm}}$, 82

$\mathbf{Sch}/k$ schemes of finite type over k, 28

$s_d(X)$ symmetric characteristic class, 11

$\widetilde{S}^G(X)$ reduced quotient variety for X_+, 232

$S^G(X)$ quotient variety X^m/G, 232

$S^i(N)$ i^{th} symmetric power of the motive N, 66

$\mathbf{Sm}/k$ category of smooth schemes over k, xii

$\mathbf{Sm}/k$ category of smooth schemes over k, 14

$S^m(X), S^G X$ symmetric power of X, equal to X^m/Σ_m, 232

Sq^i motivic Steenrod operations when $\ell = 2$, 218

$Sq^i_{\text{top}}, P^i_{\text{top}}$ topological Steenrod operations, 216

$\Sigma X, \widetilde{\Sigma} X$ simplicial suspension and reduced suspension of X, 18

$\text{Sym}^m(H)$ invariant subring of $H^{\otimes m}$, 269

$t^{\mathbb{C}}$ topological realization on $\mathbf{Ho}(\mathbf{Sm}/\mathbb{C})$, 211

τ element of $H^{0,1}(k, \mathbb{Z}/2)$, 263

$t_E, \bar{t}_E$ Thom class of vector bundle E, 223

$Th(V)$ Thom space of affine space V, 264

$Th_Y(E)$ Thom space of vector bundle E on Y, 223

$Th(V)/G$ quotient of Thom space by G, 264

$T^{\amalg m}$ G-scheme $T \amalg \cdots \amalg T$, 233

$T(n)$ truncated motivic complex, 39

$t_{d,r}$ Todd numbers, 110

$U \boxtimes K$ simplicial presheaf $n \mapsto \coprod_{K_n} X_n$, 176

$U \otimes K$ radditive version of $U \boxtimes K$, 186

$\mathfrak{X}$ 0-coskeleton of X, 17

y^D twisted dual of y, 59

Y_r tower of varieties in symbol chain, 134

$\mathbb{Z}(i), \mathbb{Z}/\ell(i)$ motivic complex in weight i, 4

$\mathbb{Z}_{\mathbf{Sm}}[X]$ restriction of $\mathbb{Z}[X]$ to $\mathbf{Sm}/k$, 28

Bibliography

[Ada74]   J. F. Adams, *Stable homotopy and generalised homology*, University of Chicago Press, Chicago, Ill., 1974, Chicago Lectures in Mathematics. MR53 #6534

[AGV73]   M. Artin, A. Grothendieck, and J.-L. Verdier (eds.), *Théorie des topos et cohomologie étale des schémas*, Springer-Verlag, Berlin, 1972-1973, Séminaire de Géométrie Algébrique du Bois-Marie 1963–1964 (SGA 4), Dirigé par M. Artin, A. Grothendieck, et J.-L. Verdier. Avec la collaboration de N. Bourbaki, P. Deligne et B. Saint-Donat, Lecture Notes in Mathematics, vol. 269, 270, 305.

[Ati61a]   M. F. Atiyah, *Bordism and cobordism*, Proc. Cambridge Philos. Soc. **57** (1961), 200–208. MR0126856

[Ati61b]   _____ , *Thom complexes*, Proc. London Math. Soc. (3) **11** (1961), 291–310. MR0131880

[Ayo07]   J. Ayoub, *Les six opérations de Grothendieck et le formalisme des cycles évanescents dans le monde motivique. I*, Astérisque (2007), no. 314, x+466 pp. (2008). MR2423375

[Beĭ87]   A. A. Beĭlinson, *Height pairing between algebraic cycles*, K-theory, arithmetic and geometry (Moscow, 1984–1986), Lecture Notes in Math., vol. 1289, Springer, Berlin, 1987, 1–25. MR923131 (89h:11027)

[BG73]   K. S. Brown and S. M. Gersten, *Algebraic K-theory as generalized sheaf cohomology*, Algebraic K-theory, I: Higher K-theories (Proc. Conf., Battelle Memorial Inst., Seattle, Wash., 1972), Springer, Berlin, 1973, 266–292. Lecture Notes in Math., vol. 341. MR50 #442

[BK72]   A. K. Bousfield and D. M. Kan, *Homotopy limits, completions and localizations*, Lecture Notes in Mathematics, vol. 304, Springer-Verlag, Berlin, 1972. MR0365573 (51 #1825)

[Bla01]   B. Blander, *Local projective model structures on simplicial presheaves*, K-Theory **24** (2001), no. 3, 283–301. MR1876801 (2003c:18014)

[Blo80]   S. Bloch, *Lectures on algebraic cycles*, Duke University Math. Series, IV, Duke University Math. Department, Durham, N.C., 1980. MR558224 (82e:14012)

[Car54]   H. Cartan, *Sur les groupes d'Eilenberg-Mac Lane. II*, Proc. Nat. Acad. Sci. U. S. A. **40** (1954), 704–707. MR0065161 (16,390b)

[CD13]   D.-C. Cisinski and F. Déglise, *Triangulated categories of mixed motives*, Preprint. Archived at arXiv:0912.2110, 2013.

[Dég03]   F. Déglise, *Modules de cycles et motifs mixtes*, C. R. Math. Acad. Sci. Paris **336** (2003), no. 1, 41–46. MR1968900 (2004c:14036)

[Del74]   P. Deligne, *Théorie de Hodge. III*, Inst. Hautes Études Sci. Publ. Math. (1974), no. 44, 5–77. MR0498552 (58 #16653b)

[Del09]   ———, *Lectures on motivic cohomology 2000/2001*, Abel Symposium, Springer, 2009, 355–409.

[dJ96]   A. J. de Jong, *Smoothness, semi-stability and alterations*, Inst. Hautes Études Sci. Publ. Math. (1996), no. 83, 51–93. MR1423020 (98e:14011)

[DT58]   A. Dold and R. Thom, *Quasifaserungen und unendliche symmetrische Produkte*, Ann. of Math. (2) **67** (1958), 239–281. MR0097062 (20 #3542)

[Flo71]   E. E. Floyd, *Actions of $(Z_p)^k$ without stationary points*, Topology **10** (1971), 327–336. MR0283820 (44 #1050)

[FS02]   E. M. Friedlander and A. Suslin, *The spectral sequence relating algebraic K-theory to motivic cohomology*, Ann. Sci. École Norm. Sup. (4) **35** (2002), no. 6, 773–875. MR1949356 (2004b:19006)

[FV00]   E. M. Friedlander and V. Voevodsky, *Bivariant cycle cohomology*, in book [VSF], 138–187.

[GJ99]   P. Goerss and J. F. Jardine, *Simplicial homotopy theory*, Progress in Mathematics, vol. 174, Birkhäuser, Basel, 1999.

[GL01]   T. Geisser and M. Levine, *The Bloch-Kato conjecture and a theorem of Suslin-Voevodsky*, J. Reine Angew. Math. **530** (2001), 55–103. MR1807268 (2003a:14031)

[Gro61]   A. Grothendieck, *Techniques de construction et théorèmes d'existence en géométrie algébrique. IV. Les schémas de Hilbert*, Séminaire Bourbaki, vol. 6, Soc. Math. France, Paris, 1960/1961, Exp. no. 221, 249–276. MR1611822

[Har77]   R. Hartshorne, *Algebraic geometry*, Springer-Verlag, New York, 1977, Graduate Texts in Mathematics, no. 52. MR57 #3116

[Hir03]   P. S. Hirschhorn, *Model categories and their localizations*, Mathematical Surveys and Monographs, vol. 99, American Mathematical Society, Providence, RI, 2003. MR1944041 (2003j:18018)

[Hov99]   M. Hovey, *Model categories*, Mathematical Surveys and Monographs, vol. 63, American Mathematical Society, Providence, RI, 1999. MR1650134 (99h:55031)

[HW09]   C. Haesemeyer and C. Weibel, *Norm varieties and the chain lemma (after Markus Rost)*, Abel Symposium, Springer, 2009, 95–130.

[Ill09]   L. Illusie, *On Gabber's refined uniformization*, preprint available at www.math.u-psud.fr/~illusie/refined_uniformization3.pdf, 2009.

[Jar86]   J. F. Jardine, *Simplicial objects in a Grothendieck topos*, Applications of algebraic K-theory to algebraic geometry and number theory, Part I, II (Boulder, Colo., 1983), Contemp. Math., vol. 55, Amer. Math. Soc., Providence, RI, 1986, 193–239. MR862637 (88g:18008)

[Jar87]   ———, *Simplicial presheaves*, J. Pure Appl. Algebra **47** (1987), no. 1, 35–87. MR906403 (88j:18005)

[Jar00]   ———, *Motivic symmetric spectra*, Doc. Math. **5** (2000), 445–553 (electronic). MR1787949 (2002b:55014)

[Jar03]   ———, *Presheaves of chain complexes*, K-Theory **30** (2003), no. 4, 365–420, special issue in honor of Hyman Bass on his seventieth birthday. Part IV. MR2064245 (2005d:18016)

[Kat80]   K. Kato, *A generalization of local class field theory by using K-groups. II*, J. Fac. Sci. Univ. Tokyo Sect. IA Math. **27** (1980), no. 3, 603–683. MR603953 (83g:12020a)

[Kel13]   S. Kelly, *Triangulated categories of motives in positive characteristic*, preprint (Ph.D. thesis). Archived at arXiv:1305.5349, 2013.

[KM13]   N. A. Karpenko and A. S. Merkurjev, *On standard norm varieties*, Ann. Sci. Éc. Norm. Supér. (4) **46** (2013), no. 1, 175–214. MR3087392

[KMRT98]   M.-A. Knus, A. Merkurjev, M. Rost, and J.-P. Tignol, *The book of involutions*, American Mathematical Society Colloquium Publications, vol. 44, American Mathematical Society, Providence, RI, 1998, with a preface in French by J. Tits. MR1632779

[Lan73]   P. Landweber, *Annihilator ideals and primitive elements in complex bordism*, Illinois J. Math. **17** (1973), 273–284. MR0322874 (48 #1235)

[Las63]   R. Lashof, *Poincaré duality and cobordism*, Trans. Amer. Math. Soc. **109** (1963), 257–277. MR0156357 (27 #6281)

[Laz55]   M. Lazard, *Lois de groupes et analyseurs*, Ann. Sci. Ecole Norm. Sup. (3) **72** (1955), 299–400. MR0077542 (17,1053c)

[Lic84]   S. Lichtenbaum, *Values of zeta-functions at nonnegative integers*, Number theory, Noordwijkerhout 1983, Lecture Notes in Math., vol. 1068, Springer, Berlin, 1984, 127–138. MR756 089

[LM07]   M. Levine and F. Morel, *Algebraic cobordism*, Springer Monographs in Mathematics, Springer, Berlin, 2007. MR2286826 (2008a: 14029)

[Man68]   Ju. Manin, *Correspondences, motifs and monoidal transformations*, Mat. Sb. (N.S.) **77 (119)** (1968), 475–507. MR41 #3482

[Mar83]   H. R. Margolis, *Spectra and the Steenrod algebra*, North-Holland Mathematical Library, vol. 29, North-Holland Publishing Co., Amsterdam, 1983, Modules over the Steenrod algebra and the stable homotopy category. MR738973 (86j:55001)

[Me81]   A. S. Merkur'ev, *On the norm residue symbol of degree* 2, Dokl. Akad. Nauk SSSR **261** (1981), no. 3, 542–547. MR638926

[MeS82]   A. S. Merkur'ev and A. A. Suslin, *K-cohomology of Severi-Brauer varieties and the norm residue homomorphism*, Izv. Akad. Nauk SSSR Ser. Mat. **46** (1982), no. 5, 1011–1046, 1135–1136. MR675529 (84i:12007)

[MeS90]   A. S. Merkur'ev and A. A. Suslin, *Norm residue homomorphism of degree three*, Izv. Akad. Nauk SSSR Ser. Mat. **54** (1990), no. 2, 339–356. MR1062517

[Milg]   R. J. Milgram, *The homology of symmetric products*, Trans. Amer. Math. Soc. **138** (1969), 251–265. MR0242149 (39 #3483)

[Mil58]   J. Milnor, *The Steenrod algebra and its dual*, Ann. of Math. (2) **67** (1958), 150–171. MR0099653 (20 #6092)

[Mil60]   _____, *On the cobordism ring Ω^* and a complex analogue. I*, Amer. J. Math. **82** (1960), 505–521. MR0119209 (22 #9975)

[Mil70]   J. Milnor, *Algebraic K-theory and quadratic forms*, Invent. Math. **9** (1969/1970), 318–344. MR0260844 (41 #5465)

[Mil71] _____, *Introduction to algebraic K-theory*, Princeton University Press, Princeton, N.J., 1971, Annals of Mathematics Studies, no. 72. MR0349811 (50 #2304)

[Mil80] J. Milne, *Étale cohomology*, Princeton University Press, Princeton, N.J., 1980. MR81j:14002

[Mor04] F. Morel, *On the motivic π_0 of the sphere spectrum*, Axiomatic, enriched and motivic homotopy theory, NATO Sci. Ser. II Math. Phys. Chem., vol. 131, Kluwer Acad. Publ., Dordrecht, 2004, 219–260. MR2061856 (2005e:19002)

[MS74] J. Milnor and J. Stasheff, *Characteristic classes*, Princeton University Press, Princeton, N.J., 1974, Annals of Mathematics Studies, no. 76. MR0440554 (55 #13428)

[MV99] F. Morel and V. Voevodsky, *$\mathbb{A}^1$-homotopy theory of schemes*, Inst. Hautes Études Sci. Publ. Math. (1999), no. 90, 45–143 (2001). MR1 813 224

[MVW] C. Mazza, V. Voevodsky, and C. Weibel, *Lecture notes on motivic cohomology*, Clay Mathematics Monographs, vol. 2, American Mathematical Society, Providence, RI, 2006. MR2242284 (2007e:14035)

[Nak57] M. Nakaoka, *Cohomology mod p of the p-fold symmetric products of spheres*, J. Math. Soc. Japan **9** (1957), 417–427. MR0121787 (22 #12519a)

[NS89] Y. Nesterenko and A. Suslin, *Homology of the general linear group over a local ring, and Milnor's K-theory*, Izv. Akad. Nauk SSSR Ser. Mat. **53** (1989), no. 1, 121–146. MR90a:20092

[OVV07] D. Orlov, A. Vishik, and V. Voevodsky, *An exact sequence for $K_*^M/2$ with applications to quadratic forms*, Ann. of Math. (2) **165** (2007), no. 1, 1–13. The 2000 preprint is archived at www.math.uiuc.edu /K-theory/0454. MR2276765 (2008c:19001)

[Pon38] L. S. Pontryagin, *A classification of continuous transformations of a complex into a sphere*, Doklady Akad. Nauk SSR **19** (1938), 147–149 and 361–363.

[Qui67] D. Quillen, *Homotopical algebra*, Lecture Notes in Mathematics, no. 43, Springer-Verlag, Berlin, 1967. MR0223432 (36 #6480)

[Qui73] _____, *Higher algebraic K-theory. I*, Algebraic K-theory, I: Higher K-theories (Proc. Conf., Battelle Memorial Inst., Seattle, Wash.,

1972), Springer, Berlin, 1973, 85–147. Lecture Notes in Math., vol. 341. MR0338129 (49 #2895)

[Rav86] D. Ravenel, *Complex cobordism and stable homotopy groups of spheres*, Pure and Applied Mathematics, vol. 121, Academic Press Inc., Orlando, FL, 1986. MR860042 (87j:55003)

[RG71] M. Raynaud and L. Gruson, *Critères de platitude et de projectivité. Techniques de "platification" d'un module*, Invent. Math. **13** (1971), 1–89. MR0308104 (46 #7219)

[Rio07] J. Riou, *Catégorie homotopique stable d'un site suspendu avec intervalle*, Bull. Soc. Math. France **135** (2007), no. 4, 495–547. MR2439197

[Rio12] ———, *Opérations de Steenrod motiviques*, preprint, 61 pp. Available at arXiv:1207.3121, 2012.

[Rio14] ———, *La conjecture de Bloch-Kato (d'après M. Rost et V. Voevodsky)*, Astérisque (2014), no. 361, Exp. no. 1073, x, 421–463. MR3289290

[RØ06] O. Röndigs and P. A. Østvær, *Motives and modules over motivic cohomology*, C. R. Math. Acad. Sci. Paris **342** (2006), no. 10, 751–754. MR2227753 (2007d:14043)

[Ros88] M. Rost, *On the spinor norm and $A_0(X, K_1)$ for quadrics*, preprint available at www.math.uni-bielefeld.de/~rost/spinor.html, 1988.

[Ros90] ———, *Some new results on the Chow groups of quadrics*, preprint available at www.math.uni-bielefeld.de/~rost/chow-qudr.html, 1990.

[Ros96] ———, *Chow groups with coefficients*, Doc. Math. **1** (1996), no. 16, 319–393 (electronic). MR1418952 (98a:14006)

[Ros98a] ———, *Chain lemma for splitting fields of symbols*, preprint available at www.math.uni-bielefeld.de/~rost/chain-lemma.html, 1998.

[Ros98b] ———, *Construction of splitting varieties*, preprint available at www.math.uni-bielefeld.de/~rost/chain-lemma.html, 1998.

[Ros98c] ———, *The motive of a Pfister form*, preprint available at www.math. uni-bielefeld.de/~rost/motive.html, 1998.

[Ros99] ———, *The chain lemma for Kummer elements of degree 3*, C. R. Acad. Sci. Paris Sér. I Math. **328** (1999), no. 3, 185–190. MR1674602 (2000c:12003)

[Ros02]   ———, *Norm varieties and algebraic cobordism*, Proceedings of the International Congress of Mathematicians, vol. II (Beijing, 2002) (Beijing), Higher Ed. Press, 2002, 77–85. MR1957022 (2003m: 19003)

[Ros06]   ———, *On the basic correspondence of a splitting variety*, preprint available at http://www.math.uni-bielefeld.de/~rost /basic-corr.html, 2006.

[Ser52]   J.-P. Serre, *Sur les groupes d'Eilenberg-Mac Lane*, C. R. Acad. Sci. Paris **234** (1952), 1243–1245. MR0046047 (13,675c)

[Ser63]   ———, *Cohomologie galoisienne*, Cours au Collège de France, vol. 1962, Springer-Verlag, Berlin-Heidelberg-New York, 1962/1963. MR0180551

[SGA71]   *Revêtements étales et groupe fondamental (SGA 1)*, Lecture Notes in Mathematics, vol. 224, Springer-Verlag, Berlin, 1971, Séminaire de géométrie algébrique du Bois Marie 1960–1961 (SGA 1), Dirigé par Alexandre Grothendieck. Augmenté de deux exposés de M. Raynaud, Lecture Notes in Mathematics, vol. 224. MR0354651 (50 #7129)

[SJ06]   A. Suslin and S. Joukhovitski, *Norm varieties*, J. Pure Appl. Algebra **206** (2006), no. 1-2, 245–276. MR2220090 (2008a:14015)

[Spi01]   M. Spitzweck, *Operads, algebras and modules in model categories and motives*, Ph.D. thesis, Bonn, 2001.

[Ste62]   N. E. Steenrod, *Cohomology operations*, Lectures by N. E. Steenrod written and revised by D.B.A. Epstein, Annals of Mathematics Studies, no. 50, Princeton University Press, Princeton, N.J., 1962. MR0145525 (26 #3056)

[Ste72]   ———, *Cohomology operations, and obstructions to extending continuous functions*, Advances in Math. **8** (1972), 371–416, Originally distributed in August 1957 as AMS Colloquium Lecture Notes. MR0298655 (45 #7705)

[Sto68]   R. E. Stong, *Notes on cobordism theory*, Mathematical notes, Princeton University Press, Princeton, N.J., 1968. MR0248858 (40 #2108)

[Sus87]   A. Suslin, *Algebraic K-theory of fields*, Proceedings of the International Congress of Mathematicians, vol. 1 (Berkeley, Calif., 1986) (Providence, RI), Amer. Math. Soc., 1987, 222–244. MR934225 (89k:12010)

[Sus91]   _____, *K-theory and K-cohomology of certain group varieties*, Algebraic *K*-theory, Adv. Soviet Math., vol. 4, Amer. Math. Soc., Providence, RI, 1991, 53–74. MR1124626 (92g:19004)

[Sus03]   _____, *On the Grayson spectral sequence*, Tr. Mat. Inst. Steklova **241** (2003), 218–253. MR2024054 (2005g:14043)

[SV96]   A. Suslin and V. Voevodsky, *Singular homology of abstract algebraic varieties*, Invent. Math. **123** (1996), no. 1, 61–94. MR97e:14030

[SV00a]   _____, *Bloch-Kato conjecture and motivic cohomology with finite coefficients*, The Arithmetic and Geometry of Algebraic Cycles (B. Gordon, J. Lewis, S. Müller-Stach, S. Saito, and Yui, N., eds.), Nato ASI series C, vol. 548, Kluwer, 2000, 117–189.

[SV00b]   _____, *Relative cycles and Chow sheaves*, in book [VSF], 10–86.

[Tat76]   J. Tate, *Relations between K_2 and Galois cohomology*, Invent. Math. **36** (1976), 257–274. MR0429837 (55 #2847)

[tD70]   T. tom Dieck, *Actions of finite abelian p-groups without stationary points*, Topology **9** (1970), 359–366. MR0285029 (44 #2253)

[Tho54]   R. Thom, *Quelques propriétés globales des variétés différentiables*, Comment. Math. Helv. **28** (1954), 17–86. MR0061823

[Tot92]   B. Totaro, *Milnor K-theory is the simplest part of algebraic K-theory*, K-Theory **6** (1992), no. 2, 177–189. MR1187705 (94d:19009)

[Voe96]   V. Voevodsky, *The Milnor conjecture*, preprint at http://www.math .uiuc.edu/K-theory/170, 1996.

[Voe99]   _____, *Voevodsky's Seattle lectures: K-theory and motivic cohomology*, Algebraic *K*-theory (Seattle, WA, 1997), Proc. Sympos. Pure Math., vol. 67, Amer. Math. Soc., Providence, RI, 1999, notes by C. Weibel, 283–303. MR1743245 (2001i:14029)

[Voe00a]   _____, *Δ-closed classes*, preprint, preliminary version of [Voe10d]. Archived at http://www.math.uiuc.edu/K-theory/0442, 2000.

[Voe00b]   _____, *Triangulated categories of motives over a field*, in book [VSF], 188–254.

[Voe03a]   _____, *Motivic cohomology with $\mathbf{Z}/2$-coefficients*, Publ. Math. Inst. Hautes Études Sci. (2003), no. 98, 59–104. MR2031199 (2005b: 14038b)

[Voe03b]   _____, *On motivic cohomology with $\mathbb{Z}/l$ coefficients*, preprint. Archived at http://www.math.uiuc.edu/K-theory/0639, 2003. Revised version [Voe11] published in 2011, 2003.

[Voe03c] _____, *Reduced power operations in motivic cohomology*, Publ. Math. Inst. Hautes Études Sci. (2003), no. 98, 1–57. MR2031199 (2005b:14038a)

[Voe10a] _____, *Cancellation theorem*, Documenta Math. Extra volume (2010), 671–685, preprint, 2002.

[Voe10b] _____, *Motives over simplicial schemes*, J. K-theory 5 (2010), 1–38, 2003 preprint is archived at www.math.uiuc.edu/K-theory /0638.

[Voe10c] _____, *Motivic Eilenberg-Mac Lane spaces*, Publ. Math. Inst. Hautes Études Sci. (2010), no. 112, 1–99. MR2737977

[Voe10d] _____, *Simplicial radditive functors*, J. K-theory 5 (2010), 201–244, Rewritten version of [Voe00a], posted in 2007. Archived at http://www.math.uiuc.edu/K-theory/0863.

[Voe11] _____, *On motivic cohomology with* $\mathbb{Z}/l$ *coefficients*, Annals of Math. (2011), no. 174, 401–438, based on 2003 preprint [Voe03b].

[VSF] V. Voevodsky, A. Suslin, and E. M. Friedlander, *Cycles, transfers, and motivic homology theories*, Annals of Mathematics Studies, vol. 143, Princeton University Press, Princeton, NJ, 2000.

[Wan50] S. Wang, *On the commutator group of a simple algebra*, Amer. J. Math. **72** (1950), 323–334. MR0034380 (11,577d)

[Wei56] A. Weil, *The field of definition of a variety*, Amer. J. Math. **78** (1956), 509–524. MR0082726 (18,601a)

[Wei82] _____, *Adeles and algebraic groups*, Progress in Mathematics, vol. 23, Birkhäuser, Boston, Mass., 1982, with appendices by M. Demazure and T. Ono, originally published as Lecture Notes in 1961, Institute for Advanced Study. MR670072 (83m:10032)

[Wei94] C. Weibel, *An introduction to homological algebra*, Cambridge Studies in Advanced Mathematics, vol. 38, Cambridge University Press, Cambridge, 1994. MR1269324 (95f:18001)

[Wei04] _____, *A road map of motivic homotopy and homology theory*, Axiomatic, enriched and motivic homotopy theory, NATO Sci. Ser. II Math. Phys. Chem., vol. 131, Kluwer Acad. Publ., Dordrecht, 2004, 385–392. MR2061859 (2005e:14033)

[Wei08] _____, *Axioms for the norm residue isomorphism*, K-theory and Noncommutative Geometry, European Math. Soc. Pub. House, 2008, 427–436.

[Wei09]   _____, *The norm residue isomorphism theorem*, J. Topology **2** (2009), 346–372.

[Wei13]   _____, *The K-book*, Graduate Studies in Mathematics, vol. 145, American Mathematical Society, Providence, RI, 2013, an introduction to algebraic K-theory. MR3076731

[Whi78]   G. W. Whitehead, *Elements of homotopy theory*, Springer-Verlag, New York, 1978. MR80b:55001

Index

A boldface page number indicates that the term is defined on that page.